Environmental Contamination Remediation and Management

There are many global environmental issues that are directly related to varying levels of contamination from both inorganic and organic contaminants. These affect the quality of drinking water, food, soil, aquatic ecosystems, urban systems, agricultural systems and natural habitats. This has led to the development of assessment methods and remediation strategies to identify, reduce, remove or contain contaminant loadings from these systems using various natural or engineered technologies. In most cases, these strategies utilize interdisciplinary approaches that rely on chemistry, ecology, toxicology, hydrology, modeling and engineering.

This book series provides an outlet to summarize environmental contamination related topics that provide a path forward in understanding the current state and mitigation, both regionally and globally.

Topic areas may include, but are not limited to, Environmental Fate and Effects, Environmental Effects Monitoring, Water Re-use, Waste Management, Food Safety, Ecological Restoration, Remediation of Contaminated Sites, Analytical Methodology, and Climate Change.

Raman Kumar • Rajeev Kumar
Savita Chaudhary
Editors

Advanced Functional Nanoparticles "Boon or Bane" for Environment Remediation Applications

Combating Environmental Issues

Editors
Raman Kumar
Department of Biotechnology
Maharishi Markandeshwar University
Mullana, India

Rajeev Kumar
Department of Environment Studies
Panjab University
Chandigarh, India

Savita Chaudhary
Department of Chemistry and Center
of Advanced Studies in Chemistry
Panjab University
Chandigarh, India

ISSN 2522-5847 ISSN 2522-5855 (electronic)
Environmental Contamination Remediation and Management
ISBN 978-3-031-24418-6 ISBN 978-3-031-24416-2 (eBook)
https://doi.org/10.1007/978-3-031-24416-2

This Springer imprint is published by the registered company Springer Nature Switzerland AG
The registered company address is: Gewerbestrasse 11, 6330 Cham, Switzerland

Preface

The applications of advanced nanomaterials have immensely affected human progress. In material science, nanomaterials are the most inventive products of the chemical industry, and useful in varied products. The last decade has seen the tremendous growth of nanomaterials in high-technology areas such as electronic printing, magnetic recording, biotechnology, microelectronics, drug delivery, and much more. Nanomaterials are currently customized to overcome the commercial drawbacks of bulk materials. The usage of advanced nanomaterials in optical devices, biomedicine, and environmental remediation's, memory and storage devices has brought a significant revolution in the industrial sector. Nanotechnology has also facilitated wastewater management by creating detection and monitoring equipment with high sensitivity and selectivity. The existence of a highly active surface over nanomaterials has further induced the variations in optical and luminescence properties of particles. In comparison to the commercially used technologies such as chemical oxidation, bioremediation, and sedimentation, the usage of nanotechnology in waste cleaning is found to be less expensive and more effective.

However, the high-end applications of nanoparticles have further enhanced their production rate and made living beings more prone to their exposure. Water resources, soil, and air ecosystem are believed as an drainage source for these nanomaterials. The discharge of sewage into rivers or dumping yards is severely damaging the health of living beings. Here, we present new tools and advanced methodologies for the risk assessment of advanced nanomaterials in environmental remediation applications. We also discuss the harmful impacts of uncontrolled exposure of nanoparticles towards living systems by taking metal oxide–based nanomaterials as model particles.

The purpose of this book to make student aware about the application of functionalized nanoparticles as a new approach to supplement traditional treatment methods in cost- and time-effective manner. The field of nanotechnology has undergone fast revolution over the past two decades and produced considerable impact in scientific and industrial sectors. The advancement in synthetic procedures with controlled morphologies and optimized physicochemical properties of diverse range of nanomaterials have made them accessible tool in environmental protection science.

The simplistic means to assemble nanoparticles to the constituents of next-generation technologies in environment cleanup and sensing are the main objectives of the book. The book has also intended to present before the readers the toxicological footprinting of released advanced functional nanomaterials in ecosystem. In addition, the reflection of advancements in the field of nanotechnology and related concerns of released nanomaterials in environment is the central theme the editors bring readers. This book provides an overview of applications of advanced nanomaterials, basic lab set up and requirements in for their synthesis, techniques and career scope of nanotechnology in industries, and research and their related concerns.

This book is a comprehensive reference that provides a whole and systematic role of nanotechnology in biomedicines, and agricultural and environmental sectors, discussing the fundamental role of different types of nanomaterials and their applications. The sources, risk assessment, and safety aspect of nanomaterials in environment has also been discussed in detail. The interlinked confronts of advanced nanomaterials with environmental safety have been thoroughly discussed in this book. Different types of fate descriptors for screening nanotoxicity and pollutant sensing have been extensively discussed in this book. The present achievements and upcoming challenges of nanotechnology in the area of agriculture and biomedicine have been discussed in detail. The new perspectives application and hazardous impact of nanomaterials in aquatic environment have been thoroughly investigated in this book. The details regarding the risk governance policies for sustainable use of nanomaterials have been discussed in this book for the safe use of nanomaterials in technological applications.

This book is a supplement to graduate and postgraduate students of environmental science, nanotechnology, nano-biotechnology, and biochemistry for their theory and practical solutions in treating environmental remediation application. Along with academia, this book provides fundamentals test and methods used in industry for treating harmful toxins and producing effective sensory devices. The book has been organized to facilitate its use as quick reference as well as a general information source. It is the earnest anticipation of the authors that the compiled work will effectively convey the basic concepts of nanotechnology to readers. Our greatest source of contentment lies in the possible stimulation of interest in the minds of young learners and researchers for whom this book is primarily meant.

Mullana, India Raman Kumar

Chandigarh, India Rajeev Kumar

Chandigarh, India Savita Chaudhary

Contents

About the Editors

Raman Kumar is working as a senior associate professor in Department of Biotechnology, Maharishi Markendeshwar (Deemed to be University), Mullana (Ambala), Haryana, India, since 2011. He has also worked as a senior research fellow in the Division of Soil and Crop Management at the Central Soil Salinity Research Institute, Karnal, India, from 2007 to 2010. He has experience of about 17 years in the field of teaching and research in the area of microbial and environmental biotechnology. Raman has published more than 65 publications and 3 books in various national and international peer-reviewed journals. He is also the associate editor for the research journal *Current Trends in Biotechnology and Chemical Research* (eISSN 2321-0265; pISSN 2249-4073), which is currently indexed in many reputed databases including CAS and Medical Journal Links. He has been working in the area of environmental biotechnology and also got a major project from Haryana DST in the area of environmental microbiology. Raman developed many bacterial and fungal strains for removal of heavy metals from wastewater. His current research areas are bioremediation and biodegradation of toxic pollutants, pesticides, and azo dyes using microbial consortium from the industrial effluents, mechanism involved in the heavy metal bioremediation.

Rajeev Kumar is a permanent faculty in the Department of Environment Studies, Panjab University, Chandigarh, since July 2010. Dr. Kumar obtained his PhD in chemistry from the Department of Chemistry, Panjab University, Chandigarh, during which he published various research articles in journals of international repute. His laboratory focuses on the eco-friendly removal of environmental pollutants, specifically on the fungal remediation of recalcitrant compounds. His research group makes use of different modern technologies based on organometallics and nano- chemistry for this purpose. Dr. Kumar is recipient of research awards from the different funding agencies of the Government of India *viz.* Science and Engineering Research Board (SERB) – a statutory body of the Department of Science & Technology (DST) and the University Grants Commission (UGC) for his research endeavors. He has also been the project director of the Indo-US Partnership on Green Chemistry/ Engineering and Technologies Education, Research and Outreach for Sustainable Development funded by UGC and the Ministry for Human Resource and Development (MHRD), India. He has published 62 international research papers with more than 180 citations in journals of international repute, making his H-index of 21. His publications also include 17 international book chapters with Springer International; CRC, London; and Wiley Interscience, among others. Dr. Rajeev Kumar has visited several countries for his research work and has international collaborators from various countries including the USA, Israel, Singapore, Vietnam, and Australia.

Savita Chaudhary is working as an assistant professor in the Department of Chemistry and the Centre of Advanced Studies, Panjab University, Chandigarh. She has made significant contributions to the visionary mission of the central government under Swachata Abhiyan by the application of advanced nanomaterials. The current state of her work offers a sustainable use of nanoparticles in treating realistic water and harmful radiation problems. Her research methodology provides the better option of waste biomass conversion into useful advanced carbon dots. Therefore, her research work overcomes the problem of waste disposal and burning of crop stubble to a greater extent. The fabrication of chemodosimeters, fluorescence, and electrochemical-based sensors by Chaudhary et al. has unwrapped colossal range of prospects of their utilization from laboratory study to realistic wastewater treatment. Dr. Chaudhary is credited with 150 publications in reputed journals with h-index of 29, i_{10} index of 73, and 2858 citations. The consolidated impact factor of her research work is >600. Her work has been acknowledged at various national and international platforms. She has authored 2 book and 14 book chapters. She has been conferred with "Haryana Yuva Vigyan Ratan Award (2014–15)", "DST-DAAD fellowship (2008)," and best posters and presentation awards. Dr. Chaudhary has handled five independent major research projects.

Chapter 1
Advanced Nanomaterials: From Properties and Perspective Applications to Their Interlinked Confronts

Chitven Sharma, Deepika Bansal, Dhruv Bhatnagar, Sanjeev Gautam, and Navdeep Goyal

Abstract In today's era, nanotechnology has undoubtedly revolutionized a major part of modern-day techniques. It is gradually becoming an inevitable part of human life. From electronics to the energy storage/consumption sector, from pollution sensing to water purification systems, from food to the biomedical sector, this technology has certainly proved its potential everywhere. The materials at such nanoscale range begin to show some unexpected results, which may either prove to be fruitful or may pose risks at times. However, their application areas, methods of formation, and atmospheric conditions are some of the leading factors that influence the consequences of their usage. This chapter includes a discussion on various advanced nanomaterials and their properties including several metals, metal oxides, and carbon-based NPs. Based on their characteristic features, their usage in numerous sectors essential for sustaining human life has been considered. In addition, a detailed analysis regarding the involved risks of these nanoparticles has been included in this chapter.

Keywords Metals · Metal oxides · Carbon-based NPs · Advanced functional nanomaterials · Public risks

C. Sharma · D. Bansal · D. Bhatnagar · S. Gautam (✉)
Advanced Functional Materials Lab., Dr. S. S. Bhatnagar University Institute of Chemical Engineering & Technology, Panjab University, Chandigarh, India
e-mail: sgautam@pu.ac.in

N. Goyal
Department of Physics, Panjab University, Chandigarh, India
e-mail: ngoyal@pu.ac.in

R. Kumar et al. (eds.), *Advanced Functional Nanoparticles "Boon or Bane" for Environment Remediation Applications*, Environmental Contamination Remediation and Management, https://doi.org/10.1007/978-3-031-24416-2_1

1.1 Introduction to Advanced Nanomaterials

Nanotechnology has been regarded as the fastest growing sector in terms of developing novel and promising techniques in a vast range of application areas. It deals with materials at the nanometer scale (1–100 nm) including organic and inorganic nanoparticles, nanorods, nanotubes, nanowires, nanosheets, quantum dots, and their composites – termed as advanced nanomaterials (NMs). Since most of the properties are size dependent, these NMs greatly differ from their bulk counterparts as surface effects also predominate at this scale). For example, gold in its bulk form appears gold in color; however, when it is reduced to nanoscale, it changes to red. This is because of the electron confinement, due to which they react differently with light. Now with the introduction of these materials in various fields, it has become feasible to develop miniaturized, portable devices with rapid response timings. Their remarkable physical, mechanical, electrical, magnetic, chemical, and optical properties have gained them huge popularity in numerous industries. Fields like agriculture/food, biomedical, wastewater treatment, pollution sensing, and defense are some of the major areas where these nanomaterials are currently being studied (Kumar et al. 2020; McNamara and Tofail 2017; Hong et al. 2019). Many metals (such as Ag, Au, Zn, and Cu), metal oxides (e.g., Fe_2O_3, ZnO, TiO_2, CuO), carbon-based nanomaterials (including GO, rGO, and CNTs), and their nanocomposites have shown great potential in improving the existing techniques along with devising numerous modern-day technologies. However, these nanomaterials possess certain limitations among which their toxicity is the major cause of concern. Researchers are working in this direction, and several techniques have been proposed from time to time to counteract their risk factors. Their synthesis techniques, atmospheric conditions, and area of application also play an important role in determining their toxicity. It is mainly their small size that may pose risk to human health as they can easily penetrate inside the cells and it may result in fatal consequences. Thus, proper and prior knowledge of all the characteristic features of nanomaterials is highly recommended when working with them.

This chapter brings forth characteristic features of several advanced nanomaterials that are being extensively studied by researchers worldwide. Also, their usage in various sectors has been critically analyzed and discussed along with the risk factors involved in using them.

1.2 Properties of Advanced Nanomaterials

1.2.1 Metal Nanoparticles

Nano-sized inorganic particles of either simple or complex nature display unique physical and chemical properties and are an important material in the development of novel nanodevices that may be used in numerous physical, biological, biomedical, and pharmaceutical applications (Khan et al. 2019). In various industrial applications, metallic nanoparticles have attracted significant attention due to their different physical and chemical properties from bulk metals like high mechanical strength, large surface area, low melting point, and good optical and magnetic properties.

Silver Nanoparticles

Silver nanoparticles are one of the most prominent metal nanoparticles, as they are being widely utilized in various fields, including medical, food, health care, and other industrial purposes, because of their unique physical and chemical properties. These include optical, electrical, thermal, high electrical conductivity and biological properties (Zhang et al. 2016). The advantage of the nanosilver over the other noble metals in their physicochemical properties is the small loss of the optical frequency during the surface-plasmon propagation, non-toxicity, high electrical and thermal conductivity, stability at ambient conditions, high-primitive character, wide absorption of visible and far IR region of the light, chemical stability, and nonlinear optical behavior. Moreover, they exhibit a broad spectrum of high antimicrobial activity (bactericidal and fungicidal activity) attracting scientists and technologists with much interest to develop nanosilver-based disinfectant products.

Currently, nanosilver technologies have appeared in a variety of manufacturing processes and end products. There are many consumer products and applications where nanosilver is being utilized such as in soap, shampoo, textile, disinfecting medical devices, and residential appliances to water treatments, with the highest degree of business value (Chouhan 2018). Figure 1.1 shows the various applications of commonly used silver and gold nanoparticles (Agrawal et al. 2018; Madkour 2018).

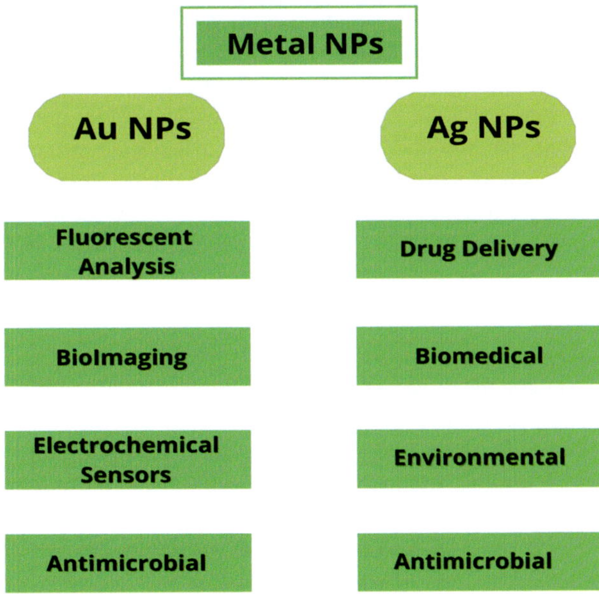

Fig. 1.1 A simple diagram showing the usual applications of silver and gold nanoparticles

Gold Nanoparticles
Gold nanoparticles (Au NPs) have also gained the attention of researchers. These are the most extensively studied NPs and are the center of fascination among the scientific community due to their excellent biocompatibility, large effective surface area, and distinctive size-dependent optical and electronic properties. Owing to their excellent conductivity, Au NPs are also successful in the area of electrochemical biosensors where these NPs amplify biorecognition signals and improve the analytical performance of biosensors. Gold nanoparticles are widely utilized in the field of radiation medicine as radiation enhancers and also provide therapeutic enhancement in radiotherapy because of the efficient and targeted drug delivery to the tumor site (Das et al. 2019). Gold nanoparticles have various applications as platform nanomaterials for biomolecular ultrasensitive detection, killing cancer cells by hyperthermal treatment, labeling for cells and proteins, and delivering therapeutic agents within cells. These are also being used in the tableware industry, which uses gold for decoration purposes, and in the electronic industry where specialized gold nanoparticle inks are used to provide metalized gold tracks. New uses envisaged in the electronics industry include gold nanomaterials to enhance solar cells, liquid crystal displays, and even nonvolatile storage devices (Falahati et al. 2020).

1.2.2 Metal Oxide Nanoparticles

Metal oxide nanomaterials have certainly gained much recognition in a wide range of industrial applications owing to their exceptional magnetic, optical, electronic, and catalytic properties (Liu and Liu 2019). They have found their applicability in various technologies including batteries, capacitors, biosensors, gas sensors, fuel cells, solar cells, anticorrosion coatings, ceramics, catalysts, etc. (Falcaro et al. 2016). Their limited size and high density greatly affect their structural and chemical properties. This is why a proper understanding of their several features is highly needed, concerning their synthesis and application area. As of now, various metal oxide nanomaterials have been explored for developing novel techniques or to enhance the efficiency of the existing technologies. Among them, a few of the materials have been elaborated in this section.

Zinc Oxide (ZnO)
ZnO is considered to be the most prominent metal oxide nanomaterial having excellent semiconducting properties. It has a wurtzite structure and consists of a wide band gap (~3.3 eV) with correspondingly large exciton binding energy, i.e., ~60 meV. It can be synthesized using various techniques and in a variety of morphologies. Figure 1.2 shows FESEM images of sol-gel synthesized thin films and nanowires of ZnO. It was reported that with the rise in annealing temperature, the grain size becomes larger and denser, which results in increasing the diameter of ZnO nanowires (Huang and Lin 2008). Moreover, its unique electrical and optical

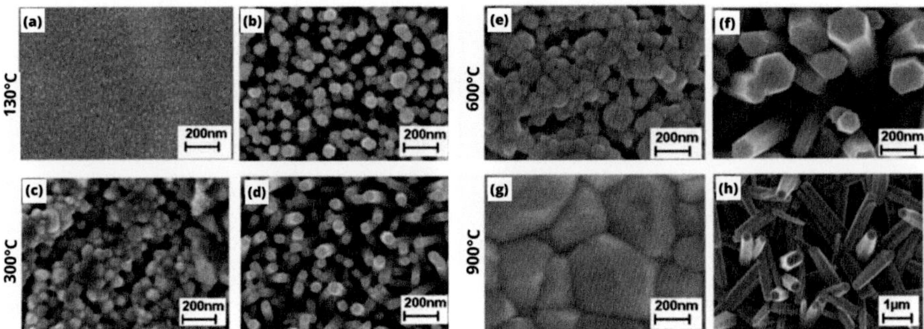

Fig. 1.2 FESEM images of ZnO thin films at (**a**) 130 °C, (**c**) 300 °C, (**e**) 600 °C, and (**g**) 900 °C annealing temperatures. (**b**, **d**, **f**, and **h**) describe ZnO nanowire arrays formed at 90 °C each. (Reprinted with permission from Huang and Lin 2008; J. Appl. Phys. Copyright 2008 American Institute of Physics)

properties including high thermal and chemical stability, excellent radiation absorption capability, and high electrochemical coupling coefficient along with low cost and easy synthesis techniques have led to a huge interest in various fields ranging from optoelectronics to antimicrobial agents (Djurišić et al. 2012; Kumar et al. 2017). However, its wide band gap largely affects its ability to absorb UV radiation, which limits its functionality as photoelectrodes, antimicrobial agents, and photocatalysts. So to counteract this problem, various intrinsic defects are introduced in their structure. Among them, oxygen vacancy (OV) has been identified as the most crucial and commonly found defect in their structure (Wang et al. 2018). OV helps in forming additional energy levels, which eventually lessens the band gap of ZnO. A study was reported where the optical properties of ZnO_{1-x} films were analyzed. It was observed that with the rise in oxygen vacancies, the light absorption range of ZnO was eventually improved and the band gap was reduced to 2.15 eV (Zhang et al. 2017). Now with the increase in the usage of ZnO NPs, the concerns related to their toxicity and threat to the ecosystem and mankind are also rising. Since their applicability has been extended in the biomedical, food, as well as cosmetic industry, it becomes essential to monitor their harmful effects (Sruthi et al. 2018).

The production of reactive oxygen species (ROS) and Zn^{+2} ions are mainly held responsible for the toxic effects of ZnO. However, the same attributes are also helpful in carrying antibacterial mechanisms or photocatalytic dye degradation (Djurišić et al. 2012). Thus, a proper insight into the various outcomes of using these materials is highly needed, and accordingly, different ways can be employed to neutralize their negative effects.

Titanium Dioxide (TiO_2)

TiO_2 is another commonly exploited inorganic nanomaterial. In nature, it is found in three crystal forms (polymorphs), which include anatase, rutile, and brookite where rutile is its most stable form and brookite has very rare usage due to its difficult synthesis process (Hanaor and Sorrell 2011). TiO_2 has gained attention in diverse

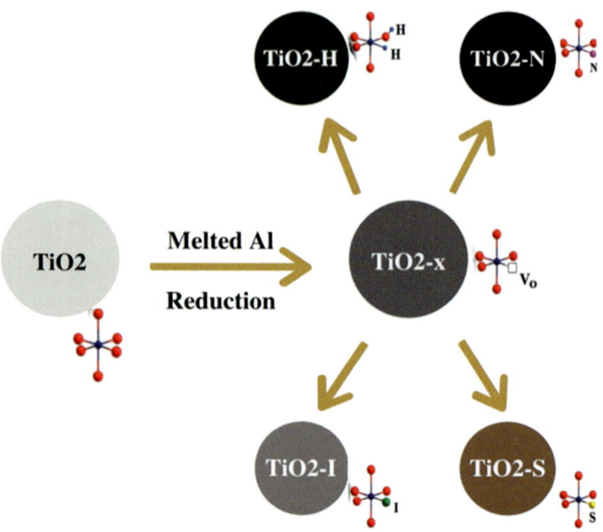

Fig. 1.3 A typical reaction mechanism showing the formation of oxygen-deficient TiO_{2-x} from white pristine TiO_2 and then conversion to TiO_2-H, N, S, I

application areas including skincare products such as sunscreens, in biomedical applications such as drug delivery and biosensing, for environment protection such as water purification and photocatalytic degradation of pollutants, and as a white pigment in paints, food, plastics, etc., all thanks to its remarkable properties (Qiu et al. 2011; Ziental et al. 2020). Recently, black TiO_2 has been reported, which tends to have better optical properties and shows greater solar energy absorption capacity ranging from ultraviolet (UV) to infrared (IR) region when compared to white TiO_2 nanomaterials. It turns black on reducing white TiO_2 with Ti^{+4} to Ti^{+3} through oxygen vacancies, self-doping, Ti-H bonds, or surface hydroxyl groups (Ullattil et al. 2018). Lin et al. (2014) reported Al-reduced TiO_2 nanomaterial where oxygen vacancies were created in the crystal structure. In place of vacant oxygen sites, nonmetals such as H, N, S, and I were incorporated to form corresponding black titania as shown in Fig. 1.3.

Black titania has certainly found its usage in hydrogen generation, photocatalytic water splitting, solar cells, batteries, photodegradation, etc. (Ullattil et al. 2018). In literature, several conflicting reviews have been reported related to the toxicity of TiO_2 NMs, where some state that they are nontoxic and biocompatible while others have shown them to be carcinogenic and genotoxic. This variation in studies could be due to several reasons like size differences, coatings, variation in assays used for toxicity assessment, or concentration of NMs used. Hence, a definite conclusion cannot be drawn regarding their toxic effects.

Iron Oxide NPs

Magnetic nanomaterials have considerably attracted much attention among researchers due to the vast array of constructive characteristics exhibited by them.

Their large surface area, high coercivity, large magnetic susceptibility, superparamagnetic behavior, and low Curie temperature are some of the important aspects for which they have shown promising results in numerous application areas. The controlled size and the synthesis technique used to fabricate the iron-based nanoparticles has a huge impact on its efficiency. Chemical methods are generally used to as they produce efficient and controlled size NPs (Gautam et al. 2023). Among them, iron oxide nanoparticles (IONPs) have emerged as a promising candidate in the biomedical and health care sector, agro-food industry, defense sector, fuel cells and batteries, wastewater treatment, etc. (Ali et al. 2016). Various forms of iron oxides have been investigated so far including Fe_3O_4, γ-Fe_2O_3, α-Fe_2O_3, β-Fe_2O_3, ε-Fe_2O_3, and FeO; however, among them, magnetite (Fe_3O_4) has been regarded as more prominent. It has a cubic inverse spinel structure where oxygen exhibits an fcc arrangement while Fe cations occupy tetrahedral and octahedral sites (Kefeni and Mamba 2020). These NMs are highly susceptible to oxidation and easily get oxidized into Fe_2O_3 NPs, which can result in the loss of their magnetism as well as dispersibility. Also, they have the tendency to agglomerate owing to the high energy surface and strong forces of attraction that exist between their particles. Therefore, to protect them from external factors and to keep them stable, surface coatings can be done, which helps them to improve their hydrophilicity and biocompatibility by maintaining their functionality. Zhu et al. (2018) reported surface modifications of IONPs on coating with inorganic materials like SiO_2, carbon NPs, Au and Ag NPs, and various metal oxides/sulfides and with organic materials like polymers and surfactants, as shown in Fig. 1.4. It was thus concluded that surface coatings largely contribute to enhancing the properties of IONPs, which has certainly improved their scope in diverse application fields.

1.2.3 Carbon-Based Nanomaterials

Nanofabrication methods have been improving over the years, which has made carbon-based nanomaterials (CBNs) famous for use in different application spheres. CBNs possess an amalgamation of various physical and chemical properties such as thermal and electrical conductivity, high mechanical strength, and optical properties. CBNs generally include carbon nanotubes, both single-walled and multi-walled, graphenes, fullerenes, nanodots, nanodiamonds, etc. Figure 1.5 shows the different types of carbon-based nanomaterials used for applications in modern-day technologies.

Carbon Nanotubes (CNTs)

Ever since CNTs were discovered by Sumio Iijima in the 1990s, these materials have found their usage in biomedical areas, water desalination and purification, dye removal, and many more such areas. CNTs possess more tensile strength, flexibility, and rigidity than well-known materials. It is interesting to note that graphene, by itself, can be characterized as either a zero-gap semiconductor or a metal, since the

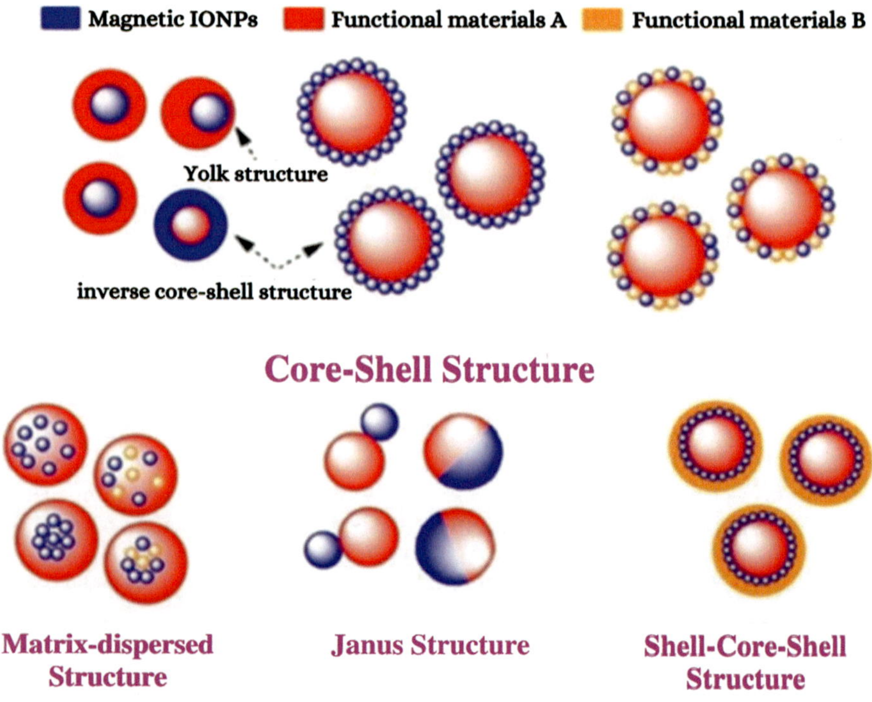

Fig. 1.4 Various morphologies of surface functionalized IONPs. (Adapted from Zhu et al. 2018; Nanomaterials, 8(10), 810. Copyright 2018 MDPI)

density of states (DOS) is zero at the Fermi energy (EF), and imparts those properties to a nanotube. It is also well known that the fundamental conducting properties of a graphene tubule depend on the nature of wrapping (chirality) and the diameter (typically, single-walled carbon nanotubes (SWCNTs) have diameters in the range of 0.4–2 nm). Multi-walled carbon nanotubes (MWCNTs) are composed of coaxial nanotube cylinders, of different helicities, with a typical spacing of ~0.34 nm, which corresponds closely to the interlayer distance in graphite of 0.335 nm. These adjacent layers are generally non-commensurate (different chiralities) with a negligible inter-layer electronic coupling and could alternate randomly between metallic and semiconducting varieties. The layers, constituting the individual cylinders, are found to close in pairs at the very tip of an MWCNT, and the detailed structure of the tips plays an important role, say, in the electronic and field emission properties of nanotubes (Bandaru 2007). Carbon nanotubes can also be chemically modified with functional groups to make them suitable for biological applications like enhanced material compatibility, cellular responsiveness, and injectable drug delivery systems. Due to oxidative stress in CNTs, there have been reports of cytotoxicity. Mostly, these NPs are cleared out from the body, but small amounts may still get accumulated in some organs and can cause inflammation. Furthermore, toxicity can be reduced by using CNTs as nanocomposites with other materials (Cha et al. 2013).

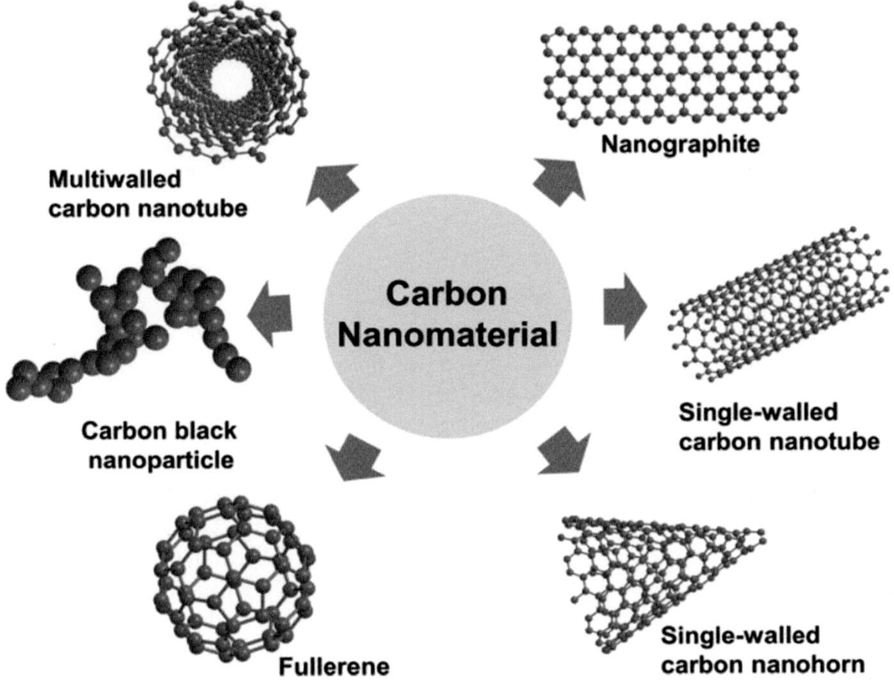

Fig. 1.5 Various types of carbon-based nanomaterials used for applications in modern technologies. (Adapted from Yuan et al. 2019; Part. Fibre Toxicol., 16(1), 1–27)

Graphene

Graphene is one of the strongest known materials to date. It identifies itself as a building block of many graphitic entities that are formed as a result of the selective removal of hydrogen from organic molecules. Graphene possesses many qualities such as high electron mobility, high conductivity, and strong mechanical strength that make them suitable for use in building membranes for purification purposes, energy harvesting mechanisms, sensors, and electronics. The quality of graphene plays a crucial role as the presence of defects, impurities, grain boundaries, multiple domains, structural disorders, and wrinkles in the graphene sheet can have an adverse effect on its electronic and optical properties. Graphene remains capable of conducting electricity even at the limit of nominally zero carrier concentration because the electrons do not seem to slow down or localize. Due to their excellent electron-transport properties and extremely high carrier mobility, graphene and other direct band-gap monolayer materials such as transition metal dichalcogenides (TMDCs) and black phosphorus show great potential to be used for low-cost, flexible, and highly efficient photovoltaic devices. A relatively new method of purifying brackish water is *capacitive deionization (CDI) technology* that uses graphene-like nanoflakes as electrodes for capacitive deionization. The advantages of CDI are that it has no secondary pollution, is cost-effective, and is energy efficient. CDI is energy efficient because it aims to remove only the salt ions, which are a small percentage of the feed solution, as compared to most other technologies that aim to separate water, which accounts for 90% of the feed solution.

1.3 Application Areas

1.3.1 Biomedical Areas

Nanomaterials are being used in different sectors such as wastewater treatment, food packaging, and dye removal, and one of its main application areas is in the biomedical field. In recent years, nanomaterials have emerged as important players in modern medicine, with many clinical applications starting from contrast agents in imaging to carriers for drug and gene delivery into tumors (McNamara and Tofail 2017). Metals and nanoparticles are also used in the fabrication of biomedical implants due to high mechanical strength and biocompatibility (Gautam et al. 2022). Indeed, there are some instances where nanoparticles enabled analyses and therapies were performed that merely cannot be performed otherwise. Mainly the nanomaterials that show antibacterial properties are getting used in the medical field.

Some of the uses of nanomaterials in the biomedical sphere are shown in Fig. 1.6.

Among various antimicrobial agents, silver has been employed most extensively since decades to fight infections and control spoilage as it possesses a broad spectrum of antibacterial, antifungal, and antiviral properties. Silver nanoparticles (Ag NPs) have received special interest, especially in biomedicine. Ag NPs are famous for their broad-spectrum and highly efficient antimicrobial and anticancer activities (Xu et al. 2020). Other biological activities of Ag NPs are also explored, including promoting bone healing and wound repair and enhancing the immunogenicity and antidiabetic effects. Nanosilver is widely used for diagnostic and therapeutic applications (e.g., in wound healing, arthritic disease, etc.) and also as an antiseptic to treat dermal problems. It is observed that Ag NPs can anchor and then penetrate the bacterial membrane and subsequently trigger the destruction of semipermeable membrane and leakage of content (Tang and Zheng 2018). Figure 1.7 displays the mechanism of the antibacterial action of Ag NPs.

Gold nanoparticles have also gained attention because of their application in medicine due to their unique physical and chemical properties. Au NPs have attracted tremendous interest from different fields of science because of their particular features: high X-ray coefficient of absorption, ease of synthetic manipulation, strong binding affinity, and unique optical and distinct electronic properties (Dykman and Khlebtsov 2011). Gold nanoparticles are especially utilized in genomics, biosensors, immune analysis, clinical chemistry, and detection and photothermolysis of microorganisms and cancer cells; the targeted delivery of drugs, DNA, and antigens; optical bioimaging; and the monitoring of cells and tissues using modern registration systems. The intrinsic optical properties of Au NPs provide the

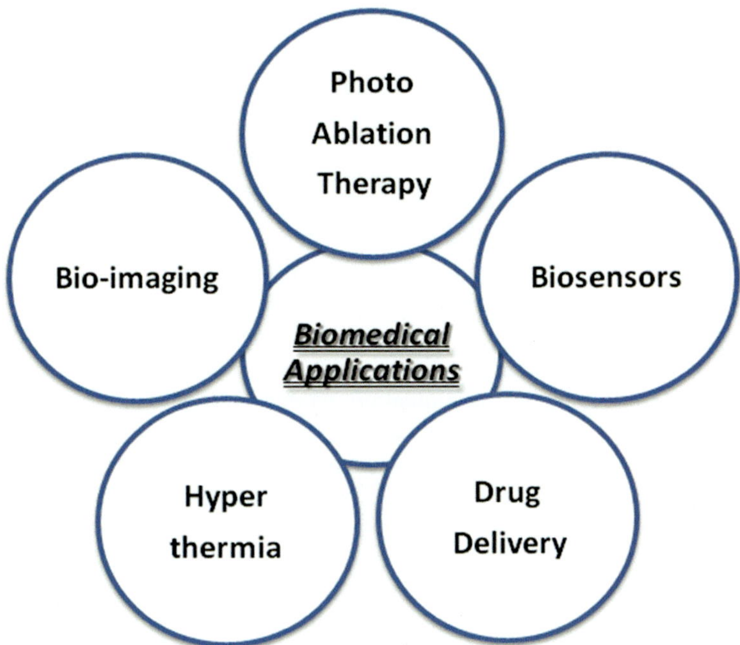

Fig. 1.6 Schematic diagram showing important biomedical applications of nanoparticles

Fig. 1.7 Proposed mechanism of action of silver nanoparticles against bacterial growth

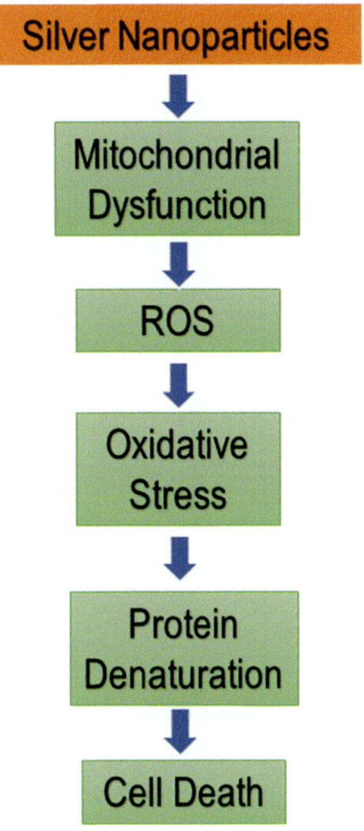

opportunity of being composite therapeutic agents in the clinic (Chhour et al. 2016). The use of gold nanoparticles for the photothermal therapy of chemotherapy-resistant forms of cancers seems to be the most promising direction. ZnO NPs have become one of the most popular metal oxide nanoparticles in biological applications because of their excellent biocompatibility, cost-effectiveness, and low toxicity. They are increasingly utilized in personal care products like cosmetics and sunscreen due to their strong UV absorption properties (Zhang et al. 2013). Additionally, ZnO NPs have superior antibacterial, antimicrobial, and excellent UV-blocking properties. It is generally known that zinc as a vital trace element extensively exists in all body tissues, including the brain, muscle, bone, skin, and so on. ZnO NPs with comparatively inexpensive and relatively less toxic properties exhibit excellent biomedical applications, such as anticancer, drug delivery, antibacterial, diabetes treatment, anti-inflammation, wound healing, and bioimaging. ZnO NPs are widely used in cancer therapy and reported to induce a selective cytotoxic effect on cancer cell proliferation. The results have suggested that ZnO NPs could selectively induce neoplastic cell apoptosis, which may be further served as a promising candidate for cancer therapy (Rajeshkumar et al. 2019).

Carbon nanotubes (CNTs) represent one of the most studied allotropes of carbon. The unique physicochemical properties of CNTs make them among prime candidates for various applications in biomedical fields including drug delivery, gene therapy, biosensors, and tissue engineering applications, as shown in Fig. 1.8. CNTs, having unique physical properties, can be manipulated to be subjected to different methods of biomedical imaging to analyze and improve their functionalities and response to their environment. Photoacoustic imaging (which allows deeper tissues to be imaged) (Alshehri et al. 2016) and magnetic resonance imaging are two additional methods that may be performed on CNTs, because of their high absorbance and also the impurities in the form of metallic nanoparticles they contain.

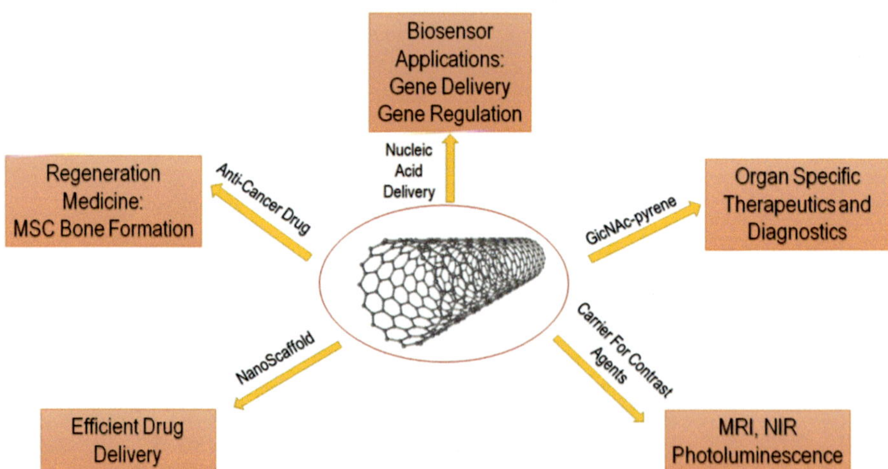

Fig. 1.8 Schematic drawing of applications of carbon nanotubes in biomedical areas

1.3.2 Water Purification and Dye Removal

Nanomaterials have posed great candidates for water purification, water desalination and dye removal. They have high adsorption capacities, and the mobility of the nanoparticles in solution form is also high. The materials discussed above hold a large potential in water treatment technologies. Silver nanoparticles possess antimicrobial properties that may be used to disinfect water. When Ag NPs come in contact with the bacteria, they interfere with the structure of the cell membrane, which causes the death of the cell eventually. Another perspective that explains the mechanism of Ag NPs on their contact with bacteria is that the dissolution of Ag NPs will release antimicrobial Ag^+ ions, which can interact with the thiol groups of many vital enzymes, inactivate them, and disrupt normal functions in the cell. Ag NPs have been used along with certain filter membranes due to their cost-effectiveness and high antimicrobial properties. However, these nanoparticles do not stand suitable for long-term use as they might get aggregated in aqueous media (Lu et al. 2016).

Compounds constituting carbon also show great attributes in wastewater treatment. A popular example of this is carbon nanotubes (CNTs). Their properties like nanosize diameter and hydrophobicity make them suitable for water purification and salt removal. Fast water transport through carbon nanotube pores has raised the possibility to use in the next generation of water treatment technologies. The tip-functionalized nonpolar interior home of carbon nanotubes provides a strong invitation to polar water molecules and rejects salts and pollutants. Low energy consumption, antifouling, and self-cleaning functions have made CNT membranes extraordinary over conventional ones. Freshwater is not only important to human health but also serves as a crucial feedstock for several industries. The reuse, recycling, and recovery of water have proven to be fruitful in creating a new and reliable water supply while not compromising public health. Membrane filtration is considered to be among the most promising and widely used processes for the purification of wastewater, seawater, and brackish water. Carbon nanotubes are promising media for seawater desalination due to their high water flow rate. Currently, narrow CNT membranes cannot be mass-produced for salt rejection purposes. Hence, introducing chemical functional groups at the rims of comparatively wider CNTs is an alternative strategy to enhance the ion exclusion performance (Hong et al. 2019). Compared to their bare CNT counterparts, CNTs with a low dipole moment at the rim can slightly increase the water flow and salt rejection rate, and CNTs with a high dipole moment can significantly improve the salt rejection efficiency, although water passage is slowed. In general, introducing functional groups with a high dipole moment on wider CNT rims can filter more than 95% of salt ions while maintaining a high water permeability of 10.2 L cm^{-2} day^{-1} MPa^{-1} (Hong et al. 2019).

The dyeing industry, including the textile industry, is one of the largest industries in the world. The effluents from these industries need proper treatment before they are discharged into water bodies. The dyes, namely, cationic, anionic, and non-ionic, are highly variable in their composition and are difficult to remove from the wastewater

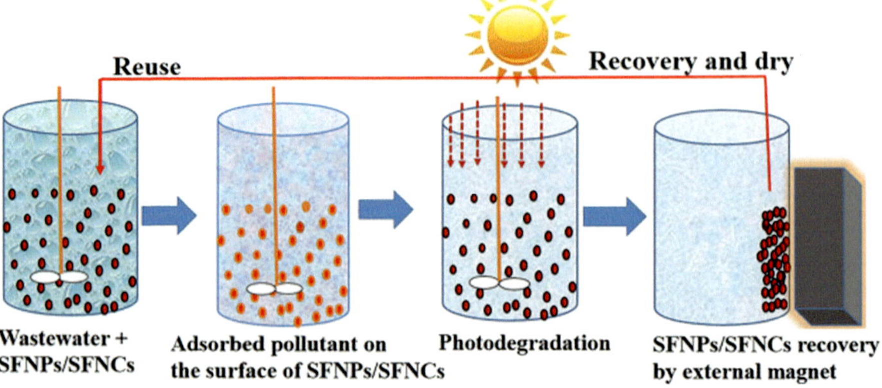

Fig. 1.9 A typical separation process of ferrite NC or NPs from the medium using an external magnet. (Adapted from Kefeni and Mamba 2020; Sustain. Mater. Techno., 23, e00140. Copyright 2020 Elsevier)

(Sivamani and Leena 2009). Dyes possess high stability due to complex aromatic molecular structures. This makes it difficult to degrade them. Moreover, the cost is an important factor that needs to be considered during the process of dye removal. Materials like fly ash, coal, and CNTs have been used for this purpose, but there always remains a disadvantage of separation. Ferrites have magnetic properties that make them suitable for separation at the end of the process. Figure 1.9 shows the separation process of ferrite nanoparticles from the medium using a commercially available external magnet. For example, sol-gel synthesized Congo Red (CR) was removed using manganese ferrites, nickel ferrites, and zinc ferrites. After various characterization methods and adsorption tests, it was concluded that nickel ferrites showed the best adsorption properties, followed by cobalt ferrites (Samoila et al. 2015).

When the large dye particles are brought into the UV radiation environment, they tend to break down into relatively smaller particles. This is the degradation of dyes. To fasten this process, some catalysts are used that promote the quicker breaking down of the dye molecules. This is what we call the photocatalytic degradation of dyes in wastewater. This can be brought about by certain materials like ZnO, TiO_2, CNTs, etc. However, again, spinel ferrites play an excellent role here due to their large surface area, low band gap, and magnetic properties. Out of all the ferrites, nickel ferrites show the best degradation properties for methylene blue dye with a rate of 2.417 per min, which gradually reduced to 2.065 per min for zinc oxide (Gupta et al. 2020).

Agriculture and Food Technology

Nanotechnology has emerged as a valuable asset for the agriculture and food sector in recent times, as it has always provided numerous opportunities to exploit the NMs in a way to find novel solutions and techniques (Kumar et al. 2020). The world population is said to reach 8.5 billion by 2030, which will eventually raise the food demands. This has certainly posed huge pressure on the food sector, as natural resources are also depleting at a faster rate. In this situation, it became essential to improve the existing techniques, agricultural productivity, post-harvest processing, food processing practices, and packaging methods to meet the rising demands (Kim et al. 2018). It is here that the role of advanced nanomaterials becomes prominent. These materials play a significant role in food science, as they constitute the monitoring of crop production, improvement in the yield and quality of the produce, safeguarding the nutritional value of food, and protection against microbial contamination, which in turn enhances the shelf life of the food commodity (Khot et al. 2012; Socas-Rodríguez et al. 2017). Figure 1.10 shows the various agricultural as well as the food safety applications of nanomaterials.

Every year almost 40% of food crops get destroyed worldwide due to plant diseases and pests, which ultimately leads to food shortage (Mesterházy et al. 2020). Most of the commercially used pesticides and other chemicals have been reportedly shown to destroy the natural mineral content of soil and cause water pollution as excess amount washes away into water bodies (Rekha et al. 2006). This calls for a great need for modern technologies to develop new protocols in this area. Researchers have recently come up with an effective solution in terms of nano-pesticides, nano-fertilizers, and other nano-chemicals to confront the situation. Due to the small size of these NMs, they can spread evenly on the surface of the pests. Moreover they are required in a lesser amount which gives them a greater edge over conventionally

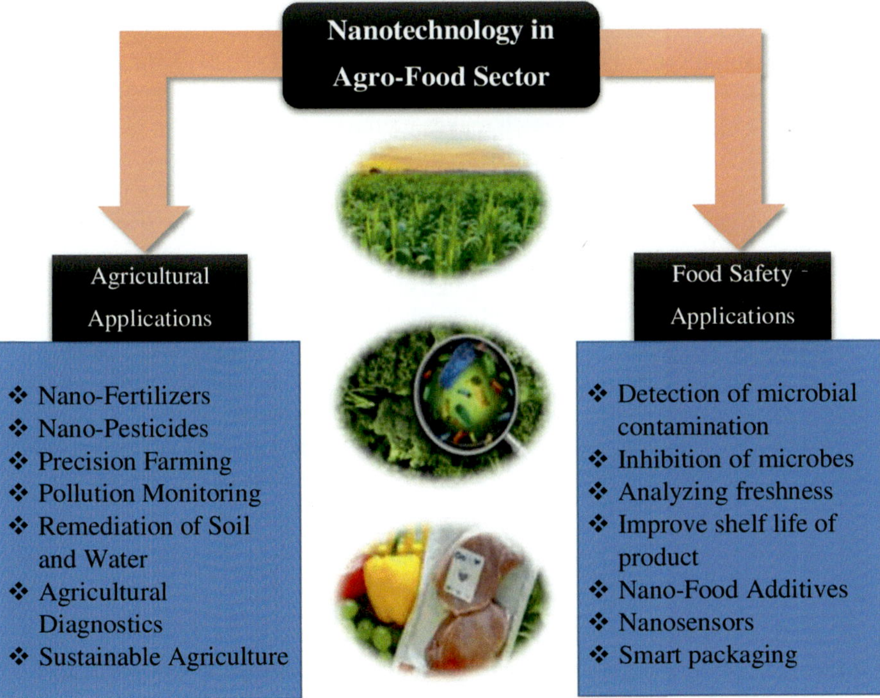

Fig. 1.10 Agricultural and food safety applications of nanotechnology in the agro-food sector

used pesticides (Rajna et al. 2019). Metal oxide NMs such as TiO_2, ZnO, and Al_2O_3 as well as CNTs have been extensively explored to improve the crop yield (Gogos et al. 2012; Yausheva et al. 2017). They have reportedly shown promising results in increasing the rate of photosynthesis, germination, and efficiency to utilize nutrients in crops. Bala et al. (2019) studied the efficiency of ZnO NMs as nano-fertilizers for rice crops grown in soil deficient in Zn. It was observed that the yield and growth certainly improved. It also increased the Zn content in the crop. Apart from this, materials like SiO_2, Cu, Ag, ZnO, TiO_2, etc., are even applied in pesticides to inhibit microbial growth (Chhipa 2017). These NMs can either be employed as nanocarriers for pesticides or as the active ingredient (Kumar et al. 2019). Now to monitor and detect the toxicity, soil pollution, bacterial activity, and nutrient deficiency in crops, an analyzing tool called nanosensor has been developed (Srivastava et al. 2018). Moreover, it even helps to reduce the loss of water during irrigation, pesticides, and fertilizers by sensing the right amount required. Certainly, Cu, Ag, Au, GO, and various metal oxide quantum dots have found their usage in these units due to their excellent properties (Arora 2018). A study reported Ag-doped Au core-shell nanorods as nanosensors for detecting bacteria using one-step assembling technique (Qiu et al. 2016).

After the agriculture sector, these NMs even have their applicability in food processing and food packaging units as antimicrobials, food additives, freshness indicators, and sensors for smart/intelligent packaging system (Mandal and Banerjee 2020). Various studies have been put forth regarding the application of Ag, Au, ZnO, CuO, GO, and carbon-based NMs such as CNTs in food safety due to their remarkable antimicrobial activity (Sánchez-López et al. 2020; Kim et al. 2020; Azizi-Lalabadi et al. 2020). These materials are embedded in the packaging material to form a nanocomposite, which makes it convenient for them to carry out bacterial inhibition mechanism. Apart from this, various food nanosensors have been fabricated in recent times to monitor the freshness and quality of the product. For example, carbon dots (CDs)-based fluorescent nanosensors have been designed with good sensitivity and selectivity for a wide range of food contaminants (Luo et al. 2020). Others include MnO, Au, CuO, and CNT/GO-based nanomaterials that are also being widely used (Can et al. 2020; Kumar and Guleria 2020; Hari et al. 2020; Fahim et al. 2020). However, with the rise in the usage of this technology in the food sector, the concerns related to their toxicity are also rising. Although some NMs are toxic, how they are designed and utilized also plays a great role. Thus, the current need is to address the prevailing misconceptions and knowledge gap related to these NMs in the food sector; only then will it be able to scale up its pace.

1.3.3 Energy Storage

If the available resources are used beyond their rate of getting replenished, they may deplete eventually. There is an immediate need for resources that are cost-effective, environment friendly, and, of course, can be stored for a longer period. Energy storage can be defined as a process where an extra amount of energy is stored in a material and can be used whenever required. This surplus energy can be used to generate electricity; or for use in vehicles, fuel cells, and rocket fuels. Energy production can be done through various materials, one of which is graphene. Using nano-sized graphene, energy can be produced through pressure and friction. This energy can be used to power fuel cells or can be burned in combined cycle gas power plants. Graphite in its nanofiber form serves as a great candidate for hydrogen storage (Nechaev et al. 2020; Lotoskyy et al. 2018). It has a specific surface area of 1400 m^2 g^{-1}. It was also seen that the decoration of graphite nanofibers with Pt caused the specific surface area to increase, which in turn decreased the hydrogen storage capacity of GNFs (Kostoglou et al. 2021). On the contrary, another study revealed that the hydrogen storage capacity increased up to a certain value and then decreased. Various samples of metal-doped graphite were considered for data analysis. When nickel was used as a dopant at 298 K, 2 MPa, the hydrogen storage capacity turned out to be 6.5%, theoretically. When doped with Fe, the capacity turned out to be 3.9%, experimentally. Theoretically, when doped with Al at 300 K, 100 MPa, the capacity is 3.48% (Züttel et al. 2002).

The properties of CNTs like structural defects, pressure, and temperature affect their hydrogen storage capacities. According to various experiments, it was concluded that the highest hydrogen storage capacity of 10 wt% can be achieved experimentally in SWCNT whereas a storage capacity of 11.2 wt% is also reported through Monte Carlo simulation. If MWCNTs are doped with lithium and potassium, the hydrogen storage capacity of 20 wt% and 14 wt% can be obtained, respectively. However, later it was reported that such a high hydrogen storage capacity was because of the presence of water impurities in the experiments. It was also observed that the synthesis reaction conditions play an important role in the tube diameter, which in turn affects hydrogen storage capacity. Defects in the structures of CNTs also affect hydrogen storage efficiency. For example, the presence of defects in CNTs may increase the adsorption binding energy of the order of 50%, which may further affect the H_2 adsorption criteria. Coiled CNTs have in-built defects and show good adsorption binding energies. Hence, hydrogen molecules get more space to get engaged. Certain additions to the original CNTs can improve the H_2 storage capacities. The addition of alkali metal to improve the diffusibility of H_2 can increase the interlayer distancing. The H_2 storage capacities get changed when CNTs are used with certain metals like Ca, K, Li, Ag, Ni, Pd, Fe, Pt, Ru, and Ti. This is mainly due to both crystal structure and electronic structure change (Froudakis 2011). Different adsorption values are compared in Table 1.1 (Mohan et al. 2019).

1.3.4 Pollution Sensing

According to the WHO reports, around seven million people die every year due to air pollution. Usually, the underdeveloped countries are affected the most. Vehicular and industrial pollution has been causing adverse effects on human health. Not only

Table 1.1 A comparison between the hydrogen storage capacities of different materials (Mohan et al. 2019)

Adsorbent	Reaction conditions	Hydrogen storage capacity
AC	303 K, 10 MPa	0.67%
AC	77 K, 0.1 MPa	1.4%
AC	296 K, 13 MPa	1.6%
AC-Ni	77 K, 0.1 MPa	1.8%
AC-Pt	298 K, 10 MPa	2.3%
Graphite-Cu	–	0.97% (experimental)
Graphite-Co	–	1.8% (experimental)
Graphite-Li	298 K, 2 MPa	6.5% (theoretical)
MWCNT-Li	–	20%
MWCNT-K	–	14%
CNT-Ti	–	8%
CNT-Ni	–	10%

outdoor pollution but indoor pollution is also an area of concern these days. There is a need to sense pollution levels and make suitable changes so to keep the levels under control. Air pollution sensors are devices that monitor the presence of air pollution in particular areas. Nanomaterial-based gas sensors have gained attention due to their easy-to-use gas sensing approach. Materials like WO_3, SnO_2, and ZnO are some of the commonly used gas-sensing materials. ZnO has a large band gap and is used in multiple spheres such as spin electronics, gas sensors, and piezoelectric devices. Bare ZnO and doped ZnO serve as great materials for gas sensing. Using indium as a dopant, ZnO showed a doping ratio of 95:5 at a temperature of about 150–250 °C (Hassan et al. 2014).

1.4 Risks and Interlinked Attributes

Although advanced nanomaterials have numerous applications in today's technologies, there may be some properties that make them less suitable for use under different reaction conditions. Nanomaterials are being increasingly introduced to the biosystem. The use of different synthesis and functionalization processes makes these NPs possess varying morphologies, sizes, and compositions. In some instances, the toxic behavior of NPs may overshadow their positive aspects. For evaluating different properties of nanomaterials, we need to have a deep knowledge of both their physical and chemical properties. These nanomaterials when disposed of may cause human health concerns. The disposal of nano-waste is a major topic of discussion and has been explained in further chapters.

High levels of distributed aromatics, including PAHs, and the formation of chlorinated furan are catalyzed by a broad spectrum of metal and metal oxide NPs, such as titanium, nickel oxide, silver, cerium, iron oxide, as well as fullerenes and quantum dots compared to their bulk counterparts. Synthesis and designing of NPs play

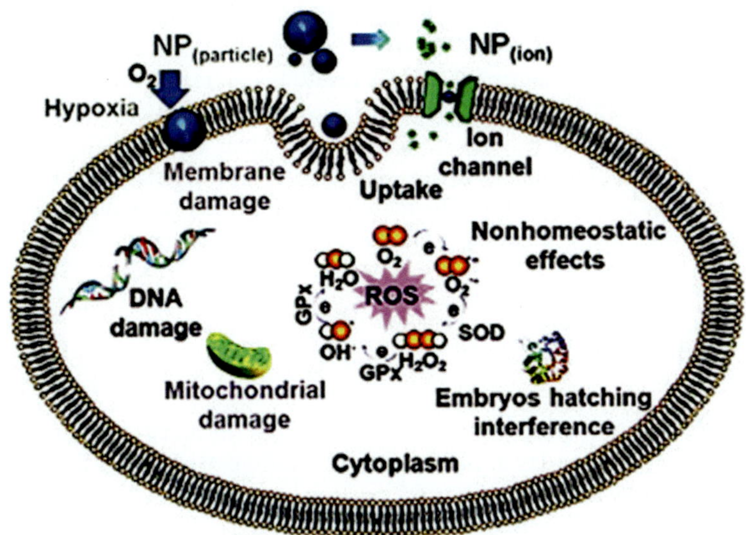

Fig. 1.11 Common risks of advanced nanomaterials as a result of ROS mechanism. (Reprinted with permission from Nasrollahzadeh and Sajadi, Interface Science and technology, 28, 2019. Copyright 2019 Academic Press)

an important role in their toxic traits. Not all NPs possess the same level of toxicity. The main interlinked attribute of all NPs is their size and morphology. On being compared to their bulk equivalents, these NPs due to their smaller size penetrate the ecosystem easily. Due to their small size, they have larger surface areas and hence increased reactivity (Nasrollahzadeh et al. 2019).

Nanoparticles may enter the human body through inhalation or skin. As a result, the body starts producing immune responses that give result in inflammatory responses. When inhaled, these NPs may cause lung tumors too. When NPs enter the soil after their disposal, they get accumulated in the food chain, thus affecting the flora and fauna. The formation of reactive oxygen species (ROS) usually results in cell death. Figure 1.11 shows different risks related to the usage of nanomaterials.

Taking into consideration carbon nanotubes, these are carcinogenic due to their fiber length. Human respiratory tracts have mechanisms to keep foreign particles away, but the small sizes of NPs make them travel to other organs. This causes swelling or apoptosis. Zinc oxide is also a great material for use in food technology. ZnO NPs are known for causing apoptosis or cell death in microorganisms to stop microbial growth. However, the same thing if caused in human nerve cells may be fatal. Doping of ZnO with Mg makes it more stable. Titanium dioxide is an important material used in photocatalysis. But its cell rupture mechanism causes damage to a large extent. CuO NPs disintegrate the membrane of the neurons. Once the NPs overcome the resistance of the cell membrane, they enter the cell and generate ROS, which eventually causes cell death. It is known that cationic gold nanoparticles can cause damage to the cell membrane when they come in the vicinity of a negatively charged cellular membrane. Au NPs with a size around 1.4 nm can cause high oxidative stress on cells and damage them as a result. In contrast to this, those NPs that had a size of about 15 nm did not damage the cell wall. This many suggest that the toxicity of gold nanoparticles may be dependent on their size. Toxicity of quantum dots causes bioaccumulation in organs and tissues, cytotoxicity, genetic material damage, disturbed cell viability, disordered immune cells (Nasrollahzadeh et al. 2019). More of this will be explained in the chapters that follow.

1.5 Conclusions

This chapter has given a brief overview of the advanced functional nanomaterials that are used in modern-day technologies. The different physical, chemical, and structural properties of commonly used metal nanoparticles, metal oxide nanoparticles, and carbon-based nanomaterials have been discussed in detail. These materials are increasingly being used in many spheres such as biomedical areas, food and agriculture, wastewater treatment, pollution sensing, as well as energy harvesting. These fields are some of the flourishing fields of nanotechnology. In the biomedical field, NPs are used for drug delivery, bioimaging, and other such procedures. In wastewater treatment, NPs are used to degrade dyes and remove unwanted ions. Similarly, these particles play an important role in the food industry for developing antibacterial food packaging films. This chapter explains these applications in a very descriptive manner. The risks linked with these NPs that tend to overweigh their advantages have been mentioned. The causes of their toxicity and other related aspects have been thoroughly explained. The measures that can be taken to reduce the toxicity of NPs to make them suitable for use have been explained in detail in the further chapters.

Multiple Choice Questions

1. Nanotechnology deals with materials with the size range of
   (i) 1–10 nm
   (ii) 1–100 nm
   (iii) 1–1000 nm
2. Gold in its nanoparticle size appears to be
   (i) Red
   (ii) Yellow
   (iii) Orange

3. Which of these nanomaterials exhibits a broad spectrum of high antimicrobial activity?
   (i) Ag
   (ii) CNT
   (iii) Graphene
4. Which of these nanomaterials amplifies biorecognition signals and improves the analytical performance of biosensors?
   (i) Ag
   (ii) CNT
   (iii) Au
5. SWCNTs have diameters in the range
   (i) 0.4–2 nm
   (ii) 4–20 nm
   (iii) 0.8–2 nm
6. Cytotoxicity is caused due to
   (i) Reductive stress
   (ii) Oxidative stress
   (iii) Cell death
7. Which of these is capable of conducting electricity even at the limit of nominally zero carrier concentration?
   (i) Au NPs
   (ii) Ag NPs
   (iii) Graphene
8. Using nano-sized graphene, energy can be produced through
   (i) Pressure and friction
   (ii) Fermentation
   (iii) Coal gasification
9. Coiled _______ have in-built defects and show good adsorption binding energies.
   (i) Graphene oxide
   (ii) Graphene
   (iii) CNTs
10. What is meant by "apoptosis"?
    (i) Cell death
    (ii) Cell regeneration
    (iii) Corona formation

Answers

1. (ii) 1–100 nm
2. (i) Red
3. (i) Ag
4. (iii) Au
5. (i) 0.4–2 nm
6. (ii) Oxidative stress
7. (iii) Graphene
8. (i) Pressure and friction
9. (iii) CNTs
10. (i) Cell death

Short Questions and Answers

1. *What is capacitive deionization (CDI) technology?*

   Answer: A relatively new method of purifying brackish water is *capacitive deionization (CDI) technology* that uses graphene-like nanoflakes as electrodes for capacitive deionization. The advantages of CDI are that it has no secondary pollution, is cost-effective, and is energy efficient. CDI is energy efficient because it aims to remove only the salt ions, which are a small percentage of the feed solution, as compared to most other technologies that aim to separate water, which accounts for 90% of the feed solution.

2. *What properties make gold nanoparticles suitable for use in biomedical areas?*

   Answer: Au NPs have attracted tremendous interest from different fields of science because of their particular features: high X-ray coefficient of absorption, ease of synthetic manipulation, strong binding affinity, and unique optical and distinct electronic properties (Dykman and Khlebtsov 2011). Gold nanoparticles are especially utilized in genomics, biosensors, immune analysis, clinical chemistry, and detection and photothermolysis of microorganisms and cancer cells; the targeted delivery of drugs, DNA, and antigens; optical bioimaging; and the monitoring of cells and tissues using modern registration systems.

3. *Why are Ag NPs used in membranes for wastewater treatment?*

   Answer: Silver nanoparticles possess antimicrobial properties that may be used to disinfect water. When Ag NPs come in contact with the bacteria, they interfere with the structure of the cell membrane, which causes death of the cell eventually. Another perspective that explains the mechanism of Ag NPs on their contact with bacteria is that the dissolution of Ag NPs will release antimicrobial Ag^+ ions, which can interact with the thiol groups of many vital enzymes, inactivate them, and disrupt normal functions in the cell. Ag NPs have been used along with certain filter membranes due to their cost-effectiveness and high antimicrobial properties.

4. *Why are ferrite nanocomposites preferred over other materials for dye removal?*

   Answer: Ferrites have magnetic properties that make them suitable for separation at the end of the process. For example, Congo Red (CR) was removed using manganese ferrites, nickel ferrites, and zinc ferrites. They were synthesized using the sol-gel method. After various characterization methods and adsorption tests, it was concluded that nickel ferrites showed the best adsorption properties, followed by cobalt ferrites.

5. *Which nanomaterials can be used to inhibit microbial growth?*

   Answer: Materials like SiO_2, Cu, Ag, ZnO, and TiO_2 are applied in pesticides to inhibit microbial growth. These NMs can either be employed as nanocarriers for pesticides or as an active ingredient.

References

Agrawal S, Bhatt M, Rai SK, Bhatt A, Dangwal P, Agrawal PK (2018) Silver nanoparticles and its potential applications: a review. J Pharmacogn Phytochem 7:930–937

Ali A, Hira Zafar MZ, ul Haq I, Phull AR, Ali JS, Hussain A (2016) Synthesis, characterization, applications, and challenges of iron oxide nanoparticles. Nanotechnol Sci Appl 9:49

Alshehri R, Ilyas AM, Hasan A, Arnaout A, Ahmed F, Memic A (2016) Carbon nanotubes in biomedical applications: factors, mechanisms, and remedies of toxicity: miniperspective. J Med Chem 59(18):8149–8167

Arora K (2018) Advances in nano based biosensors for food and agriculture. In: Nanotechnology, food security and water treatment. Springer, Cham, pp 1–52

Azizi-Lalabadi M, Hashemi H, Feng J, Jafari SM (2020) Carbon nanomaterials against pathogens; the antimicrobial activity of carbon nanotubes, graphene/graphene oxide, fullerenes, and their nanocomposites. Adv, Colloid Interface Sci., p 102250

Bala R, Kalia A, Dhaliwal SS (2019) Evaluation of efficacy of ZnO nanoparticles as remedial zinc nanofertilizer for rice. J Soil Sci Plant Nutr 19(2):379–389

Bandaru PR (2007) Electrical properties and applications of carbon nanotube structures. J Nanosci 7(4-5):1239–1267

Can K, Üzer A, Apak R (2020) A manganese oxide (MnO x)-Based colorimetric nanosensor for indirect measurement of lipophilic and hydrophilic antioxidant capacity. Anal Methods 12(4):448–455

Cha C, Shin SR, Annabi N, Dokmeci MR, Khademhosseini A (2013) Carbon-based nanomaterials: multifunctional materials for biomedical engineering. ACS Nano 7(4):2891–2897

Chhour P, Naha PC, Cheheltani R, Benardo B, Mian S, Cormode DP (2016) Gold nanoparticles for biomedical applications: synthesis and In vitro evaluation. In: Nanomaterials in Pharmacology. Humana Press, New York, NY, pp 87–111

Chouhan N (2018) Silver nanoparticles: synthesis, characterization and applications

Das P, Fatehbasharzad P, Colombo M, Fiandra L, Prosperi D (2019) Multifunctional magnetic gold nanomaterials for cancer. Trends Biotechnol 37(9):995–1010

Djurišić AB, Chen X, Leung YH, Ng AMC (2012) ZnO nanostructures: growth, properties and applications. J Mater Chem A 22(14):6526–6535
Dykman LA, Khlebtsov NG (2011) Gold nanoparticles in biology and medicine: recent advances and prospects. Acta Naturae (англоязычная версия) 3(2):9
Fahim IS, Hassanein AM, Said LA, Madian AH (2020) Design and fabrication of CNT/graphene-based polymer nanocomposite applications in nanosensors. In: Nanofabrication for Smart Nanosensor Applications. Elsevier, pp 281–294
Falahati M, Attar F, Sharifi M, Saboury AA, Salihi A, Aziz FM et al (2020) Gold nanomaterials as key suppliers in biological and chemical sensing, catalysis, and medicine. Biochimica et Biophysica Acta (BBA)-General Subjects 1864(1):129435
Falcaro P, Ricco R, Yazdi A, Imaz I, Furukawa S, Maspoch D et al (2016) Application of metal and metal oxide nanoparticles@ MOFs. Coord Chem Rev 307:237–254
Froudakis GE (2011) Hydrogen storage in nanotubes & nanostructures. Mater Today 14(7-8):324–328
Gautam S, Bhatnagar D, Bansal D, Batra H, Goyal N (2022) Recent advancements in nanomaterials for biomedical implants. Bio Engg Adv 100029
Gautam S, Bansal D, Bhatnagar D, Sharma C, Goyal N (2023) Synthesis of iron-based nanoparticles by chemical methods and their medical applications. In: Kumar P, Kandasamy G, Singh JP, Maurya P (eds). Oxides for Medical Applications. Woodhead Publishing. https://doi.org/10.1016/B978-0-323-90538-1.00009-1
Gogos A, Knauer K, Bucheli TD (2012) Nanomaterials in plant protection and fertilization: current state, foreseen applications, and research priorities. J Agric Food Chem 60(39):9781–9792
Gupta NK, Ghaffari Y, Kim S, Bae J, Kim KS, Saifuddin M (2020) Photocatalytic degradation of organic pollutants over MFe_2O_4 (M= Co, Ni, Cu, Zn) nanoparticles at neutral pH. Sci Rep 10(1):1–11
Hanaor DA, Sorrell CC (2011) Review of the anatase to rutile phase transformation. J Mater Sci 46(4):855–874
Hari N, Radhakrishnan S, Nair AJ (2020) Nanosensors as potential multisensor systems to ensure safe and quality food. Biotechnol Approach Food Adulte 204
Hong Y, Zhang J, Zhu C, Zeng XC, Francisco JS (2019) Water desalination through rim functionalized carbon nanotubes. J Mater Chem A 7(8):3583–3591
Huang JS, Lin CF (2008) Influences of ZnO sol-gel thin film characteristics on ZnO nanowire arrays prepared at low temperature using all solution-based processing. Int J Appl Phys 103(1):014304
Kefeni KK, Mamba BB (2020) Photocatalytic application of spinel ferrite nanoparticles and nanocomposites in wastewater treatment. SM & T 23:e00140
Khan I, Saeed K, Khan I (2019) Nanoparticles: properties, applications and toxicities. Arab J Chem 12(7):908–931
Khot LR, Sankaran S, Maja JM, Ehsani R, Schuster EW (2012) Applications of nanomaterials in agricultural production and crop protection: a review. Crop Prot 35:64–70
Kim DY, Kadam A, Shinde S, Saratale RG, Patra J, Ghodake G (2018) Recent developments in nanotechnology transforming the agricultural sector: a transition replete with opportunities. J Sci Food Agric 98(3):849–864
Kim I, Viswanathan K, Kasi G, Thanakkasaranee S, Sadeghi K, Seo J (2020) ZnO nanostructures in active antibacterial food packaging: preparation methods, antimicrobial mechanisms, safety issues, future prospects, and challenges. Food Rev Int:1–29
Kostoglou N, Liao CW, Wang CY, Kondo JN, Tampaxis C, Steriotis T et al (2021) Effect of Pt nanoparticle decoration on the H2 storage performance of plasma-derived nanoporous graphene. Carbon 171:294–305
Kumar V, Guleria P (2020) Application of DNA-nanosensor for environmental monitoring: recent advances and perspectives. Curr Pollut Rep:1–21
Kumar R, Umar A, Kumar G, Nalwa HS (2017) Antimicrobial properties of ZnO nanomaterials: a review. Ceram Int 43(5):3940–3961

Kumar S, Nehra M, Dilbaghi N, Marrazza G, Hassan AA, Kim KH (2019) Nano-based smart pesticide formulations: Emerging opportunities for agriculture. J Control Release 294:131–153
Kumar P, Mahajan P, Kaur R, Gautam S (2020) Nanotechnology and its challenges in the food sector: a review. Mater Today Chem 17:100332
Lin T, Yang C, Wang Z, Yin H, Lü X, Huang F et al (2014) Effective nonmetal incorporation in black titania with enhanced solar energy utilization. Energy Environ Sci 7(3):967–972
Liu B, Liu J (2019) Sensors and biosensors based on metal oxide nanomaterials. TrAC Trends Anal Chem 121:115690
Lotoskyy M, Denys R, Yartys VA, Eriksen J, Goh J, Nyamsi SN et al (2018) An outstanding effect of graphite in nano-MgH 2–TiH 2 on hydrogen storage performance. J Mater Chem A 6(23):10740–10754
Lu H, Wang J, Stoller M, Wang T, Bao Y, Hao H (2016) An overview of nanomaterials for water and wastewater treatment. Adv Mater Sci 2016
Luo X, Han Y, Chen X, Tang W, Yue T, Li Z (2020) Carbon dots derived fluorescent nanosensors as versatile tools for food quality and safety assessment: a review. Trends Food Sci Technol 95:149–161
Madkour LH (2018) Applications of gold nanoparticles in medicine and therapy. Pharm Pharmacol Int J 6(3):157–174
Mandal A, Banerjee ER (2020) Nanotechnology and functional food. In: Nanomaterials and biomedicine. Springer, Singapore, pp 85–112
McNamara K, Tofail SA (2017) Nanoparticles in biomedical applications. Adv Phys X 2(1):54–88
Mesterházy Á, Oláh J, Popp J (2020) Losses in the grain supply chain: causes and solutions. Sustainability 12(6):2342
Mohan M, Sharma VK, Kumar EA, Gayathri V (2019) Hydrogen storage in carbon materials – a review. Energy Storage 1(2):e35
Nasrollahzadeh M, Sajadi MS, Atarod M, Sajjadi M, Isaabadi Z (2019) An introduction to green nanotechnology. Academic
Nechaev YS, Alexandrova NM, Cheretaeva AO, Kuznetsov VL, Öchsner A, Kostikova EK, Zaika YV (2020) Studying the thermal desorption of hydrogen in some carbon nanostructures and graphite. Int J Hydrog 45(46):25030–25042
Qiu J, Zhang S, Zhao H (2011) Recent applications of TiO2 nanomaterials in chemical sensing in aqueous media. Sensors Actuators B Chem 160(1):875–890
Qiu L, Wang W, Zhang A, Zhang N, Lemma T, Ge H et al (2016) Core–shell nanorod columnar array combined with gold nanoplate–nanosphere assemblies enable powerful in situ SERS detection of bacteria. ACS Appl Mater Interfaces 8(37):24394–24403
Rajeshkumar S, Lakshmi T, Naik P (2019) Recent advances and biomedical applications of zinc oxide nanoparticles. In: Green synthesis, characterization and applications of nanoparticles. Elsevier, pp 445–457
Rajna S, Paschapur AU, Raghavendra KV (2019) Nanopesticides: its scope and utility in pest management. Indian Farmer 6:17–21
Rekha S, Naik SN, Prasad R (2006) Pesticide residue in organic and conventional food-risk analysis. J Chem Health Saf 13(6):12–19
Samoila P, Cojocaru C, Cretescu I, Stan CD, Nica V, Sacarescu L, Harabagiu V (2015) Nanosized spinel ferrites synthesized by sol-gel autocombustion for optimized removal of azo dye from aqueous solution. J Nanomater 2015
Sánchez-López E, Gomes D, Esteruelas G, Bonilla L, Lopez-Machado AL, Galindo R et al (2020) Metal-based nanoparticles as antimicrobial agents: an overview. Nanomaterials 10(2):292
Sivamani S, Leena GB (2009) Removal of dyes from wastewater using adsorption-a review. Int J Biosci Technol 2(4):47–51
Socas-Rodríguez B, Gonzalez-Salamo J, Hernandez-Borges J, Rodríguez-Delgado MÁ (2017) Recent applications of nanomaterials in food safety. TrAC Trends Anal Chem 96:172–200
Srivastava AK, Dev A, Karmakar S (2018) Nanosensors and nanobiosensors in food and agriculture. Environ Chem Lett 16(1):161–182

Sruthi S, Ashtami J, Mohanan PV (2018) Biomedical application and hidden toxicity of Zinc oxide nanoparticles. Mater Today Chem 10:175–186

Tang S, Zheng J (2018) Antibacterial activity of silver nanoparticles: structural effects. Adv Healthc Mater 7(13):1701503

Ullattil SG, Narendranath SB, Pillai SC, Periyat P (2018) Black TiO2 nanomaterials: a review of recent advances. Chem Eng J 343:708–736

Wang J, Chen R, Xiang L, Komarneni S (2018) Synthesis, properties and applications of ZnO nanomaterials with oxygen vacancies: a review. Ceram Int 44(7):7357–7377

Xu L, Wang YY, Huang J, Chen CY, Wang ZX, Xie H (2020) Silver nanoparticles: synthesis, medical applications and biosafety. Theranostics 10(20):8996

Yausheva EV, Sizova EA, Gavrish IA, Lebedev SV, Kayumov FG (2017) Effect of Al2O3 nanoparticles on soil microbiocenosis, antioxidant status and intestinal microflora of red Californian worm (Eisenia foetida). Agric Biol 52(1):191–199

Yuan X, Zhang X, Sun L, Wei Y, Wei X (2019) Cellular toxicity and immunological effects of carbon-based nanomaterials. Part Fibre Toxicol 16(1):1–27

Zhang Y, Nayak TR, Hong H, Cai W (2013) Biomedical applications of zinc oxide nanomaterials. Curr Mol Med 13(10):1633–1645

Zhang XF, Liu ZG, Shen W, Gurunathan S (2016) Silver nanoparticles: synthesis, characterization, properties, applications, and therapeutic approaches. Int J Mol Sci 17(9):1534

Zhang C, Geng X, Liao H, Li CJ, Debliquy M (2017) Room-temperature nitrogen-dioxide sensors based on ZnO1− x coatings deposited by solution precursor plasma spray. Sensors Actuators B Chem 242:102–111

Zhu N, Ji H, Yu P, Niu J, Farooq MU, Akram MW et al (2018) Surface modification of magnetic iron oxide nanoparticles. Nanomaterials 8(10):810

Ziental D, Czarczynska-Goslinska B, Mlynarczyk DT, Glowacka-Sobotta A, Stanisz B, Goslinski T, Sobotta L (2020) Titanium dioxide nanoparticles: prospects and applications in medicine. Nanomaterials 10(2):387

Züttel A, Sudan P, Mauron P, Kiyobayashi T, Emmenegger C, Schlapbach L (2002) Hydrogen storage in carbon nanostructures. Int J Hydrog Energy 27(2):203–212

Chapter 2
Advanced Nanoparticles: A Boon or a Bane for Environmental Remediation Applications

Deepak Rohilla and Savita Chaudhary

Abstract The potential usage of nanotechnology in diverse range of applications has found promising alternative answers to the crucial problems associated with industrial and commercial sectors. The nanomaterials are currently customized to overcome the commercial drawbacks of bulk materials. The usage of advanced nanomaterials in optical devices, biomedicine, and environmental remediations, memory, and storage devices has brought a significant revolution in the industrial sector. However, the tremendous enhancement in the production of nanoparticles has raised the questions of potential risks associated with nanotechnology. The detailed overviewing of wellbeing, security, and ecological effects during fabrication, usage in different applications, and their removal from the environment are the major features for the growth and development of safe usage of nanotechnology. The partial understanding of the ecological providence of nanotechnology has further posed threats and hazards to human wellbeing. This chapter presents the new tools and advanced methodologies for the risk assessment of advanced nanomaterials in environmental remediation applications. We also discuss the harmful impacts of uncontrolled exposure of nanoparticles on living systems.

Keywords Nanotechnology · Optical devices · Environmental remediation · Exposure · Risk assessment

D. Rohilla · S. Chaudhary (✉)
Department of Chemistry and Centre of Advanced Studies in Chemistry, Panjab University, Chandigarh, India
e-mail: schaudhary@pu.ac.in

R. Kumar et al. (eds.), *Advanced Functional Nanoparticles "Boon or Bane" for Environment Remediation Applications*, Environmental Contamination Remediation and Management, https://doi.org/10.1007/978-3-031-24416-2_2

2.1 Introduction

The current era has brought significant attention to the miniaturization of materials in both technical and scientific areas (Zhang et al. 2018a, b, c). The nanotechnology is considered as one of the hottest topics in commercial sectors and possessed considerable attention in the biomedical field (Ravichandran 2010; Holm et al. 2002). The potential usages of nanomaterials in the ecological and food technological sector are well documented in the literature (Vambol et al. 2017; Kasaai 2020). The applications of advanced nanoparticles in optical devices, screen display, textile, cosmetics, and memory devices hold high consideration among industries (Winardi et al. 2020; Dréno et al. 2019). Nanotechnology has also facilitated the wastewater management by creating detection and monitoring equipment with high sensitivity and selectivity (Matlochová et al. 2013; Crane and Scott 2012). These advanced nanomaterials comprise both organic and inorganic materials with small size, usually lesser than 100 nm (Ye et al. 2019; Wang et al. 2020). The significant size reduction has induced quantum effect and further enhanced the surface to volume ratio in nanomaterials as compared to their bulk counterparts (Jancik Prochazkova et al. 2020; Modena et al. 2019). The existence of a highly active surface over nanomaterials has further induced the variations in optical and luminescence properties of particles (Fahmy et al. 2019; Zhang et al. 2020).

These unique and outstanding properties of nanomaterials have further enhanced their scope in therapeutic and disease prevention area (Tricoli and Bo 2020). In comparison with the commercially used technologies such as chemical oxidation, bioremediation, and sedimentation, the usage of nanotechnology in waste water clean is found to be less expensive and more effective (Sarong et al. 2020; Patil et al. 2016). The usage of nanomaterials has allowed the treatment of containments in a time-effective manner and reduced the application of additional chemicals during processing (Singh et al. 2020). However, the high-end applications of nanoparticles have further enhanced their production rate and made living beings more prone to their exposure (Singh 2019; Prajitha et al. 2019). The different water resources, soil, and air ecosystem are believed as an ultimate descend for these nanomaterials. The uncontrolled discharge in sewage, river, or dumping yards has created a severely damaging impact of the health of living beings (Oliveira et al. 2019; Chaudhary et al. 2019). Their environmental fate is characterized by their size, surface characteristic properties, coating, and bioavailability of nanomaterials in a different medium (Zhang et al. 2019). For instance, have studied that the environmental fates of different types of nanomaterials are dependent not only on their initial composition but also based on their environmental transformation in the ecosystem. A thorough understanding of these conversions is crucial for defining the ultimate fate and impact of nanomaterials during their application usage and clearance points.

This chapter presents the new tools and advanced methodologies for the risk assessment of advanced nanomaterials in environmental remediation applications. We also discuss the harmful impacts of uncontrolled exposure of nanoparticles towards living systems by taking metal oxide-based nanomaterials as model particles.

2.2 Classification and Properties of Advanced Nanoparticles

In general, the advanced nanomaterials are classified into the following main categories:

- Organic-based nanoparticles
- Inorganic nanoparticles
- Fullerenes or carbon-based nanoparticles
- Polymeric nanoparticles

However, it has been found from the available data that the organic- and polymeric-based nanoparticles are highly biodegradable. The toxicity imparted by these nanoparticles is minimum, and they are unstable under the heat and electromagnetic exposure (Naahidi et al. 2013). In addition, C-dots have also possessed better control over the toxicity and are considered as a good alternative material in the biomedical application (Yi et al. 2019; Bhamore et al. 2019). Metal oxide nanoparticles have a broad spectrum of applications in electrochemistry, catalysis, energy storage systems, detectors, lubrication, coatings and climate remediation, etc. (Narayanan and El-Sayed 2005; Zhang et al. 2016; Gusain and Khatri 2013), whereas the metal and metal oxide-based nanoparticles are highly effective in conducting and semiconducting material and useful for electronic devices. The advanced reactivity and amendable properties of metal oxide nanoparticles made them quite useful as compared to their metal counterparts (Akbari et al. 2018; Degler et al. 2019). These nanoparticles have possessed superior physicochemical properties, which made them superior as compared to their bulk counterparts.

The variations in the physicochemical properties of nanoparticles can be easily achieved by modifying the available surface atoms. The higher ratio of surface area to volume in nanoparticles has further affected their size, shape, and optical and fluorescence properties (Leong et al. 2019; Shahbazi et al. 2020). In addition, the external oxide coating over the nanoparticles has further affected their dispersity rate in aqueous media and enhanced their applications in water as dispersion media. The fabrication of different types of nanoparticles is grouped into two main processes, namely, top-down method and bottom-up approaches. The fundamental properties of nanoparticles are determined by analysing their specific physiochemical properties. For instance, the corresponding reactivity and stability of formed nanoparticles are easily checked by measuring the particle size of nanoparticles. The nature of the surface charge has provided an idea about the homogeneity of used nanoparticles. Also, the size of used nanoparticles calculated from dynamic light scattering and different types of microscopic techniques such as transmission electron microscope (TEM), scanning electron microscope (SEM), and atomic force microscope (AFM) has provided the idea about the bioavailability of the used nanoparticles. The retention and distribution rate has also been estimated by getting information about the particle size of fabricated nanoparticles.

2.3 Environmental Remediation Applications of Advanced Nanoparticles

Modern industrialization, increasing population, growing urbanization, and widespread use of unsustainable assets and uncontrolled exploitation of natural resources are affecting the climate. The anthropogenic activities and waste discharge from multiple sectors have affected our natural resources and are making them unfit for living beings. Every day, the fabrics, paper, fibre, fertilizers, chemicals, pesticides, batteries, metal plating, refineries, food processing, and pharmaceutical industries have released millions of gallons of wastewater that polluted our land and water bodies (Low and Isserman 2009; Mason et al. 2016; Islam and Tanaka 2004). For instance, persistent organic pollutants (POPs) are incredibly resistant to degradation and triggered congenital disabilities, cancer, reproductive and immune dysfunctional mechanisms, and impairing infant growth (Ogata et al. 2009; Goerke et al. 2004). Therefore the non-biodegradable nature and toxic impact on human health and the climate have made the researchers think better means to treat these toxic wastes.

On the other hand, organic dyes are complex molecules and a quite useful component in a different range of industries (Nigel Corns et al. 2009; Barnett 2007). Currently, thousands of distinct kinds of dyes are accessible worldwide, with estimated manufacturing of 7 to 105 tons (Carneiro et al. 2010). The water-soluble nature with slow biodegradability of dyes has possessed a challenge of high-level detection in aqueous media (Castillo and Barceló 2001). The release of organic dyes in the hydrosphere generates undesirable colour and restricts sunlight absorption, which compromises the photochemical and biological activities of aquatic species (Parasuraman et al. 2012). Another important class of organic pollutants in the water bodies is agrochemicals such as pesticides, herbicides, fungicides, and different ranges of synthetic fertilizers (Dai et al. 2019; Alharbi et al. 2018). The uncontrolled and excessive usage of pesticide can contaminate soil (farm drainage) and groundwater and adversely affects living beings. These existences of water contaminants are considered as a significant source of water scarcity in the domestic, agricultural, and industrial sectors. Therefore, extensive attempts have been aimed towards the preservation of water resources. Wastewater treatment via employing economically viable and environmentally friendly methods (Fig. 2.1) has attracted significant concern among the researchers.

To date, several methods such as physical, biological, and chemical methods have been used to remove and minimize the impact of organic contaminants. These methods are used to detoxify wastewater either separately or in combination with others. In addition to this, adsorption, biological degradation, liquid-liquid

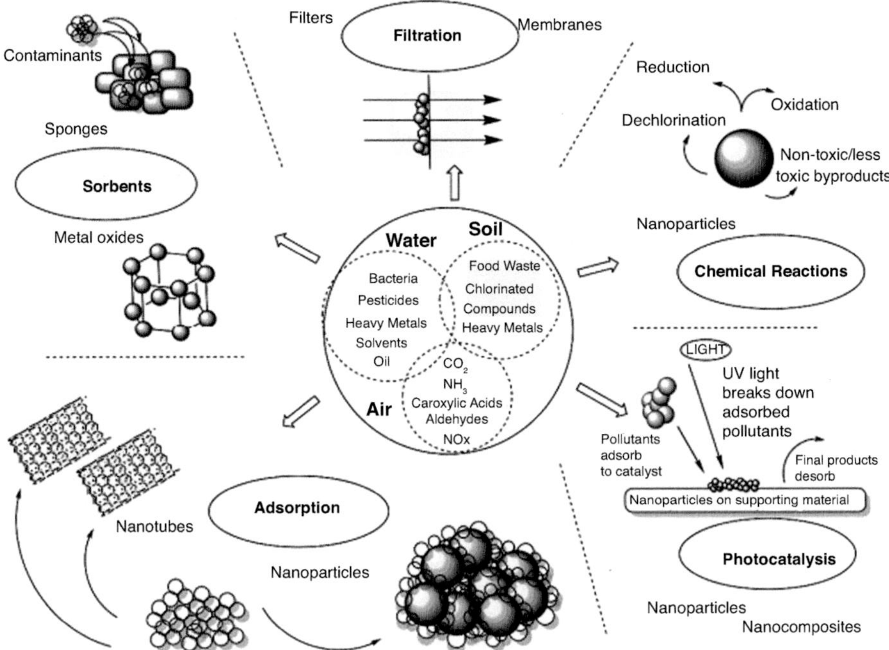

Fig. 2.1 Applications of environmental nanotechnologies involving adsorption, chemical reactions, photocatalysis, and filtration, particularly for the removal of contaminants from water, soil, and air. (Adapted figure from Nascimento et al. 2019, with permission from copyright, Elsevier publishing company)

extraction, membrane filtration, photocatalytic degradation, oxidation, nanofiltration, UV irradiation, and reverse osmosis (RO) have also been used for decontaminating the water resources. Out of all available methods, nanoparticles come out to be a potential and futuristic candidate in adsorption and photocatalytic degradation processes of organic pollutants, in a cost-effective manner (Liu et al. 2020a, b; Islam et al. 2020).

In comparison with adsorption, the photocatalytic degradation process is found to be more eco-friendly although adsorption is found to be the rapid, cost-effective, and scalable approach towards the removal of organic pollutants. However, the production of secondary waste is one of the significant concerns of adsorptive removal strategy. Another key challenge in the adsorption process is the recyclability of adsorbent species for subsequent cycles. On the contrary, the photocatalytic method of degradation has contributed to the transformation of pollutants into less poisonous or intermediate chemicals having simpler molecular structure in comparison with their parent chain, which ultimately are transformed to mineral-based useful products like H_2O, CO_2, etc. In addition to the bandgap engineering and light source, the method of photo-degradation is primarily based on organic pollutant adsorption ability over the photocatalyst surface. The photocatalyst with higher adsorption surface makes the degradation process easier as compared to other remediation processes. On the other hand, the adsorption process has created the interaction between organic molecules and adsorbate species (Tachikawa et al. 2007). Guo et al. (2016) have shown that fluorinated species of TiO_2 (f-TiO_2) are more capable towards photo-degradation of methyl orange and rhodamine B dyes. As a result, the photocatalytic degradation strategy becomes more advantageous than adsorption in terms of doing complete degradation of pollutants. However, effective adsorption of pollutants is also required to attain an outstanding photo-degradation rate. Nanostructured metal oxides have the highest adsorption abilities towards organic pollutants. The development of smart materials having modified designs with the advantage of environmental friendliness and cost-effectiveness in addition to high specific surface area and structure of a crystal is the need of the time for the removal of harmful toxins. The revival of photocatalysts and adsorbents has further proved to be an economically practical way for proper management of secondary waste generated during processing (Ballav et al. 2018). Hence, there is a significant demand for economically viable, environmentally benign, and more effective approaches to the elimination or degradation of organic pollutants for clean and safe water resources.

The nanoparticles of transition series metal oxide have displayed notable surface characteristics such as high surface area and advanced microstructural features, which made them appealing applicants for the adsorption of harmful pollutants (Henglein 1989). Metal oxide nanomaterials have possessed a specific adsorbent capacity for the effective extraction of pollutants from wastewater sources (Li et al. 2015). Transitional metal oxides and their respective composites have displayed outstanding photocatalytic activities for the degradation of harmful pollutants (Ahmed et al. 2011). The existence of controllable morphology, crystalline structure, and bandgap values in semiconducting nanomaterials have enhanced their potential in environmental remediation activities. The existence of oxygen over the exterior surface of semiconducting nanomaterials has further affected the photocatalytic degradation rate for organic contaminants. The photon intake (energy) through semiconducting material has further been stimulated the oxygen and promoted water clean-up by turning organic contaminants into less toxic components and ultimate conversion to water, CO_2, and inorganic ions (Gupta et al. 2015). The high surface area of nanoparticles has further promoted the adsorption rate of incident radiations. Also, the variations in the surface to volume ratio have easily been attained by decreasing the bulk surface. Furthermore, the volume reduction in nanomaterials has brought the catalytic site's interaction on their surface for better interaction with organic molecules. On comparing with the bulk materials, the higher surface areas of nanomaterials have made them effective for adsorptive removal of harmful toxins from contaminated resources. Biomass-derived activated carbon dots, nanotubes, grapheme, and carbon fibres have possessed the potential to remove organic pollutants from contaminated water sources.

These nano-photocatalysts have possessed the ability to absorb the visible range of the solar spectrum and delayed the recombination rate of electron-hole pairs and act as effective photocatalysts (Zhao et al. 2005). Up to now, numerous types of metal oxide-based nanomaterials such as Al_2O_3, ZnO, TiO_2, CeO_2, CuO, and iron oxides, have displayed significant interest as adsorbents and photocatalysts in environmental remediation activities (Gaya and Abdullah 2008; El-Hankari et al. 2016). Furthermore, the efficiency of nano-catalysts has been enhanced by compositing the particles with different kinds of metal oxides to produce metal-metal oxides, graphene-metal oxides, and magnetic-metal oxides particles (Zhu et al. 2016; Zhang et al. 2013a, b). These modified nanocomposites have better control over the magnitude, surface, and texture characteristics properties, which further regulate the adsorption behaviour towards toxins (Mclaren et al. 2009).

2.3.1 Advanced Nanoparticles as an Effective Adsorbent for Toxic Pollutants

The usage of nanoparticles has provided a low cost and sustainable adsorbent resource with high effectiveness, low maintenance, and multi-pollutant concurrent adsorption ability for the effective management of wastewater resources (Yang et al. 2019a, b; Tang et al. 2019). Carbon-based nanomaterials such as carbon nanotubes, graphene, biomass-derived activated carbons, clay-based materials, chitosan, metal nano-chalcogenides, inorganic complexes, layered double hydroxides (LDHs), organic-inorganic nanocomposites, metal nanocomposites and metal oxide nanomaterials have been widely used as adsorbents for removing a diverse range of organic pollutant from aqueous media (Gupta et al. 2019; Ma et al. 2016). The utilities of a different form of iron (Fe)-based nanoparticles such as magnetite (Fe_3O_4), hematite (α-Fe_2O_3), and maghemite (γ-Fe_2O_3) have been used as effective adsorbents for harmful toxins (Cornell and Schwertmann 2003). The existence of high surface area to volume ratio and superior magnetism in Fe-based nanoparticles (Table 2.1) have made them quite effective for the removal of lethal toxins from wastewater resources (Bystrzejewski et al. 2009; Selvan et al. 2010). For instance, the chlorinated pesticides like hexachlorocyclohexane (α-HCH and γ-HCH), lindane (1,2,3,4,5,6-hexachlorocyclohexane), 2,4,5-trichlorophenoxyacetic acid (2,4,5-T), dieldrin, and 2,4-dichlorophenoxyacetic acid (2,4-D) have been easily removed from water samples by using the magnetite-based nanoparticles (Mac Rae 1985). These nanomaterials have been extensively used in the form of pure, composite, and doped form for the adsorption of pollutants. The ease of separation after adsorption of toxins from water resources via the presence of their inbuilt magnetism has further supported the popularity of iron-based nanoparticles in wastewater remediation (Ambashta and Sillanpaa 2010).

The organic dyes, namely, bromophenol blue, Eriochrome Black T, fluorescein, and bromocresol green with hydroxyl groups, have a higher ability to absorb with magnetite nanoparticles as compared to the non-hydroxylated dye from aqueous media (Saha et al. 2011). The application of external surface modifiers or surface acting agents such as polyacrylic acid, glutamic acid, organosilane, chitosan, ionic

Table 2.1 List of commonly used Fe-based nanoparticles for adsorptive removal of harmful toxic dyes

Nature of adsorbent	Morphology	Type of toxin removed	Reference
Ordered Fe_2O_3 microspheres	Microspheres	Methyl orange, congo red	Kang et al. (2018)
Magnetic porous carbon-doped Fe_2O_3 spheres	Ordered spongy spheres	Methyl orange	Siyasukh et al. (2018)
Activated carbon-functionalized Fe_3O_4 nanoparticles	Hollow nanospheres	Reactive black 5	Saroyan et al. (2017)
Suspension of magnetic nanoparticles of Fe_2O_3 modified by using chitosan	Ellipsoidal and quasi-spherical-shaped Fe_3O_4 nanoparticles	Congo red	Hui et al. (2018)
Double-layered Mg/Al hydroxide of Fe_2O_3	Nanocrystalline particles	Fulvic acid	Fang et al. (2018)
Magnetically activated carbon with Fe_2O_3	Porous solid nanomaterial	Methylene blue	Altýntýg et al. (2017)
Fe_3O_4-gelatine nanoparticles	Spherical nanoparticles	Direct yellow 12	Mir et al. (2018)
Magnetic Fe/Ni-doped carbon mesoporous nanoparticles	Bimodal porous geometry	Methyl orange, methylene blue	Liu et al. (2015)
Organic-functionalized silica-Fe_3O_4-rGO nanocomposite	Silica matrix decorated over Fe_3O_4-rGO sheets	Rhodamine B	Sahu et al. (2017)
Metal-organic framework of porous cubic Fe_2O_3 nanoparticles	Rod-like aggregates	Methyl blue	Li et al. (2018a, b)

liquids, surfactants, and polymers has further enhanced the adsorption rate of iron oxide nanoparticles (Wang et al. 2012). For example, humic acid-functionalized iron oxide nanoparticles have displayed higher adsorption efficiency towards methylene blue dye as compared to bare counterparts (Zhang et al. 2013a, b). In addition, a novel hierarchical magnetic nanorods of core-shell iron oxide@magnesium silicate (HIO@MgSi) prepared by hydrothermal flexible sol-gel methodology have possessed higher adsorption kinetics towards methylene blue dye (Zhang et al. 2013a, b), whereas self-assembled layered double hydroxide colloidal nanohybrids of Fe_3O_4 have possessed 505 mgg^{-1} adsorption ability towards congo red dye (Chen et al. 2011). Additionally, the surface reactivity with favourable zero-point charge ranges from 6 to 6.8, and controlled structural and textural characteristics of titanium dioxide (titania) nanomaterials have also made it useful adsorbent materials for the removal of toxins from aqueous media. The amendable size, shape, phase structure, and crystallinity of titania have further inspired its application in adsorption and photocatalytic degradation of pollutants. For instance, the 1D nanorods and 3D microspheres of TiO_2 have displayed morphologically modified featured-dependent photocatalytic phenol adsorption, and degradation behaviour in aqueous media (Liu et al. 2008). The flower-like nanostructure of sodium titanate (TMF) with one-dimensional nanoribbon morphology has indicated the higher adsorption

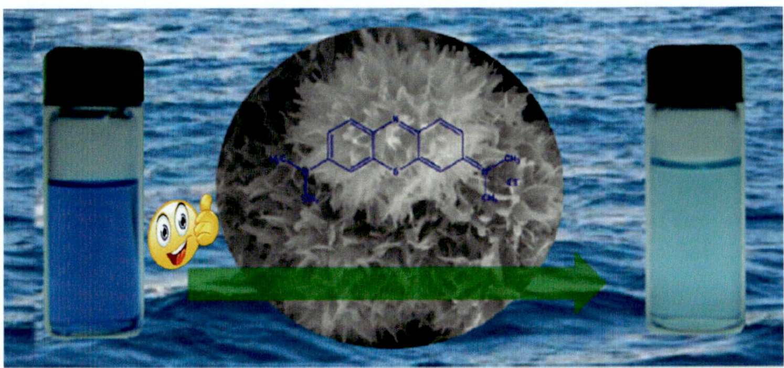

Fig. 2.2 Schematic illustration of change in methylene blue colour dye adsorption over the flower-like surface of designed nanostructures. (Adapted figure from Feng et al. 2013, with permission from copyright, American Chemical Society, Washington, DC, USA)

ability towards methylene blue from the wastewater sample (Feng et al. 2013). The electrostatic association between positively charged methylene blue and negatively charged titanate is mainly responsible for the higher efficacy of sodium titanate towards wastewater treatment (Fig. 2.2).

The higher surface area and extensive surface-active sites over the mesoporous titanium have further facilitated the adsorption process and made the removal fast and effective for different types of dyes molecules (Xiong et al. 2010). In addition, carbon nanotube (CNT)-stacked TiO_2 nanomaterials can adsorb more than 95% of methyl orange dye in 10 min. The existence of hydrogen bonding, hydrophobic interactions, and π-π interactions between the modified CNT-stacked TiO_2 and dye molecules can improve the efficacy of used nanoparticles (Ahmad et al. 2017). In addition, the surface adsorption behaviour of nanoparticles has also been enhanced by coating with potential surfactants. For instance, the adsorption of acid red 1 and acid blue 9 dyes were promoted by functionalized titanium nanotubes with hexadecyltrimethylammonium chloride as a coating agent (Lee et al. 2007). The degradation efficiency of particles has also been modified by doping the nanoparticles with metal ions such as Zn and Cu (Khairy and Zakaria 2014).

Additionally, the ease of preparation, economically viable synthesis, non-toxicity with controlled surface, and textural properties of ZnO have made it a suitable adsorbent for wastewater remediation applications. Chitosan-modified ZnO nanoparticle can adsorb direct blue 78 and acid black 2 dye from aqueous media (Salehi et al. 2010). The comparative studies of pure chitosan and chitosan@ZnO nanoparticles have displayed a higher efficacy of nanoparticles towards pesticides (Dehaghi et al. 2014), whereas the effectiveness of ZnO nanoparticles has also been enhanced by doping the surface with external metal ions such as Cr^{3+} ions for the effective removal of anionic dyes from the aqueous media. The presence of Cr^{3+} ions can affect the surface defects in ZnO nanoparticles (Meng et al. 2015). Moreover, the hollow microsphere of ZnO-ZIF-8 [$Zn(MeIM_2)$] has further improved the selective dye adsorption behaviour of ZnO. The incorporation of $NiFe_2O_4$ in

ZnO nanoparticles has further reduced the photo corrosion chances in nanoparticles and increased their adsorption ability towards congo red (Zhu et al. 2016).

A range of other nanomaterials such as CeO_2, Cu_2O, MoO_2, MoO_3, SnO_2, and ThO_2 has also been used for the adsorptive removal of different types of pollutants from aqueous media (Brock et al. 1998). For instance, the applications of hierarchic nanoporous MnO_2 have been used for the removal of methylene blue dye from aqueous media. The application of three different types of sugar (glucose, fructose, and sucrose) molecules during the synthesis has induced the morphological and structural changes in particles for preparing better adsorbent for dye molecules (Kim et al. 2017). The temperature has also induced the adsorption variations in MnO_2 nanoparticles and facilitated the dye removal from aqueous media (Chen and He 2008; Murray 1974). The electrostatic force of the attraction from methylene blue to negative MnO_2 promotes the effective removal of cationic dyes. The zero point charge value of ~2.4 has further contributed to the negative surface charge of MnO_2 over a broad pH range and their efficacy towards cationic dye (Kim and Saito 2018). The morphological changes such as cubic or spherical hollow nanostructures of MnO_2 have also displayed variations in the adsorption capacity of nanoparticles towards congo red (Fei et al. 2008). The hierarchical porous $Ni(OH)_2$ and NiO particles have applications for removing organic dyes (Cheng et al. 2011). The nanomaterials of ThO_2 are highly isoelectric (pH = 9–11), making them ideal for adsorbent dyes (Parks 1965). For instance, the monodisperse ThO_2 nanoparticles demonstrated an excellent potential to remove congo red dye from wastewater. In comparison with the spherical ThO_2, the monodisperse asymmetrical nanoparticles have advanced adsorption ability for the congo red (Wang et al. 2014). The CeO_2 hydroxyapatite nanoparticles showed good adsorption behaviour towards Eriochrome Black T dye (Wu et al. 2016). The bidirectional connections of the Ce^{4+} cation and the group of sulfonate dye molecules further improved the adsorption capacity (Chaudhary et al. 2016).

The superior surface properties of graphene made it an attractive alternative for dye removal application from aqueous media (Xiao et al. 2016). The adsorption ability for the cationic dyes was significantly higher in comparison with anionic dyes in the case of graphene molecules. This behavioural aspect has been explained due to the adequate electrostatic bonding between the organic dyes with the negative oxygen residual functionalities of the rGO (Gupta and Khatri 2017). The adsorption of malathion, chlorpyrifos, and endosulfan pesticides has been taken care of by using GO and rGOs. The water-mediated interaction of graphene with pesticides is mainly responsible for the effective removal of toxins (Maliyekkal et al. 2013). Metal oxide nanoparticles over the external surface of graphene and GO with variable structure, texture, and composition and surface characteristics have further increased the adsorption capacity towards toxins and expedited the adsorption and recovery rate (Upadhyay et al. 2014). For instance, Fe_3O_4-rGO nanocomposite was utilized for the removal of harmful pesticides, namely, ametryn, simazine, simeton, prometryn, and atrazine from the contaminated water resources. The inherent magnetic properties of iron oxide-graphene nanocomposites have facilitated the separation of used adsorbent from the water by applying the external

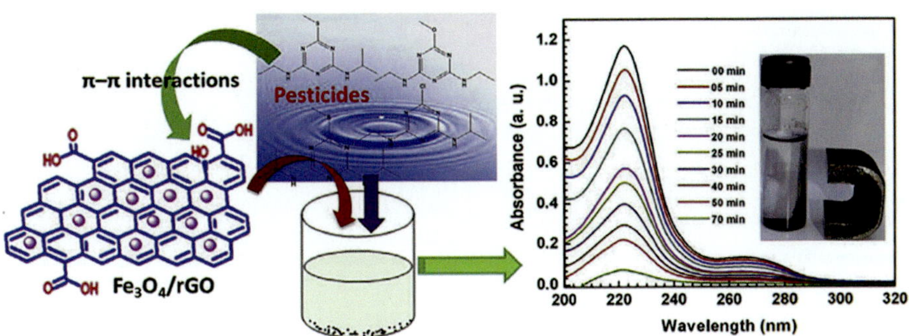

Fig. 2.3 Schematic illustration is showing pesticide removal by using Fe_3O_4/rGO. The gradual decrease in absorption was observed over time. (Adapted figure from Boruah et al. 2017, with permission from copyright, Elsevier publishing company)

magnetic force (Fig. 2.3). The used nanoparticles can be easily segregated and retrieved during the process and employed for the next adsorption cycle (Boruah et al. 2017). The strategy involves phase transfer without decomposition of current pollutants and converts the solution into the active phase over adsorbent.

Industrialization bloom in the twenty-first century produces a large number of pollutants to the environment. Among them, the medical industry is the new alarming pollutant generator (Tiwari et al. 2020; Bharathiraja et al. 2019). With the course of time, the usage of antibiotics for treating several microbial diseases increases day by day, and one of the major drawbacks of these antibiotics is that they are not completely absorbed on ingestion and 70% of these antibiotics are excreted via urine and faeces to the environment in active form (Bao et al. 2020; Zheng et al. 2019; Zhou et al. 2019). No doubt that the current concentration of these antibiotics in the surface water is about ng L^{-1} to μg L^{-1} range, but their long time exposure can cause drug-resistant microbial stains. On taking concern of these pollutants, Song et al. (2020) and his co-workers synthesized nickel (Ni) (II)-modified porous boron nitride nanomaterial as a potential adsorbent for the antibiotic pollutant (tetracycline) removal from aqueous solution. In this respect, Ni (II) was successfully attached with a porous boron nitride (BN) surface to improve the adsorption capacity for tetracycline (TC) (Fig. 2.4).

The synthesized Ni (II)-modified porous BN nanostructure has shown outstanding percentage removal of 99.76% for TC. Consequently, this analysis shows the excellent ability of the Ni-modified porous BN as an effective adsorbent for wastewater treatment. In this context, Xiang et al. (2020) work on the fabrication of viable manganese ferrite-customized biochar from vinasse for improved adsorption of antibiotic (fluoroquinolone) pollutants. They performed a batch adsorption experiment to check the adsorptive behaviour of fluoroquinolone antibiotics ciprofloxacin (CIP) and pefloxacin (PEF) over fabricated nanomaterials. While performing this experiment, they work on various experimental factors such as adsorbent dose, pH, temperature, ionic strength, and contact time because these factors affect the adsorption of PEF and CIP. These excellent porous structures show outstanding adsorption

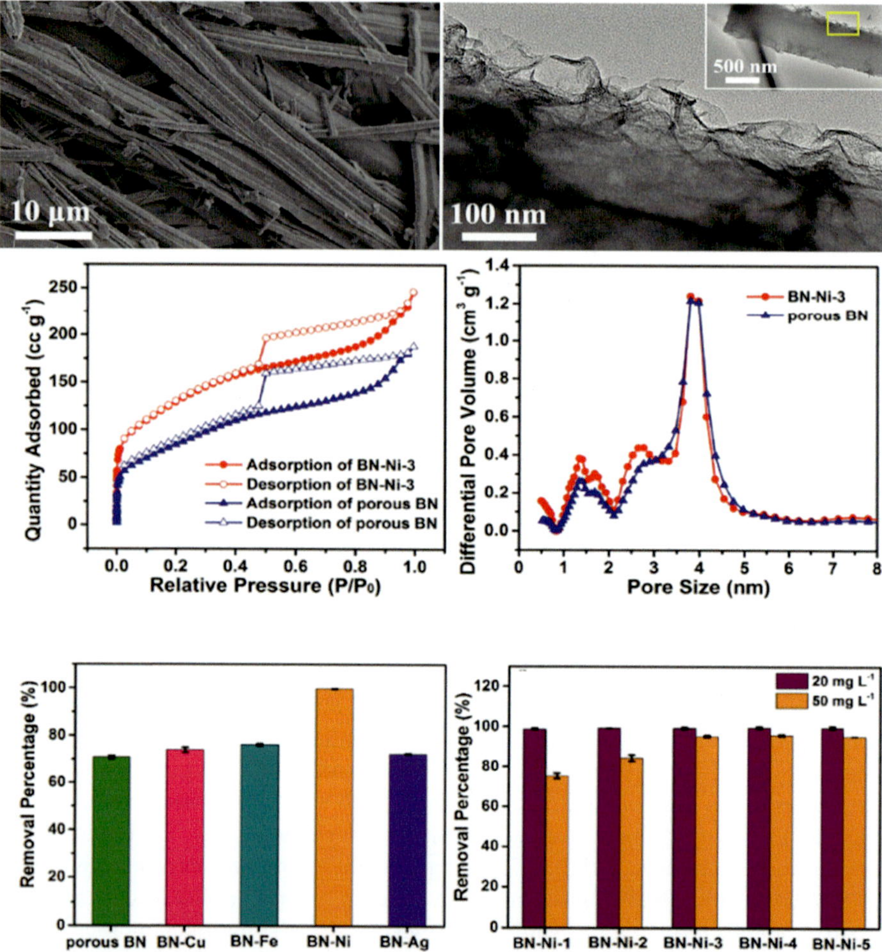

Fig. 2.4 (**a**) SEM image, (**b**) TEM image, and compared removal percentages of (**c**) porous BN and various metal-modified BN and (**d**) Ni (II)-modified porous BN with different loading. (Adapted figure from Song et al. 2020, with permission from copyright, Elsevier publishing company)

ability for CIP and PEF was 135 and 270 mg g^{-1}, respectively, and obeyed the pseudo-second-order kinetics. The controlled adsorption mechanisms comprised of pore filling effect, hydrogen bonding, π–π EDA, hydrophobicity, and π–π stacking interaction (Fig. 2.5). Fabricated manganese ferrite-modified biochar (FMB) nanostructures not only solved the problem of antibiotic pollutant but also utilized the vinasse wastes and used them as a raw material of biochar.

Adsorption of antibiotic pollutants has considered one of the ways to separate them from the aqueous medium, but the major challenge is to degrade them completely. Qiao et al. (2020) made advanced nanomaterials that are not able to adsorb

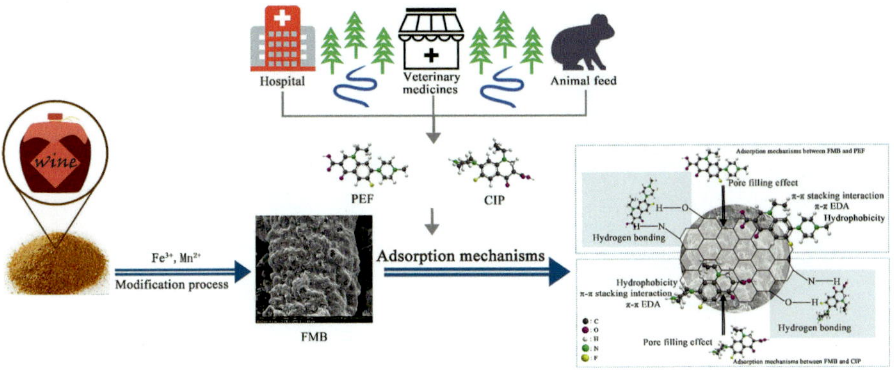

Fig. 2.5 Schematic diagram showing collection of antibiotics from different sources, synthesis of FMB, and possible mechanisms of PEF or CIP adsorption onto the FMB. (Adapted figure from Xiang et al. 2020, with permission from copyright, Elsevier publishing company)

the pollutants but also have the potential to degrade them photocatalytically. Fabricated magnetic graphene oxide/ZnO nanocomposites (MZ) show effective adsorption and photocatalytic degradation of tetracycline (TC) (Fig. 2.6). To understand the adsorption process, different kinetic models were used, among which pseudo-second-order kinetics model suits well for TC and adsorption capacity found was 1590.28 mg g^{-1}. To support this mechanism of adsorption between MZ and TC, the π–π interaction, cation exchange, electrostatic attraction, complexation, and hydrogen bonding were key factors. After successful adsorption, the authors have tried to explore the mechanism of photocatalytic degradation of TC. During the study, they found that the graphene oxide works as the scaffold and helps in electron transfer and ZnO acts as an ideal photocatalyst. While exploring the catalytic efficiency of MZ, it is found that these nanocomposites continued relatively high levels of removal after four cycles.

2.3.2 Advanced Nanoparticles as Effective Photocatalysts for Degrading Harmful Toxins

The application of nano-catalysts for decomposing organic pollutants offered a clean, viable, and environment friendly approach to treat the harmful toxins. The nano-photocatalytic strategy can eliminate harmful contaminants from wastewater without using harmful chemicals. These nano-catalysts offered sophisticated oxidative methodologies for the environmental clean-up by adsorption. This absorption has further led to the separation of charge by pushing electrons from the valence band and resulted in the formation of positive holes in the conduction band (Akpan and Hameed 2009). These generated positive holes have the tendency to oxidize the water into a hydroxyl radical ($OH^{\bullet}$) and further reduced the adsorbed oxygen

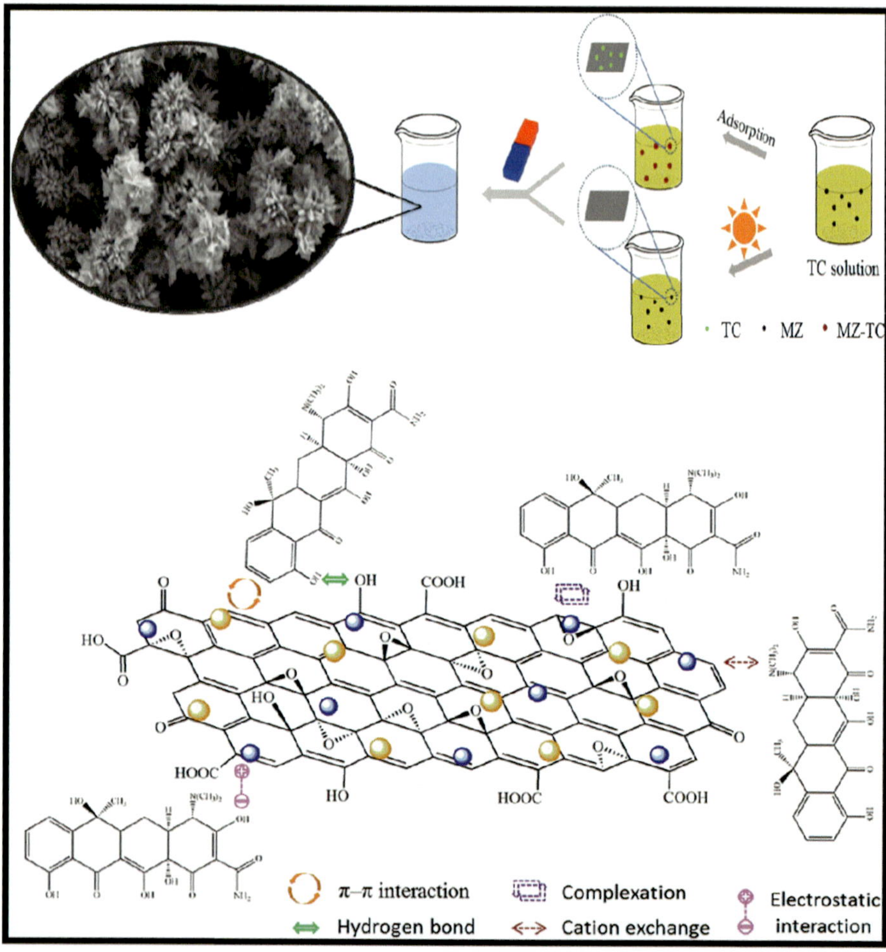

Fig. 2.6 Schematic illustration for the formation of magnetic graphene oxide/ZnO nanocomposites (MZ) and adsorption mechanism of TC over MZ. (Adapted figure from Qiao et al. 2020, with permission from copyright, Elsevier publishing company)

molecules present at the surface. Generated radicals have further employed for the transformation of organic contaminants or pollutants into non-toxic degraded form. However, the recombination rate of electron and hole must be controlled for producing the effective photocatalytic performance of nanomaterials. Also, the nature of the photocatalyst, light, intensity, type of organic pollutants, response temperature, pH, sacrificial reagents, and solvent have affected the effectiveness of photocatalytic degradation (Konstantinou and Albanis 2004; Rauf et al. 2011). Generally, the rate of mono azo dyes photo-degradation is higher than anthraquinone dyes (Khataee and Kasiri 2010). Till now, a diverse range of nano-catalysts such as ZnO, TiO_2, ZrO_2, SnO_2, Fe_2O_3, CdS, and WO_3 have been employed for the photocatalytic degradation of toxic pollutants (Zhao et al. 2005).

Out of all, low price, chemical inertness, extensive availability, photocatalytic stability and non-toxic environmental conduct have made TiO_2 as an extraordinarily acceptable and promising photocatalyst for harmful toxins (Akpan and Hameed 2009; Schneider et al. 2014). The high bandgap value of TiO_2 has further enabled the photons to absorb radiation of wavelength equivalent to 387 nm (Nassar et al. 2017). In the presence of solar radiation, electrons from the valence band get excited and jump into the conducting band and generate pairs of electron-hole in TiO_2 nanoparticles (Khataee and Kasiri 2010). The hole (h^+) in the valence band in the alkaline medium can convert water molecules to hydroxylic radicals. However, the molecular oxygen can react with H_2O molecules and form a strong oxidizer, i.e., H_2O_2, which is further reduced by photo-generated electrons inside the conduction band (Fig. 2.7). These formed oxidizers then degrade organic pollutants into harmless and non-toxic components, i.e., CO_2 and H_2O (Khataee and Kasiri 2010).

TiO_2 nanoparticles can degrade large varieties of organic dyes such as Orange G, malachite green, alizarin green, rhodamine 6G, Solvent Red 23, indigo carmine, and Sudan III, in the presence of UV light. It has also been observed that the wide bandgap values of anatase and rutile forms of TiO_2 (3.0–3.2 eV) have contributed towards the catalytic activity of TiO_2 (Kim et al. 2013). In addition, substantial research activities have been carried out by researchers to make TiO_2-based nanostructures degrade harmful toxins in visible light and become active photocatalysts. Budarz et al. (2019) explored the photo-degradation capacities of TiO_2 nanoparticles for the degradation of organophosphorus pesticide (chlorpyrifos) in the presence of visible light. It has been observed that 80% of pesticide was degraded within 24 h in the presence of TiO_2 nanoparticles. The composites of TiO_2 with ceramic have further enhanced the porosity and photocatalytic activities of nanocomposites for

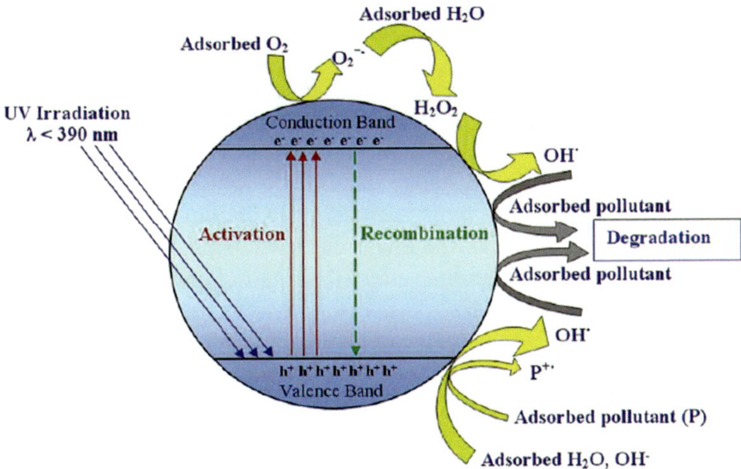

Fig. 2.7 Photocatalytic reaction schematic explanation. The electron-hole pairs generated through photon make organic contaminants more degradable. (Adapted figure from Khataee and Kasiri 2010, with permission from copyright, Elsevier publishing company)

photo-degradation of harmful pesticides, including lindane, atrazine, dipterex, thiobencarb, malathion dimethoate, and bentazone in aqueous media. In addition, TiO_2-supported zeolitic composites (HBETA) have also been used for the degradation of dichlorvos pesticide. The TiO_2 contents in the HBETA have possessed higher photocatalytic activities. TiO_2 loading in HBETA has been significantly increased by 20% for dichlorvos as compared to bare TiO_2 (Gomez et al. 2015). However, one associated disadvantage with TiO_2 was its higher rate of replication of photogenerated pairs, which further reduced the catalytic potential of nanoparticles. This problem can be easily overcome by adding co-adsorbents as effective dopants in TiO_2. This addition has delayed the combination of electron-hole pairs with photogenerated particles and improved their photocatalytic efficiency towards harmful toxins. For instance, the presence of MWCNT with titanium dioxide has enhanced the surface properties of nanoparticles and found to be effective for methyl orange dye (Saleh and Gupta 2012).

Sarkar et al. (2016) have used the efficacy of biocomposites of amylopectin-functionalized TiO_2 with Au nanoparticles (NPs) for removing toxic organic dyes. The synergistic effects of both adsorption and photocatalytic degradation activities of nanoparticles are responsible for the degradation of dye molecules under sunlight. The developed photocatalyst is quite selective for the degradation of methyl violet dye from the mixture of dye solution by employing plasmonic photocatalysis phenomena. The degradation of dye molecules mainly occurred due to the photoexcitation of TiO_2 particles in the presence of UV light present in sunlight. This will further lead to the self-photosensitization of methyl violet dye. The photoexcitation of dye molecules to excited state (MV*) has infused an electron into the conduction band of TiO_2 molecules. Moreover, the presence of UV light can excite the TiO_2 particles and produced electron in the conduction band (Fig. 2.8). These formed electrons in the conduction band of TiO_2 can react with the adsorbed water and oxygen molecules to form reactive oxygen radicals, which are mainly responsible for the degradation of dye molecules.

In addition, the dopant molecules can modify the surface characteristics such as surface acidity, active site, and defects of TiO_2. Such modifications are further useful for active photocatalytic activities of harmful toxins under visible light (Zhao et al. 2005; Nassar et al. 2017). For instance, lanthanide ion-doped TiO_2 nanoparticles have displayed higher photocatalytic activities for the degradation of direct blue dye. This mainly occurred due to the lowering of the bandgap of TiO_2 in the presence of lanthanide ions, which further affect their particle size and surface area and pore volume (El-Bahy et al. 2009).

Subsequently, the application of zinc oxide nanomaterials as potential photocatalysts has been well documented in literature due to their high quantity effectiveness and low fabrication cost with excellent stability and biocompatibility towards living beings (Daneshvar et al. 2004; Sakthivel et al. 2003). The higher potential of ZnO nanoparticles was further associated with the more absorption capacity of ZnO nanoparticles towards solar radiation. Also, the fluorescence emission features of the ZnO have been contributed towards the photocatalytic degradation of numerous types of pollutants (Hariharan 2006). The relation of fluorescence intensity of ZnO

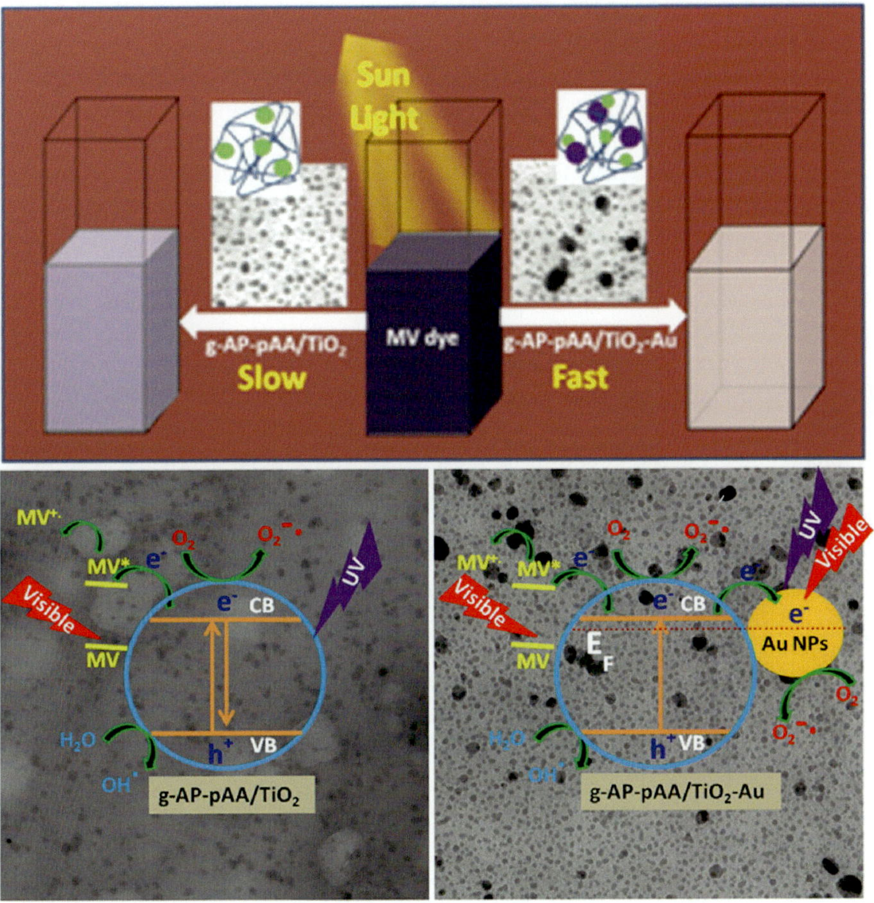

Fig. 2.8 Schematic illustration showing the photocatalytic degradation of methyl violet dye in the presence of TiO_2 with proposed scheme for photocatalytic degradation of dyes under solar light irradiation. (Adapted figure from Sarkar et al. 2016, with permission from copyright, American Chemical Society, Washington, DC, USA)

with the recombination method of electron-hole has contributed towards the effectiveness of ZnO. The presence of organic compounds over the external surface of ZnO can scavenge photo-generated holes and compete for the fluorescence emissions charging and reduced their emission intensity. The higher sensitivity of ZnO towards the pH of reaction media has further contributed towards the photocatalytic activity of prepared nanoparticles. The strong acidic nature, i.e., less than pH 4, has further supported the photo corrosion (Daneshvar et al. 2004). For example, the photocatalytic functions of nanocrystalline ZnO for the degradation of diazinon pesticide have been compared with the commercially available ZnO material. It has been observed from the data that the small crystalline size of ZnO nanoparticles has the potential to enhance the active area for the photocatalytic reaction and affected the number of discrete ZnO particles per volume in the solution. As a result, the photon absorption efficiency of ZnO nanoparticles has been enhanced tremendously towards the degradation of toxins (Daneshvar et al. 2007). As an effective photo-oxidant, the nanocomposites of $ZnO/Na_2S_2O_8$ can photo-degrade various types of pesticides by using natural sunlight (Navarro et al. 2009). However, the regular retardation of photoactivity of ZnO and rapid rate of electron-hole pairs recombination are mainly responsible for limiting the long-term usage of ZnO as photocatalyst (Chen et al. 2015). To resolve this problem, various synthetic strategies have been developed for hindering the recombination rate in ZnO particles and enhancing their stability towards photo corrosion. For instance, an electric spun strategy can synthesize hetero-nanofibers of ZnO-SnO_2 with improved photocatalytic degradability efficiency towards eosin red, methylene blue, methyl orange, and congo red dyes. These hetero-nanofibers tend to separate photo-generated electron-hole pairs and provided high adsorption ability to degrade harmful toxins (Chen et al. 2015). The developed catalyst was found to be recyclable, with efficiency greater than 90%. The presence of external dopants can change the optical and electronic characteristics of nanostructures and their sensitivity to degrade toxins in the visible spectral range. Additionally, Mn^{2+}-doped ZnO nanostructure has also enhanced the photocatalytic activity of bare ZnO nanoparticles under visible light (Ullah and Dutta 2008). On the other hand, Ag@ZnO microstructures have possessed enhanced photocatalytic activity towards methylene blue dye under UV light exposure. Ag nanoparticles can act as an electron sink during the process and have the ability to catch the electrons generating during synthesis and increasing the life span of the electron-hole pair and enhanced their efficiency towards the degradation of harmful toxin (Rafaie et al. 2017).

Similarly, Cu@ZnO nanorods were applied by deferring the recombination of electron-hole pairs for degrading organophosphorus compounds (Shirzad-siboni et al. 2017). Pd-doped ZnO particles can reduce borohydride with efficiency greater than 96% (Fig. 2.9) (Guy et al. 2016). Er-doped ZnO nanoparticles with a varying percentage amount of Er can degrade toxins under a visible spectrum of light. The presence of Er ions can modify the bandgap values of ZnO particles and also modified the half-life span of the electron-hole pair. These variations have a direct influence over the photocatalytic degradation towards direct red-31 dye (Bhatia et al. 2016).

Additionally, the wide bandgap of cerium oxide (3.1 eV) has also made it useful nanomaterial in catalytic activities. The higher biocompatibility of nanoparticles with high stability towards heat and light has further supported their efficacy towards catalysis (Park et al. 2000; Shchukin and Caruso 2004). The photocatalytic capacity of ceria has mainly depended on the calcination temperature during the fabrication of nanoparticles. The positively charged phosphorus-containing pesticides are more effective in getting adsorbed over the surface of ceria via the electrostatic interaction between the hydroxyl functionality of ceria and pesticide molecules. Also, the particles synthesized at lower temperatures have possessed more number of hydroxyl groups and found to be more effective towards the adsorption of organophosphate pesticides (Janos et al. 2014). The efficacy has further been enhanced by

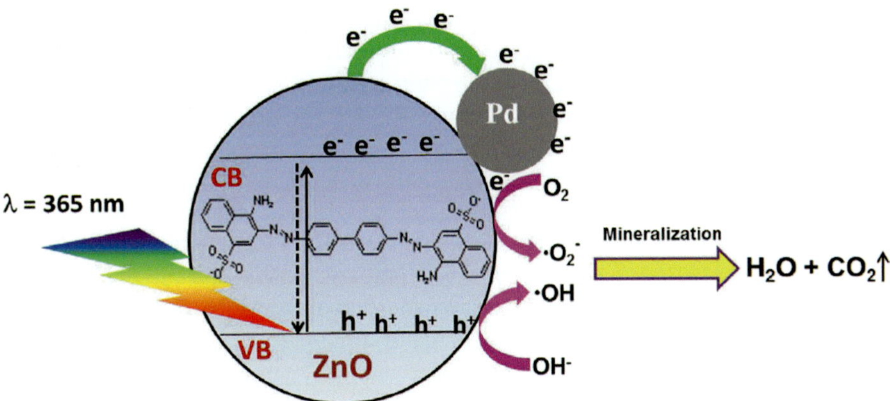

Fig. 2.9 Schematic illustration showing the photocatalytic degradation of congo red dye using Pd@ZnO photocatalyst. The diagram shows the possible photocatalytic mechanism for congo red dye degradation using Pd-doped ZnO photocatalyst. (Adapted figure from Guy et al. 2016, with permission from copyright, Elsevier publishing company)

co-doping the ceria particles with magnetically separable Fe_2O_3 particles. The formed composite was quite stable and displayed maximum degrading efficacy with a temperature range of 300–400 °C. On further enhancing, the calcination temperature over 500 °C has decreased the activity due to the reduction in the surface area and pore size of ceria particles (Janoš et al. 2015). The single-dimensional CeO_2 nanotubes with a high-pitched interior have possessed superior photocatalytic degradation efficiency towards phenols in comparison with commercially available CeO_2 nanoparticles (Tang et al. 2011). The degradation efficacy of ceria particles has also been modified by doping the particles with CuO. The H_2O_2-generated hydroxyl radicals have also accelerated the degradation process towards organic pollutants (Xu et al. 2014). The different molar ratio of the binary metal oxide of CeO_2 with Y_2O_3 has displayed degradation efficiency of 98% towards rhodamine B. The improvement in photocatalytic activity has been explained due to the enhancement in the active surface and increased oxygen vacancies in particles (Magdalane et al. 2017).

The low bandgap values of CuO particles have also enabled the absorption of visible light and enhanced the scope of advanced materials towards photocatalysis. The electronic composition and surface activity of CuO nanoparticles have also been modified by doping the same with MoS_2. The presence of graphene oxide and other metal oxides such as SiO_2, TiO_2, and ZnO has affected the light absorption charges in CuO and further improved their photocatalytic efficiency. For instance, the hierarchical 3D-ordered structures of ZnO with Fe_3O_4 particles have improved the photocatalytic activity towards the degradation of congo red dye (Malwal and Gopinath 2016). The engineering of bandgaps has further improved the charge separation and improved the photocatalytic effectiveness towards congo red dye under visible light conditions. The photocatalytic activity of CuO@ZnO nanoparticles gets doubled under the visible light-irradiation condition as compared to ZnO for the degradation of a wide range of textile dyes. The presence of CuO with ZnO can produce structural flaws and stoichiometric impairment in particles and shifting their energy values and enhancing their photocatalysis activities. In addition, the semiconductor p-n joint of CuO-TiO_2 with natural zeolite can degrade azo dyes under visible light exposure of 60 min (Zhao et al. 2015). The enhancement of interfacial charge transfer from the active surface of particles, as well as bandgap modification, is primarily responsible for enhancing CuO's photocatalytic activity under visible light. The heterogeneous composite of MoS_2-CuO nanoparticles has a better efficiency in degrading methylene blue in comparison with the pristine form of MoS_2. The presence of tiny band offsets in nano-flowers of MoS_2-CuO particles has provided the existence of stalled type II band alignment in nanomaterials. As a result, no electron or holes have been trapped and easily migrated over the surface and degrade the adsorbed dye (Hossain et al. 2016). Therefore, these particles have possessed an outstanding option for environmental remediation towards a wide range of organic pollutants for wastewater management.

Other metal oxides such as n-type semiconducting nanoparticles of SnO_2 with broad bandgap value of ~3.6 eV have also favoured the degradation of toxins at a wide range of pH media (Long et al. 2011). Modifying the surface of SnO_2 with MgO particles has displayed a negative impact over the combination of photogenerated electrons and enhanced the photocatalytic properties of prepared nanocomposites (Bayal and Jeevanandam 2013). The chitosan-functionalized SnO_2 have possessed superior degradation activity towards methyl orange dye as compared to rhodamine B. This behavioural aspect has been explained due to the effective adsorption of dye molecules over the chitosan-functionalized SnO_2 (Gupta et al. 2017). In addition, the degrading response in the presence of light has further been modified by varying the ratio of chitosan and SnO_2. The higher stability of the nanostructure of Cu_2O in a broad pH range has also supported their efficacy towards the degradation of toxins (Mao et al. 2007). SnO_2 doping in Cu_2O has prevented the oxidation of Cu_2O oxidation and encouraged the photocatalytic action towards textile dyes (Bai et al. 2014; Bai et al. 2015). The heterostructures of WO_3-TiO_2, Fe_2O_3-WO_3, and MoO_3-TiO_2 nanocomposites have also displayed higher dye removal ability as compared to single counterparts. The potential energy variations between Fe_2O_3 and WO_3 have allowed superior photocatalytic activity towards organic pollutants (Bayal and Jeevanandam 2013). The p-n heterojunction structures of Co_3O_4/$BiVO_4$ composites have displayed excellent stability and the effective ability for degrading organic toxins. This aspect has been associated with the decreased recombination rate of photoluminescence-endorsed photo-generated charge carriers in bi-composites (Long et al. 2006). Currently, Bi_2O_3 particles with a bandgap value of 2.8 eV have possessed higher oxidation capacity in the valence pair and degraded acetaminophen (Xie et al. 2013). The direct oxidation of acetaminophen has been suggested to degrade the toxin in the presence of β-Bi_2O_3. The doping of Bi_2O_3 with transition metals oxides has significantly increased its photocatalytic ability towards organic pollutants (Bai et al. 2015).

2.3.3 Biomedical Applications of Advanced Nanomaterials

The demand for highly functional biomaterials has increased, as biomedical engineering has become more accurate and sophisticated (Shan et al. 2019; Yousaf et al. 2020). Novel hybrid nano-biomaterials (Fig. 2.10) have unique chemical, physical, and optical properties that are now being developed and implemented in different areas of biomedicine (Park et al. 2020; Montanheiro et al. 2020). Polymer-, metal-, and composite-based biomaterials have a significant context of biomedical use in the diagnosis and effective in the treatment of diseases (Zare et al. 2019; Dubey et al. 2020). Different kinds of biocompatible polymers have been improved because of the individualities of polymers with biodegradability and biocompatibility (Asghari et al. 2017). Hybrid nanocomposites are usually described as a

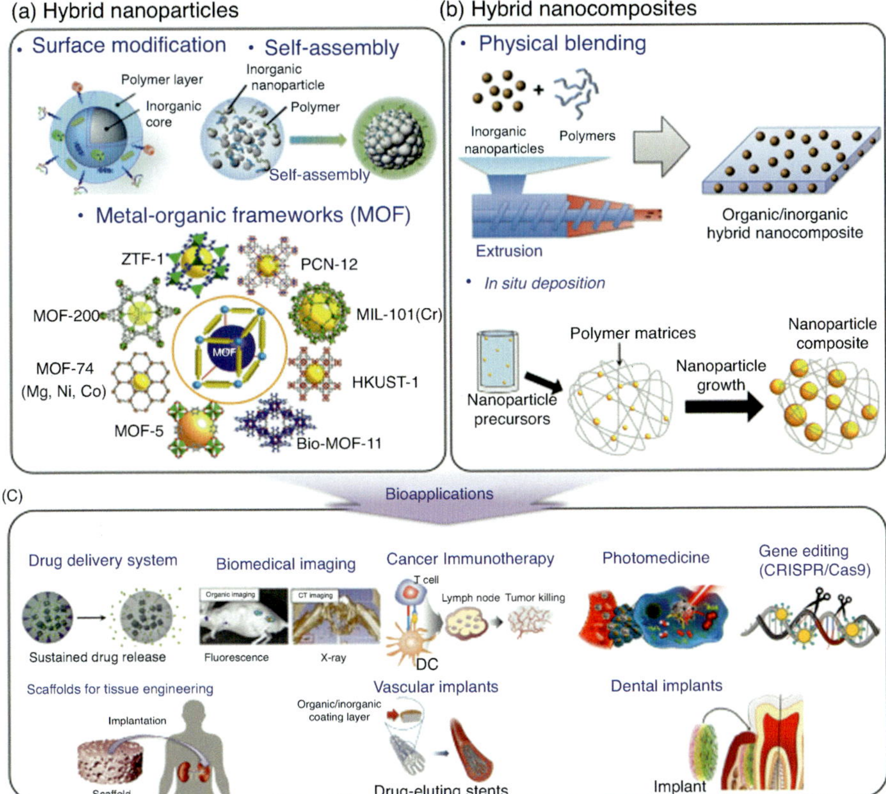

Fig. 2.10 The schematic representative of hybrid nanomaterials: (**a** and **b**) types and synthesis of representative organic/inorganic hybrid (**a**) nanoparticles and (**b**) nanocomposites and (**c**) bioapplications of the organic/inorganic hybrid materials. (Adapted figure from Park et al. 2020, with permission from copyright, Elsevier publishing company)

combination of submicron-sized organic and inorganic materials (Azhar et al. 2018; Rasheed et al. 2020).

First of all, we will discuss the core/shell nanoparticles; they are widely described as two-part hybrid materials, a core and a shell where both can be composed of either organic or inorganic material (Yang et al. 2019a, b). Over time, researchers have developed different core/shell structures for biomedical applications, in which inorganic core parts have silica nanoparticles, gold nanoparticles, quantum dots, or magnetic nanoparticles. In contrast, organic shell parts have biomolecules, polymers, or lipids (Katz 2019; Fanizza et al. 2020; Zhuang et al. 2017). In general, the inorganic components of core/shell nanostructures contribute to both imaging and therapeutic implications due to their unique physicochemical, electrical, optical, and magnetic properties (Zhang et al. 2018a, b, c; Ju et al. 2019). In addition, the biocompatibility and stability of inorganic nanostructures are improved by the organic part of the core/shell nanostructures, which allow large numbers of therapeutic agents to be loaded simultaneously to exploiting theranostic abilities (Lin et al. 2017; Hameed et al. 2018). Inorganic nanomaterial surfaces typically need to be modified to deal with biological conditions; for this, the surface of inorganic nanomaterials can be fabricated with organic materials to overcome their toxicity (Mao et al. 2019; Dong et al. 2019). Mangadlao et al. (2018) fabricated prostate-specific membrane antigen-targeted gold nanoparticles (AuNP-5kPEG-PSMA-1-Pc4) and characterized them by spectroscopic and surface imaging techniques. Results show that the fabricated AuNPs were stable for a wide range of buffers, solvents, and media. Theranostic agents (AuNP-5kPEG-PSMA-1-Pc4) have prostate-specific membrane antigen (PSMA-1) and consist of fluorescent photodynamic therapy (PDT) drug, Pc4 (Fig. 2.11a).

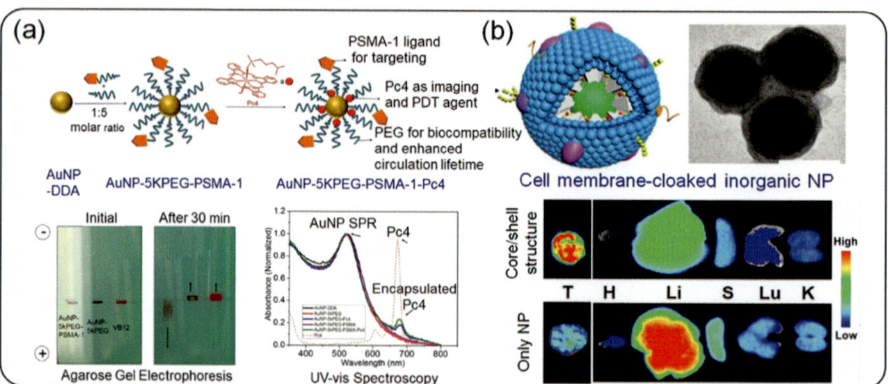

Fig. 2.11 (**a**) Fabricated theranostic agent (AuNP-5kPEG-PSMA-1-Pc4) showing successful binding (5kPEG-PSMA-1) and loading (Pc4) to AuNPs. (**b**) Cell membrane-cloaked core/shell nanostructures for enhanced in vivo bio-distribution. (Adapted figure from Park et al. 2020, with permission from copyright, Elsevier publishing company)

Further to this, they demonstrated the potential binding of 5kPEG-PSMA-1 to AuNPs, with the help of agarose gel electrophoresis and the successful incorporation of Pc4 into the nanoparticle system, and confirmed that using UV-vis absorbance spectroscopy. In vitro and in vivo studies reveal that the fabricated nanostructures can provide surgical guidance and treatment in cases where surgery is ineffective for prostate tumours. In some studies, it is found that the nanoparticle surfaces have been coated with membrane from cancer cells, stem cells (Fig. 2.11b), and macrophages. The coating of these membranes provides a simple top-down approach for the fabrication of complex and different functions to the nanocarriers required for effectual bio-interfacing. The features of source cells are naturally imitated by nanoparticles surrounded by cell membranes and help to perform a broad range of functions like disease-related targeting and tissue tropism. The next important application of these hybrid nanomaterials has its theranostic properties.

The term theranostics intends (Lee et al. 2017) a concept of uniting both diagnoses (through MRI, magnetic resonance imaging; CT, computed tomography; MPI, magnetic particle imaging, etc.) and therapeutic therapy (like drug delivery and controlled release, etc.) together into a single stage that allows us to monitor and treat disease in the real-time environment (Fig. 2.12). The targeted drug delivery to the

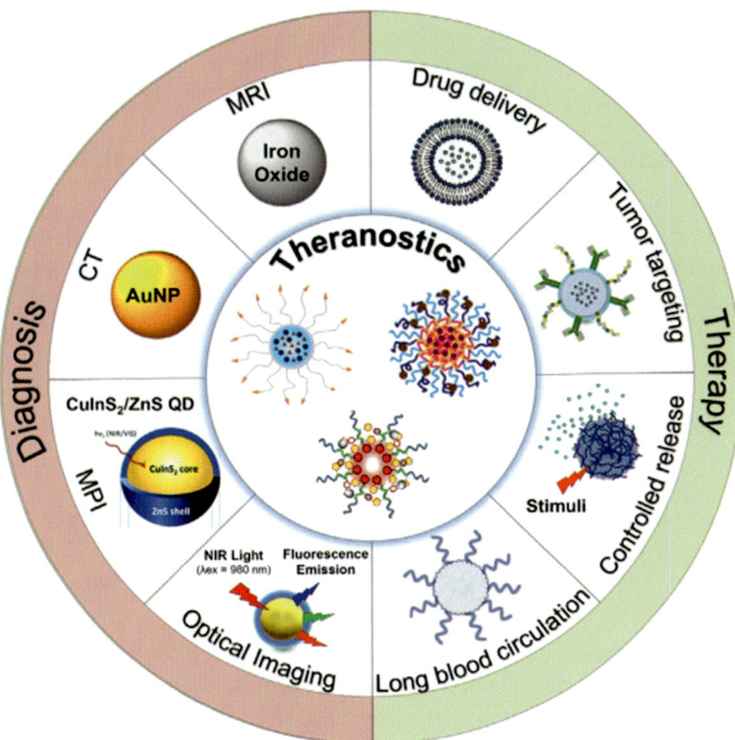

Fig. 2.12 Schematic illustration of the theranostic concept. (Adapted figure from Park et al. 2020, with permission from copyright, Elsevier publishing company)

specific tissue and therapeutic responses capability of designed nanomaterials inspires us to develop personalized medicines for accurate disease control (Tahara 2020; Mu et al. 2018). Due to these amendable properties of hybrid nanomaterials, the world is moving towards the era of nanomedicine (Sharma et al. 2018; Bhise et al. 2017).

These nanomedicines have the potential to specifically deliver the required amount of drug to the infected area without delay of time. In the last part of this section, we will discuss the recently known photodynamic therapy (PDT), which is an assuring non-invasive method for the treatment of numerous diseases. Li et al. (2018a, b) fabricate the tumour pH-sensitive photodynamic nano agents (PPNs) comprised of self-built photosensitizers (PSs) fasten to pH-responsive polymeric ligands (PLLs) and upconversion nanoparticles (UCNPs) and examine an enhanced PDT effect. Quenching was observed in fluorescence at blood pH (~7.4); however, it gets unquenched rapidly at tumour microenvironment pH (~6.5) because of the ionization of the imidazole groups (pKa = 6.8). Consequently, the lanthanide centre of PPNs absorbs these photons after irradiation by NIR light at pH 6.5. It emits high energy visible light, thus exerting a PDT phenomenon by stimulating chlorin e6 (Ce6) of PLLs. PPNs showed improved antitumor results against innate tumours in A549 tumour-bearing nude BALB/c mice, signifying that they could surmount the limitations on the PDT effect due to the low tissue-penetration capability of visible light (Fig. 2.13).

2.4 Environmental Impact of Advanced Nanoparticles

The current era has witnessed the excessive utilization of advanced nanomaterials in different types of applications. Several research groups have developed innovative approaches for water rehabilitation by using advanced nanoparticles for supplying safe drinking water. Various strategies have been developed so far for producing environmentally viable nanoparticles for safe usage in water treatment processes. Although the prospective benefits of nanomaterials in water treatments have been well established at the same time, their related health hazards towards living beings need a better understanding of their widespread usage in wastewater management.

2.4.1 Environmental Fate of Advanced Nanomaterials

Natural nanomaterials came into existence by the triggering of volcano explosions, woodland flames, humidity, stone weathering, precipitation reaction, and biological processing in the environment. The released gases from open street smoke, combustion of gasoline in vehicles, jet planes, and agricultural products are also responsible for the emission of harmful nanosized particulate materials in the atmosphere. The free nanoparticulate in the atmosphere can react chemically with other existing

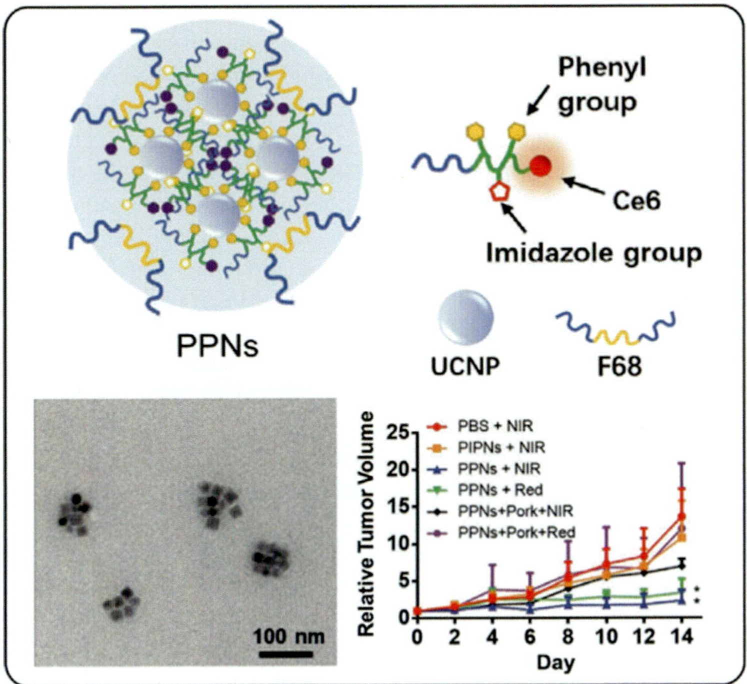

Fig. 2.13 Hybrid nanoparticles for advanced photodynamic therapy (PDT): upconversion nanoparticle enhancing PDT effect by increasing the tissue-penetration depth of light (Li et al. 2018a, b). (Adapted figure from Park et al. 2020, with permission from copyright, Elsevier publishing company)

pollutants and create a secondary source of pollution (Shi et al. 2001). Daughton (2004) has described the toxic impact of nanomaterials and their ineradicable xenobiotic counterparts in the air. Their wide-spreading ability in the air has also enhanced their concentration compared to their source. Therefore, better knowledge of their diffusion, agglomeration, wet-dry processing, and gravitational alteration is vital for understanding the toxic impact of nanomaterials (Wiesner et al. 2006). At the reduced limit of 1–100 nm size range, dispersive pressure and electron conductivity have produced a dominant impact with decreased grain volume (Mädler and Friedlander 2007). However, the inertial and gravitational forces are more effective at the micrometre scale. However, the Brownian movement of particles that controls the diffusion of particles and flux rates is inversely proportional to the diameter of the particles. This high-speed motion of nanoscaled particles is essential in conversion procedures, where the velocity of agglomeration of particles is controlled by fluid motion in the cell. As a result, the small concentration of nanomaterials in the aerosolized state has possessed a higher agglomeration ability. However, the deposition ability of nanomaterials is depended on the gravitational adjustment speed in the media, which is further equivalent to the particulate diameter. The agglomeration will substantially improve the deposition ability of nanomaterial.

In addition, the nanomaterials can enter the marine atmosphere via an instant waste heat released from soil fluid-containing nanomaterials or through soil fluid or air conversion (Weinberg et al. 2011). Interestingly, multiple technologies such as nanodegradation, nanoadsorption, and nanofiltration, with the widespread usage of nanoparticles, are few of the sources to affect the quality of water resources. Therefore, the application of nanomaterials should be made with the utmost care to the atmosphere. Consumer goods containing a diverse range of nanomaterials are one of another source to add nanomaterials in water resources. The excessive usage of nanomaterials in environmental remediation processes has also transformed these materials into water sources and further reached soil and underground water resources (Brar et al. 2010). The solubility rate of nanomaterials is found to be more as compared to their mass equivalent (Müller 2007). In addition, the seafloor microlayers composed of lipids, carbohydrates, and DNA with humic acid colloids can transport the dissolved nanomaterials in the aquatic environment. These interactions have further supported the higher rate of solubility of nanomaterials in aqueous media. The presence of organic natural materials in aquatic media has also supported the water regulation of nanomaterials. The existence of high surface to mass proportions of nanomaterials has also supported the capturing efficiency of sedimentary soil and partial solid components. The photo-degradation has also affected the physical and chemical characteristics of nanomaterials in aquatic environments and altered their performance in the environment (Nurmi et al. 2005). Nanomaterials have further interacted with the particulates in aquatic media and lead to the aggregation of particles and lead to the sedimentation of particles (Tiede et al. 2016). Therefore, the fate of nanomaterials in the aquatic environment has been influenced by distinct processes like biotic, abiotic, aggregation, and distribution relationships between nanomaterials and water components. General water chemistry has also played a substantial role in changing the size and velocity of nanomaterials. In such processes, redox reactions, soluble sorption, aggregation, and pH of water have also played an outstanding role in defining the fate of nanomaterials.

On the other hand, soil is considered as a multilayered food web context matrix and displayed the complicated connection between organic/inorganic gas and wildlife (Mukhopadhyay 2014). Soil conditions are highly complicated and varying in nature. Therefore, it is highly hard to produce a clear prediction of the final destiny of nanomaterials in soil. It has been found that the soil matrix has affected the spreading power of nanoparticles and affected their movement in soil samples. The transportation of nanomaterials has also been affected by the modifications of the physical and chemical characteristics of nanomaterials in soil samples (Li et al. 2006). The small-sized nanomaterials can reach the plants grown in the soil media and affect their growth (Fig. 2.14) (Conway et al. 2015). The high surface area of nanomaterials has made them able to stick strongly to the soil matrix and made them inert and immobilized in nature. The size of NPs, chemical composition, and soil features have also influenced the soil ability towards nanoparticles.

Yusefi-Tanha et al. (2020) have demonstrated the effect of copper oxide nanoparticle (CuONP) size and concentration on soybean (*Glycine max* cv. Kowsar) plant during its lifecycle. They examined potential phytotoxicity of CuONPs mainly

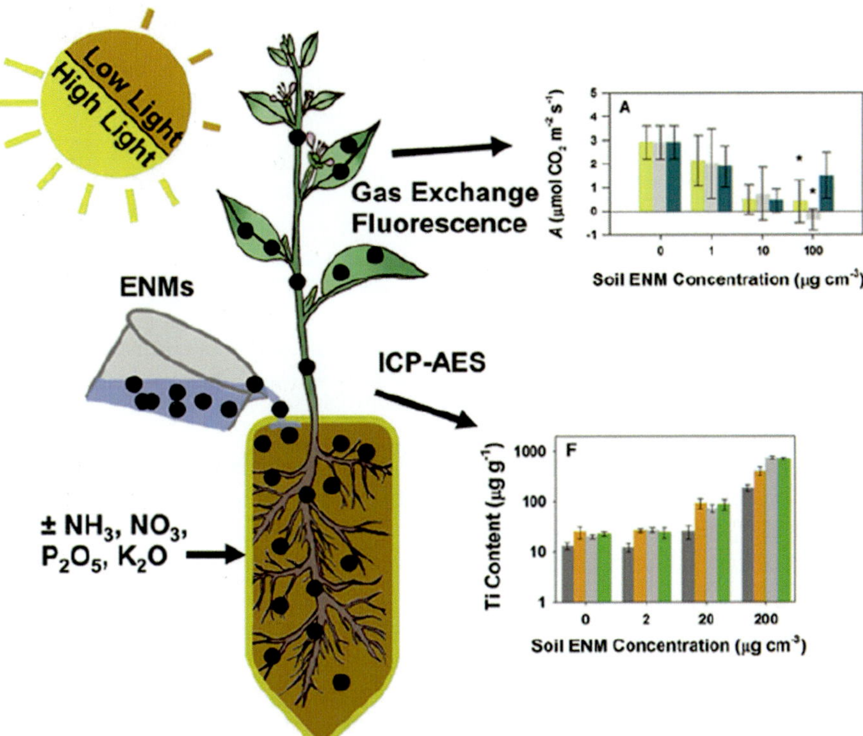

Fig. 2.14 Fate of engineered nanomaterials (ENMs) from soil to a different part of plants and their quantitative analysis in terms of Ti content and photosynthetic rate (*A*). (Adapted figure from Conway et al. 2015, with permission from copyright, American Chemical Society, Washington, DC, USA)

focused on particle size and concentration and studied the response of various antioxidant defence biomarkers (SOD, CAT, POX, APX). The research team made an effort to fabricate three distinct sizes of CuONPs – 25, 50, and 250 nm (CuONP-25, CuONP-50, CuONP-250, respectively), which have a monoclinic crystal structure. Seeds were sown in pot soil having N-fixing bacteria (*Rhizobium japonicum*), and each pot has two seeds examined for 120 days (Fig. 2.15). Results show that the smallest size (CuONP-25) has higher toxicity for the majority of antioxidant biomarkers in comparison with bigger size (CuONP-50, CuONP-250) or dissolved free Cu^{2+} ion treatments. The concentration-dependent curves for CuONP-25 and Cu^{2+} ions show linear response, whereas the bigger size CuONPs (CuONP-50, CuONP-250) have a nonlinear response for the majority of antioxidant biomarkers examined. Different forms of metal nanoparticles have a different impact on plant growth and antioxidant markers. To study this phenomenon, Elhaj Baddar and Unrine (2018) investigate the effects of ZnO NPs and $ZnSO_4$ on seed germination and plant growth of wheat (*Triticum aestivum* L.) with pot experiments. Findings from this experiment suggest that the ZnO NPs have more potential to increase Zn content significantly as compared to $ZnSO_4$ in *Triticum aestivum* L. plant.

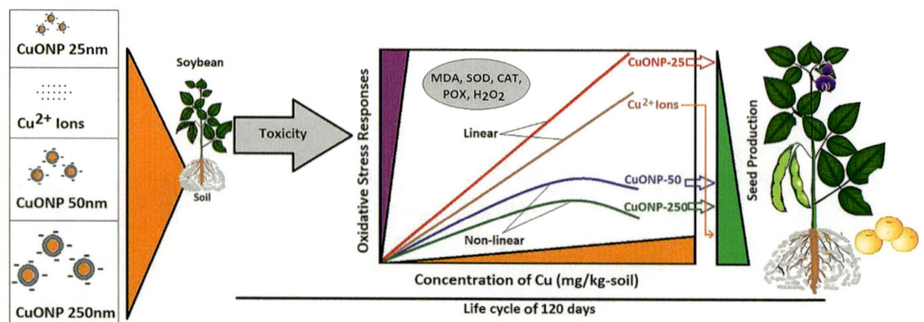

Fig. 2.15 Schematic illustration of the effect of copper oxide nanoparticle (CuONPs) size and concentration on soybean (*Glycine max* cv. Kowsar) plant during its lifecycle. (Adapted figure from Yusefi-Tanha et al. 2020, with permission from copyright, Elsevier publishing company)

Moderate doses of both ZnO NPs and $ZnSO_4$ help to gain seed germination and biomass of the plant. One interesting thing that comes while examining the results was that at higher doses, $ZnSO_4$ was found to be more toxic in comparison with ZnO NPs. Higher doses of $ZnSO_4$ show an inhibitory effect on seed germination, shoot length, and root length. So from the study, they conclude that the ZnO NPs do not have nano-specific risk and can be explored further for better results. Apart from studying the effect of nanoparticles on seed germination and plant growth, a research team (Ke et al. 2020) from the Netherlands and China studied the toxicity of silver nanoparticles (AgNPs) to *Arabidopsis thaliana* plant flowering and floral development. To examine the effect of AgNP dose (12.5 mg/kg) by employing on parental (P-AgNPs) and offspring (O-AgNPs) generation of *Arabidopsis thaliana* flowering and floral development (Fig. 2.16), the observation that comes out from the experiments shows that the presence of P-AgNPs considerably reduces the petal and pollen viability, which leads to a decrease in pod production. This negative impact further affected the offspring and became highly severe in the O-AgNP group. During this research, they found that the negative impact generated in the floral part also transferred to the offspring and showed that these NPs create a great risk to our food safety and security. In vivo studies reveal the actual picture of the toxicity intensity of NPs to the different parts of the plant. Ullah et al. (2020) and their research team studied the in vivo phytotoxicity of PbS nanoparticles in maize (*Zea mays* L.) plants (Fig. 2.17). They studied the phytotoxicity generated in plants due to plant taking and translocation of PbS nanoparticles. To study the whole mechanism of toxicity, they synthesized PbS NPs of size range 15 ± 6 nm and concentration ranging from 5 to 50 mg/L of PbS NPs and 1.5 mg/L for Pb^{2+} ions. All of these concentrations inhibit the germination rate and root elongation in maize plants due to chlorotic effects. After taking up of PbS NPs by maize plant, STEM mapping was done to investigate the effect of NPs and found that NPs easily crossed the cell wall and enter into the cytoplasm and intercellular space of the cortical cell of maize seedlings by symplastic and apoplastic pathways. The bioaccumulation rate of PbS NPs was found to be more in comparison with the Pb ions in root and shoot of the

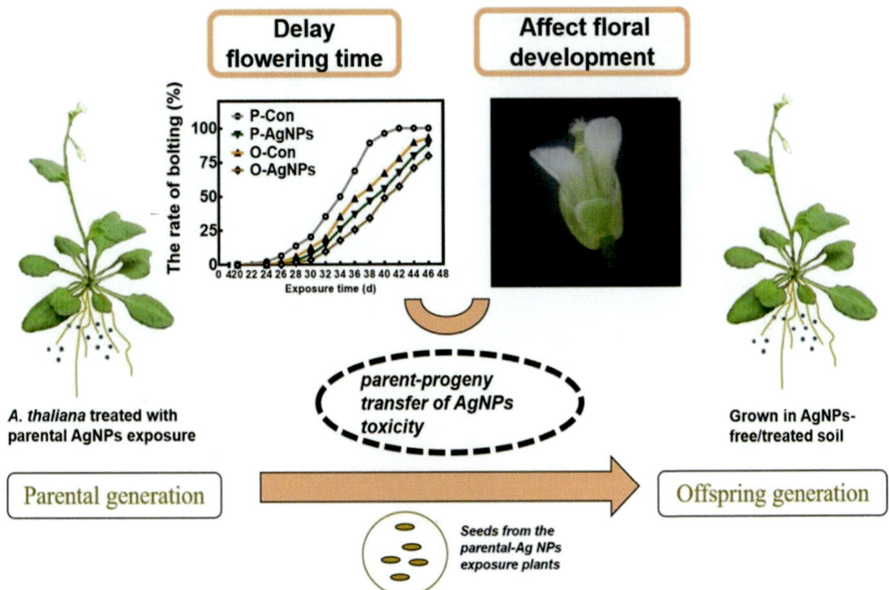

Fig. 2.16 Diagrammatic illustration of the toxicity of silver nanoparticles (AgNPs) to *Arabidopsis thaliana* plant flowering and floral development. (Adapted figure from Ke et al. 2020, with permission from copyright, Elsevier publishing company)

maize plant. So this study reveals how the accumulation of NPs not only affects the plants, but their consumption by humans and animals can create a serious threat to their life. The flexible nature of nanoparticle has also influenced the soil properties, such as pressure, porosity, and seed size. Surface photo-generated reactions have further changes the physicochemical characteristics of nanoparticles.

It has further been observed that the nanomaterials in soil sample act as an ordinary colloid and affect the pore characteristics of nanomaterials. The power of the sorption of soil has also been affected in the presence of nanomaterials (Lecoanet and Wiesner 2004). Like mineral colloids, the presence of electrical charging modifications in soil and sediment has produced a substantial effect on the circulation of nanomaterials in soil samples.

2.4.2 *Related Risk and Toxic Impact*

Several studies have found that nanomaterials are carcinogenic and toxic. However, it is still not clear whether the effect of poisoning is mainly based on the characteristic properties or the chemical structure of nanomaterial. The excessive usage has generated the release of a diverse range of nanomaterials in environmental resources such as air, water, and soil (Zhu et al. 2019). The disposal of nanomaterials has been

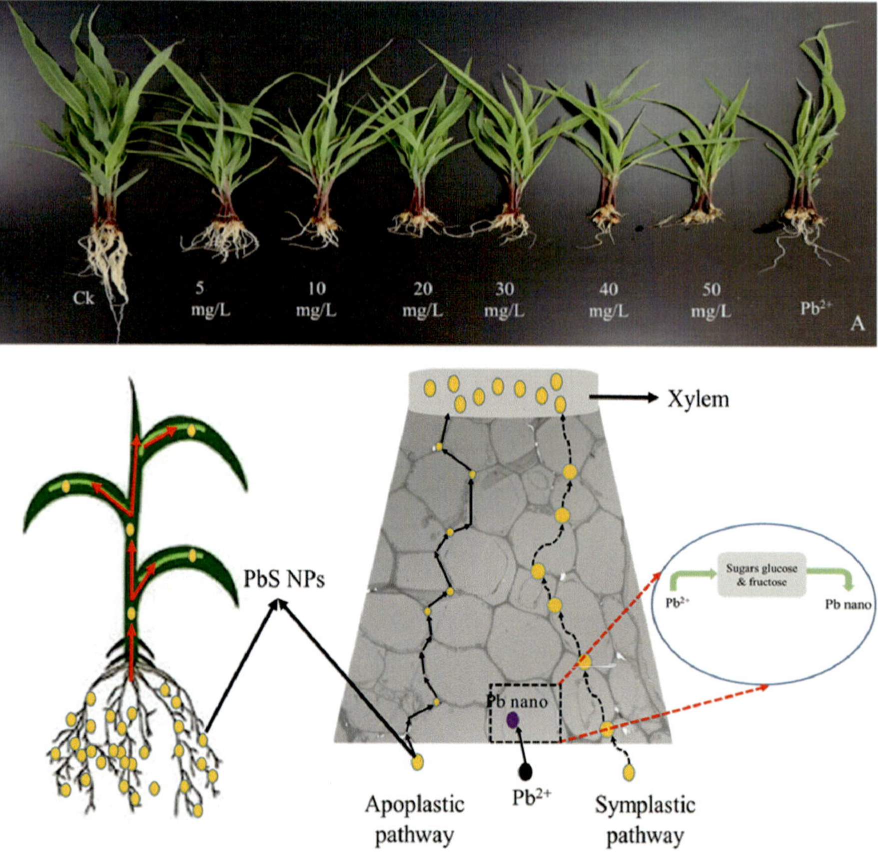

Fig. 2.17 Schematic illustration of the effects of PbS NPs and its uptake and translocation mechanisms in maize (*Zea mays* L.) plants. (Adapted figure from Ullah et al. 2020, with permission from copyright, Elsevier publishing company)

mainly dependent on their characteristic properties. These nanomaterials can leech out to water and soil surface and act as a source of pollution to plants, animals, and microbes (Fig. 2.18). In conjunction with the current growth of synthesis and implementation policies, it is essential to discuss the knowledge gap of their toxic impact on living beings.

Better understanding of bioavailability and toxicity imparted by nanomaterials has been assessed by measuring their impact on biotic flora and fauna (Alengebawy et al. 2021). The exposure rate of the living being is started during all phases of nanoparticle synthesis and their related application. The production, storage, usage, and disposal of advanced nanoparticles are few of the phases where living beings came across to the exposure of these particles. The released nanoparticles into the atmosphere have posed a dangerous impact on living beings. After the desired application, all employed nanomaterials have been inadvertently rounded up the air,

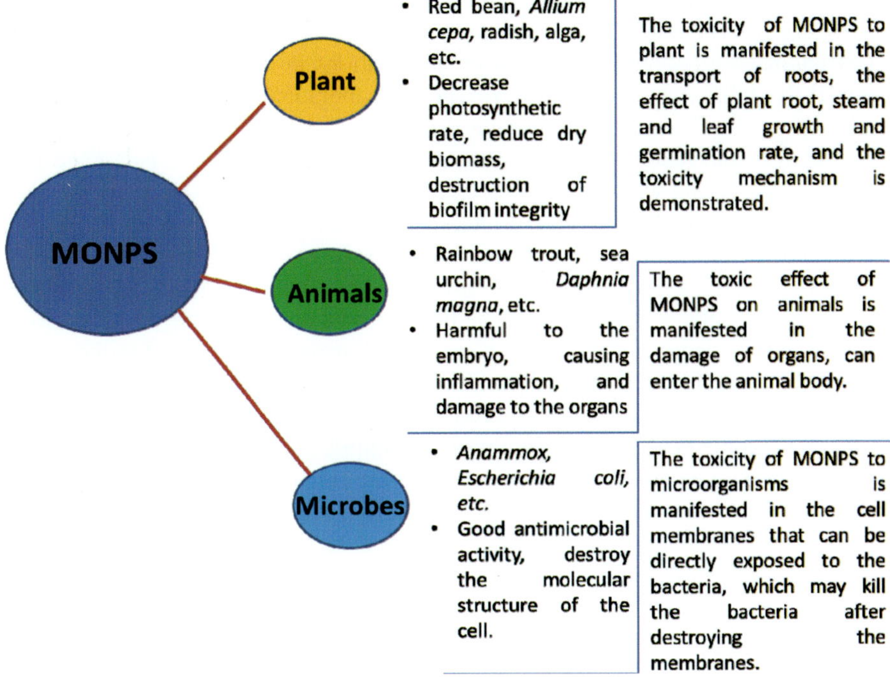

Fig. 2.18 Toxic effects of metal oxide nanoparticles (MONPs) on plants, microorganisms, and animals

water, and soil. Human exposure has also produced skin allergies, inhalation problems, and cancer in the extreme case (Chrishtop et al. 2021)). Therefore, a better understanding of the bioavailability and toxicity imparted by used nanomaterials is necessary to overcome the impact of the danger associated with these advanced materials.

2.4.2.1 Toxic Impact of Nanomaterials on Human Beings

The proposed mechanisms of toxicity imparted by different types of nanomaterials must be understood to evaluate the periling effect of nanomaterials on human beings. The overexposure of living beings with nanomaterials has affected the normal functioning of vital organs. The thorough investigation of both the amount of exposure and the particular time of ingestion of nanomaterials has given a piece of direct evidence for the toxicity evaluation of nanomaterials (Bengalli et al. 2021). The distribution of absorbed nanomaterials in the living organism is an important variable to study the toxicity assessment of nanomaterials. The overexposure of nanomaterials generally occurs to the employee during synthesis and their utilization in hi-tech products. Heavy metal exists naturally in the environment, but it has been introduced into water bodies, air, and soil as a result of human and natural activity and has become a serious global issue. Aquatic systems are the main recipient of engineered nanomaterials, and transformations in these environments have received the most attention. Transformations in the aquatic environment include various physical, chemical, and biological processes, such as aggregation, dissolution, sorption, and redox reactions (Zhang et al. 2018a, b, c). Toxic nanomaterials contained in drinking water are posing a global concern that is growing by the day. They are responsible for a variety of fatal diseases in humans. For example, Al can cause neurotoxicity by creating reactive oxygen species (ROS) and blocking DNA repair enzymes. They are responsible for causing damage to organs such as the brain, liver, and bones within human bodies. At the same time, they cause serious illnesses such as cancer, Alzheimer's disease, Parkinson's disease, and reduction in fertility both in male and in female (Rehman et al. 2020). The handling of raw materials during the fabrication has also produced ill effects over the health of human beings (Rafieepour et al. 2021). In addition, the shipping and transportation of nanomaterials are the few other sources of the exposure of nanomaterials. The overusing of consumer products is another source of contact of nanomaterials to living beings that can lead to damaging and destructive impacts on living beings (Resnik 2019). The nanoparticles can enter the lung areas via inhalation and get accumulated by macrophages in the alveolar region that move to the bronchial and bronchial areas, depending on their magnitude (Cho et al. 2012). The quantity of particulate matter inside the living system is alarming and transferring the materials to the main bloodstream and affecting the normal functioning of the body. On inhaling these small-sized particulates, these particulates can reach up to the deep tissue due to the considerably smaller size of the chest area (Fig. 2.19). Kreyling et al. (2012) have observed that the gold nanoparticles on inhaling have the ability to overcome an air-blood barrier. The particles of size in the range of 1.4 nm can easily get translocated inside the body and affect the living system. Their research has also supported that the translocation rate of nanoparticles is dependent on their size and aggregation rate. These particles have further been reached to the main bloodstream and end up in secondary organs. Another study has shown that high metal oxides or other nanomaterials that form part of the granular part of bio-persistent substances could lead to lung inflammation. Also, the elastic form of the electrons is a major factor determining pulmonary toxicity imparted by nanomaterials. The organizational variations between mineral substrates and produced nanofibers, such as nanotubes, nanorods, and nanowire, were noted to produce a significant impact on the normal functioning of human being.

In addition, the dermal skin layer is the source of exposure to these nanomaterials. It has been observed that these particles can either affect the living system locally or have the ability to pass through the body and distributed inside the living organs through the bloodstream (Warheit and Donner 2015). The other usage of cosmetic products such as creams, lotions, and sunscreens that contain ZnO and TiO_2 particles has made the skin more allergic and highly sensitive to these nanomaterials (Lee et al. 2020). The additional sources of contact to these nanomaterials

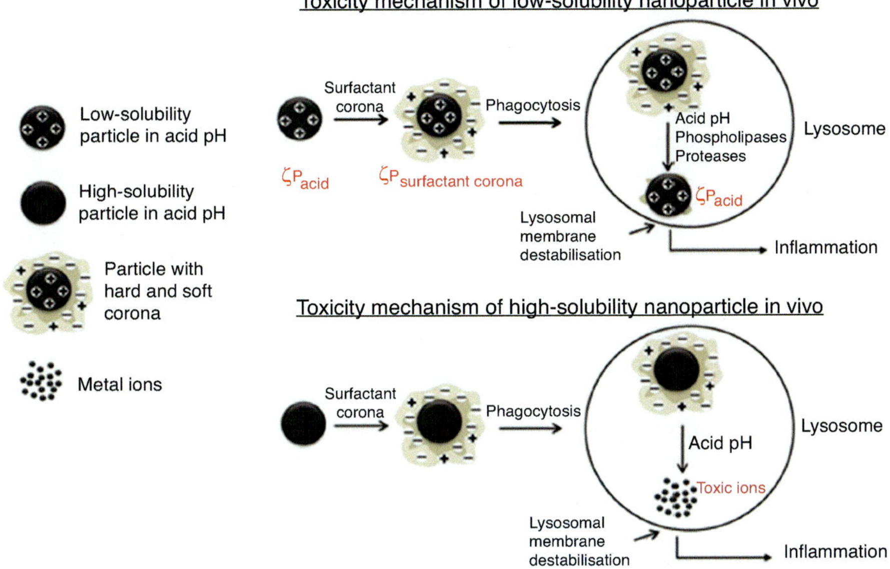

Fig. 2.19 Schematic illustration showing the mechanism of lung inflammation caused by metal/metal oxide nanoparticles. (Adapted figure from Cho et al. 2012, with permission from copyright, Oxford University Press)

include emission of car tailpipe, ultra-fine waxing particulate, and natural gasoline products (Sun et al. 2017). The solder fumes, organic oil, and coal emissions have also made human beings to encounter these particulate matters. The entry of nanomaterials to the living nervous system mainly occurred via the intercellular epidermal transition or through the pores of the skin and hair cavities. The lipid-soluble electrons reaching corneal strata cells can lead to the initial intake of nanoparticles via intercellular lipid constructions, trans tissue process, and skin. Therefore the toxicity imparted by overexposure to nanomaterials is highly debatable among scientists and environmentalists. Excessive exposure has produced hair cytotoxicity, long-term hair storage deficiency, and photoactivation in the body. It has been observed that the skin is usually less susceptible to the entry of nanomaterials. However, the hair follicles become more susceptible to the toxicity imparted by nanomaterials.

Additionally, the overexposure of nanoparticles has caused reduced viability, inhibition of mitochondrial activities, and initiation of apoptosis and cell death in living beings (Wani and Shadab (2021). For instance, carbon-based nanomaterials can induce mitochondrial dysfunction, oxidizing phosphorylation, and reactive oxygen species when incubated with keratinocytes into tissue culture. Inflammatory injuries may also be caused by nanomaterials in the body. It has also been observed that the proteins could be denatured and epitopes become unmasked in the presence of nanoparticles (Caruso et al. 2021).

2.4.3 Bioaccumulation of Advanced Nanomaterials

One of the important aspects of analysing the potential toxicity of nanomaterial is to study the extent to which these particles reach and build up tissues within an organism. From the studies, it has been observed that the interactions of nanomaterials with aquatic habitats (Malhotra et al. 2020) have influenced differently biofilms, bio-plants, and aquatic and terrestrial multicellular animals. For instance, TiO_2 nanoparticles (Zhu et al. 2010) can transmute from low-trophic animals such as biofilm and soil to high-tropical environments such as nematodes and sand. The toxic impact of nanomaterials (Vignardi et al. 2020) has further been enhanced if these materials are employed for the development of pesticides, crop additives, fertilizers, and crop security goods. The higher adsorption rate of these materials has further affected their transportation rate and imparts a more harmful impact on the living system. However, the impact of nanomaterials on the production of water and the consequences for natural and economic safety have been poorly understood in the literature (Resnik 2019). The applications of nanomaterials into nutrition technology have further produced an effect on the living system. The physicochemical properties of nanomaterials in particularly surface effect, concentration, and quantum tunnelling has produced a damaging impact on the living system. These characteristic properties have further influenced their conduct in the biosystem. The integrated use of polymers with nanomaterials in packaging industries has further caused the adverse effects on the living system (Chaudhary et al. 2020). Understanding the poisoning regime for nanomaterials can provide needed data on nanomaterials that will serve as instructions for their proper use. This also contributes to more creative ideas for solving future food-technology problems.

2.5 Conclusion

Engineered nanomaterials with intrinsic features, intended for a specific purpose, provide distinct functions to these nano-objects, whose manufacture and use are growing at an exponential rate. Nanotechnology offers new advanced techniques, methods, and products that bring huge advantages for medicine and industry, effective adsorbent for toxic pollutants, photochemical degradation of harmful toxins, and in agriculture field most importantly in environmental remediation. As in every human domain, evidence is mounting that, in addition to the good advances brought about by nanotechnologies, they may also have detrimental impacts on human health and environmental security. Although the mechanisms of nanotoxicity are not fully known at this time, toxic effects caused by nanoparticles have been observed at both flora and fauna. Nanomaterials can also cause air, water, and most important a soil-persistent form of pollution, which is too small to be detected easily, making nanopollution another man-made unwanted environmental impact, with uncertain effects in the long term. Larger and multicentre studies are needed to determine the human reactivity and the fate of the nanoparticles in the environment before large-scale nanotechnology is completely settled. In this chapter, we tried to cover all the aspects which can justify the advantages and disadvantages of these nanoparticles. A detail research is still in need to justify the actual role of fabricated advanced nanoparticles in environmental remediation.

Multiple-Choice Question (MCQ)

1. The term "nano" in nanotechnology stands for:
   (a) A nanometre is one-billionth of a metre
   (b) A nanometre is one-millionth of a metre
   (c) A nanometre is one-thousand of a metre
   (d) A nanometre is one-trillionth of a metre

Answer: (a)

2. The term nanotechnology was first coined by which scientist and when?
   (a) Norio Taniguchi, 1974
   (b) Eric Drexler, 1986
   (c) Richard Feynman, 1959
   (d) Sumio Iijima, 1991

Answer: (c)

3. What do you understand by the term carbon quantum dots?
   (a) Carbon quantum dots are carbon nanoparticles of 10–100 nm range.
   (b) Carbon quantum dots are small carbon nanoparticles with some form of surface passivation.
   (c) Carbon quantum dots are buckminsterfullerene, composed of 60 carbon atoms.
   (d) All of the above.

Answer: (b)

4. Nanofiltration is widely used in water purifiers; what does it mean?
   (a) A technique used macro-sized channels for filtration
   (b) A technique used for the filtration of nanoparticles
   (c) Membrane-based filtration method that uses nanometre-sized pores for filtration
   (d) None of the above

Answer: (c)

5. Quantum confinement effect in nanomaterials is observed because of:

   (a) The particles follow rules of quantum chemistry.
   (b) The size of the particle is too small to be comparable to the wavelength of the electron.
   (c) The property of both bulk and nanomaterials.
   (d) Both (b) and (c).

Answer: (b)

6. The term nanocomposites means:

   (a) Materials that incorporate nanosized particles into a matrix of standard material
   (b) A mixture of micron-sized nanoparticles
   (c) Particles having a size in sub-micron level
   (d) All of the above

Answer: (a)

7. Which one of these statements at the nanoscale is NOT true?

   (a) Aluminium at the nanoscale is highly combustible.
   (b) Silicon at the nanoscale is an insulator.
   (c) Gold at the nanoscale is red.
   (d) None of the above

Answer: (b)

8. The term graphene used to for:

   (a) A one-atom-thick sheet of carbon
   (b) A new material made from carbon nanotubes
   (c) Gold at the nanoscale
   (d) Both (a) and (b)

Answer: (b)

9. What is the term related to the study of the toxic effect of nanoparticles?

   (a) Nano-remediation
   (b) Nano-robotics
   (c) Neurotoxicology
   (d) Nanotoxicology

Answer: (d)

Short Questions

Q1. How do nanomaterials differ as compared to their bulk counterparts?

Answer: The significant size reduction has induced quantum effect and further enhanced the surface to volume ratio in nanomaterials as compared to their bulk counterparts.

Q2. Why is nanotoxicity a threat to living beings?

Answer: The high-end applications of nanoparticles have further enhanced their production rate and made living beings more prone to their exposure. With time slow deposition results into various health hazards in living beings.

Q3. Explain the properties of photocatalyst used in the degradation of organic pollutants.

Answer: The method of photo-degradation is primarily based on organic pollutant adsorption ability over the photocatalyst surface. The photocatalyst with higher adsorption surface makes the degradation process easier as compared to other remediation processes.

Q4. Why are semiconducting nanomaterials used in pollutant degradation?

Answer: The existence of controllable morphology, crystalline structure, and bandgap values in semiconducting nanomaterials have enhanced their potential in environmental remediation activities. The existence of oxygen over the exterior surface of semiconducting nanomaterials has further affected the photocatalytic degradation rate for organic contaminants.

Q5. In various literature, zinc oxide (ZnO) nanoparticles act as a potential photocatalyst. Explain.

Answer: The application of zinc oxide nanomaterials as potential photocatalyst has been well documented in literature due to their high quantity effectiveness and low fabrication cost with excellent stability and biocompatibility towards living beings. The higher potential of ZnO nanoparticles was further associated with the more absorption capacity of ZnO nanoparticles towards solar radiation.

References

Ahmad A, Razali MH, Mamat M, Mehamod FSB, Amin KAM (2017) Adsorption of methyl orange by synthesized and functionalized-CNTs with 3-aminopropyltriethoxysilane loaded TiO_2 nanocomposites. Chemosphere 168:474–482

Ahmed S, Rasul MG, Brown R, Hashib MA (2011) Influence of parameters on the heterogeneous photocatalytic degradation of pesticides and phenolic contaminants in wastewater: a short review. J Environ Manag 92:311–330

Akbari A, Amini M, Tarassoli A, Eftekhari-Sis B, Ghasemian N, Jabbari E (2018) Transition metal oxide nanoparticles as efficient catalysts in oxidation reactions. Nano-Struct Nano-Objects 14:19–48

Akpan UG, Hameed BH (2009) Parameters affecting the photocatalytic degradation of dyes using TiO_2-based photocatalysts: a review. J Hazard Mater 170:520–529

Alengebawy A, Abdelkhalek ST, Qureshi SR, Wang MQ (2021) Heavy metals and pesticides toxicity in agricultural soil and plants: Ecological risks and human health implications. Toxics 9:42–75

Alharbi OM, Khattab RA, Ali I (2018) Health and environmental effects of persistent organic pollutants. J Mol Liq 263:442–453

Altýntýg E, Altundag H, Tuzen M, Sarý A (2017) Effective removal of methylene blue from aqueous solutions using magnetic loaded activated carbon as novel adsorbent. Chem Eng Res Des 122:151–163

Ambashta RD, Sillanpaa M (2010) Water purification using magnetic assistance: a review. J Hazard Mater 180:38–49

Asghari F, Samiei M, Adibkia K, Akbarzadeh A, Davaran S (2017) Biodegradable and biocompatible polymers for tissue engineering application: a review. Artif Cells Nanomed Biotechnol 45:185–192

Azhar MR, Vijay P, Tadé MO, Sun H, Wang S (2018) Submicron sized water-stable metal organic framework (bio-MOF-11) for catalytic degradation of pharmaceuticals and personal care products. Chemosphere 196:105–114

Bai S, Zhang K, Sun J, Luo R, Li D, Chen A (2014) Surface decoration of WO_3 architectures with Fe_2O_3 nanoparticles for visible-light-driven photocatalysis. Cryst Eng Comm 16:3289–3294

Bai S, Liu H, Sun J, Tian Y, Chen S, Song J (2015) Improvement of TiO_2 photocatalytic properties under visible light by WO_3/TiO_2 and MoO_3/TiO_2 composites. Appl Surf Sci 338:61–68

Ballav N, Das R, Giri S, Muliwa AM, Pillay K, Maity A (2018) L-cysteine doped polypyrrole (PPy@ L-Cyst): a super adsorbent for the rapid removal of Hg^{+2} and efficient catalytic activity of the spent adsorbent for reuse. Chem Eng J 345:621–630

Bao LP, Chong CO, Mohamed SMS, Pau-Loke S, Jo-Shu C, Tau CL, Su SL, Joon CJ (2020) Conventional and emerging technologies for removal of antibiotics from wastewater. J Hazard Mater:122961–122968

Barnett JC (2007) Synthetic organic dyes, 1856–1901: an introductory literature review of their use and related issues in textile conservation. Stud Conserv 52:67–77

Bayal N, Jeevanandam P (2013) Sol–gel synthesis of SnO_2–MgO nanoparticles and their photocatalytic activity towards methylene blue degradation. Mater Res Bull 48:3790–3799

Bengalli R, Colantuoni A, Perelshtein I, Gedanken A, Collini M, Mantecca P, Fiandra L (2021) In vitro skin toxicity of CuO and ZnO nanoparticles: application in the safety assessment of antimicrobial coated textiles. NanoImpact 21:100282–100295

Bhamore JR, Jha S, Park TJ, Kailasa SK (2019) Green synthesis of multi-color emissive carbon dots from *Manilkara zapota* fruits for bioimaging of bacterial and fungal cells. J Photochem Photobiol B Biol 191:150–155

Bharathiraja B, Selvakumari IAE, Iyyappan J, Varjani S (2019) Itaconic acid: an effective sorbent for removal of pollutants from dye industry effluents. Curr Opin Environ Sci Health 12:6–17

Bhatia S, Verma N, Bedi RK (2016) Optical application of Er-doped ZnO nanoparticles for photodegradation of direct red – 31 dye. Opt Mater 62:392–398

Bhise K, Sau S, Alsaab H, Kashaw SK, Tekade RK, Iyer AK (2017) Nanomedicine for cancer diagnosis and therapy: advancement, success and structure–activity relationship. Ther. Deliv 8:1003–1018

Boruah PK, Sharma B, Hussain N, Das MR (2017) Magnetically recoverable Fe_3O_4/graphene nanocomposite towards efficient removal of triazine pesticides from aqueous solution: investigation of the adsorption phenomenon and specific ion effect. Chemosphere 168:1058–1067

Brar SK, Verma M, Tyagi RD, Surampalli RY (2010) Engineered nanoparticles in wastewater and wastewater sludge-Evidence and impacts. Waste Manag 30(3):504–520

Brock SL, Duan N, Tian ZR, Giraldo O, Zhou H, Suib SL (1998) A review of porous manganese oxide materials. Chem Mater 10:2619–2628

Budarz JF, Cooper EM, Gardner C, Hodzic E, Ferguson PL, Gunsch CK, Wiesner MR (2019) Chlorpyrifos degradation via photoreactive TiO2 nanoparticles: assessing the impact of a multi-component degradation scenario. J Hazard Mater 372:61–68

Bystrzejewski M, Pyrzyñska K, Huczko A, Lange H (2009) Carbon-encapsulated magnetic nanoparticles as separable and mobile sorbents of heavy metal ions from aqueous solutions. Carbon 47:1201–1204

Carneiro PA, Umbuzeiro GA, Oliveira DP, Zanoni MVB (2010) Assessment of water contamination caused by a mutagenic textile effluent/dye house effluent bearing disperse dyes. J Hazard Mater 174(1–3):694–699

Caruso G, Fresta CG, Costantino A, Lazzarino G, Amorini AM, Lazzarino G, Tavazzi B, Lunte SM, Dhar P, Gulisano M, Caraci F (2021) Lung surfactant decreases biochemical alterations and oxidative stress induced by a sub-toxic concentration of carbon nanoparticles in alveolar epithelial and microglial cells. Int J Mol Sci 22:2694–2707

Castillo M, Barceló D (2001) Characterisation of organic pollutants in textile wastewaters and landfill leachate by using toxicity-based fractionation methods followed by liquid and gas chromatography coupled to mass spectrometric detection. Analytica Chimica Acta 426(2):253–264

Chaudhary S, Sharma P, Renu R, Kumar R (2016) Hydroxyapatite doped CeO_2 nanoparticles: impact on biocompatibility and dye adsorption properties. RSC Adv 6:62797–62809

Chaudhary RG, Bhusari GS, Tiple AD, Rai AR, Somkuvar SR, Potbhare AK, Lambat TL, Ingle PP, Abdala AA (2019) Metal/Metal Oxide Nanoparticles: toxicity, Applications, and Future Prospects. Curr Pharm Des 25(37):4013–4029

Chaudhary P, Fatima F, Kumar A (2020) Relevance of nanomaterials in food packaging and its advanced future prospects. J Inorg Organomet Polym Mater:1–13

Chen H, He J (2008) Facile synthesis of monodisperse manganese oxide nanostructures and their application in water treatment. J Phys Chem C 112:17540–17545

Chen C, Gunawan P, Xu R (2011) Self-assembled Fe_3O_4-layered double hydroxide colloidal nanohybrids with excellent performance for treatment of organic dyes in water. J Mater Chem 21:1218–1225

Chen X, Zhang F, Wang Q, Han X, Li X, Liu J (2015) The synthesis of ZnO/SnO_2 porous nanofibers for dye adsorption and degradation. Dalton Trans 44:3034–3042

Cheng B, Le Y, Cai W, Yu J (2011) Synthesis of hierarchical $Ni(OH)_2$ and NiO nanosheets and their adsorption kinetics and isotherms to Congo red in water. J Hazard Mater 185:889–897

Cho WS, Duffin R, Thielbeer F, Bradley M, Megson IL, MacNee W, Poland CA, Tran CL, Donaldson K (2012) Zeta potential and solubility to toxic ions as mechanisms of lung inflammation caused by metal/metal oxide nanoparticles. Toxicol Sci 126(2):469–477

Chrishtop VV, Mironov VA, Prilepskii AY, Nikonorova VG, Vinogradov VV (2021) Organ-specific toxicity of magnetic iron oxide-based nanoparticles. Nanotoxicology 15:167–204

Conway JR, Beaulieu AL, Beaulieu NL, Mazer SJ, Keller AA (2015) Environmental stresses increase photosynthetic disruption by metal oxide nanomaterials in a soil-grown plant. ACS nano 9(12):11737–11749

Cornell RM, Schwertmann U (2003) The iron oxides: structure, properties, reactions, occurrences and uses. Wiley

Crane RA, Scott TB (2012) Nanoscale zero-valent iron: future prospects for an emerging water treatment technology. J Hazard Mater 211:112–125

Dai Y, Zhang N, Xing C, Cui Q, Sun Q (2019) The adsorption, regeneration and engineering applications of biochar for removal organic pollutants: a review. Chemosphere 223:12–27

Daneshvar N, Salari D, Khataee AR (2004) Photocatalytic degradation of azo dye acid red 14 in water on ZnO as an alternative catalyst to TiO_2. J Photochem Photobiol A Chem 162:317–322

Daneshvar N, Aber S, Seyeddorraji M, Khataee A, Rasoulifard M (2007) Photocatalytic degradation of the insecticide diazinon in the presence of prepared nanocrystalline ZnO powders under irradiation of UV-C light. Sep Purif Technol 58:91–98

Daughton CG (2004) Non-regulated water contaminants: Emerging research. Environ. Impact Assess. Rev. 24(7):711–732

Degler D, Weimar U, Barsan N (2019) Current understanding of the fundamental mechanisms of doped and loaded semiconducting metal-oxide-based gas sensing materials. ACS Sens 4(9):2228–2249

Dehaghi SM, Rahmanifar B, Moradi AM, Azar PA (2014) Removal of permethrin pesticide from water by chitosan–zinc oxide nanoparticles composite as an adsorbent. J Saudi Chem Soc 18:348–355

Dong F, Lu L, Ha CS (2019) Silsesquioxane-containing hybrid nanomaterials: fascinating platforms for advanced applications. Macromol Chem Phys 220(3):1800324–1800329

Dréno B, Alexis A, Chuberre B, Marinovich M (2019) Safety of titanium dioxide nanoparticles in cosmetics. J Eur Acad Dermatol 33:34–46

Dubey N, Kushwaha CS, Shukla SK (2020) A review on electrically conducting polymer bionanocomposites for biomedical and other applications. Int J Polym Mater PO 69(11):709–727

El-Bahy ZM, Ismail AA, Mohamed RM (2009) Enhancement of titania by doping rare earth for photodegradation of organic dye (Direct Blue). J Hazard Mater 166:138–143

Elhaj Baddar Z, Unrine JM (2018) Functionalized-ZnO-nanoparticle seed treatments to enhance growth and zn content of wheat (Triticum aestivum) seedlings. J Agric Food Chem 66(46):12166–12178

El-Hankari S, Aguilera-Sigalat J, Bradshaw D (2016) Surfactant-assisted ZnO processing as a versatile route to ZIF composites and hollow architectures with enhanced dye adsorption. J Mater Chem A 4:13509–13518

Fahmy HM, Mosleh AM, Abd Elghany A, Shams-Eldin E, Serea ESA, Ali SA, Shalan AE (2019) Coated silver nanoparticles: synthesis, cytotoxicity, and optical properties. RSC Adv 9(35):20118–20136

Fang L, Hou J, Xu C, Wang Y, Li J, Xiao F (2018) Enhanced removal of natural organic matters by calcined Mg/Al layered double hydroxide nanocrystalline particles: adsorption, reusability and mechanism studies. Appl Surf Sci 442:45–53

Fanizza E, Zhao H, Zio SD, Depalo N, Rosei F, Vomiero A, Curri ML, Striccoli M (2020) Encapsulation of dual emitting giant quantum dots in silica nanoparticles for optical ratiometric temperature nanosensors. Appl Sci 10(8):2767–2772

Fei JB, Cui Y, Yan XH, Qi W, Yang Y, Wang KW (2008) Controlled preparation of MnO2 hierarchical hollow nanostructures and their application in water treatment. Adv Mater 20:452–456

Feng M, YouW WZ, Chen Q, Zhan H (2013) Mildly alkaline preparation and methylene blue adsorption capacity of hierarchical flower-like sodium titanate. ACS Appl Mater Interfaces 5:12654–12662

Gaya UI, Abdullah AH (2008) Heterogeneous photocatalytic degradation of organic contaminants over titanium dioxide: a review of fundamentals, progress and problems. J Photochem Photobiol C Photchem Rev 9:1–12

Goerke H, Weber K, Bornemann H, Ramdohr S, Plötz J (2004) Increasing levels and biomagnification of persistent organic pollutants (POPs) in Antarctic biota. Mar Pollut Bull 48(3-4):295–302

Gomez S, Marchena CL, Renzini MS, Pizzio L, Pierella L (2015) In situ generated TiO_2 over zeolitic supports as reusable photocatalysts for the degradation of dichlorvos. Appl Catal B Environ 162:167–173

Guo J, Yuan S, Jiang W, Yue H, Cui Z, Liang B (2016) Adsorption and photocatalytic degradation behaviors of rhodamine dyes on surface-fluorinated TiO_2 under visible irradiation. RSC Adv 6:4090–4100

Gupta K, Khatri OP (2017) Reduced graphene oxide as an effective adsorbent for removal of malachite green dye: plausible adsorption pathways. J Colloid Interface Sci 501:11–21

Gupta A, Saurav JR, Bhattacharya S (2015) Solar light based degradation of organic pollutants using ZnO nanobrushes for water filtration. RSC Adv 5:71472–71481

Gupta VK, Saravanan R, Agarwal S, Gracia F, Khan MM, Qin J, Mangalaraja RV (2017) Degradation of azo dyes under different wavelengths of UV light with chitosan-SnO_2 nanocomposites. J Mol Liq 232:423–430

Gupta K, Gupta D, Khatri OP (2019) Graphene-like porous carbon nanostructure from Bengal gram bean husk and its application for fast and efficient adsorption of organic dyes. Appl Surf Sci 476:647–657

Gusain R, Khatri OP (2013) Ultrasound assisted shape regulation of CuO nanorods in ionic liquids and their use as energy efficient lubricant additives. J Mater Chem A 1:5612–5616

Guy N, Cakar S, Ozacar M (2016) Comparison of palladium/zinc oxide photocatalysts prepared by different palladium doping methods for Congo red degradation. J Colloid Interface Sci 466:128–137

Hameed S, Bhattarai P, Dai Z (2018) Cerasomes and bicelles: hybrid bilayered nanostructures with silica-like surface in cancer theranostics. Mater Chem Front 6:127–132

Hariharan C (2006) Photocatalytic degradation of organic contaminants in water by ZnO nanoparticles: revisited. Appl Catal A Gen 304:55–61

Henglein A (1989) Small-particle research: physicochemical properties of extremely small colloidal metal and semiconductor particles. Chem Rev 89:1861–1873

Holm BA, Bergey EJ, De T, Rodman DJ, Kapoor R, Levy L, Friend CS, Prasad PN (2002) Nanotechnology in biomedical applications. Mol Cryst Liq Cryst 374(1):589–598

Hossain M, Shima H, Islam MA, Hasan M, Lee M (2016) Novel synthesis process for solar light-active porous carbon-doped CuO nanoribbon and its photocatalytic application for the degradation of an organic dye. RSC Adv 6:4170–4182

Hui M, Shengyan P, Yaqi H, Rongxin Z, Anatoly Z, Wei C (2018) A highly efficient magnetic chitosan "fluid" adsorbent with a high capacity and fast adsorption kinetics for dyeing wastewater purification. Chem Eng J 345:556–565

Islam MS, Tanaka M (2004) Impacts of pollution on coastal and marine ecosystems including coastal and marine fisheries and approach for management: a review and synthesis. Mar Pollut Bull 48(7-8):624–649

Islam MR, Ferdous M, Sujan MI, Mao X, Zeng H, Azam MS (2020) Recyclable Ag-decorated highly carbonaceous magnetic nanocomposites for the removal of organic pollutants. J Colloid Interface Sci 562:52–62

Jancik Prochazkova A, Salinas Y, Yumusak C, Scharber MC, Brüggemann O, Weiter M, Sariciftci NS, Krajcovic J, Kovalenko A (2020) Controlling quantum confinement in luminescent perovskite nanoparticles for optoelectronic devices by the addition of water. ACS Appl Nano Mater 3(2):1242–1249

Janos P, Kuran P, Kormunda M, Stengl V, Grygar TM, Dosek M, Stastny M, Ederer J, Pilarova V, Vrtoch L (2014) Cerium dioxide as a new reactive sorbent for fast degradation of parathion methyl and some other organophosphates. J Rare Earths 32:360–370

Janoš P, Kuráň P, Pilařová V, Trögl J, Šťastný M, Pelant O, Henych J, Bakardjieva S, Životský O, Kormunda M, Mazanec K (2015) Magnetically separable reactive sorbent based on the CeO2/γ-Fe2O3 composite and its utilization for rapid degradation of the organophosphate pesticide parathion methyl and certain nerve agents. Chem Eng J 262:747–755

Ju Y, Dong B, Yu J, Hou Y (2019) Inherent multifunctional inorganic nanomaterials for imaging-guided cancer therapy. Nano Today 261:08–122

Kang D, Hu C, Zhu Q (2018) Morphology controlled synthesis of hierarchical structured Fe_2O_3 from natural ilmenite and its high performance for dyes adsorption. Appl Surf Sci 459:327–335

Kasaai MR (2020) Biopolymer-based nanomaterials for food, nutrition, and healthcare sectors: an overview on their properties, functions, and applications. In: Handbook of functionalized nanomaterials for industrial applications, pp 167–184

Katz E (2019) Synthesis, properties and applications of magnetic nanoparticles and nanowires – a brief introduction. Magnetochemistry 5(4):61–65

Ke M, Li Y, Qu Q, Ye Y, Peijnenburg WJGM, Zhang Z, Xu N, Lu T, Sun L, Qian H (2020) Offspring toxicity of silver nanoparticles to Arabidopsis thaliana flowering and floral development. J Hazard Mater 386:121975–121979

Khairy M, Zakaria W (2014) Effect of metal-doping of TiO_2 nanoparticles on their photocatalytic activities toward removal of organic dyes. Egypt J Pet 23:419–426

Khataee AR, Kasiri MB (2010) Photocatalytic degradation of organic dyes in the presence of nanostructured titanium dioxide: influence of the chemical structure of dyes. J Mol Catal A Chem 328:8–26

Kim H, Saito N (2018) One-pot synthesis of purple benzene-derived MnO_2-carbon hybrids and synergistic enhancement for the removal of cationic dyes. Sci Rep 8:4342–4348

Kim HS, Lee JW, Yantara N, Boix PP, Kulkarni SA, Mhaisalkar S, Grätzel M, Park NG (2013) High efficiency solid-state sensitized solar cell-based on submicrometer rutile TiO_2 nanorod and $CH_3NH_3PbI_3$ perovskite sensitizer. Nano Lett 13:2412–2417

Kim H, Watthanaphanit A, Saito N (2017) Simple solution plasma synthesis of hierarchical nanoporousMnO2for organic dye removal. ACS Sustain Chem Eng 5:5842–5851

Konstantinou IK, Albanis TA (2004) TiO_2-assisted photocatalytic degradation of azo dyes in aqueous solution: kinetic and mechanistic investigations: a review. Appl Catal B Environ 49:1–14

Kreyling WG, Semmler-Behnke M, Takenaka S, Muller W (2012) Differences in the biokinetics of inhaled nano- versus micrometer-sized particles. Acc Chem Res 46(3):714–722

Lecoanet HF, Wiesner MR (2004) Velocity effects on fullerene and oxide nanoparticle deposition in porous media. Environ Sci Technol 38(16):4377–4382

Lee CK, Liu SS, Juang LC, Wang CC, Lyu MD, Hung SH (2007) Application of titanate nanotubes for dyes adsorptive removal from aqueous solution. J Hazard Mater 148:756–760

Lee ST, Kulkarni HR, Singh A, Baum RP (2017) Theranostics of neuroendocrine tumors. Visc Med 33:358–366

Lee CC, Lin YH, Hou WC, Li MH, Chang JW (2020) Exposure to ZnO/TiO_2 nanoparticles affects health outcomes in cosmetics salesclerks. Int J Environ Res Public Health 17:6088–6100

Leong KY, Rahman MRA, Gurunathan BA (2019) Nano-enhanced phase change materials: a review of thermo-physical properties, applications and challenges. J Energy Storage 21:18–31

Li XQ, Elliott DW, Zhang WX (2006) Zero-valent iron nanoparticles for abatement of environmental pollutants: materials and engineering aspects. Crit Rev Solid State 31(4):111–122

Li L, Xiao J, Liu P, Yang G (2015) Super adsorption capability from amorphousization of metal oxide nanoparticles for dye removal. Sci Rep 5:9028

Li F, Du Y, Liu J, Sun H, Wang J, Li R, Kim D, Hyeon T, Ling D (2018a) Responsive assembly of upconversion nanoparticles for pH-activated and near-infrared-triggered photodynamic therapy of deep tumors. Adv Mater 30(35):1802808–1802812

Li X, Liu Y, Zhang C, Wen T, Zhuang L, Wang X, Song G, Chen D, Ai Y, Hayat T, Wang X (2018b) Porous Fe_2O_3 microcubes derived from metal organic frameworks for efficient elimination of organic pollutants and heavy metal ions. Chem Eng J 336:241–252

Lin LS, Yang X, Zhou Z, Yang Z, Jacobson O, Liu Y, Yang A, Niu G, Song J, Yang HH, Chen X (2017) Yolk–Shell nanostructure: an ideal architecture to achieve harmonious integration of magnetic–plasmonic hybrid theranostic platform. Adv Mater 29(21):1606681–1606687

Liu L, Liu H, Zhao YP, Wang Y, Duan Y, Gao G, Ge M, Chen W (2008) Directed synthesis of hierarchical nanostructured TiO_2 catalysts and their morphology-dependent photocatalysis for phenol degradation. Environ Sci Technol 42:2342–2348

Liu Y, Zeng G, Tang L, Cai Y, Pang Y, Zhang Y, Yang G, Zhou Y, He X, He Y (2015) Highly effective adsorption of cationic and anionic dyes on magnetic Fe/Ni nanoparticles doped bimodal mesoporous carbon. J Colloid Interface Sci 448:451–459

Liu J, Chen H, Shi X, Nawar S, Werner JG, Huang G, Ye M, Weitz DA, Solovev AA, Mei Y (2020a) Hydrogel microcapsules with photocatalytic nanoparticles for removal of organic pollutants. Environ Sci Nano 7(2):656–664

Liu M, Anderson RC, Lan X, Conti PS, Chen K (2020b) Recent advances in the development of nanoparticles for multimodality imaging and therapy of cancer. Med Res Rev 40(3):909–930

Long M, Cai W, Cai J, Zhou B, Chai X, Wu Y (2006) Efficient photocatalytic degradation of phenol over $Co_3O_4/BiVO_4$ composite under visible light irradiation. J Phys Chem B 110:20211–20216
Long J, Xue W, Xie X, Gu Q, Zhou Y, Chi Y, Chen W, Ding Z, Wang X (2011) Sn^{2+} dopant induced visible-light activity of SnO2 nanoparticles for H_2 production. Catal Commun 16:215–219
Low SA, Isserman AM (2009) Ethanol and the local economy: industry trends, location factors, economic impacts, and risks. Econ Dev Q 23(1):71–88
Ma L, Wang Q, Islam SM, Liu Y, Ma S, Kanatzidis MG (2016) Highly selective and efficient removal of heavy metals by layered double hydroxide intercalated with the MoS_4^{2-} ion. J Am Chem Soc 138:2858–2866
Mac Rae I (1985) Removal of pesticides in water by microbial cells adsorbed to magnetite. Water Res 19:825–830
Mädler L, Friedlander SK (2007) Transport of nanoparticles in gases: overview and recent advances. Aerosol Air Qual Res 7(3):304–342
Magdalane CM, Kaviyarasu K, Vijaya JJ, Siddhardha B, Jeyaraj B, Kennedy J, Maaza M (2017) Evaluation on the heterostructured CeO_2/Y_2O_3 binary metal oxide nanocomposites for UV/Vis light induced photocatalytic degradation of Rhodamine – B dye for textile engineering application. J Alloys Compd 727:1324–1337
Malhotra N, Villaflores OB, Audira G, Siregar P, Lee JS, Ger TR, Hsiao CD (2020) Toxicity studies on graphene-based nanomaterials in aquatic organisms: current understanding. Molecules 25(16):3618–3623
Maliyekkal SM, Sreeprasad TS, Krishnan D, Kouser S, Mishra AK, Waghmare UV, Pradeep T (2013) Graphene: a reusable substrate for unprecedented adsorption of pesticides. Small 9:273–283
Malwal D, Gopinath P (2016) Enhanced photocatalytic activity of hierarchical three dimensional metal oxide@CuO nanostructures towards the degradation of Congo red dye under solar radiation. Cat Sci Technol 6:4458–4472
Mangadlao JD, Wang X, McCleese C, Escamilla M, Ramamurthy G, Wang Z, Govande M, Basilion JP, Burda C (2018) Prostate-specific membrane antigen targeted gold nanoparticles for theranostics of prostate cancer. ACS Nano 12(4):3714–3725
Mao Y, Park TJ, Zhang F, Zhou H, Wong SS (2007) Environmentally friendly methodologies of nanostructure synthesis. Small 3:1122–1139
Mao LC, Zhang XY, Wei Y (2019) Recent advances and progress for the fabrication and surface modification of AIE-active organic-inorganic luminescent composites. Chin J Polym Sci 37(4):340–351
Mason SA, Garneau D, Sutton R, Chu Y, Ehmann K, Barnes J, Fink P, Papazissimos D, Rogers DL (2016) Microplastic pollution is widely detected in US municipal wastewater treatment plant effluent. Environ Pollut 218:1045–1054
Matlochová A, Plachá D, Rapantová N (2013) The application of nanoscale materials in groundwater remediation. Pol J Environ Stud 22(5):1401–1410
Mclaren A, Valdes-Solis T, Li G, Tsang SC (2009) Shape and size effects of ZnO nanocrystals on photocatalytic activity. J Am Chem Soc 131:12540–12541
Meng A, Xing J, Li Z, Li Q (2015) Cr-doped ZnO nanoparticles: synthesis, characterization, adsorption property, and recyclability. ACS Appl Mater Interfaces 7:27449–27457
Mir AA, Amooey AA, Ghasemi S (2018) Adsorption of direct yellow 12 from aqueous solutions by an iron oxide-gelatin nanoadsorbent; kinetic, isotherm and mechanism analysis. J Clean Prod 170:570–580
Modena MM, Rühle B, Burg TP, Wuttke S (2019) Nanoparticle characterization: what to measure? Adv Mater 31(32):1901556–1901562
Montanheiro TLDA, Ribas RG, Montagna LS, Menezes BRCD, Schatkoski VM, Rodrigues KF, Thim GP (2020) A brief review concerning the latest advances in the influence of nanoparticle reinforcement into polymeric-matrix biomaterials. J Biomater Sci Polym Ed:1–25
Mu J, Lin J, Huang P, Chen X (2018) Development of endogenous enzyme-responsive nanomaterials for theranostics. Chem Soc Rev 47(15):5554–5573

Mukhopadhyay SS (2014) Nanotechnology in agriculture: prospects and constraints. Nanotechnol Sci Appl 7:63–71

Müller N (2007) Nanoparticles in the environment: Risk assessment based on exposure modelling: what concentrations of nano titanium dioxide, carbon nanotubes and nano silver are we exposed to? Doctoral dissertation, Department of Environmental Sciences, ETH Zurich, Zurich

Murray JW (1974) The surface chemistry of hydrous manganese dioxide. J Colloid Interface Sci 46:357–371

Naahidi S, Jafari M, Edalat F, Raymond K, Khademhosseini A, Chen P (2013) Biocompatibility of engineered nanoparticles for drug delivery. J Control Release 166(2):182–194

Narayanan R, El-Sayed MA (2005) Catalysis with transition metal nanoparticles in colloidal solution: Nanoparticle shape dependence and stability. J Phys Chem B 109:12663–12676

Nascimento R, Ferreira OP, De Paula AJ, Neto VD (eds) (2019) Nanomaterials applications for environmental matrices: water, soil and air. Elsevier

Nassar MY, Mohamed TY, Ahmed IS, Samir I (2017) MgO nanostructure via a sol-gel combustion synthesis method using different fuels: an efficient nano-adsorbent for the removal of some anionic textile dyes. J Mol Liq 225:730–740

Navarro S, Fenoll J, Vela N, Ruiz E, Navarro G (2009) Photocatalytic degradation of eight pesticides in leaching water by use of ZnO under natural sunlight. J Hazard Mater 172:1303–1310

Nigel Corns S, Partington SM, Towns AD (2009) Industrial organic photochromic dyes. Color Technol 125(5):249–261

Nurmi JT, Tratnyek PG, Sarathy V, Baer DR, Amonette JE, Pecher K, Wang C, Linehan JC, Matson DW, Penn RL, Driessen MD (2005) Characterization and properties of metallic iron nanoparticles: spectroscopy, electrochemistry, and kinetics. Environ Sci Technol 39(5):1221–1230

Ogata Y, Takada H, Mizukawa K, Hirai H, Iwasa S, Endo S, Mato Y, Saha M, Okuda K, Nakashima A, Murakami M (2009) International pellet watch: global monitoring of persistent organic pollutants (POPs) in coastal waters. 1. Initial phase data on PCBs, DDTs, and HCHs. Mar Pollut Bull 58(10):1437–1446

Oliveira ML, Saikia BK, da Boit K, Pinto D, Tutikian BF, Silva LF (2019) River dynamics and nanopaticles formation: a comprehensive study on the nanoparticle geochemistry of suspended sediments in the Magdalena River, Caribbean Industrial Area. J Clean Prod 213:819–824

Parasuraman D, Leung E, Serpe MJ (2012) Poly (N-isopropylacrylamide) microgel based assemblies for organic dye removal from water: microgel diameter effects. Colloid Polym Sci 290(11):1053–1064

Park S, Vohs JM, Gorte RJ (2000) Direct oxidation of hydrocarbons in a solid-oxide fuel cell. Nature 404:265–269

Park W, Shin H, Choi B, Rhim WK, Na K, Han DK (2020) Advanced hybrid nanomaterials for biomedical applications. Prog Mater Sci:100686–100692

Parks GA (1965) The isoelectric points of solid oxides, solid hydroxides, and aqueous hydroxo complex systems. Chem Rev 65:177–198

Patil SS, Shedbalkar UU, Truskewycz A, Chopade BA, Ball AS (2016) Nanoparticles for environmental clean-up: a review of potential risks and emerging solutions. Environ Technol Innov 5:10–21

Prajitha N, Athira SS, Mohanan PV (2019) Bio-interactions and risks of engineered nanoparticles. Environ Res 172:98–108

Qiao D, Li Z, Duan J, He X (2020) Adsorption and photocatalytic degradation mechanism of magnetic graphene oxide/ZnO nanocomposites for tetracycline contaminants. Chem Eng Trans:125952–125957

Rafaie HA, Nor RM, Azmina MS, Ramli NIT, Mohamed R (2017) Decoration of ZnO microstructures with Ag nanoparticles enhanced the catalytic photodegradation of methylene blue dye. J Environ Chem Eng 5:3963–3972

Rafieepour A, Azari MR, Khodagholi F, Jaktaji JP, Mehrabi Y, Peirovi H (2021) Interactive toxicity effect of combined exposure to hematite and amorphous silicon dioxide nanoparticles in human A549 cell line. Toxicol Ind Health 37:289–302

Rasheed T, Rizwan K, Bilal M, Iqbal H (2020) Metal-organic framework-based engineered materials – fundamentals and applications. Molecules 25(7):1598–1603

Rauf MA, Meetani MA, Hisaindee S (2011) An overview on the photocatalytic degradation of azo dyes in the presence of TiO_2 doped with selective transition metals. Desalination 276:13–27

Ravichandran R (2010) Nanotechnology applications in food and food processing: innovative green approaches, opportunities and uncertainties for global market. Int J Green Nanotechnol Phys Chem 1(2):72–96

Rehman AU, Nazir S, Irshad R, Tahir K, Rehman K, Islam RU, Wahab Z (2020) Toxicity of heavy metals in plants and animals and their uptake by magnetic iron oxide nanoparticles. J Mol Liq 1:114455–114480

Resnik DB (2019) How should engineered nanomaterials be regulated for public and environmental health? AMA J Ethics 21:363–369

Saha B, Das S, Saikia J, Das G (2011) Preferential and enhanced adsorption of different dyes on Iron oxide nanoparticles: a comparative study. J Phys Chem C 115:8024–8033

Sahu TK, Arora S, Banik A, Iyer PK, Qureshi M (2017) Efficient and rapid removal of environmental malignant arsenic (III) and industrial dyes using reusable, recoverable ternary Iron oxide-ORMOSIL-reduced graphene oxide composite. ACS Sustain Chem Eng 5:5912–5921

Sakthivel S, Neppolian B, Shankar MV, Arabindoo B, Palanichamy M, Murugesan V (2003) Solar photocatalytic degradation of azo dye: comparison of photocatalytic efficiency of ZnO and TiO_2. Sol Energy Mater Sol Cells 77:65–82

Saleh TA, Gupta VK (2012) Photo-catalyzed degradation of hazardous dye methyl orange by use of a composite catalyst consisting of multi-walled carbon nanotubes and titanium dioxide. J Colloid Interface Sci 371:101–106

Salehi R, Arami M, Mahmoodi NM, Bahrami H, Khorramfar S (2010) Novel biocompatible composite (chitosan-zinc oxide nanoparticle): preparation, characterization and dye adsorption properties. Colloids Surf B: Biointerfaces 80:86–93

Sarkar AK, Saha A, Tarafder A, Panda AB, Pal S (2016) Efficient removal of toxic dyes via simultaneous adsorption and solar light driven photodegradation using recyclable functionalized amylopectin–TiO_2–Au nanocomposite. ACS Sustain Chem Eng 4(3):1679–1688

Sarong MM, Orge RF, Eugenio PJG, Monserate JJ (2020) Utilization of rice husks into biochar and nanosilica: for clean energy, soil fertility and green nanotechnology. Int J Des Nat Ecodyn 15(1):97–102

Saroyan HS, Giannakoudakis DA, Sarafidis CS, Lazaridis NK, Deliyanni EA (2017) Effective impregnation for the preparation of magnetic mesoporous carbon: application to dye adsorption. J Chem Technol Biotechnol 92:1899–1911

Schneider J, Matsuoka M, Takeuchi M, Zhang J, Horiuchi Y, Anpo M, Bahnemann DW (2014) Understanding TiO_2 photocatalysis: mechanisms and materials. Chem Rev 114:9919–9986

Selvan ST, Tan TT, Yi DK, Jana NR (2010) Functional and multifunctional nanoparticles for bioimaging and biosensing. Langmuir 26:11631–11641

Shahbazi MA, Faghfouri L, Ferreira MP, Figueiredo P, Maleki H, Sefat F, Hirvonen J, Santos HA (2020) The versatile biomedical applications of bismuth-based nanoparticles and composites: therapeutic, diagnostic, biosensing, and regenerative properties. Chem Soc Rev 49(4):1253–1321

Shan D, Ma C, Yang J (2019) Enabling biodegradable functional biomaterials for the management of neurological disorders. Adv Drug Deliv Rev 148:219–238

Sharma M, Pandey C, Sharma N, Kamal M, Sayee U, Akhtar S (2018) Cancer nanotechnology-an excursion on drug delivery systems. Anti-Cancer Agents Med Chem 18(15):2078–2092

Shchukin DG, Caruso RA (2004) Template synthesis and photocatalytic properties of porous metal oxide spheres formed by nanoparticle infiltration. Chem Mater 16:2287–2292

Shi JP, Evans DE, Khan AA, Harrison RM (2001) Sources and concentration of nanoparticles (<10nm diameter) in the urban atmosphere. Atmos Environ 35(7):1193–1202

Shirzad-Siboni M, Jonidi-Jafari A, Farzadkia M, Esrafili A, Gholami M (2017) Enhancement of photocatalytic activity of Cu-doped ZnO nanorods for the degradation of an insecticide: kinetics and reaction pathways. J Environ Manag 186:1–11

Singh S (2019) Zinc oxide nanoparticles impacts: cytotoxicity, genotoxicity, developmental toxicity, and neurotoxicity. Toxicol Mech Methods 29(4):300–311

Singh R, Behera M, Kumar S (2020) Nano-bioremediation: an innovative remediation technology for treatment and management of contaminated sites. In: Bioremediation of industrial waste for environmental safety, pp 165–182. Springer, Singapore

Siyasukh A, Chimupala Y, Tonanon N (2018) Preparation of magnetic hierarchical porous carbon spheres with graphitic features for high methyl orange adsorption capacity. Carbon 134:207–221

Song Q, Liang J, Fang Y, Guo Z, Du Z, Zhang L, Liu Z, Huang Y, Lin J, Tang C (2020) Nickel (II) modified porous boron nitride: an effective adsorbent for tetracycline removal from aqueous solution. Int J Chem Eng:124985–124990

Sun TY, Mitrano DM, Bornhöft NA, Scheringer M, Hungerbühler K, Nowack B (2017) Envisioning nano release dynamics in a changing world: using dynamic probabilistic modeling to assess future environmental emissions of engineered nanomaterials. Environ Sci Technol 51:2854–2863

Tachikawa T, Fujitsuka M, Majima T (2007) Mechanistic insight into the TiO_2 photocatalytic reactions: design of new photocatalysts. J Phys Chem C 111:5259–5275

Tahara K (2020) Pharmaceutical formulation and manufacturing using particle/powder technology for personalized medicines. Adv Powder Technol 31(1):387–392

Tang ZR, Zhang Y, Xu YJ (2011) A facile and high-yield approach to synthesize one-dimensional CeO_2 nanotubes with well-shaped hollow interior as a photocatalyst for degradation of toxic pollutants. RSC Adv 1:1772–1778

Tang Y, Li M, Mu C, Zhou J, Shi B (2019) Ultrafast and efficient removal of anionic dyes from wastewater by polyethyleneimine-modified silica nanoparticles. Chemosphere 229:570–579

Tiede K, Hanssen SF, Westerhoff P, Fern GJ, Hankin SM, Aitken RJ, Chaudhry Q, Boxall AB (2016) How important is drinking water exposure for the risks of engineered nanoparticles to consumers? Nanotoxicology 10(1):102–110

Tiwari B, Drogui P, Tyagi RD (2020) Removal of emerging micro-pollutants from pharmaceutical industry wastewater. In: Current developments in biotechnology and bioengineering, pp 457–480

Tricoli A, Bo R (2020) Nanoparticle-based biomedical sensors. Front Nanosci 15:247–269

Ullah R, Dutta J (2008) Photocatalytic degradation of organic dyes with manganese-doped ZnO nanoparticles. J Hazard Mater 156:194–200

Ullah H, Li X, Peng L, Cai Y, Mielke H (2020) In vivo phytotoxicity, uptake, and translocation of PbS nanoparticles in maize (*Zea mays* L.) plants. Sci Total Environ 737:136994–136998

Upadhyay RK, Soin N, Roy SS (2014) Role of graphene/metal oxide composites as photocatalysts, adsorbents and disinfectants in water treatment: a review. RSC Adv 4:3823–3851

Vambol S, Vambol V, Suchikova Y, Deyneko N (2017) Analysis of the ways to provide ecological safety for the products of nanotechnologies throughout their life cycle. Eastern-European J Enterp Technol 1(10):27–36

Vignardi CP, Muller EB, Tran K, Couture J, Means JC, Murray J, Ortiz C, Keller AA, Sanchez NS, Lenihan HS (2020) Conventional and nano-copper pesticide are equally toxic to the estuarine amphipod Leptocheirus plumulosus. Aquat Toxicol:105481–105486

Wang Y, Cheng R, Wen Z, Zhao L (2012) Facile preparation of Fe_3O_4 nanoparticles with cetyltrimethylammonium bromide (CTAB) assistant and a study of its adsorption capacity. Chem Eng J 181–2:823–827

Wang L, Zhao R, Wang XW, Mei L, Yuan LY, Wang SA, Chai ZF, Shi WQ (2014) Size-tunable synthesis of monodisperse thorium dioxide nanoparticles and their performance on the adsorption of dye molecules. CrystEngComm 16:10469–10475

Wang H, Liang X, Wang J, Jiao S, Xue D (2020) Multifunctional inorganic nanomaterials for energy applications. Nanoscale 12(1):14–42

Wani MR, Shadab GHA (2021) Low doses of thymoquinone protect isolated human blood cells from TiO2 nanoparticles induced oxidative stress, hemolysis, cytotoxicity, DNA damage and collapse of mitochondrial activity. Phytomedicine Plus 1:100056–100067

Warheit DB, Donner EM (2015) Risk assessment strategies for nanoscale and fine-sized titanium dioxide particles: recognizing hazard and exposure issues. Food Chem Toxicol 85:138–147

Weinberg H, Galyean A, Leopold M (2011) Evaluating engineered nanoparticles in natural waters. TrAC 30(1):72–83

Wiesner MR, Lowry GV, Alvarez P, Dionysiou D, Biswas P (2006) Assessing the risks of manufactured nanomaterials. Environ Sci Technol 40(14):4336–4345

Winardi S, Qomariyah L, Widiyastuti W, Kusdianto K, Nurtono T, Madhania S (2020) The role of electro-sprayed silica-coated zinc oxide nanoparticles to hollow silica nanoparticles for optical devices material and their characterization. Colloids Surf A Physicochem Eng Asp:125327–125332

Wu J, Wang J, Du Y, Li H, Jia X (2016) Adsorption mechanism and kinetics of azo dye chemicals on oxide nanotubes: a case study using porous CeO_2 nanotubes. J Nanopart Res 18:191–196

Xiang Y, Yang X, Xu Z, Hu W, Zhou Y, Wan Z, Yang Y, Wei Y, Yang J, Tsang DC (2020) Fabrication of sustainable manganese ferrite modified biochar from vinasse for enhanced adsorption of fluoroquinolone antibiotics: effects and mechanisms. Sci Total Environ 709:136079–136083

Xiao J, Lv W, Xie Z, Tan Y, Song Y, Zheng Q (2016) Environmentally friendly reduced graphene oxide as a broad-spectrum adsorbent for anionic and cationic dyes via π–π interactions. J Mater Chem A 4:12126–12135

Xie T, Liu C, Xu L, Yang J, Zhou W (2013) Novel heterojunction $Bi_2O_3/SrFe_{12}O_{19}$ magnetic photocatalyst with highly enhanced photocatalytic activity. J Phys Chem C 117:24601–24610

Xiong L, Yang Y, Mai J, Sun W, Zhang C, Wei D, Chen Q, Ni J (2010) Adsorption behavior of methylene blue onto titanate nanotubes. Chem Eng J 156:313–320

Xu D, Cheng F, Lu Q, Dai P (2014) Microwave enhanced catalytic degradation of methyl orange in aqueous solution over CuO/CeO_2 catalyst in the absence and presence of H_2O_2. Ind Eng Chem Res 53:2625–2632

Yang G, Liu Y, Wang H, Wilson R, Hui Y, Yu L, Wibowo D, Zhang C, Whittaker AK, Middelberg AP, Zhao CX (2019a) Bioinspired core–shell nanoparticles for hydrophobic drug delivery. Angew Chem Int 58(40):14357–14364

Yang S, Chen S, Fan J, Shang T, Huang D, Li G (2019b) Novel mesoporous organosilica nanoparticles with ferrocene group for efficient removal of contaminants from wastewater. J Colloid Interface Sci 554:565–571

Ye X, Wang J, Fan M (2019) Evaluating tribological properties of the stearic acid-based organic nanomaterials as additives for aqueous lubricants. Tribol Int 140:105848–105853

Yi C, Pan Y, Fang Y (2019) Surface engineering of carbon nanodots (C-Dots) for biomedical applications. In: Novel nanomaterials for biomedical, environmental and energy applications. Elsevier, pp 137–188

Yousaf SS, Houacine C, Khan I, Ahmed W, Jackson MJ (2020) Importance of biomaterials in biomedical engineering. In: Advances in medical and surgical engineering, pp 149–176

Yusefi-Tanha E, Fallah S, Rostamnejadi A, Pokhrel LR (2020) Particle size and concentration dependent toxicity of copper oxide nanoparticles (CuONPs) on seed yield and antioxidant defense system in soil grown soybean (*Glycine max cv.* Kowsar). Sci Total Environ 715:136994–136999

Zare EN, Makvandi P, Ashtari B, Rossi F, Motahari A, Perale G (2019) Progress in conductive polyaniline-based nanocomposites for biomedical applications: a review. J Med Chem 63(1):1–22

Zhang S, Xu W, Zeng M, Li J, Li J, Xu J, Wang X (2013a) Superior adsorption capacity of hierarchical iron oxide@magnesium silicate magnetic nanorods for fast removal of organic pollutants from aqueous solution. J Mater Chem A 1:11691–11696

Zhang X, Zhang P, Wu Z, Zhang L, Zeng G, Zhou C (2013b) Adsorption of methylene blue onto humic acid-coated Fe_3O_4 nanoparticles. Colloids Surf A Physicochem Eng Asp 435:85–90

Zhang H, Wang X, Chen C, An C, Xu Y, Dong Y, Zhang Q, Wang Y, Jiao L, Yuan H (2016) Facile synthesis of diverse transition metal oxide nanoparticles and electrochemical properties. Inorg Chem Front 3:1048–1057

Zhang J, Li S, Ju DD, Li X, Zhang JC, Yan X, Long YZ, Song F (2018a) Flexible inorganic core-shell nanofibers endowed with tunable multicolor upconversion fluorescence for simultaneous monitoring dual drug delivery. Chem Eng J 349:554–561

Zhang P, Wang F, Yu M, Zhuang X, Feng X (2018b) Two-dimensional materials for miniaturized energy storage devices: from individual devices to smart integrated systems. Chem Soc Rev 47(19):7426–7451

Zhang J, Guo W, Li Q, Wang Z, Liu S (2018c) The effects and the potential mechanism of environmental transformation of metal nanoparticles on their toxicity in organisms. Environ Sci Nano 5:2482–2499

Zhang W, Ke S, Sun C, Xu X, Chen J, Yao L (2019) Fate and toxicity of silver nanoparticles in freshwater from laboratory to realistic environments: a review. Environ Sci Pollut Res 26(8):7390–7404

Zhang L, Wang WX, Li A, Liu J, Li HW, Wu Y (2020) Influence of pressure on the structure and luminescence properties of AMP-protected gold nanoparticles as revealed by fluorescence spectra and 2D correlation analysis. J Mol Struct:128173–128178

Zhao J, Chen C, Ma W (2005) Photocatalytic degradation of organic pollutants under visible light irradiation. Top Catal 35:269–278

Zhao L, Cui T, Li Y, Wang B, Han J, Han L, Liu Z (2015) Efficient visible light photocatalytic activity of p–n junction CuO/TiO_2 loaded on natural zeolite. RSC Adv 5:64495–64502

Zheng X, Jiang B, Lang H, Zhang R, Li Y, Bian Y, Guan X (2019) Effects of antibiotics on microbial communities responsible for perchlorate degradation. Water Air Soil Pollut 230(10):244

Zhou LJ, Li J, Zhang Y, Kong L, Jin M, Yang X, Wu QL (2019) Trends in the occurrence and risk assessment of antibiotics in shallow lakes in the lower-middle reaches of the Yangtze River basin, China. Ecotoxicol Environ Saf 183:109511

Zhu X, Chang Y, Chen Y (2010) Toxicity and bioaccumulation of TiO2 nanoparticle aggregates in Daphnia magna. Chemosphere 78(3):209–215

Zhu HY, Jiang R, Fu YQ, Li RR, Yao J, Jiang ST (2016) Novel multifunctional $NiFe_2O_4/ZnO$ hybrids for dye removal by adsorption, photocatalysis and magnetic separation. Appl Surf Sci 369:1–10

Zhu Y, Wu J, Chen M, Liu X, Xiong Y, Wang Y, Feng T, Kang S, Wang X (2019) Recent advances in the biotoxicity of metal oxide nanoparticles: Impacts on plants, animals and microorganisms. Chemosphere 237:124403

Zhuang J, Young AP, Tsung CK (2017) Integration of biomolecules with metal–organic frameworks. Small 13(32):1700880–1700886

Chapter 3
Nanomaterials in Environment: Sources, Risk Assessment, and Safety Aspect

Ashpreet Kaur and Harmandeep Singh

Abstract Nanotechnology has been gaining so much popularity recently that it is being applied in all fields of sciences. But because it is still a new field and its full potential is not yet observed physically, there are risks and safety issues that come along while using this technology. Due to less amount of knowledge available on the safety aspects, risk management, inefficient technology, and unpredictable nature of these highly reactive nanomaterials and nanoparticles, some precautions need to be taken strictly into account. Nanotechnology can lead to adverse environmental effects if not used and regulated properly. To overcome any risks and accidents in workspace, proper risk management tools should be provided to occupants. Good lab practices, latest technology, and human health must be given priority when working with highly reactive and potentially toxic nanomaterials. In this chapter, we will talk about the sources of nanomaterials and nanoparticles, risk assessment strategies that can be used to combat possible hazard, and safety measures that can be taken as a safeguard against any unfortunate event related to nanotechnology.

Keywords Nanoparticles · Pollutant · Contamination · Environment · Nano-safety · Toxicity

A. Kaur (✉)
School of Natural Resource and Environmental Science, University of Florida, Gainesville, FL, USA

H. Singh
Panjab University, Chandigarh, Chandigarh, India

R. Kumar et al. (eds.), *Advanced Functional Nanoparticles "Boon or Bane" for Environment Remediation Applications*, Environmental Contamination Remediation and Management, https://doi.org/10.1007/978-3-031-24416-2_3

3.1 Introduction

Nanotechnology is a field of science, engineering, and technology that deals with molecules on the nanoscale (10^{-9} m) ranging from 1 to 100 nm and using nanoscience to convert into useful products (Drexler 1990; Mansoori and Soelaiman 2005). By working on an atom-to-atom level, it finds its ways in creating larger structures that have advanced or novel significance due to improved physical, chemical, and biological properties of the molecular structure of the compound. It is said that Michael Faraday made a colloid from nanoparticles of gold and phosphorus, changing the metallic state of gold into different colloidal states, and it was known as "activated gold" in 1857. This gave different colors like green, blue, ruby, etc., to the gold. But the idea of using nanotechnology is said to be first conceived by James Clerk Maxwell in 1871 when he regulated the flow of molecules passing through a door. Scientists started talking about this idea, and it was in 1959 that Richard Feynman gave a lecture on the idea of using small-scale particles from where it gained full attention. And in 1974, Taniguchi coined the term "nanotechnology" where he talked about basic concepts of nanotechnology (Sudha et al. 2018; Taniguchi 1974).

Nanoparticles have been in use even before nanotechnology fully developed. Nanoparticles of lead were used to dye hair black, and gold was used to cure ailments or make stained glass (Ramsden 2016). The early Mesopotamian era has evidence of using nanoparticles in pottery to give glittering effects on the pots, which were nanoparticles of gold and silver (Reiss and Hutten 2010). About 400 AD, Romans invented the Lycurgus Cup, which changed color when light was passed through it. The cup is made of silver and gold alloys. Around 300–1700 A.D., the middle east used steel swords with better strength, resistance, and sharpness. The swords are called Damascus swords and were found to have carbon nanotubes (Reibold, 2006). The Maya civilization, known as one of the oldest and advanced civilizations, created Maya Blue, which is a corrosion-resistant pigment made from clay mixed with indigo in 800 AD (Sudha et al. 2018). Nanoparticles are nano-objects having all the three dimensions in the range of 1–100 nm (ISO 2008). Nanomaterials (NM) are the materials that have nanoscopic dimensions and are used in nanotechnology (Martin 1994). Nanomaterials (NM) are materials that have nanoscopic dimensions (external or internal dimensions are in nanoscale) and are used in nanotechnology (Martin 1994; ISO 2017). They can be insoluble or bio-persistent and can be naturally, incidentally, or intentionally manufactured. They can be insoluble or bio-persistent, be naturally, incidentally, or intentionally manufactured, and exist in an unbound state, agglomerate, or as an aggregate having 50% or more particles in size range of 1–100 nm (EU 2012).

Nanotechnology has the potential to revolutionize, harness every sector, and make a new updated world. Nanomaterials made from nanotechnology have unique chemical and physical characteristics compared to conventional technologies, such as better elasticity, toughness, adsorption, absorption, adhesion, and agglomeration (Khan et al. 2017). Nanotechnology is emerging as future technology because of the following reasons. First, it deals with nanoparticles that provide more surface-to-volume ratio and works on a one-billionth scale that enhances reactivity, mobility, strength, and efficiency along with decreasing space requirement (Mansoori and Soelaiman 2005). Second, nanotechnology can provide the desired size, shape, weight, and other characteristics of a material by manipulating it. Third, it reduces the cost and increases the shelf-life of the material and commercialization of technology (Mobasser and Firoozi 2016). Fourth, it is an interdisciplinary field that includes almost every sector such as science, environment, industries, and engineering (Bhushan 2017). Fifth, further research and development in this field will lead to making a new world and the life of individuals easier.

Nanotechnology is now used to provide more robust insulin delivery that can measure minute fluctuations of insulin in the body and hence modulate the rate of insulin released in the body automatically. Researchers are proposing a closed-loop insulin delivery system to increase glucose sensor sensitivity in blood, leading to continuous glucose monitoring in the body. A subfield in nanotechnology called "green nanotechnology" aims at designing safe nanomaterials that have lesser environmental effects and designing process safety, energy and material efficiency, and reduction of nano-waste (Hutchison 2016).

3.2 General Applications of Nanotechnology

Nanotechnology has shown its applications in varied fields like in medicine, diagnostics, therapeutic, and drug delivery by increasing specificity and mobility for transportation to targeted sites, e.g., silver nanoparticles for cancer treatment (Fan and Alexeeff 2010; Connolly et al. 2014). It is used in sustainable energy production by decreasing the cost and increasing power generation efficiency with the help of nanoscale solar cells. (Low et al. 2015). And it has also found its route in cosmetics; for example, titanium dioxide provides protection against UV light and also reduces the whitening problem that is caused by traditional sunscreens (Fan and Alexeeff 2010; Smith et al. 2013). It is also used in display, information, and communication, e.g., nanoscale chips, carbon nanotubes, and nano-RAM. Some other sectors also use nanotechnology like catalysis, food and agriculture, textile, tissue engineering, electrical engineering, mechanical engineering, construction, automobiles, environment, etc., for performance improvement, mobility, shelf-life, efficiency, specificity, cost-effectiveness, and reactivity (Mobasser and Firoozi 2016; Rakesh et al. 2015). Nanorobotics is being used in the nuclear industry for radiation treatment to treat cancer. Microbivore nanorobots mimic the white blood cell functions to destroy bacteria at a faster rate and respirocyte nanorobots mimic the red blood cells to carry more oxygen (Sudha et al. 2018). Nanoparticles are used in creating more strong, lightweight, and multifunctional aircrafts, smart windows working on the principal of photcromaticity, nanophosphors in HDTVs (Sudha et al. 2018), in data storage, and automotive industries (miniature sensors) (Mamalis 2007).

3.3 Environmental Applications

Apart from some technical human-serving technologies, nanotechnology has a future to serve as an innovative option to remediate environmental issues. Its environmental applications are categorized into the following parts: remediation, protection, maintenance, and enhancement (Jalaja et al. 2016). In environmental protection, it provides enhanced adsorption and transformation of the pollutants like PAHs, PCBs, pesticides, and dioxins that are resistant to biodegradation and are persistent in air, soil, surface-subsurface-ground water, and sediments (Jones et al. 2007). For example, C-based nanotubes, zinc oxide, titanium dioxide, and other metal-based and magnetic nanoparticles enhance the removal of pollutants (Rickerby and Morrison 2007). In environmental remediation, nanoscale zerovalent iron and bi-metallics are proven effective with increased reductive dichlorination, beta elimination, and mobility in detecting and removing groundwater contaminants like TCA, PCBs, TCE, VC, and heavy metals (Hg, As), etc. (Elliot 2006; Bowman 2003). The removal of organic and inorganic pollutants and water purification use nanotechnology-based adsorbents and filters that are proven to be effective (Sobolev and Shah 2015; Mishra et al. 2012). Carbon- and graphene-based nanomaterials are efficient absorbents for oils and solvents; high-performance photocatalyst like TiO_2, ZnO, Cu, Fe, etc., are used to degrade refractory pollutants and voltaic organic compounds (Liu et al. 2016).

Recent advances in water treatment have used nanotechnology to remove nitrate from water. Nanomaterials have a small surface area, resulting in high catalytic properties, and hence can reduce nitrates in waste to ammonia, nitrite, ammonium, or nitrogen gas. Nanomaterials also have good adsorbing properties, and therefore, these reducing and adsorbing properties make them a good choice to remove pollutants from water (Tyagi et al. 2018). Nanotechnology have cost-effective options such as solar cells or nanofuels for sustainable energy production, (Low et al. 2015), engineered enzymes for degradation, ligands, and bio-activators with better performance than conventional processes that increase the efficiency of contaminant removal (Mobasser and Firoozi 2016).

3.4 Types of Nanomaterials

On the basis of dimension, nanomaterials can be classified into (a) 0 dimensional (0D), (b) 1 dimensional (1D), (c) 2 dimensional (2D), and (d) 3 dimensional (3D). On the basis of composition, nanomaterials are classified into (a) carbon-based NMs, (b) inorganic NMs, (c) organic NMs, and (d) composite NMs (Sudha et al. 2018). Carbon is one of the most abundant elements on earth and is used in nanomaterial manufacturing as it is present in all four dimensions. In zero dimension as fullerene molecules, 1D as carbon nanotubes, 2D as graphite sheets, and 3D as diamond crystals and graphite. One of its structures called fullerene is used a lot in nanotechnology as it has a symmetric structure and bounces back to its original

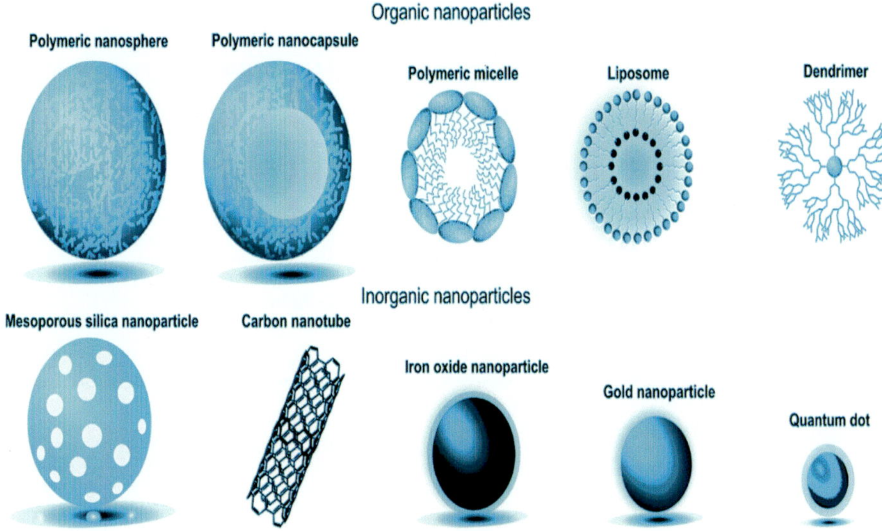

Fig. 3.1 Examples of organic and inorganic nanoparticles. (Tran Le 2018)

shape when high pressure is applied on it. Organic nanomaterials are made from lipids, carbohydrates, and polymeric substances with size ranging from 1 nanometer to 10 micrometers (Fig. 3.1). These polymers have application in drug delivery. Inorganic nanomaterials can be naturally occurring like clay and silica from volcanic eruptions or they can be engineered and converted into metal and metal oxides (Fig. 3.1). They have good optical and electronic properties, photostability, and multicolor capability and can be used in semiconductors. Composite nanomaterials are combinations of two or more different materials to get the best property out (Sudha et al. 2018; Jeevanandam et al. 2018).

3.5 Sources of Nanomaterials

Nanomaterials have been known to occur naturally before humans started engineering them. They are present as airborne particles that help the water to condense into rain., in intermetallic particles, sea plankton, biologically occurring in the form of protein, DNA, and viruses. Because of their size range, which is usually between 10 nm and 200 nm, they show unique properties like increased surface area or modified electronic states (Brabazon 2016).

They can naturally occurr during processes like forest fires, volcanic eruptions, supernovas, or form by humans during orocesses like vehicle engine exhaust, welding fumes, and combustion. They are also specially designed and manufactured to use in industrial applications (Boverhof et al. 2015; Jeevanandam et al. 2018). So sources of nanomaterials can be classified into three major categories, shown in Fig. 3.2, with examples:

NATURAL	INCIDENTAL	ENGINEERED
Natural Oxides	Magneli Phases	Quantum dots
Viruses	Nano plastic	Carbon nanotubes
Sulfides	Welding Fumes	Metals
Clay minerals	Soot, agriculture, mining	Liposomes

Fig. 3.2 Examples of nanomaterials depending on source. (Hochella et al. 2019)

1. Naturally produced nanomaterials.
2. Incidental nanomaterials.
3. Engineered nanomaterials.

3.6 Natural Nanomaterials

Nanoparticles that naturally occur in the environment like in plants and animal bodies or are naturally formed by biogeochemical or mechanical processes like dust storms, volcanic eruptions, etc., are called natural nanomaterials. These are abundantly present on earth since their formation and evolution for the past 4.54 billion years (Hochella et al. 2019).

Volcanic eruptions and soil formation: A volcanic eruption can lead to propulsion of an enormous number of aerosols, silica, bismuth oxide, heavy metals, and fine particles into the atmosphere with sizes ranging from micrometers to several nanometers. A single volcanic eruption can expel up to 30×10^6 tons of NPs mixed in ash into the atmosphere. These released NPs then spread over distances and settle in the lower layers of earth's atmosphere (Jeevanandam et al. 2018). The volcanically erupted particles may possess heavy metals that are toxic to humans. These emissions include short-term effects like nose, eye, throat, and skin irritations and bronchial symptoms and long-term effects like diseases such as podocin (a protein in kidney) aids and Kaposi's sarcoma (Blundell et al. 1989; Corachan 1988). Weathering and mineral formation in soils are the main sources of natural organically and inorganically occurring nanomaterials like sulfides, carbonates, phosphates, and metal oxides. These are usually transported to different places by rivers, glaciers, and wind movement.

Living organisms: Nanoparticles are found in organisms like nanobacteria, viruses, fungi, algae like *Chlorella vulgaris* that produces silver NPs, and magnetotactic bacteria that produce magnetic oxides, and in wings of insects like grasshopper, butterfly, and dragonfly (Jeevanandam et al. 2018). Nanocomposite particles are found in bones, abalone shells, mollusk shells, and eggshells (Sudha et al. 2018).

Nanoparticles in the human body: Bones are considered nanomaterials with composite hierarchical structures consisting of inorganic nano-hydroxyapatite and organic collagen. DNA, antibodies, enzymes, some bacteria and viruses, organelles, peptides, and polysaccharides are all composed of nanoparticles (Jeevanandam et al. 2018).

Dust storms and cosmic dust: Dust storms can transport fine particles from deserts, aerosols, and human activities to long distances and cause health problems caused by airborne particles leading to respiratory issues. Cosmic dust, meteorites, and star dust have been known to contain nanomaterials based on carbide, oxide, magnetic nanomaterial, silicate, nitride, and carbon. Collisions, electromagnetic traditions, shock waves, etc., energize these particles (Hochella et al. 2012; Rietmeijer and Nuth 2012).

Other sources where nanomaterials are found are in ocean water; for example, iron is found in fractions of 20–200 nm and manganese is found in less than 20 nm size (Fitzsimmons et al. 2014; Odlham et al. 2017).

3.7 Incidental Nanomaterials

Nanomaterials Incidental nanomaterials are unintentionally produced as by-products of any direct or indirect human activity. These are not usually present in nature but are added in nature by human activities (vehicle exhaust, combustion process, etc.) or some natural incident or event that resulted in the creation of a different compound by accident like forest fires, volcanic activities, or photochemical reactions (Jeevanandam et al. 2018).

Forest fires: Every year thousands of forest fires take place, and most are due to lightning strikes and human activities, and many forest fires have been reported to transport micro- and nanosized particles through smoke and ash. The smoke and ash from forest fires can travel long distance, degrading the standard of ambient air quality, thereby increasing the number of particulates in air. A huge number of black carbon and soot are carried and deposited over the Himalayan glaciers by clouds. This results in an increase in absorption of sun's heat, which accelerates the glacial melting (Sapkota et al. 2005). Smoke is known to contain nanoparticles that can cause or worsen the cardiopulmonary conditions in humans leading to fire-related deaths (Mott et al. 2002).

Industries and coal burning: Coal is a dominant source of energy in developing countries like India where around 54.67% of primary energy came from coal in 2019, and for the USA, it was 11.98% with a global rise in coal prediction by 1.5% (BP statistical review 2020). Burning of coal leads to the release of hazardous particles in fly ash. Elements like As, Cd, Cr, Hg, Se, B, CI, F, Mn, Mo, Ni, Pb, Be, Cu, P, Th, U, V, Zn, Ba, Co, Sb, Sn, and Tl are found to be present in traces in coal ash (Saikia et al. 2018).

Diesel and engine exhaust: In urban environment, automobile exhaust contributes to major source of micro and nanoparticles in environment especially carbon compounds like CNTs and fibers (Soto et al. 2005). Diesel engines release 20–130 nm sized particles, whereas gasoline engines release 20–60 nm sized particles like polycyclic aromatic hydrocarbons, metal oxides, nitrates, etc. (Hochella et al. 2019; Sioutas et al. 2005).

Some other sources are cigarette smoking and demolition of building material, which leads to the release of complex nanoparticles (Ning et al. 2006; Stefani et al. 2005). Mining, agricultural processes, nuclear development, and environmental changes are also some of the reasons leading to accidental formation of some particles.

3.8 Engineered Nanomaterials

Engineered nanomaterials (ENMs) are deliberately or intentionally prepared materials with a nanoscale size and have specific physical and chemical properties that are different from the bulk. They are produced to exploit the best properties of an element or to produce desired results under a controlled environment. ENMs are used in diverse fields like biology, pharmacology, medicine, biomedical imaging, biomolecular sensing, drug delivery, tissue engineering, data storage, photocatalytic pigments, cosmetics, food, etc. (Jeevanandam et al. 2018). The most abundantly engineered nanomaterials are TiO_2, Fe SiO_2, Cu, ZnO, Al_2O_3, CeO_2, Ag, carbon nanotubes, graphenes, and nanoclay composites (Hochella et al. 2019).

3.9 Risk Assessment of Nanomaterials

3.9.1 What Is Risk?

Risk is a possibility of occurrence of unwanted events that can be hazardous. It is something hazardous that can happen in the future. When we talk about things that can go wrong if a hazardous event happens, the likelihood of causality/harm of the event to occur, and consequences of a hazardous event, we are talking about risk analysis (Rausand 2011). So risk analysis is the identification of the hazard and risk to individuals, property, and environment by using the current available information in a systematic way (Rausand 2011). Risk evaluation is the process of decision-making regarding the amount of risk that is acceptable. Risk assessment is the combination of risk analysis and risk evaluation. It is a regulatory and a guiding principle for the evaluating the risks due to nanotechnological practices. Hence, it is a

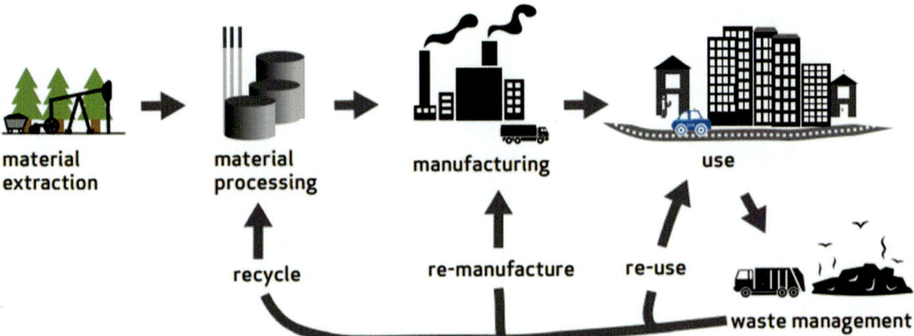

Fig. 3.3 Life cycle of nanomaterials. (In-Briefs – ECHA. https://euon.echa.europa.eu/in-brief-guides. Accessed 31 Aug. 2020)

step-by-step approach to evaluate risk by gathering the available toxicity and exposure data, quantifying it, and making decisions (Covello and Merkhoher 2013; Wilhelmsen and Ostrom 2019). The entire goal of a risk assessment is to reduce the amount of risk to an acceptable level by evaluating the hazard, exposure, dose, and risk characterization (Kuempel et al. 2012). It is a part of the overarching topic of risk management that furthers the risk assessment to track and control the risk and also document as well as communicate the information available on the risk (Popov et al. 2016; Rausand 2011). A technology is useful only if the benefits of using the technology are more than the risks related to using it. As discussed earlier, nanotechnology has lifesaving applications like disease prevention, repairing damage in the human body, clean water provision, human life longevity, maximizing agricultural productivity, and advanced applications like quick response to natural disasters, poverty reduction, improved fuel, and powerful information technology, but it also poses health and environmental risks of toxicity, exposure to unwanted chemicals, and negligence to the actual potential of using nanotechnology. Nanomaterials are produced, used, and disposed leading to the formation of nanomaterial's life cycle (Fig. 3.3). Science-based methods have been used to control and formulate approaches to risk analysis and regulations. In 2008, Environmental Health and Safety (EHS) presented a document that gathered information on various efforts done by the Nanotechnology Environmental and Health Implications (NEHI) working group. Ever since then, EHS has been inviting academia, the federal government, and experts from the industry to share information on latest technology and developments in the risk management of nanotechnology. The National Institute for Occupational Safety and Health (NIOSH) is disseminating all the latest scientific information about the potential health hazards of nanomaterials. Other agencies like OECD are developing guidelines and recommendations on characterizing the nanomaterials and using them safely. Case studies are being used to develop best management routes to manage the uncertain risk of nanotechnology (Fadel et al. 2015).

3.9.2 Health Risks

As the nanomaterials have different properties than their bulk compounds, there is less knowledge available as to how they will react in an environment in human contact. And because there is less knowledge and information about the exposure and physical and chemical reactions, there is a concern that the exposure can exceed the present understanding of nanomaterial behavior. Ultrafine nanoparticles have a size less than 100 nm and can be inhaled easily, leading to breathing problems, lung inflammation, and distal organ involvement (Dominici et al. 2006). Ultrafine particles of carbon, silica, and titanium have been reported to cause problems like mesothelioma, pulmonary toxicity, inflammation, genotoxicity and neurotoxicity, DNA damage, and human embryo development (Baeza-Squiban et al. 2013; Zhang et al. 2006; Shukla et al. 2011). Because of their nanosize, the chances of entering into human bloodstream increase, and they are electrochemically more reactive (Sudha et al. 2018).

3.9.3 Environmental Risks

The field is advancing rapidly and so are the concerns about its safety aspect. Researchers or industries working with nanomaterials should have access to personal protective equipment (PPE). Information on toxicity, regulation standards, exposure time, high precision instruments, effectiveness of PPE, usage of PPE, waste reuse and disposal flow, etc., should be clearly mentioned, explained, and easily accessible before starting to work with nanomaterials. Another way is to use engineered controls to reduce the exposure of nanomaterials, for example, source enclosure (enclosing the source of nanomaterial generation from the workers at workplace), laminar flow cabinets, or local exhaust ventilation (like using lab fume hoods) to capture airborne nanoparticles while conducting the experiment. The exposure limit of only few nanoparticles is known, hence, it gets difficult to measure their exposure limit without designing an accurate instrument. This can be overcome if we detect the overdose for nanoparticles (above the safe level for humans) through a sensor (Seal and Karn 2014).

3.9.4 Risk Assessment Frameworks

Risk assessment is the evaluation of things that can go wrong, likelihood and consequences of it happening (Kaplan and Garrick 1981). Risk assessment of chemical comprises of dose response estimation, hazard and exposure assessments, risk characterization, and uncertainty accountability of the overall estimation. Risk assessment has four components: identification of the hazard, identification of the case of the hazard, hazard evaluation, and prioritization. These components include information gathering, hazard assessment, determination of protection measures, review of the effectiveness of measures, and documentation (Heinemann and Schafer 2009). Applying these basic criteria in the field of nanotechnology needs some modifications specific to the type of nanomaterial (Sayre and Steinhauser 2016). Oomen et al. (2018) suggest different frameworks that can be used to assess the risks related to nanomaterials and the need to keep updating these frameworks. If steps are taken early in the innovation chain to weed out the nanomaterials that can cause human and environmental health problems, there are less chances of toxic nanomaterials with heavy risk to go into the market. Screening level assessment and categorization can be helpful to (Hansen et al. 2014; Godwin et al. 2015; Hund-Rinke et al. 2015) maintain records about known hazards and exposure potential of nanomaterials (to humans and environment) and prioritize those nanomaterials for further testing. Apart from this, the information should be communicated to different companies and stakeholders. Bos et al. (2015) mention MARINA risk assessment strategy focusing on the development of an efficient and flexible data collection system to assess the risk of nanomaterials. Oosterwijk et al. (2016) mention risk banding framework where the particles that are deposited in the lungs and the respiratory tract are studied to collect information regarding toxicity. Gottardo et al. (2017) mention about analyzing the European Union regulatory framework dealing with nanomaterials to provide strong practical knowledge to the industries and regulatory authorities. NANoREG (2020) provides a risk screening framework with six risk identifiers during the early stage of a new project of innovation using nanoparticles: accumulation, stability of coating, genotoxicity, inflammation, ecotoxicity, and solubility rate. Linkov et al. (2007) come up with multicriteria decision analysis (MCDA) in which a decision matrix is used to score the performance based on utility, risk, valuation, and uncertainty of the nanomaterials and hence come up with alternatives if needed. Three basic groups of people are needed to make these decisions: nanotechnology managers and policy/decision makers, scientists and engineers, and stakeholders.

Having some analytical tools to learn about the importance of the nanomaterial is helpful for risk assessment (Lenz e Silva et al. 2019). Fan and Alexeeff (2010) came up with three ways to keep a check on risks related to nanotechnology use. First is to limit the usage of particles that are known or have been known to show hazardous properties. Second is to minimize the exposure of these materials during occupational work and ultimately into the environment. And the last one is to monitor the exposure closely and at regular intervals to detect any harm to human health at initial stages. The USA has a good number of institutes like National Institutes of Health (NIH), National Institute of Standards and Technology (NIST), Environmental Protection Agency (EPA), National Institute for Occupational Safety and Health (NIOSH), and Food and Drug Administration (FDA) that are monitoring, controlling, and regulating nanotechnology-related activities and research in fields like occupational, environmental, health, and risk management (Fan and Alexeeff 2010). For risk management, steps like improving workplace practices, examining material life cycle to inform any risks and reduce them, classifying particles based on physical and chemical properties, nanomaterial use and safety information, exposure level, route nanomaterials take to enter the human body, expected uses, recycling and disposal at the end of use, and developing risk communication approaches should be taken into account (Green facts 2020). Though currently the risks created by nanopracticles are still being explored, these steps can be used as precautionary guidelines to minimize the expore of risk at workplace.

Stress is laid on providing sufficient information on handling the nanomaterials in a top-down approach (identify stakeholders and integrate their decision into risk assessment; objective driven). Creating a common database of information leading to specific links for in-depth information, guidance documents, things to keep in mind when entering the nanotechnology business, validation of this data by third parties, and integration of hazard and exposure data into the admin database (Fadel et al. 2015, Fadeel et al. 2018) can provide us with all the important information about risks related to nanoparticles discovered or being discovered. Making a big database will take time, and efforts to create the data are still ongoing. Standardized methods and test guidelines, specific analytics instruments, engagement of the stakeholders, predictability of the risk of new nanomaterials, and nanotechnology are some of the solutions recommended (Fadel et al. 2015). Biodurability it another parameter of risk assessment. Biodurability of a particle is the extent to which it can resist any physical or chemical resistance, and dissolution is the rate of release of the molecules or ions of the particles (fast rate of ion release, short-term toxic effect) (Bernstein et al. 2005).

In vitro dissolution tests can find their usage to provide information on how particles and fibers may react in a biological setting (human body) (Utembe et al. 2015). The GUIDEnano tool and SUN Decision Support System (SUNDS) are two web-based risk assessment tools that are bridging the gap of quantifying risk assessment and safety limits of nanomaterials. They provide the scope of risk assessment at different levels and scenarios and then provide a multitude of risk mitigation measures that can be put under consideration (Fadeel et al. 2018). Qualitative policy instruments like alternative assessment take into consideration a broad range of alternatives to deal with hazardous materials and activities. Quantitative instruments like multicriteria decision analysis (MCDA) are more structured and provide results for nanomaterial management by running suitable mathematical models (Linkov et al. 2009).

3.10 Safety Aspect of Nanomaterials

The field of nanotechnology is still in its rudimentary stages, and there is a lot to learn about using this technology but a lot more to learn about how to use this technology safely. The uncertainties give rise to addressing the needs of safety and toxicology of nanomaterials. The possible ways of getting exposed to nanomaterials are inhalation, dermal absorption, and ingestion. Researchers have addressed the need to make the multidisciplinary effort at the federal as well as the state level to address health and environmental risks of nanotechnology (Fan and Alexeeff 2010). The main components of occupational health and safety are information gathering, hazard assessment, determination of protection measures, review of the effectiveness of measures, and documentation. The way nanomaterials are stored, handled, and disposed can vary from lab to lab. Most often the nanomaterials under study are stored in a sealed, non-breakable container. Nanomaterials should be kept in a designated area; cleanliness should be taken care of; protective gears should be used like gloves, face masks, etc. (Dhawan et al. 2011). Sensitive nanomaterials are stored in sealed nitrogen boxes. The containers are labelled for the nanomaterial to be used, properties of the nanomaterial, etc. The sealed containers containing the nanomaterial in free form or dry nanomaterials are opened in a ventilated hood, fumed hoods, or closed glove box. Anything that meets the nanomaterial, like the container it was placed in, any wet wipes used to clean it, and gloves handling it, etc., is considered contaminated and should be disposed of properly (Dhawan et al. 2011; Seal and Karn, 2014). Steps have been taken where nanomaterials are converted into nanocomposites to immobilize their movement. Toxicity can be analyzed by using MTT (3-(4,5-dimethylthiazol-2-Yl)-2,5-diphenyltetrazolium bromide), AB, and alkaline comet assays; decreased viability indicates that the toxicity has decreased (Tokarcˇíková et al. 2014). If a nanomaterial is in solvent, it should be immobilized with 1% agar dissolved in distilled water (Dhawan et al. 2011). Methods like standard gravimetric measurement can be used to count the number of particles, however assessment based on mass only gives limited informaiton. The scanning mobility particle sizer is a better technique to measure nanoparticles but is expensive, work intensive and it is not yet standardized or validated. (Heinemann and Schafer 2009). There are different ways to handle a nanomaterial-contaminated surface like wet wiping or using a tack roll mop, which is usually unsafe for people using it as they are exposed to the nanomaterial but is cheap; HEPA vacuuming, which is expensive and has high maintenance; and using strippable decontamination agents, which are safer to use as they do not leave the nanomaterials into the air (Seal and Karn 2014). Proper disposal of these nanomaterials is one of the major issues in safety handling. Before disposal, nanomaterials should be either kept in a labelled waste container or labelled plastic bags and treated as hazardous waste (Dhawan et al. 2011).

Incineration is one of the common methods to treat nanomaterial waste or nanomaterials can be used in soil for construction purposes (Lenz e Silva et al. 2019). United Nations came up with a platform, Globally Harmonized System (GHS), to classify and communicate the hazardous properties of chemicals globally (Catalan and Norppa 2017). High-content analysis/screening (HCA/S) and high-throughput screening (HTS) are suggested to be used in drug discovery and predictive testing of chemicals. Hazard prediction tools that predict not only short-term, high-dose exposure but also long-term, low-dose exposure are needed (Fadeel et al. 2018). A list of nanoparticles has been maintained by the Project on Emerging Technologies, Woodrow Wilson International Center for Scholars (www.wilsoncenter.org/nano) (Jamil et al. 2018).

3.11 Conclusion

As the industry has decided to use nanotechnology and create nanomaterials, it should also be responsible for the health and safety risk it comes with. Detailed knowledge of using nanomaterials needs to be circulated to all the stakeholders as well as the public so that proper measures could be taken to safeguard from potential hazard. Risk assessment frameworks should be studied in detail so that nanomaterials with highest risk concern can be substituted before they enter the research field. Case studies, success stories, and research outcomes should be studied to improve these frameworks as well as handle regulations. Scientific knowledge regarding the risk and behavior of nanomaterials needs to be researched more. New techniques and instrumentation should be invented to measure the exposure level and toxicology of nanomaterials and to avoid any workplace hazards. Collaboration and cooperation of with nanotechnology industries, scientists, decision makers, and health and safety department are necessary to evolve safely in this field. A shift is needed from risk control to risk prevention and reduction.

Multiple Choice Questions

Question 1. The superior properties of nanoparticles are due to.

(a) High surface-to-volume ratio.
(b) Large size.
(c) Low melting point.
(d) No quantum effect.

Question 2. The main applications of nanotechnology in the medical field are.

(a) Drug delivery.
(b) Diagnosis.
(c) Tissue engineering.
(d) All of above.

Question 3. The word "nano" in nanotechnology means.

(a) A nanometer is one-billionth of a meter.
(b) A nanometer is one-millionth of a meter.
(c) A nanometer is one thousand of a meter.
(d) A nanometer is one-trillionth of a meter.

Question 4. Who coined the term nanotechnology?

(a) Norio Taniguchi.
(b) Richard Feynman.
(c) Eric Drexler.
(d) Sumio Iijima.

Question 5. The toxic effect of nanoparticles is termed.

(a) Nano-remediation.
(b) Nano-robotics.
(c) Neurotoxicology.
(d) Nanotoxicology.

Answers

Question 1 (a)
Question 2 (d)
Question 3 (a)
Question 4 (b)
Question 5 (d)

Short Questions and Answers

Q1. What is the basic difference between nanomaterials and bulk counterparts?

Answer: The significant size reduction has induced quantum effect and further enhanced the surface-to-volume ratio in nanomaterials as compared to their bulk counterparts.

Q2. What are the dangerous impact of nanomaterials?

Answer: The high-end applications of nanoparticles have further enhanced their production rate and made living beings more prone to their exposure. With time, slow deposition results into various health hazards in living beings.

Q3. Explain how photocatalytic properties of nanomaterials are useful?

Answer: The method of photo-degradation has been primarily based on the organic pollutant adsorption ability over the photocatalyst surface. The photocatalyst with higher adsorption surface makes the degradation process easier as compared to other remediation processes.

Q4. List out the major pollutants present in the environment.

Answer: Inorganic anions, toxic gases (nitrogen oxides, sulfur oxides, ozone, carbon oxides, etc.), suspended particulate matter, as well as volatile organic compounds.

References

Baeza-Squiban A, Vranic S, Boland S (2013) Fate and health impact of inorganic manufactured nanoparticles. In: Brayner R et al (eds) Nanomaterials: a danger or a promise? vol 9. Springer-Verlag, London ch, pp 245–268. https://doi.org/10.1007/978-1-4471-4213-3_9

Bernstein D, Castranova V, Donaldson K, Fubini B, Hadley J, Hesterberg T (2005) Testing of fibrous particles: short-term assays and strategies. Inhal Toxicol 17:497–537

Bhushan B (2017) Introduction to nanotechnology. In: Bhushan B (ed) Springer handbook of nanotechnology. Springer handbooks. Springer, Berlin, Heidelberg, pp 1–19. https://doi.org/10.1007/978-3-662-54357-3_1

Blundell G, Henderson WJ, Price EW (1989) Soil particles in the tissues of the foot in endemic elephantiasis of the lower legs. Ann Trop Med Parasitol 83(4):381–385

Bos PM, Gottardo S, Scott-Fordsmand JJ, Van Tongeren M, Semenzin E, Fernandes TF, Hristozov D, Hund-Rinke K, Hunt N, Irfan MA, Landsiedel R, Peijnenburg WJ, Sanchez Jimenez A, Van Kesteren PC, Oomen AG (2015) The MARINA risk assessment strategy: a flexible strategy for efficient information collection and risk assessment of nanomaterials. Int J Environ Res Public Health 12:15007–15021

Boverhof DR, Bramante CM, Butala JH, Clancy SF, Lafranconi M, West J, Gordon SC (2015) Comparative assessment of nanomaterial definitions and safety evaluation considerations. Regul Toxicol Pharmacol 73(1):137–150

Bowman RS (2003) Applications of surfactant-modified zeolites to environmental remediation. Microporous Mesoporous Mater 61(1):43–56

Brabazon D (2016) Nanostructured materials. Reference Module in materials science and materials engineering. In: Reference module in materials science and materials engineering. https://doi.org/10.1016/B978-0-12-803581-8.04103-5

Catalan J, Norppa H (2017) Safety aspects of bio-based nanomaterials. Bioengineering 4(4):94

Connolly JM, Raghavan V, Owens P, Wheatley A, Keogh I (2014) Nanogold-based photosensitizers probes for dual-model bioimaging and therapy of cancer. J Nanomed Nanotechnol 5:249

Corachán M, Tura JM, Campo E, Soley M, Traveria A (1988) Podoconiosis in Aequatorial Guinea. Report of two cases from different geological environments. Trop Geogr Med 40:359–364

Covello VT, Merkhoher MV (2013) Chapter 1: risk assessment methods. In: Approaches for assessing health and environmental risks. Springer, New York, pp 1–33

Dhawan A, Shanker R, Das M, Gupta K (2011) Guidance for safe handling of nanomaterials. J Biomed Nanotechnol 7(1):218–224

Dominici F, Peng RD, Bell ML, Pham L, McDermott A, Zeger SL, Samet JM (2006) Fine particulate air pollution and hospital admission for cardiovascular and respiratory diseases. JAMA 295(10):1127–1134

Drexler K (1990) Engines of creation, 1st edn. Anchor Press/Doubleday, Garden City

Elliot DW (2006) NZVI chemistry and treatment capabilities. Federal Remediation Technologies Roundtable (FRTR). Remediation technologies screening matrix and reference guide. Accessed 25 Sept.

European Commission Regulation (EU) (2012) No 528/2012 of the European Parliament and of the Council of 22 May 2012 concerning the making available on the market and use of biocidal products. Off J Eur Union L167:1–123

Fadeel B, Farcal L, Hardy B, Vazquez-Campos S, Hristozov D, Marcomini A, Lynch I, Valsami-Jones E, Alenius H, Savolainen K (2018) Advanced tools for the safety assessment of nanomaterials. Nat Nanotechnol 13:537–543

Fadel TR, Steevens JA, Thomas TA, Linkov I (2015) The challenges of nanotechnology risk management. NanoToday 10(1):6–10

Fan AM, Alexeeff G (2010) Nanotechnology and nanomaterials: toxicology, risk assessment, and regulations. J Nanosci Nanotechnol 10(12):8646–8657.

Fitzsimmons JN, Boyle EA, Jenkins WJ (2014) Distal transport of dissolved hydrothermal iron in the deep South Pacific Ocean. Proc Natl Acad Sci U S A 111:16654–16661. https://doi.org/10.1073/pnas.1418778111

Godwin H, Nameth C, Avery D, Bergeson LL, Bernard D, Beryt E, Boyes W, Brown S, Clippinger AJ, Cohen Y, Doa M, Hendren CO, Holden P, Houck K, Kane AB, Klaessig F, Kodas T, Landsiedel R, Lynch I, Malloy T, Miller MB, Muller J, Oberdorster G, Petersen EJ, Pleus RC, Sayre P, Stone V, Sullivan KM, Tentschert W, P. and Nel, A.E. (2015) Nanomaterial categorization for assessing risk potential to facilitate regulatory decision-making. ACS Nano 9(4):3409–3417

Gottardo S, Alessandrelli M, Amenta V, Atluri R, Barberio G, Bekker C, Bergonzo P, Bleeker E, Booth AM, Borges T, Buttol P, Carlander D, Castelli S, Chevillard S, Clavaguera S, Dekkers S, Delpivo C, Di Prospero Fanghella P, Dusinska M, Einola J, Ekokoski E, Fito C, Gouveia H, Grall R, Hoehener K, Jantunen P, Johanson G, Laux P, Lehmann HC, Leinonen R, Mech A, Micheletti C, Noorlander C, Olof-Mattsson M, Oomen A, Quiros Pesudo L, Letizia Polci M, Prina-Mello A, Rasmussen K, Rauscher H, Sanchez Jimenez A, Riego Sintes J, Scalbi S, Sergent JA, Stockmann-Juvala H, Simko M, Sips A, Suarez B, Sumrein A, Van Tongeren M, Vázquez-Campos S, Vital N, Walser T, Wijnhoven S, Crutzen H (2017). NANoREG framework for the safety assessment of nanomaterials, EUR 28550 EN

Greenfacts (2020) Are current risk assessment methodologies for nanoparticles adequate? https://copublications.greenfacts.org/en/nanotechnologies/l-3/8-risk-assessment.htm#4p0

Hansen FS, Alstrup Jensen K, Baun A (2014) NanoRiskCat: a conceptual tool for categorization and communication of exposure potentials and hazards of nanomaterials in consumer products. J Nanopart Res 16:2195. https://doi.org/10.1007/s11051-013-2195-z

Heinemann M, Schäfer H (2009) Guidance for handling and use of nanomaterials at the workplace. Hum Exp Toxicol 28(6–7):407–411

Hochella MF Jr, Aruguete D, Kima B, Madden AS (2012) Chapter 1: naturally occurring inorganic nanoparticles: general assessment and a global budget for one of Earth's last unexplored major geochemical components. In: Barnard AS, Guo H (eds) Nature's nanostructures. Jenny Stanford Publishing, New York, pp 1–31

Hochella MF Jr, Mogk DW, Ranville J, Allen IC, Luther GW, Marr LC, McGrail BP, Murayama M, Qafoku NP, Rosso KM, Sahai N, Schroeder PA, Vikesland P, Westerhoff P, Yang Y (2019) Natural, incidental, and engineered nanomaterials and their impacts on the earth system, review summary. Science 363:1414

Hund-Rinke K, Herrchen M, Schlich K, Schwirn K, Völker D (2015) Test strategy for assessing the risks of nanomaterials in the environment considering general regulatory procedures. Environ Sci Eur 27:24. https://doi.org/10.1186/s12302-015-0053-6

Hutchison JE (2016) The road to sustainable nanotechnology: challenges, progress and opportunities. ACS Sustain Chem Eng 4(11):5907–5914

Iso.org, 80004-1:2010, I (2017) ISO/TS 80004-1:2010—Nanotechnologies—Vocabulary—Part 1 Core terms (2017). [online] Iso.org. Available from: https://www. iso.org/standard/51240.html

ISO/TS 27867 (2008) Nanotechnologies—terminology and definitions for nano-objects—nanoparticle, nanofibre and nanoplate. Available from: https://www.iso.org/standard/44278.html

Jalaja K, Naskar D, Kundu SC, James NR (2016) Potential of electrospun core-shell structured gelatin–chitosan nanofibers for biomedical applications. Carbohydr Polym 136:1098–1107

Jamil B, Javed R, Qazi AS, Syed MA (2018) Nanomaterials: toxicity, risk management and public perception. In: Rai M, Biswas JK (eds) Nanomaterials: Ecotoxicity, safety, and public perception, pp 283–304. https://doi.org/10.1007/978-3-030-05144-0_14

Jeevanandam J, Barhoum A, Chan YS, Dufresne A, Danquah MK (2018) Review on nanoparticles and nanostructured materials: history, sources, toxicity, and regulations. Beilstein J Nanotechnol 9(1):1050–1074

Jones J, Parker D, Bridgwater J (2007) Axial mixing in a ploughshare mixer. Powder Technol 178(2):73–86

Kaplan S, Garrick BJ (1981) On the quantitative definition of risk. Risk Anal 1(1):11–27

Khan I, Saeed K, Khan I (2017) Nanoparticles: Properties, applications, and toxicities. Arab J Chem. https://doi.org/10.1016/j.arabjc.2017.05.011

Kuempel ED, Geraci CL, Schulte PA (2012) risk assessment and risk management of nanomaterials in the workplace: translating research to practice. Ann Occup Hyg 56(5):491–505. https://doi.org/10.1093/annhyg/mes040

Linkov IF, Satterstrom K, Steevens J, Ferguson E, Pleus RC (2007) Perspectives multi-criteria decision analysis and environmental risk assessment for nanomaterials. J Nanopart Res 9:543–554. https://doi.org/10.1007/s11051-007-9211-0

Linkov I, Satterstrom FK, Monica JC Jr, Hansen SF, Davis TA (2009) Nano risk governance: current developments and future perspectives. Nanotechnol Law Bus 6(2):203–220

Liu Y, Ong CN, Xie J (2016) Emerging nanotechnology for environmental applications. Nanotechnol Rev 5(1):1–2

Low J, Yu J, Ho W (2015) Graphene-based Photocatalysts for CO2 reduction to solar fuel. J Phys Chem Lett 6(21):4244–4251

Mamalis AG (2007) Recent advances in nanotechnology. J Mater Process Technol 181(1–3):52–58

Mansoori GA, Soelaiman TAF (2005) Nanotechnology – an introduction for the standards community. J ASTM Int 2(6):1–21. Paper ID JAI13110

Martine CR (1994) Nanomaterials: A Membrane-Based Synthetic Approach. Science 266(5193):1961–1966. https://doi.org/10.1126/science.266.5193.1961

Mishra Y, Chakravadhanula V, Hrkac V, Jebril S, Agarwal D, Mohapatra S, Adelung R (2012) Crystal growth behavior in Au-ZnO nanocomposite under different annealing environments and photo switchability. J Appl Phys 112(6):301–309

Mobasser S, Firoozi AA (2016) Review of nanotechnology applications in science and engineering. Journal of Civil Engineering and Urbanism 6(4):84–93

Mott JA, Meyer P, Mannino D, Redd SC, Smith EM, Gotway-Crawford C, Chase E (2002) Wildland forest fire smoke: health effects and intervention evaluation, Hoopa, California, 1999. West J Med 176(3):157–162

NANoREG D6.04 (2020) DR inventory of existing regulatory accepted toxicity tests I RIVM.https://www.rivm.nl/en/documenten/nanoreg-d604-dr-inventory-of-existing-regulatory-accepted-toxicity-tests

Ning Z, Cheung CS, Fu J, Liu MA, Schnel, l M.A. (2006) Experimental study of environmental tobacco smoke particles under actual indoor environment. Sci Total Environ 367(2–3):822–830

Oldham VE, Mucci A, Tebo BM, Luther GW III (2017) Soluble Mn (III)–L complexes are abundant in oxygenated waters and stabilized by humic ligands. Geochim Cosmochim Acta 199:238–246. https://doi.org/10.1016/j.gca.2016.11.043

Oomen AG, Steinhäuser KG, Bleeker EAJ, Van Broekhuizen F, Sipsa A, Dekkers S, Wijnhovena SWP, Sayre PG (2018) Risk assessment frameworks for nanomaterials: scope, link to regulations, applicability, and outline for future directions in view of needed increase in efficiency. NanoImpact 9:1–13

Oosterwijk MT, Feber ML, Burello E (2016) Proposal for a risk banding framework for inhaled low aspect ratio nanoparticles based on physicochemical properties. Nanotoxicology 10(6):780–793. https://doi.org/10.3109/17435390.2015.1132344

Popov G, Lyon BK, Hollcroft B (2016) Risk assessment: a practical guide to assessing operational risks. Wiley, Incorporated

Rakesh M, Divya TN, Vishal T, Shalini K (2015) Applications of nanotechnology. J Nanomed Biotherapeutic Discov 5:131. https://doi.org/10.4172/2155-983X.1000131

Ramsden J (2016) Chapter 1: what is nanotechnology? In: Ramsden J (ed) Nanotechnology – an introduction, 2nd edn. Elsevier, pp 1–18

Rausand M (2011) Risk assessment- theory, methods and applications. Wiley, pp 1–664. 9780470637647

Reibold M, et al (2006) Materials: Carbon Nanotubes in an Ancient Damascus Sabre. Nature 444(7117):286. https://doi.org/10.1038/444286a

Reiss G, Hutten A (2010) Magnetic nanoparticles. In: Sattler KD (ed) Handbook of nanophysics: nanoparticles and quantum dots. CRC press, Boca Raton, pp 2–1

Rickerby DG, Morrison M (2007) Nanotechnology and the environment: A European perspective. Sci Technol Adv Mater 8(1):19–24

Rietmeijer FJ, Nuth JA (2012) Chapter 13: nanoparticles that are out of this world. In: Barnard AS, Guo H (eds) Nature's nanostructures. Pan Stanford Publishing Pte. Ltd., Singapore, pp 329–360

Saikia BK, Saikia J, Rabha S, Silva LFO, Finkelman R (2018) Ambient nanoparticles/nanominerals and hazardous elements from coal combustion activity: implications on energy challenges and health hazards. Geosci Front 3(9):863–875

Sapkota A, Symons JM, Kleiss LJ, Wang L, Parlange MB, Ondov J, Breysse PN, Diette GB, Eggleston PA, Buckley TJ (2005) Impact of the 2002 Canadian forest fires on particulate matter air quality in Baltimore city. Environ Sci Technol 39(1):24–32

Sayre P, Steinhäuser K (2016) ProSafe – Roadmap for Members of Task Force When Reviewing Data, Protocols, Reports and Guidance Notes for Regulatory Relevance. Final version

Seal S, Karn B (2014) Safety aspect of nanotechnology-based activity. Saf Sci 63:217–225

Shukla R, Sharma V, Pandey A, Singh S, Sultana S, Dhawan, and A. (2011) ROS-mediated genotoxicity induced by titanium dioxide nanoparticles in human epidermal cells. Toxicol In Vitro 25(1):231–241

Silva GL, Viana C, Domingues D, Vieira F (2019) Risk assessment and health, safety, and environmental management of carbon nanomaterials. In: Clichici S, Filip A, Gustavo M, Nascimento D (eds) Nanomaterials – toxicity, human health and environment. IntechOpen. https://doi.org/10.5772/intechopen.85485. Available from: https://www.intechopen.com/books/nanomaterials-toxicity-human-health-and-environment/risk-assessment-and-health-safety-and-environmental-management-of-carbon-nanomaterials

Sioutas C, Delfino RJ, Singh M (2005) Exposure assessment for atmospheric ultrafine particles (UFPs) and implications in epidemiologic research. Environ Health Perspect 113(8):947–955

Smith DM, Simon JK, Baker J (2013) Applications of nanotechnology for immunology. Nat Rev Immunol 13:592–605

Sobolev K, Shah SP (2015) Nanotechnology in Construction. In *Proceedings of NICOM5*, Springer

Soto KF, Carrasco A, Powell TG, Garza KM, Murr LE (2005) Comparative in vitro cytotoxicity assessment of some manufactured nanoparticulate materials characterized by transmission electron microscopy. J Nanopart Res 7:145–169

Statistical Review of World Energy | Energy Economics | Home (2020) Bp Global. https://www.bp.com/en/global/corporate/energy-economics/statistical-review-of-world-energy.html. Accessed 27 Aug. 2020

Stefani D, Wardman D, Lambert T (2005) The implosion of the Calgary general hospital: ambient air quality issues. J Air Waste Manag Assoc 55(1):52–59

Sudha PN, Sangeetha K, Vijayalakshmi K, Barhoum A (2018) Nanomaterials history, classification, unique properties, production and market. In Emerging applications of nanoparticles and architecture nanostructures, pp 341–384. https://doi.org/10.1016/b978-0-323-51254-1.00012-9

Taniguchi N (1974) On the basic concept of "nano-technology", In: *Proceedings of the International Conference on Production Engineering*, Part II, Japan Society of Precision Engineering, Tokyo, pp 18–23

Tokarcˇíková M, Tokarský J, Cˇábanová K, Mateˇjka V, Mamulová Kutláková K, Seidlerová J (2014) The stability of photoactive kaolinite/TiO_2 composite. Compos Part B 67:262–269

Tran Le MT (2018) It's a nano party: nanoparticles and their applications in medicine. https://hippocratesmedreview.org/its-a-nanoparty-nanoparticles-and-their-applications-in-medicine/

Tyagi S, Rawtani D, Khatri N, Tharmavaram M (2018) Strategies of nitrate removal from aqueous environment using nanotechnology: a review. J Water Process Eng 21:84–95

Utembe W, Potgieter K, Stefaniak., A.B. and Gulumian, M. (2015) Dissolution and bio durability: important parameters needed for risk assessment of nanomaterials. Part Fiber toxicol 12:11. https://doi.org/10.1186/s12989-015-0088-2

Wilhelmsen CA, Ostrom LT (2019) Risk assessment: tools, techniques, and their applications. John Wiley & Sons

Zhang L, Jiang Y, Ding Y, Povey M, York, and D. (2006) Investigation into the antibacterial behavior of suspensions of ZnO nanoparticles (ZnO nanofluids). J Nanopart Res 9(3):479–489

Chapter 4
Environmental Fate Descriptors for Screening Nanotoxicity and Pollutant Sensing

Pooja Chauhan, K. K. Bhasin, and Savita Chaudhary

Abstract Herein this chapter, advanced nanoparticles have been as the most promising tool for the sensing of contaminants/pollutants present in the environment. In addition to the advanced applications, the toxicological screening and biocompatible behaviour of the nanoparticles has been investigated through various quantitative multi-assay approaches such as antibacterial, antifungal, *Allium cepa*, phytotoxicity, antialgal, photodegradation, aggregation along with their dissolution and sorption experiments. These toxicologically screened nanoparticles have possessed higher aptitude in pollutant sensing due to their high stability, selectivity, sensitivity and fast simplified response array towards toxins. An overview of various types of semiconducting nanoparticles has been reported against numerous types of contaminants such as heavy metal ions, organic dyes, inorganic anions, organic volatile compounds and gases.

Keywords Nanoparticles · Gases · Pollutant · Contamination · Dyes · Biocompatibility · Toxicity · Fate · Environment · Heavy metals

4.1 Introduction

The toxic pollutant released as industrial and household waste has been contaminating the environment at a very high level. It has become a worldwide concern in front of researchers to overcome the mortal and adverse effect of toxins on environment and save the human beings and animals from adverse effects of toxins (Fig. 4.1). The toxic pollutants involve heavy metal ions, inorganic anions, phthalates,

P. Chauhan · K. K. Bhasin · S. Chaudhary (✉)
Department of Chemistry and Centre of Advanced Studies in Chemistry, Panjab University, Chandigarh, India
e-mail: kkbhasin@pu.ac.in; schaudhary@pu.ac.in

R. Kumar et al. (eds.), *Advanced Functional Nanoparticles "Boon or Bane" for Environment Remediation Applications*, Environmental Contamination Remediation and Management, https://doi.org/10.1007/978-3-031-24416-2_4

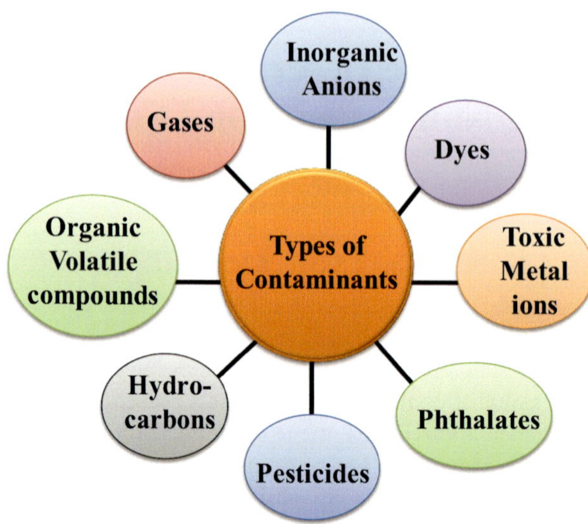

Fig. 4.1 Representation of different types of contaminants present in the atmospheric environment

pesticides, organic and inorganic volatile compounds, gases, dyes from industrial wastes and many other toxic wastes (Jiang et al. 2018; Waheed et al. 2018).

The tremendous amount of released contaminants in natural environment leads to numerous lethal health issues like kidney failure, damage to the nervous system and brain, lack of development in children, skin diseases, cancer, asthma, allergy, depression, autism, damage to DNA and, in extreme condition, death (Fig. 4.2). The environmental pollutants permeate through air, water, soil and cosmetics, whereas the primary source for the bioaccumulation of contaminants is through our food chain (Bolis et al. 2018).

A number of toxic pollutants enter into the environment through various anthropogenic sources such as industrial activities and combination of fossil fuels. These biological and non-biological conversion and permeation of harmful contaminants in the ecological system are major problems in front of living beings (Alharbi et al. 2018; Oguri et al. 2018). Therefore, according to toxicological and environmental safety point of view, it is mandatory to develop such schemes which include easy, simple, cost-effective, and high-performance strategies for the detection of harmful contaminants in a very specific and selective manner. A series of techniques are available in literature for the analysis of toxic contaminants, which helps to detect toxicity at a very low concentration. The different varieties of techniques used for the sensing of harmful toxins have been listed in Table 4.1 with their advantages and disadvantages (Clarkson et al. 2003; Wang et al. 2008; Walekar et al. 2017).

All the above-mentioned methods have very high selectivity and sensitivity towards various toxins. Simultaneously, these reported techniques have possessed series of disadvantages. Thus, there is an urgent need for the development of low-cost material, with high sensing ability useful for onsite detection and removal of

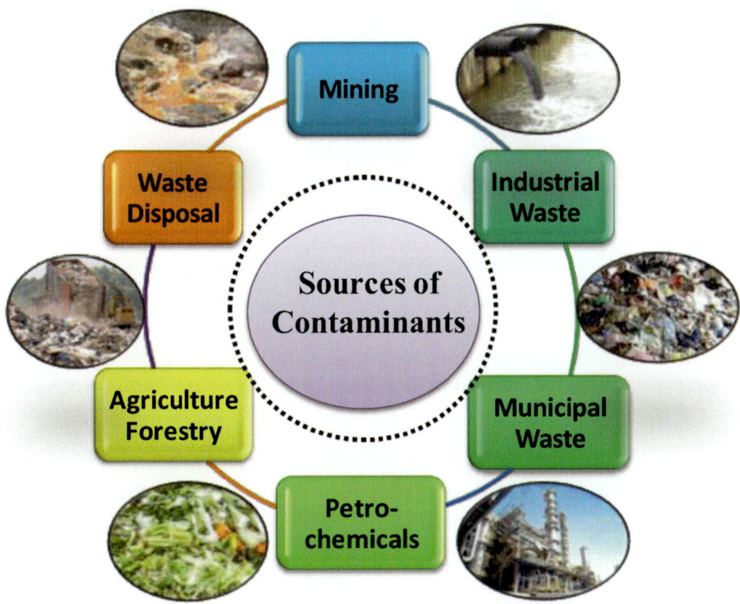

Fig. 4.2 Different sources of contaminants present in the environment

contaminants in a timely fashion. From the last few decades, nanoparticles have emerged as a new milestone in the field of environmental pollution for the detection of toxins. Meticulously nanomaterials possess excellent properties as compared to the bulk material as the former have outstanding surface area to volume ratio and quantum effect which makes them efficient sensing probe in the field of contaminant sensing. Due to their large surface area, they have been widely used in the assortment of physico-chemical processes and their potential application in nanoscience. They are capable of shaping the current needs and act as a safeguard towards environment in the assessment of novel and simpler materials and provide various remediation schemes in the handling of harmful toxins. Out of a wide variety of nanomaterials, semiconducting nanoparticles have engrossed much consideration as these are much superior to recognize and eliminate suitable contaminants from the environment. Semiconducting nanoparticles have excellent optical and photoluminescence properties, wide band gap values, tremendously biocompatible in nature and highly prone towards extensive range of toxic pollutants. The augmentation in susceptibility, discrimination, reproducibility and physico-chemical properties of the semiconducting nanomaterials has provided a platform in miniaturize sensing devices (Payan et al. 2017). In literature, diverse ranges of methods are available for the production of semiconducting nanoparticles (Fig. 4.3). (Sanchis et al. 2018; Kumar et al. 2017a, b).

This chapter mainly focuses on the different types of contaminants present in the environment and their remediation by using semiconducting nanoparticles.

Table 4.1 Representation of different sophisticated instrumental techniques used for contaminant sensing along with their advantages and disadvantages

S. no.	Techniques used	Advantages	Disadvantages
1	Stripping voltammetry	Useful for different water samples taken from different sources and has the tendency to examine environmental component out of mixture	Electrode is used and requires time-to-time modification
2	Mass spectrometry	Good resolution and capable to determine exact structure at lower concentration level	Bulky device, costly, time running process and multiple pre-treatment steps
3	Atomic absorption spectroscopy	Highly selective and sensitive towards detection	Utilization of pricey reagents, towering maintenance expenditure and time-consuming analysis process
4	Small-angle neutron scattering	Investigates surface characterization with high penetration power	Needs more complicated and superior monitoring procedure
5	Raman spectroscopy	Easy elucidation of chemical structure including fingerprint technique	Cost of laser sources is very high and works at low excitation
6	Auger electron spectroscopy	Analysis of chemical surface along with semi-quantitative data	Low surface sensitivity, high lateral resolution and low portability
7	Nuclear magnetic resonance	Crystalline and non-crystalline structure, molecular physics	Not available for higher molecular mass compounds and very expensive instrumentation
8	X-ray photoelectron spectroscopy	Analysis of surface composition	High vacuum is required and slow processing
9	High-performance chromatographic method	Analysis of various types of pollutant and proteins in real water world	Response alters with alteration in environmental exposure and low throughput
10	Amperometric-voltammetric electrochemical technique	Tendency to explore all environmental samples along with chemical detection	Prospect of intervention by alteration in stipulation and very long procedure

Semiconducting nanoparticles, such as zinc oxide (ZnO), nickel oxide (NiO), titanium dioxide (TiO_2), iron oxide (FeO), tellurium oxide (TeO_2), selenium (Se), silica oxide (SiO_2), magnesium oxide (MgO), chromium oxide (Cr_2O_3) and copper oxide (CuO), have been extensively used for the detection and removal of a wide variety of pollutants. Before utilizing the semiconducting nanoparticles in the field of environmental remediation, toxicological fate and screening of respective nanoparticles have been explored in detailed by utilizing various biological quantitative multi-assay approaches. In the end, various types of toxins, such as heavy metal ions,

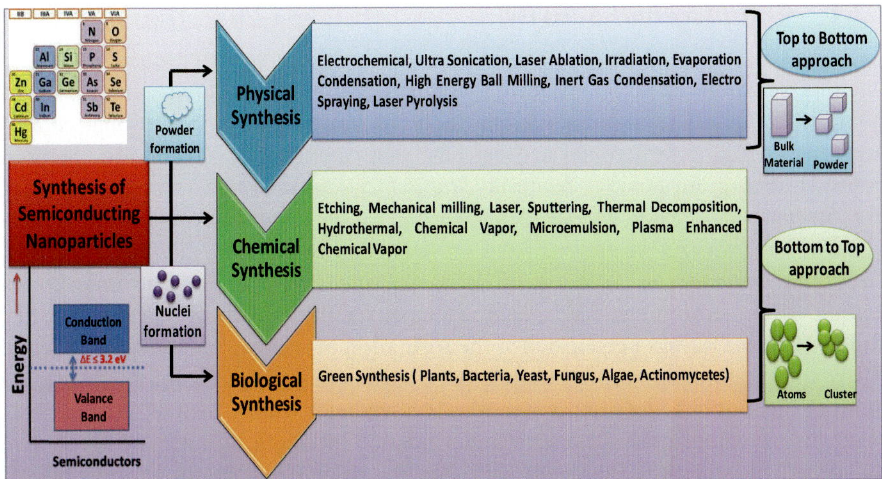

Fig. 4.3 Synthesis of semiconducting nanoparticles by using different approaches

organic pollutants, inorganic anions and organic dyes, have been studied with their origin and detection by using semiconducting nanoparticles.

4.2 Toxicological Fate of Semiconducting Nanoparticles

Semiconducting nanoparticles have been widely used in the pharmaceutical industries, material science and analytical science due to their ease of synthesis, low cost, high bioavailability and excellent response array. They can also be applied in biomedical field, such as drug delivery system, live cell imaging, tissue infection, strain resistance, vaccines and theranostic system (Rad et al. 2019). But the sensing of different types of contaminants by using semiconducting nanoparticles has engrossed much attention in the past decades. Before utilizing the synthesized semiconducting nanoparticles in the field of pollutant sensing and environmental remediation, it is mandatory to investigate the toxicological fate of developed particles (Rana et al. 2019). In literature, various types of activities have been reported to explore the toxicological fate of semiconducting nanoparticles which include antibacterial, antifungal, antialgal, anti-proliferation, *Allium cepa* chromosomal, soil sorption and photodegradation and aggregation/dissolution activities.

4.2.1 Antimicrobial Activity

Out of different toxicological profiling methods, antimicrobial activity is considered the most frequently used phenomenon due to its more standard operational parameters. In general, antimicrobial activity has been carried out on two basic bacterial strains such as gram-positive bacteria (*Staphylococcus aureus, Clostridioides difficile, Streptococcus pneumoniae, Micrococcus luteus, Bacillus clausii*) and gram-negative bacteria (*Escherichia coli, Pseudomonas aeruginosa, Salmonella typhi, Klebsiella pneumoniae, Salmonella enteric, Proteus vulgaris* and so on). Luria-Bertani (LB) medium has been generally used for the growth of bacterial strains. Muller Hinton agar has been used for the formation of solid agar plates (Gupta et al. 2016). In literature, numerous biotools have been explored to execute the respective commotion. But out of those, few methods have been widely manoeuvred such as paper disc method, optical density quantification, staining by different dyes, live and dead cell staining, agar well diffusion assay and colony formation unit for studying the antimicrobial activities in the presence of nanoparticles (Thakur et al. 2019). The permeable growth of bacterial strains was observed at solidified agar gel plate under sterilized condition in laminar air flow and incubated at 37 °C for 24 h. Semiconducting nanoparticles display excellent antibacterial activity towards both gram-positive and gram-negative bacteria and act as antiseptic agents against various infectious diseases caused by a wide range of bacterial species. From the recent reports, it has been deduced that semiconducting nanoparticles enhanced the permeability of thin bacterial cell membrane. Due to which significant functional characteristic of various proteins, fatty acids and respiratory enzymes get restricted and result in the disruption of RNA and DNA, and simultaneously protein functioning has also been influenced in the presence of nanoparticles (Chaudhary et al. 2018a). The disruption of nucleic acids results in the inhibition of ion transport process which further leads to the effectual zone of inhibition. It was observed that the development of zone of inhibition during the respective study also depends upon the size and shape of developed nanoparticles because small-sized particles have very high tendency to incorporate inside the cell wall of bacterial species. In literature, numerous authors have successfully synthesized various types of semiconducting nanoparticles, and their antimicrobial activity has been carried out against different bacterial strains and reported in Table 4.2. The biocompatible copper oxide nanoparticles form excellent nanocomposite in the presence of cellulose (CEL) and chitosan (CS). Various spectroscopic techniques were utilized to investigate the complete chemical intactness of CEL and CS with CuONPs (Tran et al. 2017). The size of developed nanoparticles was found in the range of 20–22 nm. It reflects outstanding antimicrobial activity towards a wide variety of bacterial strains both gram positive and gram negative. Iron oxide nanoparticles also serve as great antibacterial agents. However, Konwar et al. have modified iron oxide nanoparticles with chitosan (CH) and formed a nanocomposite by using graphene oxide (Konwar et al. 2016). The antimicrobial activity of chitosan iron oxide graphene nanocomposite was found to

Table 4.2 Representation of inhibition zone diameter of semiconducting nanoparticles in the presence of both gram-positive and gram-negative bacteria

S. no.	Nanoparticle	Gram-positive bacteria	ZID (mm)	Gram-negative bacteria	ZID (mm)	References
1	FeO	*Bacillus subtilis*	2	*Pseudomonas aeruginosa*	4	Rana et al. (2019)
		Staphylococcus aureus	10	*Salmonella typhi*	12	
2	TiO_2	*Bacillus subtilis*	24	*Escherichia coli*	27.67	Thakur et al. (2019)
		Staphylococcus aureus	27	*Salmonella typhi*	21.63	
3	ZnO	*Staphylococcus aureus*	14.5	*Escherichia coli*	12.7	Rad et al. (2019)
4	Se	*Staphylococcus aureus*	27.84	*Escherichia coli*	28	Chaudhary et al. (2018a, b, c)
				Salmonella typhi	27.62	
				Pseudomonas aeruginosa	26.90	
5	CuO	*Staphylococcus aureus*	10	*Enterobacter cloacae*	30	Kim et al. (2006)
				Escherichia coli	15	
6	NiO	*Staphylococcus aureus*	0	*Escherichia coli*	0	Chaudhary et al. (2018a, b, c)
7	TeO_2	*Staphylococcus aureus*	22	*Escherichia coli*	22	Gupta et al. (2016)
		Streptococcus pyogenes	0	*Klebsiella pneumoniae*	24	

be exceptionally stronger than chitosan iron oxide nanoparticles at different concentration levels and has been shown in Fig. 4.4.

4.2.2 Antifungal Activity

The antifungal activity of the semiconducting nanoparticles has been studied by using a number of various fungal species such as *E. taxodii, A. alternate, C. herbarum, C. krusei, F. oxysporum, C. albicans, A. niger, C. tropicalis, P. chrysogenum, C. glabrata* and many more. Nutrient broth and Muller Hinton agar culture medium has been utilized for the inoculation of fungal strains from the fresh colonies. The homogenous solution of agar-agar type I and malt extract prepared in equimolar ratio has been prepared for the maximum growth of fungal species (Chaudhary et al. 2017). The value of optical density for the respective culture mediums has been calculated through microdilution method, and it should not be less than 0.2 a.u. at around 600 nm wavelength (Miri et al. 2019). The maintenance of reaction condition (48–72 h) and temperature (below 17 °C) plays a very significant role for the uniform growth of fungal strains. The whole experimental procedure has been

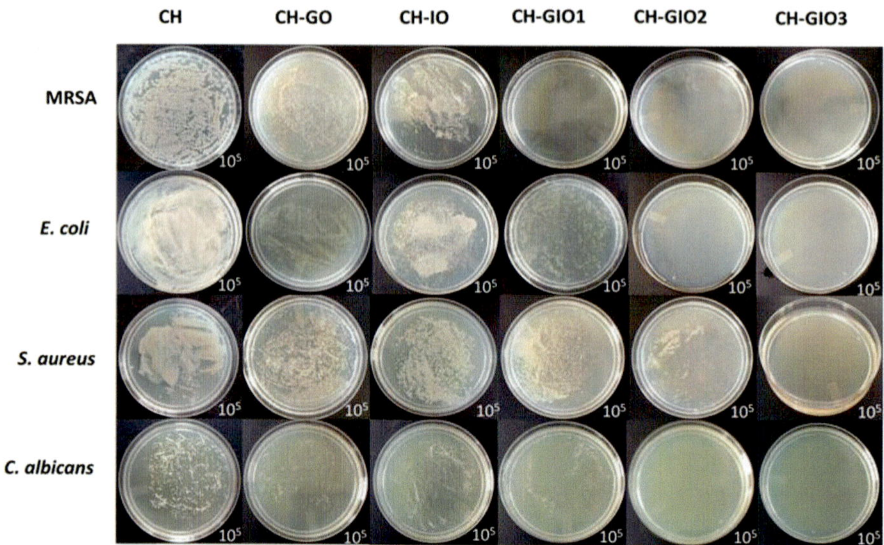

Fig. 4.4 Pictorial representation of antibacterial activity against wide variety of bacterial strains (*E. coli, S. aureus* and *C. albicans*) by using bare chitosan (CH), chitosan graphene oxide (CH-GO), chitosan iron oxide (CH-IO) and chitosan iron oxide graphene nanocomposite (CI-GIO3). (Adapted figure from Konwar et al. (2016) with permission from copyright, American Chemical Society (Washington, DC, USA))

carried out in sterilized laminar air flow under optimal condition and growth of fungal species observed below 17 °C for 4–5 days in incubator. After maintaining the standard condition, the growth of mycelium exerted by fungus was observed and compared with the standard setup (Kumar et al. 2017a, b; Sharma et al. 2018). The broth dilution, agar well diffusion and agar gel escalation activities have been performed against various fungal strains (Rajiv et al. 2013). To substantiate the reproducibility of the respective activity, the same experiment has been carried out thrice and mean of average diameter was calculated. In literature, various reports have been presented on the antifungal activity of semiconducting nanoparticles (Sharma et al. 2010). Many authors synthesized FeO, TiO_2, FeS, ZnO, SiO_2, SnO_2, Cu_2O, NiO, TeO_2 and Se nanoparticles and hence performed their antifungal activity. It was found that developed nanoparticles were highly antifungal in nature and exhibited an appreciable range of inhibition zone and spore germination. MgO-, ZnO- and Mg-doped ZnO nanoparticles have been prepared by applying chemical sol-gel synthetic method. The prepared nanoparticles have manifested excellent antifungal activity at a higher concentration level (Fernandez et al. 2017) for calcareous stone heritage against *Aspergillus niger* and *Penicillium oxalicum* fungal strains as compared to the control standard setup (Fig. 4.5). Simultaneously, the developed nanoparticles have also displayed photocatalytic activity against various dyes along with antifungal property in bioweathering of stone. ZnO nanoparticles were synthesized by Bhargav et al. (2019) by employing sol-gel, thermal decomposition and

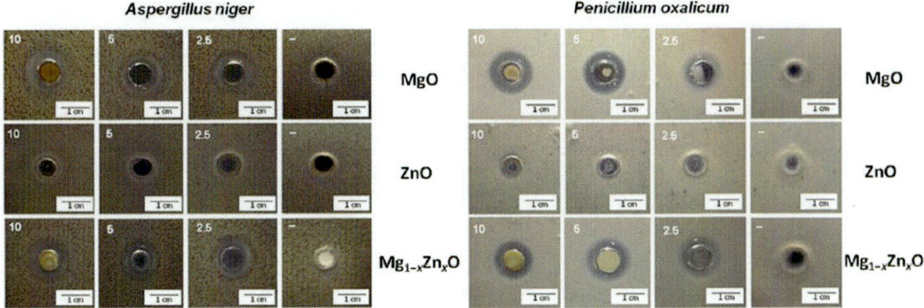

Fig. 4.5 Pictorial representation of inhibition zone carried out by MgO, ZnO and Mg1-xZnxO nanoparticles at different concentration levels with two fungal strains (*A. niger* and *P. oxalicum*). (Adapted figure from Fernandez et al. (2017) with permission from copyright, American Chemical Society (Washington, DC, USA))

direct precipitation synthetic route. The detailed characterization of the formed nanoparticles was carried out before examining their biological importance. The efficacy of developed nanoparticles was investigated on *Rhizopus stolonifer, Candida albicans* and plant *Fusarium sp.* fungal strains and therefore reflects resistance towards fungal strain.

4.2.3 Seed Germination Analysis

Nanoparticles play a very crucial role in the seed germination activity which further depends upon the size, shape, morphology and surface characteristics (surface area/mass ratio) of nanoparticles (Chaudhary et al. 2016). The environmental exposure, ageing, species and growth rate of plant play a fundamental role in the development of plant tissue. The size of nanoparticles (ranges from 10 to 50 nm) is considered as the most important factor in the analysis of proliferation because they can easily penetrate and allow accumulation into plant cell wall which further enhances the saturation stability. Aggregation, dissolution, adsorption, redox and desorption are found quite interesting processes which can transform the fate of progression and toxicity of nanoparticles on terrestrial species (Sharma and Uttam 2017). In literature, several types of seeds have been utilized to investigate the anti-proliferative effect of semiconducting nanoparticles (Fig. 4.6). Various methods have been described to carry out seed germination activity such as germination rate in petri dish, foliar spray test, agar assay, hydroponic method, germination in soil media (artificial and natural) and exposure of seeds under dark and light condition. During the measurement, various parameters are optimized for the better effect on seed germination, germination percentage, seedling growth, biomass, shape of leaf and elongation of root along weight. The structure and morphology of root leaf and shoot have displayed a significant role in understanding the mechanistic behaviour for the development and growth of plant in both positive and negative manners. The

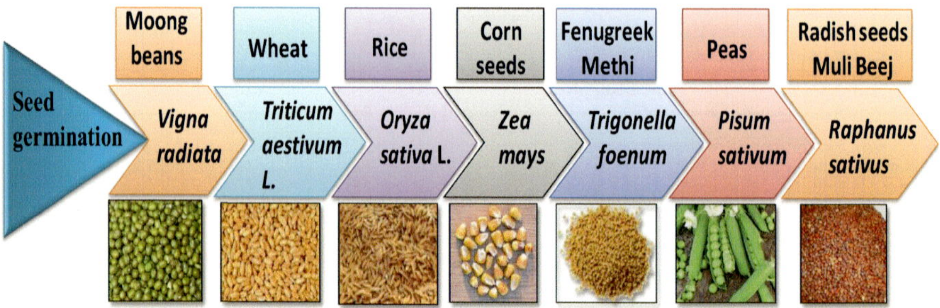

Fig. 4.6 List of different seeds used during seed germination analysis

biocompatible behaviour of semiconducting nanoparticles on the germination rate and growth of seeds highly depends upon the behaviour and fate of the utilized nanoparticles. The emergence of oxidative stress and reactive oxygen species should be measured to investigate the cytotoxic and mutagenic behaviour of semiconducting nanoparticles over plant species (Garcia Gomez and Fernandez 2019).

During the respective activity, a comparison of various semiconducting nanoparticles has provided a better platform to measure the surface physicochemistry, and it will emerge out as a new challenge in front of scientists (Chaudhary et al. 2018a, b, c). Many works have been reported in literature such as Savassa et al. (2018) that have synthesized different-sized ZnO nanoparticles by using chemical reduction method with varying concentration of the $ZnSO_4$ and employed them in agricultural field for the seed nutrient enhancement by using seed priming (Fig. 4.7). It was observed that on increasing the concentration of the treatment, the germinating ability of common bean seeds (*Phaseolus vulgaris*) starts inhibiting, whereas lower concentration of treatment provides better affinity of seeds towards germination and root elongation ability due to the slow release of zinc ions. Chaudhary et al. (2018a, b, c) selected *Vigna radiata* seeds as a potential scaffold to investigate the efficacy of NiO nanoparticles developed by simple calcination chemical synthesis process. The germination capacity of seeds incremented rapidly as the concentration of NiO nanoparticles decreases uniformly, which suggests that higher concentration of nanoparticles causes more anti-proliferation. After determining the biocompatibility of NiO nanoparticles in agricultural field, they were used in the photocatalytic degradation of methylene blue dye as a specific sensor.

4.2.4 Antialgal Activity

In literature, various methodologies have been developed to investigate the toxic aspect of semiconducting nanoparticles in terms of cellular and oxidative damage towards various environmental aqueous species through the mechanisms of dissolution, adsorption, interaction and photo-induction (He et al. 2019). Out of all those

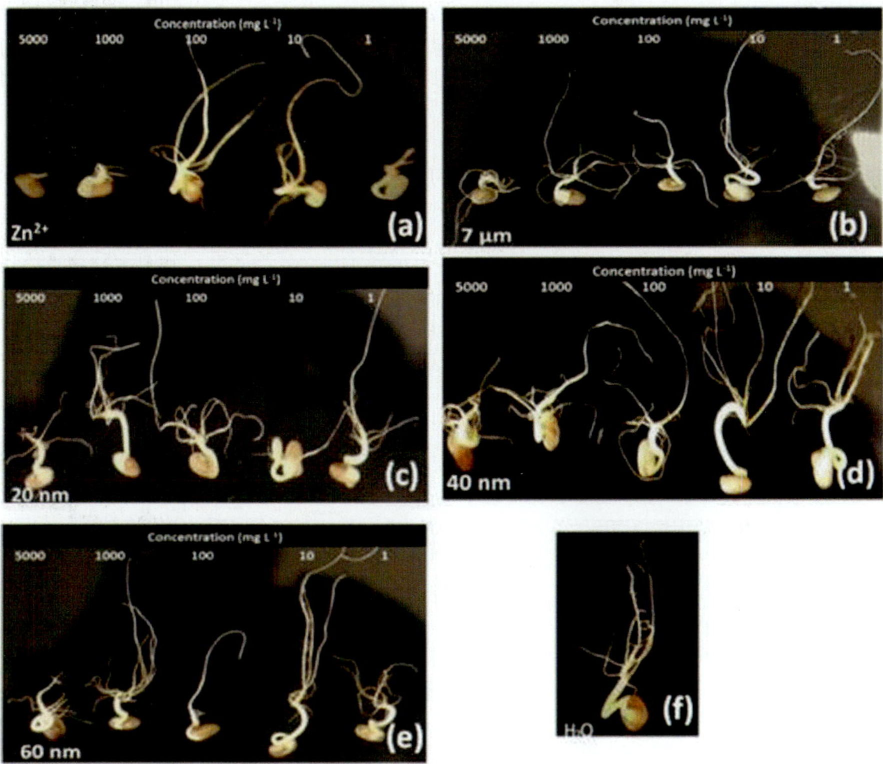

Fig. 4.7 Seed germination activity of common beans (*Phaseolus vulgaris*) in the presence of (**a**) $ZnSO_4$, (**b**) bulk ZnO, (**c**) 40 nm ZnO NP, (**d**) 60 nm ZnO NP, (**e**) 20 nm ZnO NP and (**f**) control in water. (Adapted figure from Savassa et al. (2018) with permission from copyright, American Chemical Society (Washington, DC, USA))

activities, algal activity has been considered of great importance. Algal blooms have been considered as photosynthetic organisms which have very high chlorophyll content. Algae have been present all around the environmental atmosphere and have the ability to influence the surroundings. They are most abundant, rapidly growing and untapped (unexploited) resource which have very high carbohydrate and lignin content (Kastovsky et al. 2019). Algae are divided into two categories such as macroalgae (multicellular in nature and algae have roots) and microalgae (unicellular algae without roots). Microalgae cultures are divided into four major groups, namely, green algae, red algae, cyanobacteria and chrysophyte (Rasul et al. 2017). In today's era, climate stability has become an important concern. So, it is very much important to understand how algal species interact with the environment and how they react towards climate changes. There is a diverse amount of toxicity caused by semiconducting nanoparticles (ZnO, WO_3, CoO, SnO_2, TiO_2, CuO and MgO) while interacting with aquatic species due to their water-soluble nature. So, the basic purpose of performing antialgal activity is to investigate the efficacy of semiconducting nanoparticles over aquatic species in direct and indirect manner (Luo et al. 2018). The toxicity imparted due to the uptake of nanoparticles has been depended upon the physico-chemical properties of formed material. The respective activity determines the amount of chlorophyll content present in the algal species by utilizing UV-visible and fluorescence spectroscopy (Sharan and Nara 2019). NiO and TiO_2 nanoparticles exhibit very high tendency of aggregation and ion release in seawater. The respective nanoparticles have strong capability to produce reactive oxygen species under both ultraviolet and visible light which further generates the process of oxidative stress in algal species. The adsorption photocatalytic inactivation of detrimental algal species can be visualized easily by using the developed nanoparticles (Gong et al. 2019; Wang et al. 2019).

4.2.5 Allium cepa Chromosomal Aberration Assay

Among various living beings, plants have been considered as the most fundamentally important part of life cycle because they help in maintaining the level of oxygen in the environment through photosynthesis and food chain cycle. For example, onion and garlic were dipped in nanoparticle solution and obtained rootlets used to measure the mitotic division under optical light microscope. During this activity, mitotic division during prophase, metaphase, telophase, anaphase and interphase was observed. But disorientation in mitotic division leads to the emergence of various chromosomal aberrations. Chromosomal aberration deals with the loss and breakage of chromosome which induces abnormalities at chromosomal level. The respective activity has various advantages such as trouble-free handling, highly selective and sensitive, cheap and good association with animal assay. Semiconducting nanoparticles are superior in contrast to the bulk material as they possess high surface area to volume ration, stability, small size, desired morphology and excellent physico-chemical properties (Giorgetti 2019). Nanoparticles combine with the different components of cell such as organelles, membrane, DNA and nucleus, which results in the genomic and chromosomal alteration, micronuclei formation and production of oxidative stress. During this process, highly reactive semiconducting nanoparticles dissociate into ionic forms (positively and negatively charged) and affect the DNA orientation (Roy et al. 2019). Simultaneously, NPs interact with the plasma membrane through reactive oxygen species, which results in the alteration of antioxidant enzymatic activity which causes severe damage in cell signalling. Various physiological parameters such as apical and radical growth, cellular uptake and production of biomass reflect about the knowledge of phytotoxicity. The physiological, cytotoxic and morphological effects appraise the possessions of the mitotic divisions, DNA damage and regular cell cycle caused by genotoxic activity. The described activity was found to be widely used to investigate the toxicity of any particular system. The most extensively used semiconducting nanoparticles used for the respective study are TiO_2, CeO_2, ZnO, SiO_2, CuO and Se, and the corresponding data of respective biological activity has been shown in Table 4.3.

4.2.6 Antioxidant Assay

The total antioxidant capacity of semiconducting nanoparticles such as Cr_2O_3, TiO_2, MgO, CeO_2, ZnO, SiO_2, Bi_2O_3, CuO and Se can be evaluated by simpler physiological activities as it provides better platform to understand the oxidizing and reducing behaviour of nanoparticles which can be further utilized in environmental processing. It involves paper-based colorimetric portable paper nanosensor and filter paper-based quantification. Environmental exposure and various operational parameters in the environment play a fundamental role in the determination of antioxidant content. The basic fundamental principle of antioxidant activity involves basic probe and antioxidant material. The mechanistic behaviour of antioxidant phenomena has been shown in Fig. 4.8. The antioxidant activity works in two manners: one is competitive (direct) assay and the other is non-competitive (indirect) assay (Apak 2019). In competitive assay, both probe and antioxidant get oxidized simultaneously and produce reactive oxygen species, whereas in non-competitive assay, the probe of the respective system gets oxidized by the antioxidant by utilizing simple chemical reactions in simulated manner. The antioxidant behaviour of semiconducting nanoparticles leads to the reduction of metal ion with variable kinetics (Gupta 2018; Ozkan et al. 2018). In literature, it has been found that some of the semiconducting nanoparticles such as FeO, ZnO and Se act as strong reducing antioxidant agent. They have very low redox potential values which help in the production of smaller-sized nanoparticles. In other words, weak reducing agents have a tendency to produce large-sized particles for long interval of time (Pisoschi and Negalescu 2011). Therefore, the antioxidant capacity of any type of nanoparticles can be related to the size of nanoparticles in analytical determination and further leads to reduction behaviour. The antioxidant activity involves hydrogen peroxide (H_2O_2) radical scavenging activity, superoxide anion radical scavenging activity and total reduction capacity, which are considered the most significant characteristics of antioxidant process. Various other methods have been reported in literature on the basis of their antioxidant behaviour such as crocin bleaching assay, oxygen radical absorbance capacity and so on. Different nanoprobes such as ZnO, SnO_2 and Ag_2O have been designed with high selectivity and sensitivity for the determination of hydrogen peroxide and total reduction behaviour of a wide variety of species under antioxidant activity (Rajeshkumar et al. 2018; Vidhu and Philip 2015; Vinay et al. 2019). The selectivity of hydroxyl radical offers an excellent role in the redox potential activity. Usually, UV-vis spectroscopy, fluorescence spectroscopy and colorimetric-ratiometric nanoprobe are utilized for the evaluation of antioxidant content of semiconducting nanoparticles. Out of these three, fluorescence analysis is considered as the most suitable and selective technique for the analysis purpose.v

Table 4.3 Representation of different mitotic stages and mitotic index (MI) exerted by semiconducting nanoparticles

S. no.	Material used	Prophase	Metaphase	Anaphase	Telophase	MI	References
1	ZnO NPs (25 μg/ml)	48.4	1.1	0.6	0.1	50.4	Kumari et al. (2011)
	(50 μg/ml)	39.2	0.2	0.3	0.1	40.1	
	(75 μg/ml)	33.8	0.2	0.2	0.1	35.3	
	(100 μg/ml)	29.9	0.1	0.1	0.1	29.2	
2.	Bi_2O_3 NPs	85.08	4.42	2.92	7.56	28.10	Liman (2013)
	(25 ppm)	83.66	5.37	3.1	7.85	28.44	
	(50 ppm)	86.63	5.49	2.57	5.31	30.39	
	(75 ppm)	80.73	4.69	3.42	5.17	31.01	
	(100 ppm)						
3.	MgO NPs	20.9	11.3	8.6	1.8	42.6	Mangalampalli et al. (2018)
	(25 μg/ml)	19.7	10.5	7.4	1.4	39.0	
	(50 μg/ml)	20.5	9.8	5.6	1.7	37.6	
	(100 μg/ml)						
4.	CuO NPs	50.4	1.8	0.7	0.6	54	Chaudhary et al. (2019)
5.	Cr_2O_3 NPs	28.5	26.5	0.66	2.6	29.6	Kumar et al. (2015)
	(10 μg/ml) (100 μg/ml)	20.3	17.5	0.68	1.6	20.92	

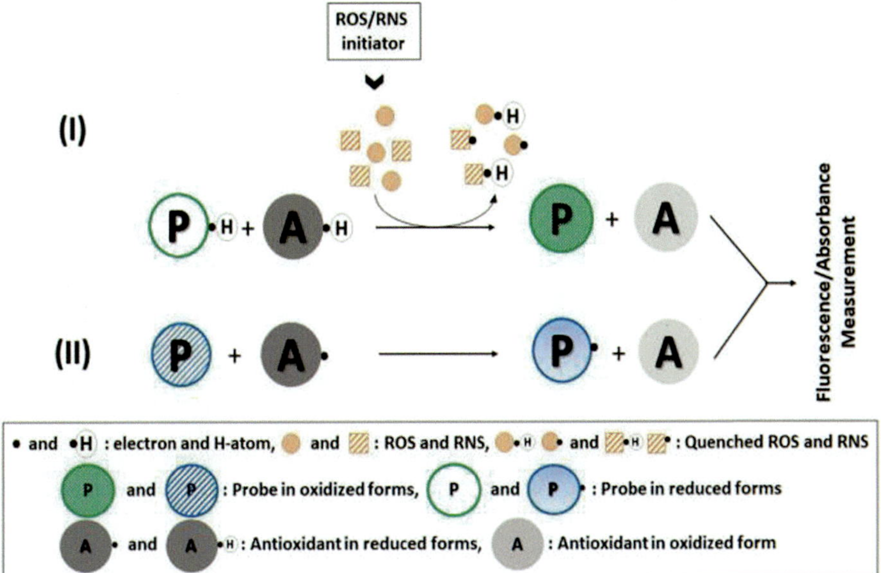

Fig. 4.8 (I) Competitive and (II) non-competitive antioxidant phenomena which involve a chromophoric or fluorophoric group with and without biologically relevant ROS/RNS. (Adapted figure from Apak (2019) with permission from copyright, American Chemical Society (Washington, DC, USA))

4.2.7 Aggregation and Dissolution Behaviour

Aggregation process deals with the combination of a series of particles which results in the formation of a large-sized particle. Aggregation is divided into two categories, namely, homo-aggregation and hetero-aggregation which depends upon pH, stearic hindrance, electrostatic interaction and ionic strength of nanoparticles along with their attachment efficiency. The aggregation tendency of nanoparticles has become beneficial according to environmental fate and prevents the toxicity of nanoparticles at their early stages. The velocity of aggregation plays a fundamental key role in nanoparticle formulation. Due to the aggregation process, nanoparticles have a very high tendency to adsorb on the surface of both organic and inorganic solid materials. It takes numerous years to reach nanoparticles at the bottom and top-down layer of earth (Markus et al. 2015; Li et al. 2018). The process of dissolution leads to splitting of molecule in their respective ionic forms and exerts toxicity on the different species present in the environment. During dissolution, the disruption of nanoparticles takes place into various fragments through either chemical or physical transformation (Zook et al. 2011). Here, oxygen content plays a fundamental role in aquatic systems as most of the water content is anoxic in nature. The oxygen works as a catalyst, and an appreciable amount of dissolved oxygen will not affect the dissolution activity. The instrumental techniques such as DLS, TEM, SEM, UV,

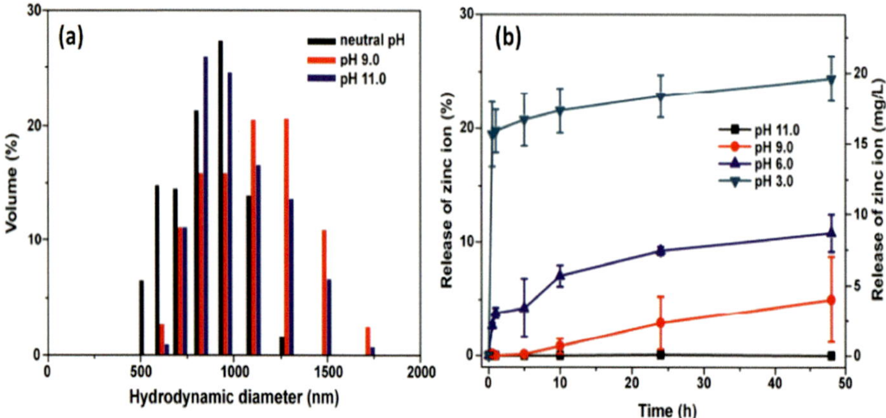

Fig. 4.9 (**a**) Representation of aggregation and (**b**) dissolution process of ZnO nanoparticles at four different pH values. (Adapted figure from Bian et al. (2011) with permission from copyright, American Chemical Society (Washington, DC, USA))

fluorescence and BET can be employed to measure aggregation and dissolution. DLS and UV-vis spectroscopy have been the most frequently used techniques due to the ease of operation parameters, high selectivity, very less time consumption and more reliable results obtained. While doing measurement through DLS instrument, diffusion-limited colloidal aggregation theory (DLCA) plays a fundamental role on the size distribution profile (Conway et al. 2015). The more the ionic strength, the more will be the adsorption of ions on the exterior surface of nanoparticles due to augmented aggregation tendency. Bian et al. (2011) have synthesized zinc oxide nanoparticles by applying chemical synthetic route with average size 4 nm. The influential behaviour of aggregation, dissolution, pH, ionic strength and adsorption was carried out on the developed nanoparticles by using standard Derjaguin-Landau-Verwey-Overbeek (DLVO) theory. From the respective theory, it was concluded that the small-sized ZnO NPs have very low tendency to aggregate and very high tendency of dissolution, whereas at higher pH, the size and aggregation tendency of nanoparticles were found to be maximum as shown in Fig. 4.9.

4.2.8 Sorption Behaviour

The contamination of soil is increasing with the greater dormant due to its harmful effect on various living beings present in the ecosystem and has become a global concern worldwide. There are various anthropogenic sources available which are polluting the soil such as mining, industrial activities, domestic sludge, land agrochemical waste and road transport that affects the landfills and water resources. Therefore it is required to understand their adverse effect on living beings and the exact impact of toxins on the living systems. The disposed off nanoparticles in water resources have also produced toxic impact on living beings (Paula and Paula 2019).

The small size, of nanoparticles can make them easily accumulate in the soil and disrupt the quality of soil along with life span of different microorganisms present in it. Therefore, before the decomposition of nanoparticles in the environment or utilizing them in agricultural field, it is very much predominant to scrutinize their sorption performance for environmental fate (Kumar et al. 2018). During this activity, different batches of soils were prepared in the presence of various types of nanoparticles, and their sorption behaviour was studied by using UV-visible spectroscopy. From the absorption spectrum, the value of optical density recorded and the amount of sorption (adsorption) of soil towards nanoparticles have been calculated out. Various reports have been published in literature over sorption activity of nanoparticles. Out of the various reports, Kah et al. (2018) observed that sorption activity of nanoparticles enhances with time of incubation (kinetic of reaction), which suggests the slow rate of diffusion of nanomaterials (sorbate) into the organic matrices of soil aggregates. If changes in the fate processes can be controlled, then nanoformulations can be a mitigating active ingredient.

4.2.9 Photodegradation Activity

In today's era, photodegradation activity is one of the most desirable activities due to the mild reaction condition and low cost of process evaluation during the degradation of harmful organic and inorganic pollutant for the wastewater remediation. Photodegradation is defined as the photo-induced reaction where the rate of reaction can be enhanced in the presence of some external agents. When compared to the additional conventional methods, photodegradation method has various advantages over other techniques. Here, the rate of reaction can be easily affected by various operational parameters such as oxygen pressure, optophysio-chemical properties, pH, light intensity and concentration of pollutant substrate. It has significant importance under UV light irradiations due to the high electron-hole recombination rate and advances their performance in wastewater treatment (Jose et al. 2018). In recent trend, it has become an admirable technique due to the degradation of harmful pollutant into non-toxic subsequent products without using any supplementary energy source (Markus et al. 2015). Degradation processes involve the entrapment of charges generated by light sources on the exterior surface, which further leads to the induction of electron transfer reaction. Semiconducting nanoparticle-based photodegradation has become the most fundamental method in industrial sector especially chemical industries and has become a widely adopted method due to NPs' excellent band gap values and redox potential. Therefore, it is very much necessary to investigate the mutagenic and carcinogenic potential of semiconducting nanomaterials on the ecosystem and human health. Kah et al. (2018) have investigated environmental fate of nanocomposites in terms of sorption, photodegradation and durability. It was inferred that the photodegradation half-life efficiency enhances up to 21% by utilizing nanoformulations in water as compared to the conventional methods. Kumar et al. (2012) have synthesized silver-titanium dioxide

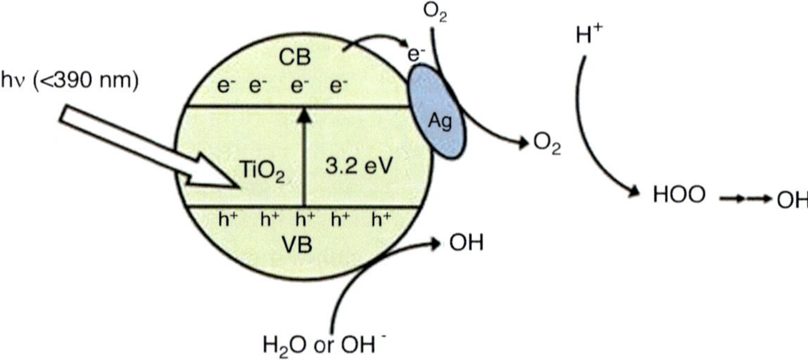

Fig. 4.10 Mechanistic schematic representation of photodegradation capacity of silver-TiO_2 nanocomposite. (Adapted figure from Kumar et al. (2012) with permission from copyright, American Chemical Society (Washington, DC, USA))

nanocomposites by sol-gel technique followed by spin coating and investigated their photodegradation efficiency. The photocatalytic efficiency of developed nanocomposite is directly related to the potential difference (Fig. 4.10). The basic aim of performing the respective study is to conclude the environmental performance, fate, new risks and benefits of developed nanomaterials. In this respect, durability and fate of nanomaterials are getting tremendous interest.

4.3 Application of Nanoparticles in Contaminant Sensing

The pollutants released in the environment are in different forms such as heavy metal ions, inorganic anions, toxic gases (nitrogen oxides, sulphur oxides, ozone, carbon oxides, etc.), suspended particulate matter (PM), as well as volatile organic compounds (VOCs). Typically, a pollutant is any toxin which is not native to the ecosystem. However, if the particular discharged chemical is indigenous to the ecosystem, then it is characterized as a pollutant if it is released into the environmental streaming in amounts that are in overabundance to the natural concentration of the organic chemical in the ecosystem, and therefore, this augmented concentration can cause damage to the flora or fauna of the ecosystem. They are able to undergo long-range atmospheric transport from the source to remote areas (Jaward et al. 2005), thereby threatening the global ecological environment and individual health. The detection and control of environmental pollutants is an essential constituent of governmental policies on the environment and of the efforts by regulatory authorities across the globe aiming to preserve and develop the eminence of our surroundings for the coming generations. It is also an issue of effervescent research, as novel procedures that are more sensitive, faster, low cost and able to sense a progressively wider range of pollutants are necessary to deal with the current challenges associated with environmental pollution. In general, different types of pollutants explained in brief are as follows: heavy metal ions, organic pollutant, organic dyes and inorganic anions.

4.3.1 *Heavy Metal Ions*

Heavy metals are a class of inorganic pollutants which can be categorized as any metal or metalloid species that exhibits toxicity, irrespective of their atomic mass or density in a form or concentration, that causes a negative human or environmental effect. The heavy metals have been referred to as persistent and bioaccumulative environmental toxins, since unlike other contaminants like pharmaceuticals drugs and dyes, these cannot be metabolized to carbon dioxide and water or easily rendered undisruptive by chemical or biological remediation processes (Khan et al. 2011). The global industrial revolution has enabled the harmful materials to spread into the environment with unprecedented speed. The carcinogenic and hazardous nature of various metal contaminants including arsenic, mercury, chromium, mercury, cadmium, copper, lead, etc. warrants a great deal of attention for detecting these heavy metals in the ecosystem. These toxic elements are present in the environment by taking diverse forms, coming from different sources and causing severe health issues. These metals once released into the atmosphere may have long-term lethal effects on human health, even with just a trace presence. Heavy metals are largely classified into essential and nonessential ions. Nonessential heavy metal ions such as lead, cadmium, mercury, antimony and arsenic are carcinogenic in nature and produce severe toxic effects to the living entities. Essential heavy metals include copper, zinc, calcium, cobalt, iron, magnesium and selenium, which are required by the body in a very less amount. Nonetheless, exceeding the quantity of essential and nonessential heavy metals than the optimal suggested level promotes various neurological and morphological disabilities in plants, animals and human beings, due to their persistent and non-biodegradable characteristics (Li and Zhang 2010; Luo et al. 2011). These metals get accumulated in the food chain in the course of biogeochemical cycles, thereby affecting the healthy individuals by means of contaminated water and food (Nabulo et al. 2011; Abuduwaili et al. 2015). Environmental attentiveness is growing within the industries due to the severe health effects of heavy metals on consumers. According to the US Environmental Protection Agency (EPA) and the International Agency for Research on Cancer (IARC), heavy metals have been classified as human carcinogens (Jaishankar et al. 2014).

4.3.1.1 Sources of Heavy Metals

Even though heavy metals are found all through the earth's crust, yet most of the environmental contamination marks from the weathering and volcanic eruptions (leaching from bedrocks and atmospheric deposition), industrial sources (metal processing in refineries, nuclear power stations, coal burning in power plants, plastics, textiles, petroleum combustion) microelectronics, paper processing plants and wood preservation, agricultural runoffs, etc. (Verma and Dwivedi 2013). Heavy metal toxicity has proven to be a major threat as there are a number of health hazards related to it. It damages the mental and central nervous system, lowers energy levels, affects blood composition as well as damages the lungs, kidneys, liver and other vital organs (Inoue 2013). Also, long-term exposure to such metals may also result in physical, muscular and neurological degenerative processes that resemble Alzheimer's disease, muscular dystrophy, Parkinson's disease, diabetes mellitus and hypertension. The most commonly reported heavy and toxic metal ions are as follows: arsenic, mercury, chromium and lead.

Arsenic

Arsenic has been considered as the most toxic and carcinogenic metal for both environment and human beings. It is present in the earth's crust and in groundwater in numerous forms such as arsenates, arsenide, etc. Direct exposure to this toxic type is unfavourable to the whole ecosystem (Yang et al. 2017). Arsenic is hazardous as it does not have a taste or odour; therefore humans can be exposed to it without knowing it. The main fundamental sources for the intake of arsenic are food and drinking water. The long-term exposure of respective metal ion in drinking water may lead to increased risks of not only skin cancer but also some other cancers and skin lesions like hyperkeratosis and pigmentation changes (Haley 2016). Arsenic pollution in groundwater has been considered as a major global concern. The primary cause of arsenic toxicity has been found to be well water contaminated by natural sources (Bhowmick et al. 2018).

Mercury

Mercury (Hg) is a toxic metal contaminant found extensively in wastewater. The prime source of contamination is modern industrial processes. This metal affects neurologic, gastrointestinal and renal systems depending upon the route of exposure. Hg (II) is extremely dangerous at low concentration level. The excess exposure has caused serious health disorders such as kidney failure, brain damage, motion disorder and a range of other cognitive diseases. The World Health Organization (WHO) has standardized the permissible toxic value for Hg (II) as 1 ppb (Liu et al. 2009).

Chromium

Chromium (Cr) is another lethal and carcinogenic element found in abundance in groundwater with Cr (III) and Cr (VI) as general oxidation states. Cr (VI) has been reported to be more toxic than Cr (III) (Shekhawat et al. 2015). Chromium contamination leads to severe health problems which includes lung and sinonasal cancer. The estimated guideline value for chromium is 50 ppb by WHO (Liu et al. 2007). The excess exposure has caused serious health disorders such as kidney failure, brain damage, motion disorder and a range of other cognitive diseases.

Lead

Like various toxic metal contaminants, lead (Pb) is well known to be carcinogenic and to create undesirable effects on the body. It is harmful to the kidneys, reproductive system and nervous system. It is extremely dangerous to foetuses and young children at the brain growth stage. Exceedingly high levels of lead contamination can result in seizures and coma and may cause death in extreme conditions. The toxic character of these metal contaminants needs precise monitoring of trace levels in water prior to use (Lopes et al. 2015).

4.3.1.2 Specialized Techniques to Remove Heavy Metals

The toxicity, persistence and bioaccumulative character of the heavy metals have raised severe public concerns across the globe. Therefore, the past decades have witnessed a growing interest towards the expansion of specialized techniques to either reduce the toxicity of these metals before their discharge or remove the metals from the waste discharges. Many analytical and industrial techniques have been developed to find the concentration level of heavy metal ions in the samples. These conventional techniques include atomic absorption spectroscopy, nuclear magnetic resonance, Raman spectroscopy, X-ray fluorescence spectroscopy and amperometric technique and neutron activation analysis. These methods can distinguish various metal ions and have fine sensing parameters. However, these methods are costly and involve tedious sample preparation (Pujol et al. 2014). Other techniques which are less expensive and easy to utilize are hence being developed to replace the conventional procedures. There has been widespread research interest in nanotechnology that provides new opportunities for the development of different portable sensors for sensing of multiple metal contaminants on site. Optical sensors, biological sensors and electrochemical sensors are the places where various nanomaterials have been used for sensing heavy metal ions. Compared to the traditional techniques, these are considered the most reliable tools for the recognition of heavy metal ions (Kailasa et al. 2019). There are various advantages of such techniques in terms of uncomplicated handling, specificity, sensitivity, portability, expeditious response time, compactness, user friendliness and reliability. In addition, these are able to determine the biotoxicity repercussions of heavy metal ions towards living beings. Mesoporous Fe_3O_4 nanoparticles are found to be highly capable for the detection and removal of heavy metal ions (Pb^{2+}, Cu^{2+}, Ni^{2+} and Cd^{2+}) from various water resources such as river water, rainwater and deionized water by utilizing adsorption and desorption phenomena as shown in Fig. 4.11. The 90–98% removal efficiency was obtained while taking consideration of mesoporous nanoparticles (Fato et al. 2019). Li et al. (2005) have synthesized water-soluble luminescent thiol-capped CdTe nanorods with an average size of 90 nm, which results in the discriminative photoluminescence enhancement of zinc ion by 68% and quenching of cobalt ion by 75%.

Silica oxide (SiO_2) nanoparticles were synthesized using simple chemical reduction method and decorated by using sensory array probes (Peng et al. 2018). The developed nanoparticles were found to be highly reactive and selective towards various metal ions (cadmium, mercury, zinc and copper) in nanomolar range by utilizing fluorescence response and sensor array technique as shown in Fig. 4.12. Wang et al. (2010) have developed highly sensitive and selective photoelectrochemical sensor for Cu^{2+} by using cadmium-sulphur nanodots with 1.0×10^{-8} M detection limit. Various other reports have been published for the detection of heavy metal ions and shown in Table 4.4.

Hung et al. have synthesized 2-mercatoethanol (2-ME)- and 4-mercaptobutanol (4-MB)-capped gold nanoparticles (Au-NPs) by applying simple and rapid chemical synthetic method. The developed nanoparticles were applied in the field of heavy metal ion sensing by using UV-visible and fluorescence spectroscopy. 2-ME-capped Au-NPs were found highly selective and sensitive towards silver and lead ions, whereas 4-MB-capped Au-NPs detected mercury in selective manner. The value of detection limit was reported 500, 70 and 45 nm for Hg^{2+}, Ag^{+} and Pb^{2+} ion by utilizing Stern-Volmer plot of fluorescence spectroscopy. The extraordinary property of respective Au-NPs is such that they can detect three metal ions simultaneously with great precision (Fig. 4.13). To carry out practical application of developed nanoparticles, real water and soil sample analysis was also carried out and 90–95% recovery was obtained.

Placido et al. (2019) have synthesized biochar-derived carbonaceous nanomaterials by using chemical depolymerization, solvent extraction and nanorefinery process. Biochar nanomaterials such as microalgae, sorghum straw and rice straw were utilized. The developed semiconducting nanomaterials have been characterized by using different analytical approaches to confirm the formation of nanoparticles. Before investigating the efficiency of developed nanomaterials in the meadow of heavy metal ions sensing, their biocompatibility, cellular localization and cellular uptake were evaluated by using three different yeast species. The developed nanomaterials were also utilized in the field of bioimaging. Out of various numerous metal ions, developed microalgae and sorghum straw biochar-derived nanomaterials have been found highly selective towards zinc and copper metal ions, whereas the value of detection limit was found to be 9 and 0.0125 μM. It has been reported that heavy metal uptake by living beings is dependent on the amount of free metal ion concentration as compared to total metal ion concentration which in turn determines the toxic strength of heavy metal ion in a free form (Verma and Singh 2006).

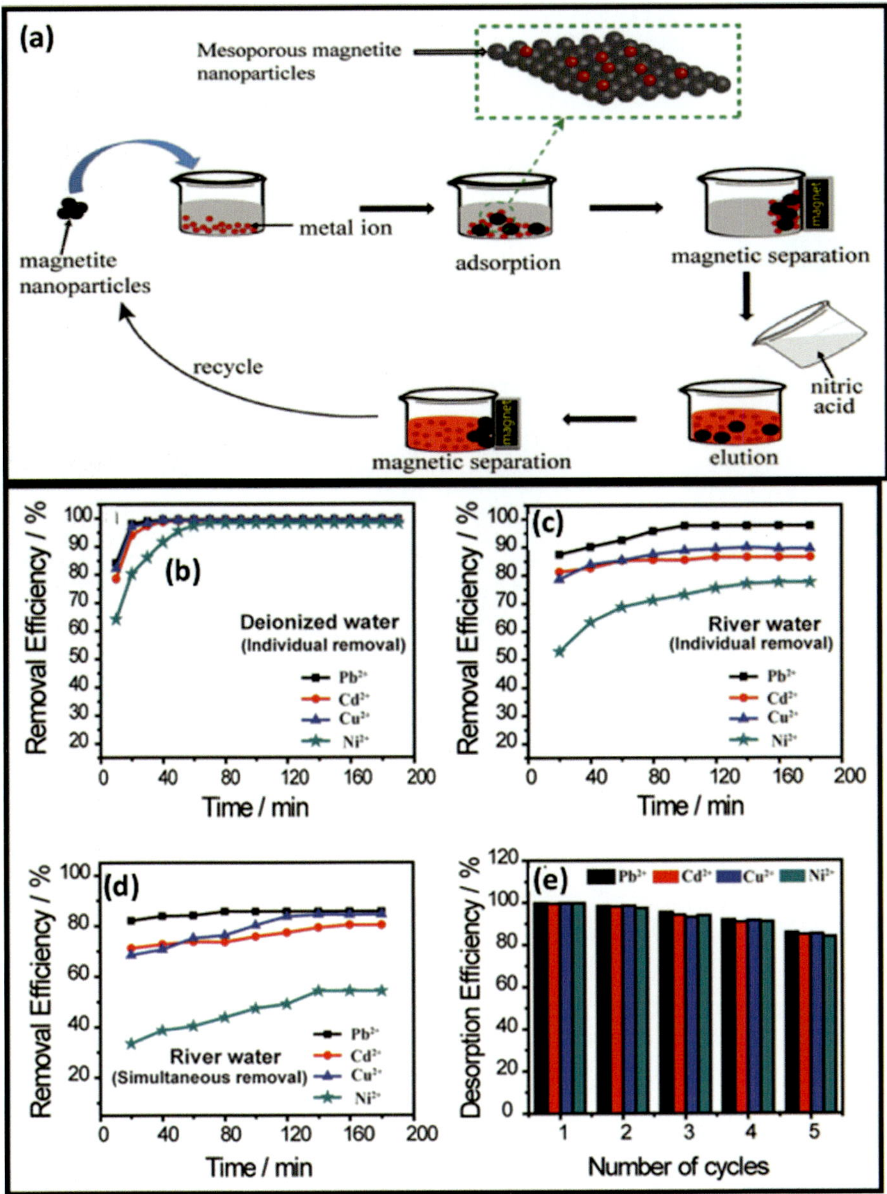

Fig. 4.11 Mechanism of desorption process for the removal of heavy metal ions from aqueous solution (**a**), removal efficiency of mesoporous nanoparticles in deionized water (**b**), removal efficiency of metal ions river water (**c**), simultaneous removal of metal ions in river water (**d**), desorption efficiency through number of cycle method using nitric acid in diluted form (**e**). (Adapted figure from Fato et al. (2019) with permission from copyright, American Chemical Society (Washington, DC, USA))

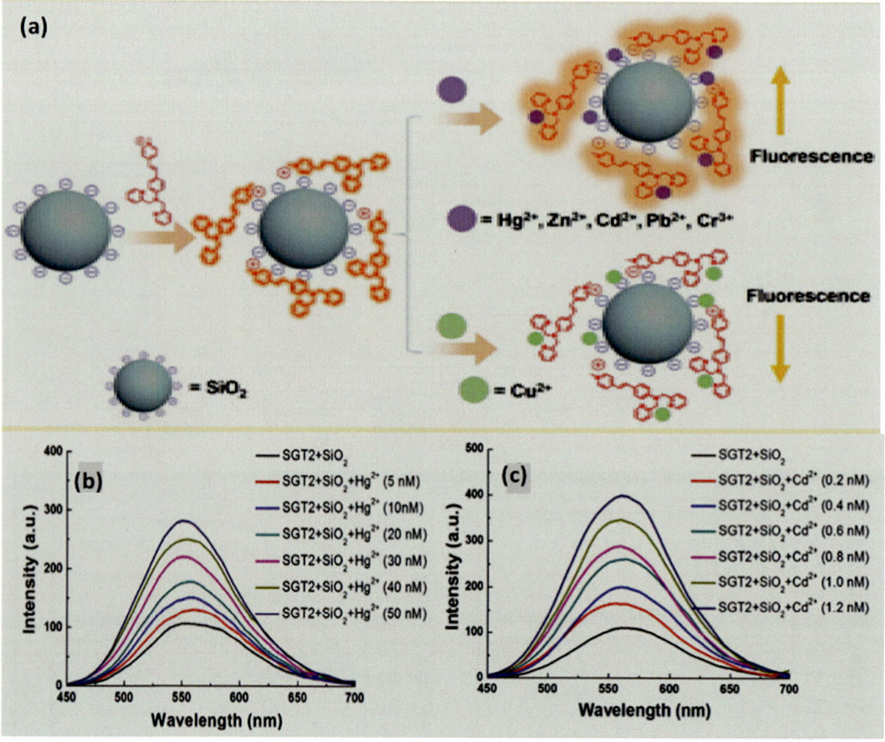

Fig. 4.12 Development of silica oxide nanoparticle and sensory array probe as a decorating agent and their application in heavy metal ion sensing (**a**) and fluorescence sensing of mercury ion (**b**) and cadmium (**c**). (Adapted figure from Peng et al. (2018) with permission from copyright, American Chemical Society (Washington, DC, USA))

4.3.2 Organic Pollutants

By way of clarification, an organic pollutant is defined as an organic chemical which enters into the environmental streams and causes pollution (however transitory or enduring) and is unsafe to the flora and fauna of the ecosystem. There are some organic chemicals which are called persistent organic pollutants (POPs) and classified as compounds that are impenetrable to environmental degradation during the different chemical and biological processes (Jacob 2013). POPs are highly non-degradable in atmospheric territory because of their extreme stability and low decomposition rates and, hence, have a long life in different ecosystems. POPs are semi-volatile, have low solubility in water, have an inherent toxicity and are lipophilic in nature. Such chemical and physical properties make them capable of long-range transport and enable them to bioaccumulate in living organisms and biomagnify in food chains. Organic chemical pollutants can enter the environment in the course of the solid phase, the liquid phase or the gas phase, can resist

Table 4.4 Representation of sensing of different heavy metal ions by using semiconducting nanoparticles

S. no.	Nanomaterials used	Heavy metal ions	Detection method	Limit of detection (LOD)	References
1	CuO	Mercury (Hg^{2+})	Fluorescence	2 μg/L	Chaudhary et al. (2019)
2	MgO	Iron (Fe^{3+})	UV-vis spectroscopy	23 μM	Jain et al. (2018)
3	ZnS	Chromium (Cr^{3+}) and cadmium (Cd^{2+})	Fluorescence	31 and 64 μM	Ayodhya and Veerabhadram (2019)
4	Ag_2S	Cadmium (Cd^{2+})	Fluorescence	546 nm	Wu et al. (2017)
5	TiO_2	Mercury (Hg^{2+})	Electrochemical	0.017 μM	Zhou et al. (2017)
6	Fe_2O_3	Lead (Pb^{2+}) and cadmium (Cd^{2+})	Electrochemical	0.17 and 0.21 ng/ml	Maleki et al. (2019)
7	CdS	Chromium (Cr^{3+})	Fluorescence	16 nM	Ahmed et al. (2015)
8	MoS_2	Cadmium (Cd^{2+})	Optical	72 nM	Yin et al. (2017)

degradation and are mobile over large distances. A huge variety of such organic pollutants originating from the massive population growth and superior agricultural and industrial activities are released into the environment which threatens the human health. Organic pollutants mainly arise from the sources like oxidation of organic particulate matter, emissions from motor vehicles, smoke from organic oils, burning of coal, etc. Therefore, these pollutants can accumulate in living organisms through the biological food chain and have destructive effects on living organisms which include acute toxicity and carcinogenicity (Abdel-Shafy and Mansour 2016). These pollutants have caused harming effect on the environment. Hence, organic pollutants are a major trouble-causing agent worldwide, especially heteroaromatic and aromatic compounds such as phenol, p-nitrophenol, pyridine, dimethyl phthalate (DMP) and trichloroethylene. These are widely used in the petrochemical, chemical and pharmaceutical industries and are therefore usually found in industrial wastewater (Essam et al. 2010). These compounds are extremely harmful and have been listed by the US Environmental Protection Agency as priority pollutants. Therefore, there is a need to remove them efficiently in order to prevent their entry into the environment. Considering their potential detrimental effects, the international community is beginning to establish particular treaties and laws to eradicate or restrict the production and usage of such pollutants. The Stockholm Convention (SC) on Persistent Organic Pollutants is one of the most famous multilateral treaties. Currently, a total of 23 chemicals or groups of halogenated compounds, including pesticides, chemicals from industries and unintentional products, are included in the list of persistent organic pollutants regulated under the Stockholm Convention. Nonetheless, despite the recognized measures and the manufacture of most POPs being ceased for over 20 years, significant levels of such pollutants remain in the environment (Mechlinska et al. 2010). Hence, there is clearly a need to develop

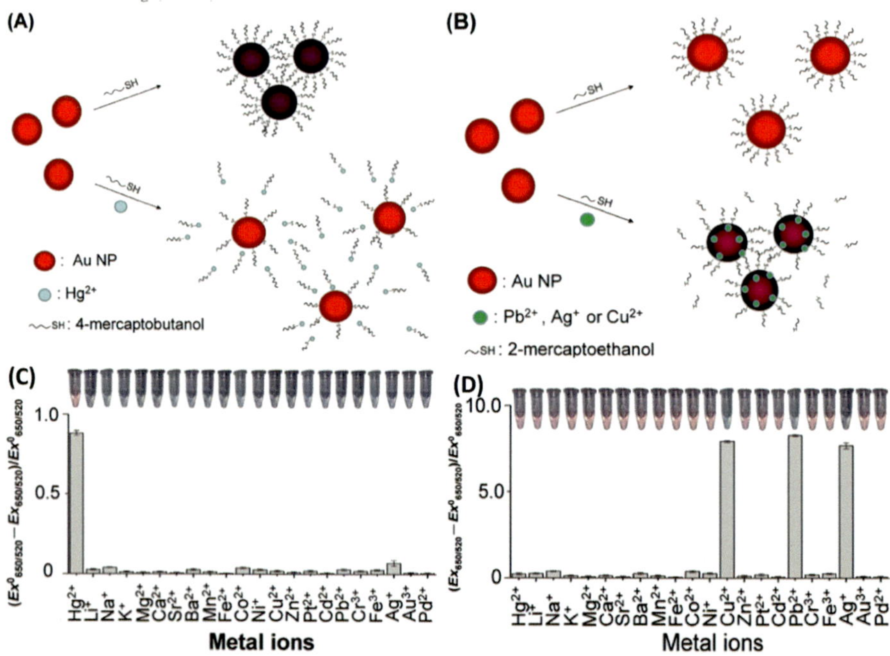

Fig. 4.13 Schematic representation of (**a**) 4-mercaptobtanol- (**b**) 2-mercaptobutanol-capped gold nanoparticles with their sensing mechanism; fluorimetric selective detection of (**c**) mercury ion in the presence of 4 MB Au-NPs and (**d**) lead, copper and silver in the presence of 2ME Au-NPs. (Adapted figure from Hung et al. with permission from copyright, American Chemical Society (Washington, DC, USA))

methods for the monitoring of pollutants in the environment to evaluate the risk to human health and the wildlife. Nanomaterials have some essential properties such as great specific surface areas, size distribution, hydrophilicity/lipophilicity, molecular structure and high reactivity. All of these physico-chemical characteristics make nanomaterials appropriate for a broad range of environmental applications like adsorbents, sensors and catalysts. For the treatment of pollutants, research is focussed on toxic-gas nanosensor, nanofilters and nanoporous film in a photocatalytic unit to remove volatile organic compounds (VOCs) and degrade residual ozone (Khan and Ghoshal 2000). In general, nanotechnology can be applied in pollution control through three major categories: (1) remediation and treatment, (2) detection and sensing and (3) pollution prevention (Yadav et al. 2017). The highly water-soluble and photocatalytic $Fe_3O_4@TiO_2$ nanoparticles act as promising agents in the photodegradation and mineralization of endocrine-disrupting organic compounds like bisphenol A (BPA) and dibutyl phthalate (DBP) under UV illumination after functionalization with carboxymethyl β-cyclodextrin (CMCD) (Fig. 4.14).

The attractive feature of the respective CMCD-functionalized $Fe_3O_4@TiO_2$ nanoparticles is such that they can be completely obtained from the organic compound suspension and reused completely without any loss of catalytic activity.

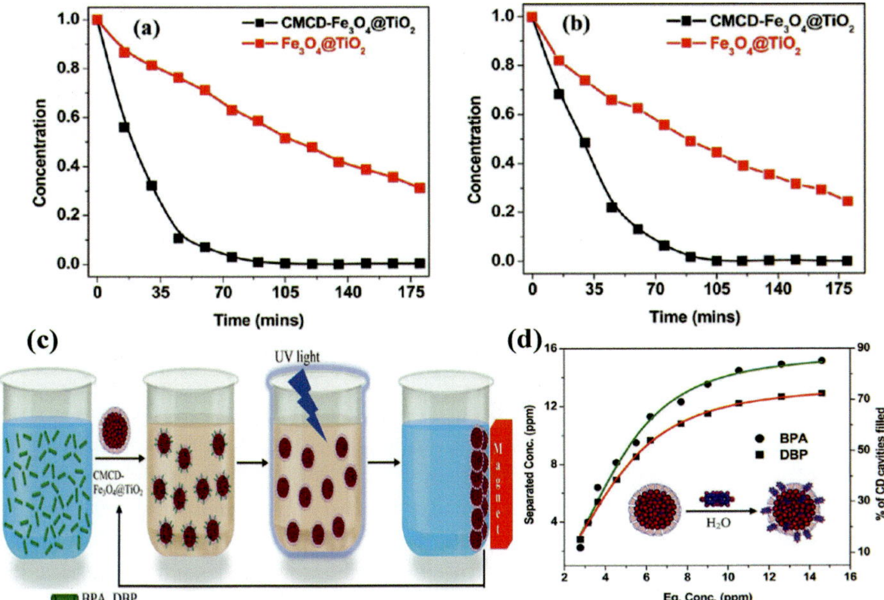

Fig. 4.14 (**a**) Photocatalytic degradation of bisphenol A and (**b**) dibutyl phthalate by using Fe_3O_4@ TiO_2 nanoparticles, (**c**) schematic representation for the mechanistic behaviour of the degradation of BP and, DBP and (**d**) sorption isotherm for the adsorption of BPA and DBP by using Fe_3O_4@ TiO_2 nanoparticles. (American Chemical Society (Washington, DC, USA))

Sharma et al. (2017) have synthesized nanohybrid of graphene nanoplatelets on the surface tin oxide nanoparticles by utilizing surface plasmon resonance method. The developed nanoparticles were found to be highly selective towards hexachlorobenzene, and the value of detection limit was found to be 8.76×10^{-13} g/L which was found to be very small as compared to other reports. Kim et al. (2017) have synthesized highly luminescent tetraphenylethylene-derived silica nanoparticles by adopting thermally reversible scheme. Diethylstilbestrol was used as sensing material in the presence of developed silica nanoparticles by using fluorescence spectroscopy. The fluorescence-quenching response calculated by utilizing Stern-Volmer plot was found to be 50 ng/ml. Biosynthesized copper oxide nanoparticles were prepared by using fruit extract of plant (Singh et al. 2017) and characterized by various analytical schemes. The electrochemical sensing of 4-nitrophenol was carried out using cyclic voltammetric method (square wave voltammetry and impedance spectroscopy). The developed nanoparticles produce significant response and degradation for 4-nitrophenol in range 10 nm–10 mM in very selective and sensitive manner.

4.3.3 Organic Dyes

Organic dyes are classified as a cluster of coloured composites, and their members may belong to a broad variety of chemical substances. Organic dyes may be aromatic, heteroaromatic or any other types of ions or molecules which have very high tendency of complex formation with variable molecular structure and properties. The basic characteristic of any dye depends upon the molecular structure which in turn is related to the temperature, ionic strength, redox potential, solvent and pH of the material. From the last few decades, enormous quantity of dyes enters into the environment due to the constant growth of printing and dyeing industries. Dyes are vital daily chemical compounds that are being widely used in many fields to fabricate colourful products such as textile, paper, rubber, leather, drug, mining, food processing industries, plastics, printing, etc. (Rafatullah et al. 2010). The textile industry discharge is one of the highest amounts of dye effluent, contributing to more than half of the existing dye effluents seen in the environment around the world. The excessive usage of reactive dyes in numerous industries around the globe causes severe contamination of water and soil (Zhang et al. 2014). Most of the organic dyes contain volatile and harmful chemical compounds such as acetic acid, nitrates, sulphur and chromium-based compounds along with various functional groups. Simultaneously, heavy metals are also present in miniature quantities such as lead, cadmium, mercury, nickel, cobalt, zinc and arsenic (Rajasimman et al. 2017). Around 50% of the commercially used dye materials are non-biodegradable and reactive in nature. Dyes are considered as pollutants that not only are lethal but also alter the colour of the water as they are visible in water bodies even in extremely low concentrations. Due to the presence of dyes, sunlight penetration gets entirely blocked and leads to imbalance in the photosynthetic reaction which further affects the aquatic biota due to oxygen deficiency. Oxygen deficiency disturbs the life cycle of various species, and hence, this may eventually lead to a process called eutrophication (Shakoor and Nasar 2016). Therefore, dyes as pollutants are deleterious to humans, aquatic animals and plants too. Various techniques are available in literature for the quantification of colour present in dyes such as spectrophotometry, chromatography and high-performance capillary electrophoresis. But, the most popular dye detection technique is UV-visible spectrophotometry which is used for monitoring the absorbance of dyes (Heidarizadi and Tabaraki 2016). In general, dyes are classified into two categories, namely, natural dyes and synthetic dyes. The natural dyes can be nontoxic because of their natural commencement as compared to the synthetic dyes. The colour of natural dyes is derived from their complex organic structures which have both chromophoric groups (including azo, polymethine, aryl methyl, nitro, nitroso, coumarin, xanthenes and anthraquinone) and auxochromic groups (like $-NH_2$, $-OH$, $-SO_3H$, etc.). These auxochromic and chromophoric groups have susceptibility to escalate the solubility and affinity towards the substrate and are electrochemically oxidizable/reducible (Hudari et al. 2017). It is coloured as it is absorbed in the visible range of the spectrum. Other properties include optical, electronic, photophysical and photochemical properties and conversion of absorbed light energy. Predominantly, a small dosage of dye in water can generate a vivid colour and contribute appreciable value of molar extinction coefficients. On the basis of their solubility and chemical structure, dyes are also classified into various categories such as acidic, basic, vat, azo, direct, mordant and so on. Among the different classes of dyes, azo dyes (contain azo group –N=N-) have been considered as the most frequently used dyes in industrial application due to their excellent and wide spectrum of colouration. Azo dyes are highly toxic and mutagenic to the ecosystem on the basis of their concentration effect in water resources (Agbe et al. 2019). Their degradation products are also toxic because of their bigger molecular structures. However, under anaerobic conditions, the formation of amines takes place due to reductive cleavage of azo linkages which are classified as toxic and carcinogenic materials (Solis et al. 2012). Azo dyes enter the human body very easily through the skin, lungs and gastrointestinal tract. After entering the human body, it destructs the haemoglobin adduct formation which in turn affects the blood formation (Karthikeyan and Jothivenkatachalam 2014). In other aspect, when these dyes enter into the agricultural field, they get easily adsorb by the soil through sorption process and assassinate the life cycle of various small organisms present in the soil. The rupturing of microorganisms (flora and fauna) affects the agricultural productivity (Imran et al. 2015). From the environmental point of view, the removal of dye from industrial waste is one of the most important concerns in today's world. So, there is an urgent need for the recognition, quantification and most importantly degradation of the organic dyes. According to various revised policies, it is expected that no synthetic chemicals should be discharged in the effluents. Since we cannot imagine a world without colours or coloured articles like clothes and objects, the manufacturers will keep on using the dyes. Therefore, new technologies must be developed to trap almost all the dyes that enter into the environmental atmosphere. Environmental scientists have adopted numerous methods such as filtration, flocculation, degradation, coagulation, adsorption and so on to overcome such hazardous threat. However, these techniques involve multiple as well as complex steps and are time-consuming, with sludge generation problems, and costly with some limitations. Hence, it is quite important to develop a competent method for the removal of such hazardous chemicals and dyes by employing sustainable approach. According to literature, zinc oxide semiconducting nanoparticles were found to be highly reactive and employed for the photocatalytic degradation of organic dyes such as methyl orange and methylene blue under the exposure of ultraviolet and sunlight (Fig. 4.15). 95–97% of degradation was observed for both dyes, and detailed mechanism of degradation was carried out on release of reactive hydroxyl species (Islam et al. 2019).

Chaudhary et al. (2018a, b, c) have synthesized NiO nanodisks by applying hydrothermal treatment at different temperatures, and average size was found to be 6–12 nm. Before investigating the degradation mechanism of developed nanomaterials, their biocompatible behaviour was evaluated by carrying out their antibacterial and seed germination analysis. The developed nanodisks degrade methylene blue dye and mixture of other dyes up to 98% and 94% within 20 mins. Zinc oxide nanoparticles were also prepared by considering microwave irradiation method

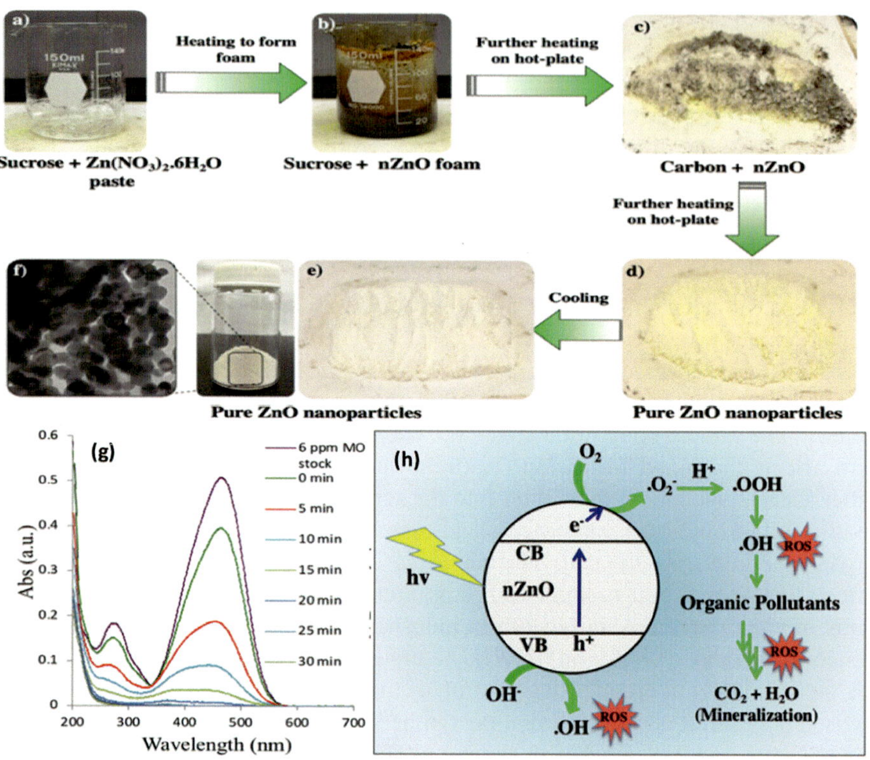

Fig. 4.15 (**a–e**) Schematic representation for the development of zinc oxide nanoparticles, (**f**) their TEM image, (**g**) photocatalytic degradation of methyl orange dye at different time intervals and (**h**) mechanism of photocatalytic generation of reactive oxygen species by zinc oxide nanoparticles and their application in degradation of harmful dyes. (Adapted figure from Islam et al. (2019) with permission from copyright, American Chemical Society (Washington, DC, USA))

with average size 4-22 nm. The developed nanoparticles were found to be highly selective towards Coomassie brilliant blue R-250 (BBR-250) dye. The adsorption and desorption studies of respective nanoparticles were performed towards BBR-250 dye. The real water sample analysis was also carried out by spiking different water sources with known concentration of dye amount. Malini et al. (2018) have synthesized C-, N- and S-doped titanium dioxide (TiO_2) nanoparticles by simple hydrolysis method. The developed nanoparticles act as efficient catalyst for the photodegradation of rose bengal dye, and 100% degradation is obtained only in 60 mins.

4.3.4 Gas Sensing

In today's scenario, the quality of air degrades at a very high rate due to the presence of various types of volatile organic compounds and gases. Simultaneously, the human health is also getting affected at a higher risk. So, the purification of air quality has become a major concern (Yin et al. 2019). The most common sources for the production of volatile gases are heat exchange arrangements, domestic commodities and chemical-manufacturing industries along with offices and printers. The most common volatile gases are acetone, n-butanol, hydrogen disulphide methanol, ethanol, carbon monoxide, hydrogen cyanide, formaldehyde and so on (Joshi et al. 2009; Chen et al. 2019). The US National Institute for Occupational Safety and Health (NIOSH) has maintained an upper and lower limit for the revelation of the volatile organic gases (Wang et al. 2016; Qu et al. 2019). Gas sensors have a variety of relevance in the field of personal safety explosive recognition. So, there is an urgent need to develop gas sensors for examining and scrutinizing the environmental pollution caused by excess of gases. In literature, different kinds of gas sensors have been reported, but out of all those, semiconducting nanoparticles have gained enormous attention and considered as the most promising materials due to their stumpy price, high selectivity and sensitivity, elevated response time, easy availability and most importantly environment-friendly nature. The band gap of the nanomaterials also plays very important role in gas sensing due to their excellent thermal and chemical steadiness (Wang et al. 2010). Out of various semiconducting nanoparticles, TiO_2, ZnO and CuO nanoparticles have very fast response time towards gases, and they have excellent band gap values (Shi et al. 2018). Huber et al. (2017) have developed zinc oxide (ZnO) nanowires on silicon by applying chemical vapour deposition (CVD) method. The developed nanowires were separated by operating dielectrophoresis and utilized in sensing of gas such as H_2S. It was observed that the prepared nanoparticles were highly selective and sensitive towards H_2S gas, and minimum 50 ppb concentration can be detected easily. Similarly, Zhao et al. (2018) also synthesized ZnO nanoparticles by cadmium doping, and highly crystalline wurtzite nanocrystal structure was revealed for the developed nanoparticles from X-ray diffraction (XRD) analysis. The developed nanoparticle acts as a promising candidate for the selective sensing of n-butanol gas sensing up to 100 ppm. Suematsu et al. (2018) developed Au-loaded zinc oxide nanoparticles by calcinations process, and the size was found in range 2–4 nm with metallic structure. The developed nanoparticles act as sensors towards various gases and show response up to 200 ppm for carbon monoxide and hydrogen and 100 ppm for toluene, ethanol, acetone and acetaldehyde in dry atmosphere at 377 °C (Fig. 4.16). The sensing response of Au-ZnO NPs was found to be 92 for toluene as compared to other combustible gases.

Yang et al. (2015) have investigated the role of copper oxide nanoparticles in sensing of gas molecules. Copper oxide nanoparticles were developed by using solvothermal analysis, and functionalization of respective nanoparticles was carried out by quartz crystal microbalance. It was observed that the developed nanoparticles

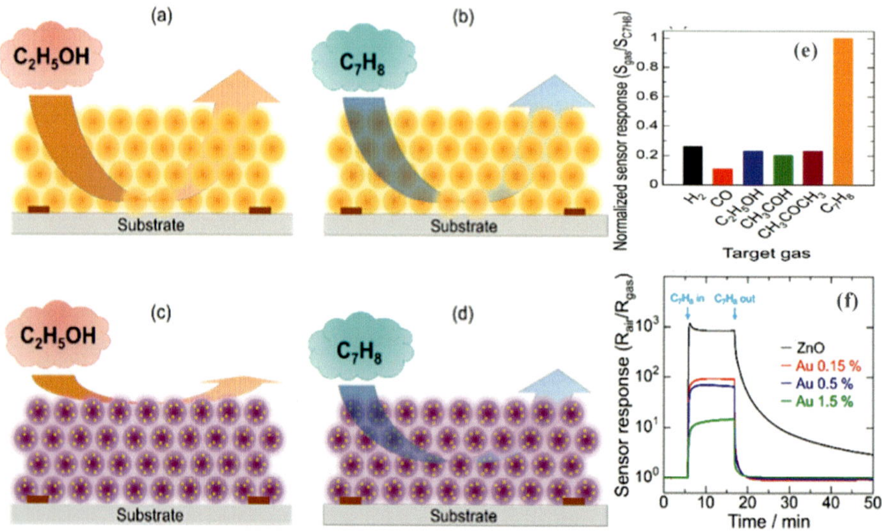

Fig. 4.16 Graphical representation of ethanol (**a**, **c**) and toluene (**b**, **d**) diffusion. Toluene gas selectivity out of various other gases (**e**) and transient response curve of ZnO and Au-ZnO nanoparticles at 377 °C for 100 ppm toluene (**f**). (Adapted figure from Suematsu et al. (2018) with permission from copyright, American Chemical Society (Washington, DC, USA))

were found highly selective towards hydrogen cyanide gas under wide-ranging humidity condition. The sensitivity, selectivity and response time of hydrogen cyanide gas increase with augmentation in the relative humidity present in the environment at 9–51%, and hydrogen cyanide gas works as a practical gas sensor. Singh et al. (2016) have synthesized Zn-doped SnO_2 polycrystalline nanoparticles by applying hydrothermal treatment with size range 9–20 nm. The developed nanoparticles were utilized in the field of gas sensing application such as methanol, ethanol and acetone. It was found that the highest concentration of dopants (zinc = 0.20%) reflects upper limit of 65% to ethanol, 75% to methanol and 62% to acetone for a concentration of 100 ppm at diverse functioning temperature.

4.3.5 Inorganic Anions

Inorganic anions such as iodide, sulphate, chloride, carbonate, sulphite and bromide have been considered as one of the major constituents of daily life cycle as they play indispensable role in various physiological and biological activities. Simultaneously, they have been widely used in the meadow of industrial processing, drug delivery, biomedical diagnostic and environmental screening (Lyu and Pu 2017; Bothra et al. 2015). The discharge of water from industries and rivers is the most common source for the release of inorganic and organic anions which in turn pollute the drinking and groundwater at a very high risk (Liu and Liu 2015). But above optimal concentration, anionic group acts as harmful pollutant towards human health and environmental living species, and sometimes its administration leads to death. Therefore, there is an urgent need to develop highly selective and sensitive sensor to stimulate the phenomenal behaviour of anionic species even at very low and higher concentration level with clinical significance (Poland et al. 2006). In literature, various techniques have been employed to evaluate the permissible limit of anions, but fluorimetric luminescent technique has been considered as the most widely used due to its ease of labour, cheap cost, simple operating software, reproducibility, reliability and high accuracy along with specificity. In recent era, semiconducting nanoparticles emerged out as the most suitable channel to investigate the performance of the anions (Ansari et al. 2016; Mehta et al. 2013). Becuwe et al. (2013) have synthesized silica nanoparticles by using sol-gel Stöber method. The developed nanoparticles are found to be spherical in nature with size 150–200 nm. During the sensing application, developed nanoparticles exhibit specific turn on (enhancement) behaviour towards perchlorate anion and turn off (quenching) behaviour towards nitrate, acetate and chloride ion. Zhang et al. (2011) have described the effect of Cu@Au nanoparticles in sensing of sulphide anion during colorimetric assay and optical absorption spectroscopy. It was observed that developed nanoparticles could detect sulphide up to 3 μM by utilizing colorimetric assay (naked eye detection) and 0.3 μM by adopting UV-vis spectroscopy.

4.4 Summary

In this book chapter, the venomous nature of different types of contaminants and toxins present in the environment was explored to a large extent. The potential aptitude of semiconducting nanoparticles has been studied here to investigate their capability in the field of pollutant sensing. Before utilizing them in the field of environmental remediation, the toxicological screening of semiconducting nanoparticles has been explored by utilizing the quantitative multi-assay approach such as antimicrobial, antifungal, antialgal, seed germination, *Allium cepa* chromosomal, photodegradation, sorption, aggregation and desorption assays. The corresponding influence of concentration and the nature of particles on different assays has provided a detailed overview regarding the biocompatibility of nanoparticles. Furthermore, the biocompatible nanoparticles were studied to explore their potential in pollutant sensing. The focus has been kept over heavy toxic metal ions, inorganic anions, organic dyes, volatile toxic gases and organic volatile compounds. The influence of corresponding interfering species on the sensing aptitude of nanoparticles has also been taken into consideration. Overall, semiconducting nanoparticles have been found free from all kinds of toxicological environmental fates and have been successfully utilized in the field of environmental remediation against various types of contaminants.

Multiple-Choice Questions

Question 1. Nanoparticles possess excellent properties as compared to bulk materials due to

(a) High surface to volume ratio
(b) Large size
(c) Low melting point
(d) No quantum effect

Question 2. How nanotechnology is used in the medical field?

(a) Drug delivery
(b) Diagnosis
(c) Tissue engineering
(d) All of the above

Question 3. Which of the statement is correct about semiconductors?

(a) Charge carrier may be positive hole or electron.
(b) They have conductivity between conductors and insulators.
(c) They have very low electrical conductivity and resistivity.
(d) They do not have properties between conductors and insulators.

Question 4. During antimicrobial activity, zone of inhibition directly depends upon?

(a) Shape and size of nanoparticles
(b) Nature of nanoparticles
(c) Concentration of nanoparticles
(d) All of the above

Question 5. While carrying out antifungal activity, what is the optimum condition for fungus growth?

(a) Low temperature range
(b) High temperature range
(c) Normal environment
(d) None of the above

Question 6. Seed germination study provides information about the

(a) Behaviour of nanoparticles
(b) Germination index
(c) Biocompatible behaviour
(d) All of the above

Question 7. *Allium cepa chromosomal* aberration assay gives knowledge of

(a) Meiotic results
(b) Mitotic division
(c) DNA rupture

(d) Nuclei formation

Question 8. Photodegradation deals with the

(a) Adsorption
(b) Desorption
(c) Degradation
(d) None of the above

Question 9. The excess concentration of heavy metal ions leads to

(a) Environmental contamination
(b) Remediation
(c) Improvement of human health
(d) No changes

Question 10. Why is sensing of different contaminants present in the environment important?

(a) To protect the environment
(b) For good health of humans
(c) For good health of animals
(d) All of the above

Short Questions

Question 1. Why do nanoparticles have different properties from bulk materials? Explain their different features.
Question 2. Name different methods utilized for the toxicological screening of semiconducting nanoparticles.
Question 3. How do nanoparticles play a crucial role during seed germination activity?
Question 4. Explain aggregation and dissolution behaviour of nanoparticles?
Question 5. What are the different pollutants present in the environment and what are their adverse effects?

Answer Sheet

Answers of MCQ

Question 1 (a)
Question 2 (d)
Question 3 (b)

Question 4 (d)
Question 5 (a)
Question 6 (d)
Question 7 (b)
Question 8 (c)
Question 9 (a)
Question 10 (d)

Short-Question Answers

Answer 1. Nanoparticles possess excellent properties as compared to the bulk material as the former have outstanding surface area to volume ratio and quantum effect which makes them efficient sensing probe in the field of contaminant sensing. Due to their large surface area, they have been widely used in the assortment of physico-chemical processes and their potential application in nanoscience. They are capable of shaping the current needs and act as a safeguard towards the environment in the assessment of novel and simpler materials and provide various remediation schemes in the handling of harmful toxins. They have excellent optical and photoluminescence properties and wide band gap values and are tremendously biocompatible in nature and highly prone towards extensive range of toxic pollutants.

Answer 2. Various types of biological and toxicity activities have been reported to explore the toxicological fate of semiconducting nanoparticles, which include antibacterial, antifungal, antialgal, anti-proliferation, *Allium cepa* chromosomal, soil sorption and photodegradation and aggregation/dissolution activities.

Answer 3. Nanoparticles play a very crucial role in the seed germination activity which further depends upon the size, shape, morphology and surface characteristics (surface area/mass ratio) of nanoparticles. The environmental exposure, ageing, species and growth rate of plant play a fundamental role in the development of plant tissue. The size of nanoparticles (ranges from 10 to 50 nm) is considered as the most important factor in the analysis of proliferation because they can easily penetrate and allow accumulation into plant cell wall which further enhances the saturation stability. Aggregation, dissolution, adsorption, redox and desorption are found quite interesting processes which can transform the fate progression and toxicity of nanoparticles on terrestrial species.

Answer 4. Aggregation process basically deals with the combination of a series of particles, which results in the formation of a large-sized particle. Aggregation is basically divided into two categories, namely, homo-aggregation and hetero-aggregation, which depends upon pH, stearic hindrance, electrostatic interaction and ionic strength of nanoparticles along with their attachment efficiency. The aggregation tendency of nanoparticles has become beneficial according to environmental fate point of view and prevents the toxicity of nanoparticles at their early stages. The process of dissolution leads to splitting of molecule in their respective ionic forms and exerts toxicity on the different species present in the environment. During dissolution, disruption of nanoparticles takes place into various fragments either through chemical or physical transformation. Here, oxygen content plays a fundamental role in aquatic systems as most of water content is anoxic in nature. Oxygen works as a catalyst, and an appreciable amount of dissolved oxygen will not affect the dissolution activity.

Answer 5. The pollutants released in the environment are in different forms such as heavy metal ions, inorganic anions, toxic gases (nitrogen oxides, sulphur oxides, ozone, carbon oxides etc.), suspended particulate matter as well as volatile organic compounds. Typically, a pollutant is any toxin which is not native to the ecosystem. However, if the particular discharged chemical is indigenous to the ecosystem, then it is characterized as a pollutant if it is emancipated into the environmental streaming in amounts that are in overabundance to the natural concentration of the organic chemical in the ecosystem, and therefore, this augmented concentration can cause damage to the flora or fauna of the ecosystem. They are able to undergo long-range atmospheric transport from source to remote areas, thereby threatening the global ecological environment and individual health.

References

Abdel-Shafy HI, Mansour MSM (2016) A review on polycyclic aromatic hydrocarbons: source, environmental impact, effect on human health and remediation. Egypt J Pet 25:107–123

Abuduwaili J, Zhang ZY, Jiang FQ (2015) Assessment of the distribution, sources and potential ecological risk of heavy metals in the dry surface sediment of Aibi Lake in Northwest China. PLoS One 10(3):1–16

Agbe H, Raza N, Dodoo-Arhin D, Kumar RV, Kim KH (2019) A simple sensing of hazardous photo-induced superoxide anion radicals using a molecular probe in ZnO-nanoparticles aqueous medium. Environmen Res 176:108424–108432

Ahmed KBA, Pichikannu A, Veerappan A (2015) Fluorescence cadmium sulfide nanosensor for selective recognition of chromium ions in aqueous solution at wide pH range. Sensor Actuat B-Chem 221:1055–1061

Alharbi OML, Basheer AA, Khattab RA, Ali I (2018) Health and environmental effects of persistent organic pollutants. J Mol Liq 228:650–657

Ansari Z, Dhara S, Bandyopadhyay B, Saha A, Sen K (2016) Spectral anion sensing and γ-radiation induced magnetic modifications of polyphenol generated Ag-nanoparticles. Spectrochim Acta A 156(2016):98–104

Apak R (2019) Current Issues in Antioxidant Measurement. J Agric Food Chem 67:9187–9202

Ayodhya D, Veerabhadram G (2019) Fabrication of Schiff base coordinated ZnS nanoparticles for enhanced photocatalytic degradation of chlorpyrifos pesticide and detection of heavy metal ions. J Mater

Becuwe M, Cazier F, Woisel P, Delattre F (2013) Turn-on/turn-off fluorescent hybrid silica nanoparticles. A new promising material for selective anions' sensing. Colloid Surf A 433:88–94

Bhargav PK, Murthy KSR, Pandey JK, Mandal P, Goyat MS, Bhatia R, Varanasi A (2019) Tuning the structural, morphological, optical, wetting properties and anti-fungal activity of ZnO nanoparticles by C doping. Nano-Structures Nano-Objects 19:10036–10044

Bhowmick S, Pramanik S, Singh P, Mondal P, Chatterjee D, Nriagu J (2018) Arsenic in groundwater of West Bengal, India: a review of human health risks and assessment of possible intervention options. Sci Total Environ 612:148–169

Bian SW, Mudunkotuwa AI, Rupasinghe T, Grassian HV (2011) Aggregation and dissolution of 4 nm ZnO nanoparticles in aqueous environments: influence of pH, ionic strength, size, and adsorption of humic acid. Langmuir 27:6059–6068

Bolis V, Garcia EC, Weder O, Hungerbuhler K (2018) New classification of chemical hazardous liquid waste for the estimation of its energy recovery potential based on existing measurements. J Clean Prod 183:1228–1240

Bothra S, Kumar R, Pati RK, Kumar A, Choi HJ, Sahoo SK (2015) Virgin silver nanoparticles as colorimetric nanoprobe for simultaneous detection of iodide and bromide ion in aqueous medium. Spectrochim Acta A 149:122–126

Chaudhary S, Sharma P, Renu KR (2016) Hydroxyapatite doped CeO_2 nanoparticles: impact on biocompatibility and dye adsorption properties. RSC Adv 6:62797–62809

Chaudhary S, Sharma P, Singh D, Umar A, Kumar R (2017) Chemical and pathogenic Cleanup of wastewater using surface-functionalized CeO_2 nanoparticles. ACS Sustain Chem Eng 5:6803–6816

Chaudhary S, Kaur Y, Jayee B, Chaudhary GR, Umar A (2018a) NiO nanodisks: highly efficient visible-light driven photocatalyst, potential scaffold for seed germination of Vigna Radiata and antibacterial properties. J Clean Prod 190:563–576

Chaudhary S, Chauhan P, Kumar R, Bhasin KK (2018b) Toxicological responses of surfactant functionalized selenium nanoparticles: a quantitative multi-assay approach. Sci Total Environ 643:1266–1277

Chaudhary S, Sharma P, Kumar S, Alex SA, Kumar R, Mehta SK, Mukherjee A, Umar A (2018c) A comparative multi-assay approach to study the toxicity behaviour of Eu_2O_3 nanoparticles. J Mol Liq 269:783–795

Chaudhary S, Rohilla D, Umar A, Kaur N, Shanavas A (2019) Synthesis and characterizations of luminescent copper oxide nanoparticles: toxicological profiling and sensing applications. Ceram Int 45:15025–15035

Chen K, Chen S, Pi M, Zhang D (2019) SnO_2 nanoparticles/TiO_2 nanofibers heterostructures: in situ fabrication and enhanced gas sensing performance. Solid State Electron 157:42–47

Clarkson TW, Magos L, Myers GJ (2003) The toxicology of mercury-current exposures and clinical manifestations. N Engl J Med 349:1731–1737

Conway RJ, Adeleye AS, Torresdey JG, Keller AA (2015) Aggregation, dissolution, and transformation of copper nanoparticles in natural waters. Environ Sci Technol 49:2749–2756

Essam T, Aly AM, El Tayeb O, Mattiasson B, Guieysse B (2010) Characterization of highly resistant phenol degrading strain isolated from industrial wastewater treatment plant. J Hazard Mater 173:783–788

Fato FP, Li DW, Zhao LJ, Qiu K, Long YT (2019) Simultaneous removal of multiple heavy metal ions from river water using ultrafine mesoporous magnetite nanoparticles. ACS Omega 4:7543–7549

Garcia-Gomez C, Fernandez MD (2019) Impacts of metal oxide nanoparticles on seed germination, plant growth and development. Compr Anal Chem 84:76–116

Giorgetti L (2019) Effects of nanoparticles in plants: phytotoxicity and genotoxicity assessment. Nanomaterials in Plants, Algae, and Microorganisms 2:65–87

Gong N, Shao K, Che C, Sun Y (2019) Stability of nickel oxide nanoparticles and its influence on toxicity to marine algae Chlorella vulgaris. Mar Pollut Bull 149:110532–110539

Gupta D (2018) Methods for determination of antioxidant capacity: a review. Int J Pharm Sci Res 6:546–566

Gupta PK, Sharma PP, Sharma A, Khan ZH, Solanki PR (2016) Electrochemical and antimicrobial activity of tellurium. Mater Sci Eng B 211:166–172

Haley R (2016) Evaluation of a colorimetric assay for the detection of arsenic in water. 1–52

He X, Xie C, Ma Y, Wang L, He X, Shi W, Liu X, Liu Y, Zhang Z (2019) Size-dependent toxicity of ThO_2 nanoparticles to green algae Chlorella pyrenoidosa. Aquat Toxicol 209:113–120

Heidarizadi E, Tabaraki R (2016) Simultaneous spectrophotometric determination of synthetic dyes in food samples after cloud point extraction using multiple response optimizations. Talanta 148:237–246

Huber F, Riegert S, Madel M, Thonke K (2017) H_2S sensing in the ppb regime with zinc oxide nanowires. Sensors Actuat B-Chem 239:358–363

Hudari FF, Brugnera MF, Zanoni MVB (2017) Advances and trends in voltammetric analysis of dyes. 75–108.

Imran M, Shaharoona BE, Crowley D, Khalid A, Hussain S, Arshad M (2015) The stability of textile azo dyes in soil and their impact on microbial phospholipid fatty acid profiles. Ecotoxicol Environ Saf 120:163–168

Inoue KI (2013) Heavy metal toxicity J Clinic Toxicol 3:2161–0495

Islam MT, Dominguez A, Alvarado-Tenorio B, Bernal RA, Montes MO, Noveron JC (2019) Sucrose-mediated fast synthesis of zinc oxide nanoparticles for the photocatalytic degradation of organic pollutants in water. ACS Omega 4:6560–6572

Jacob J (2013) A review of the accumulation and distribution of persistent organic pollutants in the environment. Int J Biosci Biochem Bioinforma 3:657–661

Jain A, Wadhawan S, Kumar V, Mehta SK (2018) Colorimetric sensing of Fe^{3+} ions in aqueous solution using magnesium oxide nanoparticles synthesized using green approach. Chem Phys Lett 706:53–61

Jaishankar M, Tseten T, Anbalagan N, Mathew BB, Beeregowda KN (2014) Toxicity, mechanism and health effects of some heavy metals. Interdiscip Toxicol 7:60–72

Jaward FM, Guardo AD, Nizzetto L, Cassani C, Raffaele F, Ferretti R, Jones KC (2005) PCBs and selected organochlorine compounds in Italian mountain air: the influence of altitude and forest ecosystem type. Environ Sci and Technol 39:3455–3463

Jiang R, Wang M, Chen W, Li X (2018) Ecological risk evaluation of combined pollution of herbicide siduron and heavy metals in soils. Sci Total Environ 626:1047–1056

Jose A, Devi KRS, Pinherio D, Narayan SL (2018) Electrochemical synthesis, photodegradation and antibacterial properties of PEG capped zinc oxide nanoparticles. J Photochem Photobiol 187:25–34

Joshi RK, Hu Q, Alvi F, Joshi N, Kumar A (2009) Au decorated zinc oxide nanowires for CO sensing. J Phys Chem C 36:16199–16202

Kah M, Walch H, Hofmann T (2018) Environmental fate of nanopesticides: durability, sorption and photodegradation of nanoformulated clothianidin. Environ Sci Nano 5:882–889

Kailasa SK, Park TJ, Rohit JV, Koduru JR (2019) Antimicrobial activity of silver nanoparticles. Nanoparticles in Pharmacotherapy:461–484

Karthikeyan KT, Jothivenkatachalam K (2014) Removal of acid Yellow-17 dye from aqueous solution using turmeric. Environ Nanotechnol 3:69–80

Kastovsky J, Fucikova K, Vesela J, Carias CB, Vegas-Vilarrubia T (2019) Algae. In: Biodiversity of Pantepui, pp 95–120

Khan FI, Ghoshal AK (2000) Removal of volatile organic compounds from polluted air. J Loss Prev Process Ind 13:527–545

Khan MS, Zaidi A, Goel R, Musarrat J (2011) Biomanagement of metal-contaminated soils, vol 20. Springer, pp 1–22

Kim YH, Lee DK, Cha HG, Kim CW, Kang YC, Kang YS (2006) Preparation and characterization of the antibacterial cu nanoparticle formed on the surface of SiO_2 nanoparticles. J Phys Chem B 110:24923–24928

Kim Y, Lee KM, Chang JY (2017) Highly luminescent tetra(biphenyl-4-yl)ethene-grafted molecularly imprinted mesoporous silica nanoparticles for fluorescent sensing of diethylstilbestrol. Sensors Actuators B 242:1296–1304

Konwar A, Kalita S, Kotoky J, Chowdhury D (2016) Chitosan–iron oxide coated graphene oxide nanocomposite hydrogel: a robust and soft antimicrobial biofilm. ACS Appl Mater Interfaces 8:20625–20634

Kumar A, Patel AS, Mohanty T (2012) Correlation of Photodegradation efficiency with surface potential of silver-TiO_2 nanocomposite thin films. J Phys Chem 116:20404–20408

Kumar D, Rajeshwri A, Jadon PS, Chaudhri G, Mukherjee A, Chandrasekaran N, Mukherjee A (2015) Cytogenetic studies of chromium (III) oxide nanoparticles on *Allium cepa* root tip cells. J Environ Sci 38:150–157

Kumar P, Kim K, Bansal V, Lazarides T, Kumar N (2017a) Progress in the sensing techniques for heavy metal ions using nanomaterials. J Ind Eng Chem 54:30–43

Kumar R, Sharma P, Bamal A, Negi S, Chaudhary S (2017b) A safe, efficient and environment friendly biosynthesis of silver nanoparticles using *Leucaena leucocephala* seed extract and its antioxidant, antimicrobial, antifungal activities. De Gruyter, pp 1–11

Kumar P, Burman U, Kaul RK (2018) Ecological risks of nanoparticles: effect on soil microorganisms. Nanomaterials in Plants, Algae, and Microorganisms 1:429–452

Kumari M, Khan SS, Pakrashi S, Mukherjee A, Chandrasekaran N (2011) Cytogenetic and genotoxic effects of zinc oxide nanoparticles on root cell of Allium cepa. J Hazard Mater 190:613–621

Li S, Zhang Q (2010) Risk assessment and seasonal variations of dissolved trace elements and heavy metals in the Upper Han River, China. J Hazard Mater 181:1051–1058

Li J, Dongsheng B, Hong X, Di L, Jinghong L, Yubai B, Tiejin L (2005) Luminescent CdTe quantum dots and nanorods as metal ion probes. J Colloid Interface Sci 257:267–271

Li S, Liu H, Gao R, Abdurahman A, Dai J, Zeng F (2018) Aggregation kinetics of microplastics in aquatic environment: complex roles of electrolytes, pH, and natural organic matter. Environ Pollut 237:126–132

Liman R (2013) Genotoxic effects of Bismuth (III) oxide nanoparticles by *Allium* and Comet assay. Chemosphere 93:269–273

Liu B, Liu J (2015) Comprehensive screen of metal oxide nanoparticles for DNA adsorption, fluorescence quenching, and anion discrimination. ACS Appl Mater Interfaces 7:24833–24838

Liu G, Lin Y-Y, Wu H, Lin Y (2007) Voltammetric detection of Cr (VI) with disposable screen-printed electrode modified with gold nanoparticles. Environ Sci Technol 41:8129–8134

Liu SJ, Nie H-G, Jiang JH, Shen GL, Yu RQ (2009) Electrochemical sensor for mercury(II) based on conformational switch mediated by interstrand cooperative coordination. Anal Chem 81:5724–5730

Lopes AC, Peixe TS, Mesas AE, Paoliello MB (2015) Lead exposure and oxidative stress: a systematic review. Rev Environ Contam Toxicol 236:193–238

Luo C, Liu C, Wang Y, Liu X, Li F, Zhang G, Li X (2011) Heavy metal contamination in soils and vegetables near an e-waste processing site, South China. J Hazard Mater 186:481–490

Luo Z, Wang Z, Yan Y, Li J, Yan C, Xing B (2018) Methods of determining titanium dioxide nanoparticles enhance inorganic arsenic bioavailability and methylation in two fresh water algae species. MethodsX 5:620–625

Lyu Y, Pu K (2017) Recent advances of activatable molecular probes based on semiconducting polymer nanoparticles in sensing and imaging. Adv Sci 4:1600481–1600495

Maleki B, Baghayeri M, Ghanei-Motlagh M, Zonoz FM, Amiri A, Hajizadeh F, Hosseinifar A, Esmaeilnezhad E (2019) Polyamidoamine dendrimer functionalized iron oxide nanoparticles for simultaneous electrochemical detection of Pb^{2+} and Cd^{2+} ions in environmental waters. Measurement 140:81–88

Malini B, Raj GAG (2018) C, N and S-doped TiO_2-characterization and photocatalytic performance for rose bengal dye degradation under day light. J Environ Chem Eng 6:5763–5770

Mangalampalli B, Dumala N, Grover P (2018) *Allium cepa* root tip assay in assessment of toxicity of magnesium oxide nanoparticles and microparticles. J Environ Sci 66:125–137

Markus AA, Parsons JR, Roex EWM, de Voogt P, Laane RWPM (2015) Modeling aggregation and sedimentation of nanoparticles in the aquatic environment. Sci Total Environ 506–507:323–329

Mechlinska A, Wolska L, Namies'nik J (2010) Isotope-labeled substances in analysis of persistent organic pollutants in environmental samples. Trends Anal Chem 29:820–831

Mehta SK, Salaria K, Umar A (2013) Highly sensitive luminescent sensor for cyanide ion detection in aqueous solution based on PEG-coated ZnS nanoparticles. Spectrochim Acta A 105:516–521

Miri A, Mahdinejad N, Ebrahimy O, Khatami M, Sarani M (2019) Zinc oxide nanoparticles: biosynthesis, characterization, antifungal and cytotoxic activity. Mater Sci Eng C 104:109981–109989

Nabulo G, Black CR, Young SD (2011) Trace metal uptake by tropical vegetables grown on soil amended with urban sewage sludge. Environ Pollut 159:368–376

Oguri T, Suzuki G, Matsukami H, Uchida N, Tue MN, Tuyen LH, Viet PH, Takahashi S, Tanabe S, Takigami H (2018) Exposure assessment of heavy metals in an e-waste processing area in northern Vietnam. Sci Total Environ 621:1115–1123

Ozkan G, Kamiloglu S, Capanoglu E, Hizal J, Apak R (2018) Use of nanotechnological methods for the analysis and stability of food antioxidants. In: Impact of nanoscience in the food industry, pp 311–350

Paula RJ, Paula AJ (2019) Treatment of soil pollutants with nanomaterials. In: Nanomaterials applications for environmental matrices, pp 355–368

Payan DAA, Koyani RD, Duhalt RV (2017) Chitosan-based biocatalytic nanoparticles for pollutant removal from wastewater. Enzym Microb Technol 100:71–78

Peng J, Li J, Xu W, Wang L, Su D, Teoh CL, Chang YT (2018) Silica nanoparticle-enhanced fluorescent sensor array for heavy metal ions detection in colloid solution. Anal Chem 3:1628–1634

Pisoschi AM, Negulescu GP (2011) Methods for total antioxidant activity determination: a review. Biochem Anal Biochem 1:1000106–1000116

Placido J, Bustamante-Lopez S, Meissner KE, Kelly DE, Kelly SL (2019) Comparative study of the characteristics and fluorescent properties of three different biochar derived-carbonaceous nanomaterials for bioimaging and heavy metal ions sensing. Fuel Process Technol 196:106163–106176

Poland K, Topoglidis E, Durrant JR, Palomares E (2006) Optical sensing of cyanide using hybrid biomolecular films. Inorg Chem Comm 9:1239–1242

Pujol L, Evrard D, Groenen-Serrano K, Freyssinier M, Ruffien-Cizsak A, Gros P (2014) Electrochemical sensors and devices for heavy metals assay in water: the French groups' contribution. Front Chem 2:1–24

Qu F, Zhou X, Zhang B, Zhang S, Jiang C, Ruan S, Yang M (2019) Fe_2O_3 nanoparticles-decorated MoO_3 nanobelts for enhanced chemiresistive gas sensing. J Alloys Compd 782:672–678

Rad SS, Sani AM, Mohensi S (2019) Biosynthesis, characterization and antimicrobial activities of zinc oxide nanoparticles from leaf extract of *Mentha pulegium* (L.). Microb Pathogenesis 131:239–245

Rafatullah M, Sulaiman O, Hashim R, Ahmad A (2010) Adsorption of methylene blue on low-cost adsorbents: a review. J Hazard Mater 177:70–80

Rajasimman M, Babu SV, Rajamohan N (2017) Biodegradation of textile dyeing industry wastewater using modified anaerobic sequential batch reactor–start-up, parameter optimization and performance analysis. J Taiwan Inst Chem Eng 72:171–181

Rajeshkumar S, Kumar SV, Ramaiah A, Agarwal H, Lakshmi T, Roopan SM (2018) Biosynthesis of zinc oxide nanoparticles using *Mangifera indica* leaves and evaluation of their antioxidant and cytotoxic properties in lung cancer (A549 cells). Enzyme Microbe Technol 117:91–95

Rajiv P, Rajeshwari S, Venckatesh R (2013) Bio-fabrication of zinc oxide nanoparticles using leaf extract of Parthenium hysterophorus L. and its size-dependent antifungal activity against plant fungal pathogens. Spectrochim Acta A 112:384–387

Rana P, Sharma S, Sharma R, Banerjee K (2019) Apple pectin supported superparamagnetic (c-Fe_2O_3) maghemite nanoparticles with antimicrobial potency. Mater Sci Energy Tech 2:15–21

Rasul I, Azeem F, Siddique HM, Muzammil S, Rasul A, Munawar A, Afzal M, Ali AM, Nadeem H (2017) Algae based polymers, blends, and composites. A Green Light for Engineered Algae, Algae Biotechnology, pp 301–334

Roy B, Krishnan SP, Chandrasekaran N, Mukherjee A (2019) Toxic effects of engineered nanoparticles (metal/metal oxides) on plants using *Allium cepa* as a model system. Compr Anal Chem 84:125–139

Sanchis A, Salvador JP, Marco MP (2018) Multiplexed immunochemical techniques for the detection of pollutants in aquatic environments. Trends Anal Chem 18:215–224

Savassa SM, Duran NM, Rodrigues ES, Almeida E, Gestel AM, Bompadre FV, Carvalho H (2018) Effects of ZnO nanoparticles on Phaseolus vulgaris germination and seedling development determined by X-ray spectroscopy. ACS Appl Nano Mater 11:6414–6426

Shakoor S, Nasar A (2016) Removal of methylene blue dye from artificially contaminated water using citrus limetta peel waste as a very low cost adsorbent. J Taiwan Inst Chem Eng 66:154–163

Sharan A, Nara S (2019) Phytotoxic properties of zinc and cobalt oxide nanoparticles in Algaes. In: Nanomaterials in plants, algae, and microorganisms, pp 1–22

Sharma S, Uttam KN (2017) Rapid analyses of stress of copper oxide nanoparticles on wheat plants at an early stage by laser induced fluorescence and attenuated total reflectance Fourier transform infrared spectroscopy. Vib Spectrosc 92:135–150

Sharma D, Rajput J, Kaith BS, Kaur M, Sharma S (2010) Synthesis of ZnO nanoparticles and study of their antibacterial and antifungal properties. Thin Solid Films 519:1224–1229

Sharma S, Usha SP, Shrivastav AM, Gupta BD (2017) A novel method of SPR based SnO2: GNP nano-hybrid decorated optical fiber platform for hexachlorobenzene sensing. Sens Actuat B-Chem 246:927–936

Sharma P, Kaur S, Chaudhary S, Umar A, Kumar R (2018) Bare and nonionic surfactant-functionalized praseodymium oxide nanoparticles: toxicological studies. Chemosphere 209:1007–1020

Shekhawat K, Chatterjee S, Joshi B (2015) Chromium toxicity and its health hazards. Int J Adv Res 3:167–172

Shi H, Li N, Sun Z, Wang T, Xu L (2018) Interface modification of titanium dioxide nanoparticles by titanium substituted polyoxometalate doping for improvement of photoconductivity and gas sensing applications. J Phys Chem Solids 120:57–63

Sierra-Fernandez A, Rosa-Garcia SC, Gomez-Villalba LS, Gomez-Cornelio S, Rabanal ME, Fort R, Quintana P (2017) Synthesis, photocatalytic, and antifungal properties of MgO, ZnO and Zn/mg oxide nanoparticles for the protection of calcareous stone heritage. ACS Appl Mater Interfaces 9:24873–24886

Singh D, Kundu VS, Maan AS (2016) Structural, morphological and gas sensing study of zinc doped tin oxide nanoparticles synthesized via hydrothermal technique. J Mol Struct 1115:250–257

Singh S, Kumar N, Kumar M, Jyoti, Agarwal A, Mizaikoff B (2017) Electrochemical sensing and remediation of 4-nitrophenol using bio-synthesized copper oxide nanoparticles. Chem Eng J 313:283–292

Solís M, Solís A, Pérez HI, Manjarrez N, Flores M (2012) Microbial decolouration of azo dyes: a review. Process Biochem 47:1723–1748

Suematsu K, Watanabe K, Tou A, Sun Y, Shimanoe K (2018) Ultraselective toluene-gas sensor: nanosized gold loaded on zinc oxide nanoparticles. Anal Chem 3:1959–1966

Thakur BK, Kumar A, Kumar D (2019) Green synthesis of titanium dioxide nanoparticles using Azadirachta indica leaf extract and evaluation of their antibacterial activity. S Afr J Bot 124:223–227

Tran DC, Makuvaza J, Munson E, Bennett B (2017) Biocompatible copper oxide nanoparticle composites from cellulose and chitosan: facile synthesis, unique structure, and antimicrobial activity. ACS Appl Mater Interfaces 9:42503–42515

Verma R, Dwivedi P (2013) Heavy metal water pollution-a case study. Recent Res Sci Technol 5:98–99

Verma N, Singh M (2006) A Bacillus sphaericus based biosensor for monitoring nickel ions in industrial effluents and foods. J Anal Meth Chem 2006:1–4

Vidhu VK, Philip D (2015) Biogenic synthesis of SnO_2 nanoparticles: evaluation of antibacterial and antioxidant activities. Spectrochim Acta A 134:372–379

Vinay SP, Udayabhanu NG, Chandrappa CP, Chandrasekhar N (2019) Novel Gomutra (cow urine) mediated synthesis of silver oxide nanoparticles and their enhanced photocatalytic, photoluminescence and antibacterial studies. J Sci Adv Mater Devices

Waheed A, Mansha M, Ullah N (2018) A nanomaterials-based electrochemical detection of heavy metals in water: current status, challenges and future direction. Trends Anal Chem 105:37–51

Walekar L, Dutta T, Kumar P, Ok YS, Pawar S, Deep A, Kim KH (2017) Functionalized fluorescent nanomaterials for sensing pollutants in the environment: a critical review. Trends Anal Chem 97:458–467

Wang H, Wang Y, Jin J, Yang R (2008) Gold nanoparticle-based colorimetric and "turn-on" fluorescent probe for mercury (II) ions in aqueous solution. Anal Chem 80:9021–9028

Wang C, Yin L, Zhang L, Xiang D, Gao R (2010) Metal oxide gas sensors: sensitivity and influencing factors. Sensors 10:2088–2106

Wang B, ,Huynh TP, Wu W, Hayek N, Do TT, Cancilla JC, (2016) Ahighly sensitive diketopyrrolopyrrole-based ambipolar transistor for selective detection and discrimination of xylene isomers. Adv Mater 28:4012–4018.

Wang X, Song J, Zhao J, Wang Z, Wang X (2019) In-situ active formation of carbides coated with NP-TiO_2 nanoparticles for efficient adsorption-photocatalytic inactivation of harmful algae in eutrophic water. Chemosphere 228:351–359

Wu Q, Zhou M, Shi J, Li Q, Yang M, Zhang Z (2017) Synthesis of water-soluble Ag_2S quantum dots with fluorescence in the second near-infrared window for turn on detection of Zn(II) and Cd(II). Anal Chem 89:6616e23

Yadav KK, Singh JK, Gupta N, Kumar V (2017) A review of nanobioremediation technologies for environmental cleanup: a novel biological approach. J Mater Environ Sci 8:740–757

Yang M, He J, Hu M, Hu X, Yan C, Cheng Z (2015) Synthesis of copper oxide nanoparticles and their sensing property to hydrogen cyanide under varied humidity conditions. Sens Actuat B Chem 213:59–64

Yang M, Jiang T-J, Wang Y, Liu J-H, Li L-N, Chen X, Huang X-J (2017) Enhanced electrochemical sensing arsenic(III) with excellent anti-interference using amino functionalized graphene oxide decorated gold microelectrode: XPS and XANES evidence. Sens Actuat B Chem 245:230–237

Yin W, Dong X, Yu J, Pan J, Yao Z, Gu Z, Zhao Y (2017) MoS_2-nanosheet-assisted coordination of metal ions with porphyrin for rapid detection and removal of cadmium ions in aqueous media. ACS Appl Mater Interfaces 9:21362–21370

Yin XT, Zhou WD, Li J, Wang Q, Wu FY, Dastan D, Wang D, Garmestani H, Wang XM, Talu S (2019) A highly sensitivity and selectivity Pt-SnO2 nanoparticles for sensing applications at extremely low level hydrogen gas detection. J Alloys Compd 805:229–236

Zhang J, Xu X, Yuan Y, Yang C, Yang X (2011) A Cu@Au nanoparticle-based colorimetric competition assay for the detection of Sulfide anion and cysteine. ACS Appl Mater Interfaces 3:2928–2931

Zhang S, Zhang Y, Bi G, Liu J, Wang Z, Xu Q, Xu H, Xi L (2014) Mussel-inspired polydopamine biopolymer decorated with magnetic nanoparticles for multiple pollutants removal. J Hazard Mater 270:27–34

Zhao R, Li K, Wang Z, Xing X, Wang Y (2018) Gas-sensing performances of Cd-doped ZnO nanoparticles synthesized by a surfactant-mediated method for n-butanol gas. J Phys Chem Solids 112:43–49

Zhou WY, Liu JY, Song JY, Li JJ, Liu JH, Huang XJ (2017) Surface-electronic-state-modulated, single-crystalline (001) TiO_2 nanosheets for sensitive electrochemical sensing of heavy-metal ions. Anal Chem 89:3386–3394

Zook JM, MacCuspie RI, Locascio LE, Halter MD, Elliott JT (2011) Stable nanoparticle aggregates/agglomerates of different sizes and the effect of their size on hemolytic cytotoxicity. Nanotoxicology 5:517–530

Chapter 5
Nanomaterials in Combating Water Pollution and Related Ecotoxicological Risk

Teenu Jasrotia, Ganga Ram Chaudhary, and Rajeev Kumar

Abstract To meet the needs of the present times, synthesis and processing of tools and resources related to nanoscience are the integral parts of the developmental industries. In this chapter, we have discussed about the emerging and evolving fields of nanotechnology in the domain of water pollution. Nanomaterial-based sensing, diagnosis, and remediation of pollutants from the aquatic ecosystems have been summarized as the state-of-the-art developments. The potentiality of different nanomaterials can tackle and remediate pollution released from diverse industrial units with greater efficacy than other outdated conventional methods. But at the same time, by keeping in view the toxicity concerns rising about the release of nanoparticles (NPs) in the water bodies, the evaluation of ecotoxicological risks on different trophic levels of aquatic ecosystems has also been discussed in this chapter. To advance the nano-field in the domain of pollution control, there is a need of designing the nanomaterials with least cons for the sustenance of aquatic life.

Keywords Nanomaterial · Pollution · Sensing · Toxicity · Aquatic ecosystem

T. Jasrotia
Department of Environment Studies, Panjab University, Chandigarh, India

Department of Chemistry, Panjab University, Chandigarh, India

G. R. Chaudhary
Department of Chemistry, Panjab University, Chandigarh, India

R. Kumar (✉)
Department of Environment Studies, Panjab University, Chandigarh, India
e-mail: rajeev@pu.ac.in

R. Kumar et al. (eds.), *Advanced Functional Nanoparticles "Boon or Bane" for Environment Remediation Applications*, Environmental Contamination Remediation and Management, https://doi.org/10.1007/978-3-031-24416-2_5

5.1 Introduction

The industrialized world of today's time has left its incarnation in every corner of the earth. The impact of fast globalization and speedy urbanization has left unforgettable scars in every sphere of the only claimed planet of life. The clenches of the materialistic world can be seen even on the final frontier of the modern era, i.e. in the abyssal plain and oceanic trenches of the ocean beds. The contribution of noxious elements by the manly activities has surpassed the geological time scale of natural nutrient and elemental cycling. These biodegradable as well as non-biodegradable waste products affect the aquatic ecosystems by degrading the aesthetic values and ecosystem services. The non-biodegradable section of the waste has a longer life cycle, so it remains persistent for the ages by playing the role of a slow poison for the ecosystem. These wastes have irreversible adverse impact on the flora and fauna of the system ranging from micro to macro in size scale (Wang et al. 2015). Natural polluting sources greatly differ from human-originated ones because of the mobility issues. Naturally released contaminants generally have a tendency of scattering and waning after a short period with the only exception of volcanic lava, whereas anthropogenic sources (except water stream-fed pollutants), owing to their constrained movement, remain intact for longer durations (Trujillo-Reyes et al. 2014). These man-made routes mainly include oil spills, fertilizers and pesticides left-outs, extraction and combustion of fossil fuels, and industrial by-products. They affect the aqueous cycle with direct as well as indirect means, which in turn affects the quality and quantity of consumable share of the hydrological cycle (Jaciw-Zurakowsky et al. 2020). The need for rectification is further boosted up by the fact that the earth holds only 2.5% water in potable form. The tripling of the human population as per the data of twentieth century has resulted in the sixfold increase of water utilization by the present human civilization (Verma et al. 2020). So, the future predictions about demographic details have provided an impetus to draw the attention of researchers and scientists towards the rectification of water pollution to meet the water needs of the fastly exploding population. The prevailing limitations in the form of quality and quantity of freshwater along with marine pollution raise the requirement of crucial and credible materials (Wang et al. 2015; Zhang et al. 2021). Thus, there is an urgent need for some modern, advanced, and efficacious solutions for combating water pollution to save and improve living standards on a global scale.

5.2 Types of Water Pollutants

Diverse forms of water contaminants (Fig. 5.1) can be efficiently treated with nanomaterials. These chemical pollutants include organic dyes, heavy metals, pesticides, phenolic compounds, pharmaceutical waste, endocrine-disrupting chemicals, and personal care products.

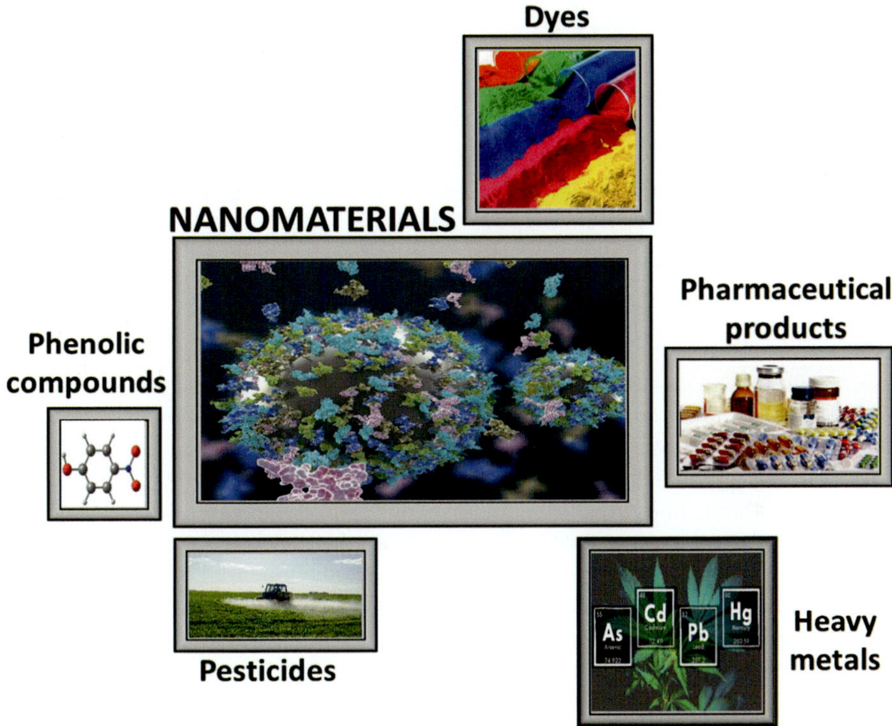

Fig. 5.1 Different types of water pollutants treated with nanomaterials

5.2.1 Organic Dyes

These are the chromophore groups containing coloured aromatic compounds, which are further categorized into anionic dyes and cationic dyes (Bagheri et al. 2020; Lu and Astruc 2020). Industries dealing with textiles and fabrics, paper, food, leather, pharmaceutical, cosmetics, and agriculture are the significant contributors of dyes in the environment (Mu and Wang 2016). As per the industrial data, every year, around 1.6 million tonnes of different coloured dyes are demanded, out of which a considerable share of 10–15% is discarded as industrial effluent waste (Hunger 2003). Health impacts such as breathing discomfort, irreparable damages to the brain cells, allergic reactions, and dermal and eye infections are reported after coming in contact with the dye-polluted water (Pan et al. 2012). The mutagenic as well as carcinogenic concerns associated with these coloured compounds have risen thoughtful processing regarding their abundant discharge in water bodies.

5.2.2 *Heavy Metals*

Metals surpassing the atomic density of 4 ± 1 g/cm³ are categorized as heavy metals. The presence of these metals in levels greater than the values set by national and international agencies such as the World Health Organization (WHO) and US Environmental Protection Agency (EPA) has become a dreadful threat for humans along with other animals (Burakov et al. 2018).

5.2.3 *Pesticides*

These are the class of chemicals used for the enhancement of food supply by restricting the growth of unwanted plants, animals, insects, and diseases. Pesticide is a broad terminology which is comprised of many terms such as weedicide, herbicide, fungicide, miticide, insecticide, and many more. Any route of exposure to pesticides like organophosphates, organochlorine, and chlorophenoxy is lethal for humans and other animals. These carcinogenic compounds have the potential of causing severe damage to body organs, reproductive system, nervous system, and genetic system. Reports suggest that annual usage of pesticides for the management of different pests in the agricultural industries marks the figure of millions of tonnes. The surface run-off and leaching phenomenons have aggravated the pesticide issue by contaminating the surface water bodies and groundwater aquifers, respectively. Thus, removal of these xenobiotics from water bodies has become an attention-seeking issue.

5.2.4 *Phenolic Compounds*

Phenol is seen as a priority water pollutant by the US EPA and National Pollutant Release Inventory (NPRI) as it stands at 11th position in the list of hazardous chemicals. Its presence even in trace quantities can aggravate serious health issues and is mostly dreadful. Resistance to degradation has resulted in their bioaccumulation and hence biomagnification in every trophic level of the food chains.

5.2.5 *Pharmaceutical Waste, Endocrine-Disrupting Chemicals, and Personal Care Products*

Detection of these products in water bodies has become an alarming issue, owing to their durability and non-biodegradability. Effluents from hospitals, industries, and municipalities dump loads and loads of pharmaceutical products like ciprofloxacin and ibuprofen in the aquatic ecosystem (Moussavi et al. 2013). Personal care products and endocrine-disrupting complexes like bisphenol A are usually characterized by higher polarity, low volatility, and non-degradability. As a result, they have been listed as harmful chemicals for humans, animals, and the environment (Archer et al. 2017). Even the trace amounts of these categories of pollutants in water can raise teratogenic, carcinogenic, and mutagenic impacts on living beings (Archer et al. 2017). The functional groups carried by these products have aided their detection in surface water bodies and groundwater table as well.

5.3 Definition and Significance of Nanomaterials

Nanomaterials belong to the class of materials that bear all or at least any one dimension in the nanoscale ranging from 1 nm to 100 nm and are characterized by dimension-dependent properties (Ray and Salehiyan 2020). The inimitable properties of these materials vary greatly from their correspondent macro-sized chemical compounds (Jeevanandam et al. 2018). According to the definition of the International Organization for Standardization (ISO), 'Nanomaterials are the materials with any external nanoscale dimension or having internal nanoscale structure'. Whereas European Commission (EC) has modified the words to define nanomaterials as 'a manufactured or natural material that possesses unbound, aggregated or agglomerated particles where external dimensions are in-between 1 to 100 nm size range'.

Nanomaterials because of their distinct and tailorable dimension-dependent properties serve the purpose of an integral helping tool in diverse evolving fields (Kagan and Murray 2015). The competence of nanomaterials has been proved in numerous fields like medicine, agriculture, food, information technology, textiles, paper and pulp, energy, concrete, transport, and cosmetics (Lugani et al. 2021). The successful exploitation of nanomaterials is achieved because of the ground principle of high surface-to-volume ratios (Bharathi et al. 2021). The effects of quantum confinement govern the enhancement of the properties (optical, structural, chemical, electronic, and magnetic) of the materials in the nanoscale regime (Scher et al. 2016).

Nanomaterials' development and implementations have proved to be more effective and better tools for water pollution monitoring and treatment as compared to their conventional counterparts. From the past few decades, nanotechnology and nanomaterials due to their higher reactivity have emerged as novel and advanced technological innovations for the remediation of water pollution with higher effective rates. The on-going advancement projects on nanomaterials have provided porous domains and surface modifiability to the structural aspects of nanomaterials. As a result, solutions like desalination of seawater, enhanced filtration potential, reduction of toxicants, and improved water quality came into existence (Lu and Astruc 2020). For combating water pollution, nanotechnology can provide promising outputs by following any one of the two approaches, namely, sensing and detection of toxic pollutants in the water bodies or degradation and remediation of different water pollutants. These attractive options for water treatment can be moulded according to customer-oriented needs and are achieved because of their remarkable catalytic nature, enhanced stability (physical, chemical, and thermal), high surface area, notable reactivity, stronger tendency of electron transfer, and many more (Fig. 5.2) (Zhang et al. 2014b; Xu et al. 2019a). On comparing nanomaterials with other traditional methods, nanomaterials proved to be more efficient, low energy consuming, less time-consuming, and environment-friendly option for the addressal of different environmental issues (Nasrollahzadeh et al. 2021).

5.4 Nanomaterials for Combating Water Pollution

5.4.1 Zero-Valent Nanoparticles

5.4.1.1 Nano Zero-Valent Iron (nZVI)

The applicability of nano zero-valent iron (nZVI) has been practised for combating water pollution since the 1990s (Lu and Astruc 2020). nZVI has a broad spectrum towards the elimination of water pollutants including both organic and inorganic contaminants. These include organic dyes, polychlorinated biphenyls, chlorinated organic compounds, pesticides, inorganic ions (like nitrates and phosphates), metalloids, personal care products, radio elements, and pharmaceutical products such as ampicillin and amoxicillin (Shahwan et al. 2011; Zhang et al. 2014a; Sadegh et al. 2014; Goodarzi et al. 2017). They have the tendency of converting toxic species of heavy metals like Cr(VI), As(III), Ni(II), As(V), Cu(II), and Pb(II) into their less toxic counterparts (Bernd 2010; Qu et al. 2013). Large surface area and shape properties result in the enhancement of reactivity of nZVI when compared with simply granular iron (Homhoul et al. 2011; Aredes et al. 2012). The reducing reaction of nZVI towards redox-labile contaminants like chlorinated solvents via reductive dehalogenation is reflected by the value '0.440 V' as standard redox potential of its elemental form (He and Zhao 2007). Because of their higher adsorption capacity, economic viability, precipitation, reduction, and oxidation properties, nZVI are the most researched zero-valent metal NPs for water remediation (Lu et al. 2016). The current scenario of the utilization of nZVI for the water combating is not restricted to the laboratorial level, but successful implementations are made at pilot-scale and at full-scale real water sampling from different contaminated sites (Lu et al. 2016). Apart from surface water remediation, nZVI is also investigated for groundwater pollution and is found to be successful for treating chlorinated hydrocarbons, halogenated organic pollutants, and perchlorates (Bernd 2010; Bishop et al. 2010; Zhan et al. 2011; Verma et al. 2020). The processes of hydrogenolysis and dehaloelimination serve as the reactive pathways of nZVI for reductive dehalogenation (Bishop et al. 2010). nZVI interacts with groundwater pollutants either by directly being injected in the stabilized form in the underwater aquifers, wherein they get adsorbed on the surfaces of aquifer solids after making a zone of iron particles, or by the

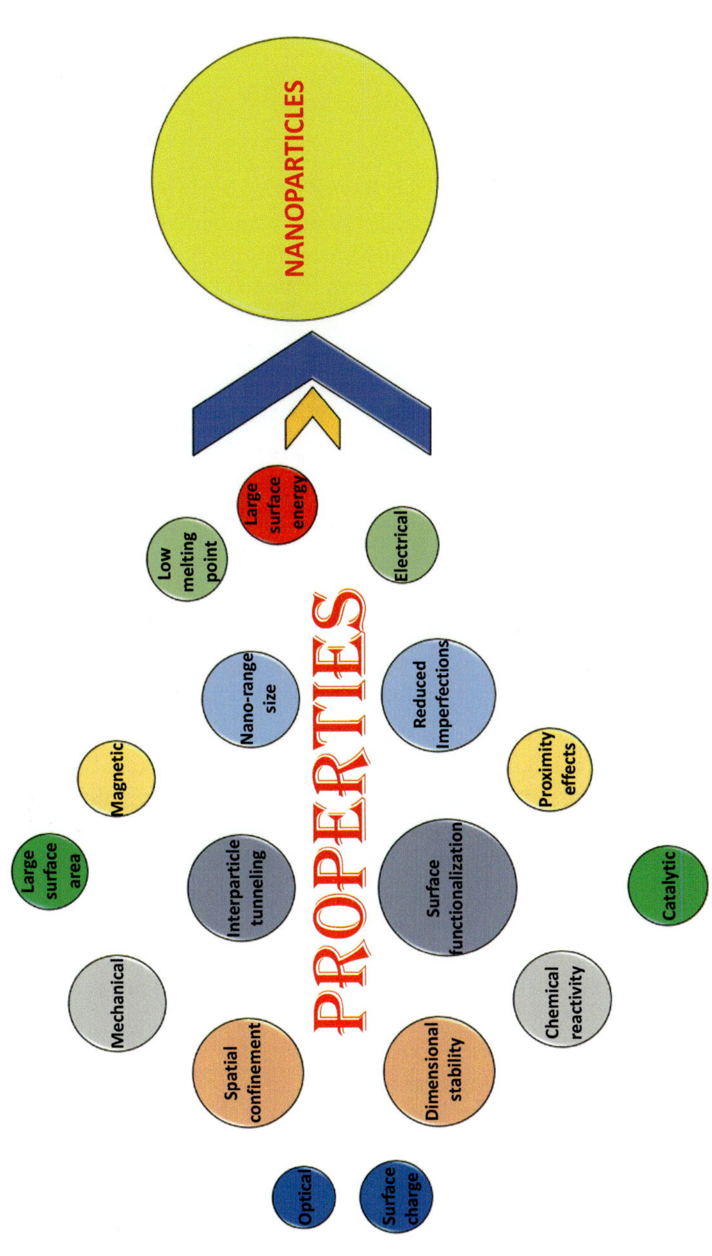

Fig. 5.2 Different properties of NPs

formation of reactive iron particles plume when mobile nZVI is incorporated in the groundwater. In the latter case, the formed plume has the tendency of destroying dissolved organic pollutants, which came into existence from the liquid source of dense non-aqueous phase present in the aquifer (Bernd 2010; Yunus et al. 2012).

5.4.1.2 Silver Nanoparticles (AgNPs)

AgNPs are known to have antimicrobial behavior towards a broad range of microbes present in water and are extensively used for water treatment. Due to their antifouling potential, many industries have installed a coating of AgNPs in their filtration membranes such as in poly sulfone membranes (Panahi et al. 2018). Many research reports also suggest AgNPs as a successful treatment for organic pollutants like 4-nitrophenol (Ismail et al. 2019). Different dyes such as crystal violet (Moghadas et al. 2020), eosin yellow, methylene blue, methyl orange, and many more are intrinsically reduced by the catalytic action of AgNPs (Marimuthu et al. 2020; Sherin et al. 2020). Many researchers are exploring the applications of AgNPs in the sensing sector. Rani et al. (2020) have utilized the AgNPs for the designing of electrochemical sensor which can detect the lowest possible concentration of a toxic compound of *p*-nitrotoluene from the water bodies (Rani et al. 2020).

5.4.1.3 Zinc Nanoparticles (ZnNPs)

For the reduction of many halogenated organic compounds, ZnNPs are always considered as an alternative to nZVI, because of their negative standard reduction potential (Bokare et al. 2013; Lu et al. 2016). Generally, ZnNPs exhibit a faster rate of degradation than their other counterparts. Dehalogenation reactions which are involved in the reduction of CCl_4 by ZnNPs are greatly influenced by the chemistry of water solutions. Complete degradation with faster rates has been achieved by ZnNPs (Tratnyek et al. 2010). A study conducted for achieving dechlorination of octachlorodibenzo-p-dioxin in aqueous solution at ambient conditions has revealed ZnNPs to be the first zero-valent metal NPs responsible for efficient degradation of this toxic compound into its lower chlorinated compounds (Bokare et al. 2013).

5.4.1.4 Gold Nanoparticles (AuNPs)

AuNPs have shown adsorption potential towards different organophosphate pesticides, serious pollutants, found in different water bodies (Das et al. 2009). AuNPs, when coated with different polyphenolic groups, can act as strong photocatalysts for a series of dyes like bromophenol blue, methylene blue, bromocresol green, and methyl orange (Choudhary et al. 2017; Kumar et al. 2019). Apart from this, many studies have reported photocatalytic interaction of AuNPs with phenol (Navalon et al. 2010), ciprofloxacin, and many more pharmaceutical products (Durán-Álvarez et al. 2016), released into the water bodies.

5.4.2 *Metal Oxide Nanoparticles*

5.4.2.1 Titanium Oxide Nanoparticles (TiO_2NPs)

TiO_2NPs are the most commonly used NPs of the semiconductor class because of their noteworthy properties like photocatalytic, chemical, biological and photostability, economic efficiency, and high reactivity in the UV range (Guesh et al. 2016). Due to their limited selectivity, they are found to be effective for a broad range of water contaminants such as heavy metals (Tahir et al. 2020), polycyclic aromatic hydrocarbons (Shaban 2019), pharmaceutical products (Mahmoud et al. 2017), pesticides (Surendra et al. 2020), dyes (Sathiyan et al. 2020), phenols (Zamri and Sapawe 2019), and cyanides (Mishra et al. 2019) along with pathogenic load of virus, harmful algae, and bacteria (Jaciw-Zurakowsky et al. 2020). The reactive oxygen species (ROS) generated on the surface of TiO_2NPs because of the formation of electron-hole pairs can degrade organic compounds in variant wavelengths of light (Budimir et al. 2020). TiO_2NPs crystallize the toxic forms of metals like Hg(II) and Ag(I) due to the generation of hydroxyl ions, when used as photocatalysts in the water (Prairie et al. 1993). The successful transformation of Cr(VI) to a less toxic form Cr(III) has been attained by these NPs in basic conditions. Food additives such as tartrazine dye undergo degradation when they come in contact with photocatalytic TiO_2NPs (Gupta et al. 2011). Many researchers support the utilization of TiO_2NPs for the eradication of pharmaceutical waste from the water bodies (Kanakaraju et al. 2014).

5.4.2.2 Zinc Oxide Nanoparticles (ZnONPs)

ZnONPs are known for their unique properties like high photocatalytic activity, direct and wide band gaps in the near UV range, and stronger oxidation potential, which renders their suitability for combating water pollution (Borysiewicz 2019). Moreover, their biocompatibility makes them one of the most efficient and demanded alternatives for being utilized in wastewater treatment (Schmidt-Mende and MacManus-Driscoll 2007). ZnONPs are well known for many plus points which involve low cost, considerable stability and reactivity, and larger intake capacity of solar spectra of wider ranges, when compared with NPs of other metal oxides (Yathisha and Nayaka 2020). ZnONPs have the potential of deactivating a wide range of pathogens found in polluted water, which makes them apt for the application of water disinfection (Gehrke et al. 2015; Wong et al. 2019). ZnONPs exhibit higher removal efficiencies of around 98% against Cu(II) released in the wastewater of electroplating industries (Rafiq et al. 2014). The toxic heavy metals like Cr(VI) are extensively eliminated from the water by getting adsorbed on the surface of these NPs (Kumar et al. 2013). The photocatalytic efficiency of ZnONPs has been reported in the UV range against dyes like methyl orange (Fu et al. 2011). The undulating fluorescent signals in the UV range after coming in contact with different chlorinated organic compounds reflect the competency of ZnONPs towards this class of water pollutants (Mohamed 2017).

5.4.2.3 Iron Oxide Nanoparticles (FeONPs+Fe_2O_3NPs + Fe_3O_4NPs)

FeONP-associated processes and properties like higher affinity towards pollutants, redox reactions, magnetic properties, and ion exchange open up the routes of their exploitation in the environmental field (Trujillo-Reyes et al. 2014). Due to large abundance, lower cost, easy accessibility, and simple synthesis, FeONPs are the preferential adsorbent for combating water pollution (Bagheri et al. 2020). FeONPs are often considered eco-friendly and biocompatible with very lower toxicity concerns associated with them (Jasrotia et al. 2020). These features aid their entry into the field of sensing and remediation of water pollutants. Nano iron hydroxide has been reported as an adsorbing substrate for arsenic present in aqueous solutions (Priyadarshni et al. 2020). Nano Fe_2O_3 and nano Fe_3O_4 are known for their superparamagnetic properties, which results in their large-scale exploitation as pollutant separators along with the additional advantage of their recovery under the influence of low-gradient magnetic field generated by them (Qu et al. 2013). From the very initial times of the nano introduction, Fe_2O_3NPs have been studied for their interaction with organic pollutants discharged in the water bodies via the phenomena of degradation, adsorption, and oxidation (Xu et al. 2019b). The removal of organic pollutants such as ethylene glycol, toluene, phenol, bisphenol A, benzene, and xylene has been successfully carried out at room temperature with the help of Fe_2O_3NPs (Yu et al. 2016). The photo-Fenton-carried degradation of dyes like black 5 (Miyashiro et al. 2021), remazol yellow (Bhuiyan et al. 2020), and rhodamine B along with other pollutants like '4-nitrophenol' has been investigated with these NPs (Piotrowski et al. 2019). Because of the magnetic properties of this class of NPs, they are easily separated after their course of action for the reusability purpose and for evading impending environmental risks. Fe_3O_4NPs have played a substantial role in the removal of organophosphorus pesticides (Fakhri et al. 2020) and polyaromatic hydrocarbons (Zhou et al. 2019). Mahanty et al. (2020) evaluated the efficacy of Fe_2O_3NPs up to five adsorption-desorption cycles against a list of heavy metals like Ni(II), Pb(II), Cu(II), and Zn(II) from aqueous solutions. Qiao et al. (2019) focused on the desired characteristics of FeONPs as magnetic oil sorbents with appreciable efficiencies. FeONPs are also involved in the removal of pharmaceutical waste like ofloxacin from the water bodies with the help of adsorptive mechanism of H-bonding and π-π electron donor-acceptor (Singh and Srivastava 2020). A study investigated not only the removal but also the efficient recovery of both the organic and inorganic species of selenium from the waste water with the help of van der Waals attractions and electrostatic interactions, respectively (Okonji et al. 2020).

Apart from these, Fe_2O_3NPs after the subjection to a certain type of treatment are efficient for the removal of radioactive materials like uranyl ions (Zhu et al. 2019).

5.4.2.4 Magnesium Oxide Nanoparticles (MgONPs)

MgONPs synthesized by different routes have the potential for the catalytic degradation of organic dyes (Nguyen et al. 2020). Sackey et al. (2020) showed the effective photocatalytic property of MgONPs against methylene blue under visible irradiation. Non-spontaneous adsorption of direct sky blue dye from the aqueous solutions has been recently reported because of these NPs (Noreen et al. 2020). MgO_2NPs are successfully applied for the removal of hydrocarbon compounds like naphthalene and toluene from the groundwater (Gholami et al. 2019). MgONPs can simultaneously act on pathogens and heavy metals like Cd and Pb to remove their considerate concentrations from the water bodies (Cai et al. 2017). The literature has demonstrated cubical- and spherical-structured MgONPs as good adsorbents with substantially higher adsorption capacities for pharmaceutical products like ciprofloxacin which are ended up in water bodies (Sivaselvam et al. 2020).

5.4.2.5 Copper Oxide Nanoparticles (CuONPs)

CuONPs are always an interesting object for combating water pollution, because of their distinctive physical and chemical properties, antimicrobial potential, and lower cost issues (Budimir et al. 2020). These NPs release small-sized copper ions into the contaminated water, which causes major disruptions in the internal structures of the bacteria after entering into their cell walls (Sunitha et al. 2020). In addition to this, CuONPs also alter the pH and conductivity of the surroundings, thus altering the chemical composition of the water. Large surface-to-volume ratio of the CuONPs provides proximity to cell membranes, which results in more intense interaction between them, thus displaying antimicrobial character in a concentration-dependent pattern towards a variety of bacteria like *S. aureus*, *E. coli*, etc. (Budimir et al. 2020). Various progressions in the photocatalytic activity of CuONPs towards pollutant degradation are very well jotted down by Raizada et al. (2020) in a review. CuONPs have applications in the treatment of different categories of pollutants such as dyes (including methyl red, methylene blue, rhodamine B, eosin yellow, methyl orange) (Sreeju et al. 2017; Rafique et al. 2020) and heavy metals (like As(III), Pb(II), Cr(VI)) (Mohan et al. 2015). Furthermore, they are also implied in the sensing field as successful sensors of phenol (Wahab et al. 2020) and as catalytic agents for the transformation of serious pollutants like 2-NP, 3-NP, and 4-NP into their analogous amino compounds within a time frame of 4–12 minutes (Sreeju et al. 2017).

5.4.2.6 Cerium Oxide Nanoparticles (CeONPs)

CeONPs are a part of emerging nanotechnological interventions for environmental applications like heterogeneous catalysis, pollutants adsorption, and contaminants sensing. Pharmaceutical wastes such as tetracycline are exothermically and spontaneously adsorbed by the cubical-structured CeONPs (Nurhasanah et al. 2020). Effective photodegradation of CeONPs is reported in literature against different dyes like congo red, methyl orange (Latha et al. 2018), rhodamine 6G (Calvache-Muñoz and Rodríguez-Páez 2020), reactive green 19, reactive yellow 81, reactive orange 84, and reactive violet 1 (Sane et al. 2018). Many reports have been published considering CeONPs as excellent adsorbents for the removal of toxic contaminants like As(III), As(V) (Feng et al. 2012), Cr(VI) (Sun et al. 2016), Pb(II), Cu(II) (Yari et al. 2015), and Hg(II) from the natural water bodies.

5.4.3 Carbon-Based Nanomaterials

5.4.3.1 Carbon Nanotubes (CNTs)

As the very name suggests, carbon nanotubes are the folded structures of carbon in the shape of a tube with size dimensions ranging in nanoscale. The inimitable behaviour of these nano-sized carbon tubes is ascribed to a list of properties like the accessibility of a larger number of adsorption sites, large surface area, high mechanical strength, remarkable electrical and thermal conductivities, and flexible surface chemistry (Alghuthaymi et al. 2020). These nanotubes can be further categorized into two types based on structural forms: single-walled carbon nanotubes and multi-walled carbon nanotubes. The former type is comprised of only a single layer of graphene folded in a cylindrical shape, whereas in the latter case, there is more than one layer with a length ranging from nano to a few microns. These CNTs are reported to be used as extracting agents for the leftovers of five N-methyl carbamate insecticides, namely, carbaryl, carbofuran, zectran, methiocarb, and aminocarb, present in different surface water bodies (Latrous El Atrache et al. 2016). CNTs are reported to have strong adsorption potential for a wide array of organic compounds and heavy metals because of their high sorption surface (Alghuthaymi et al. 2020; Verma and Balomajumder 2020). Multi-walled CNTs act as a competent and low-priced technique for the exclusion of 99.5% of unleaded gasoline hydrocarbon from the contaminated water surfaces (Lico et al. 2019). Different pollutants interact and are adsorbed differently on the internal hollow spaces of nanotubes, interstitial channels between nanotubes, outermost superficial grooves, or outside surfaces of nanotube aggregates (Panahi et al. 2018). Heavy metals like Pb(II) and Cd(II) show electrostatic interactions and chemical binding with the functional groups of the adsorption sites of CNTs, whereas organic pollutants present in the vicinity of CNTs show π-π interactions, hydrophobic effects, electrostatic attractions, hydrogen bonding, and covalent bonding (Panahi et al. 2018; Lu and Astruc 2020).

5.4.3.2 Carbon Quantum Dots (CQDs)

The attention of the scientific world has been drawn towards CQDs because of their non-toxicity and harmless green synthesis, which is further boosted up by their excellent optical and electrical properties (Si et al. 2020). CQDs have emerged as a promising alternative approach to semiconductor QDs for water remediation. These carbon-based nanomaterials are comprised of carbon NPs with a size range of less than 10 nm. CQDs are considered at par or superior to other available nano-remediation approaches because of their properties like cost-effectiveness, higher physical and chemical stability, stronger photoluminescence, good thermal and electrical conductivity, appreciable solubility in water, and lower toxicity levels (Lim et al. 2015). The higher number of ROS generation under the influence of visible light attributes a substantial amount of antibacterial potential to CQDs (Meziani et al. 2016). Degradation of dyes like rhodamine B and transformation of toxic species of heavy metals like Cr(VI) have been achieved with the participation of CQDs (Liu et al. 2020). Sensing and differentiation of four different antibiotics released in the water bodies, namely, tetracycline, metacycline, oxytetracycline, and doxycycline, have been achieved with the help of a fluorescence sensor fabricated by the conjunction of two CQDs (Xu et al. 2020).

5.4.3.3 Graphene-Based Nanomaterials

Graphene and related materials are the newest addition in the carbon family and are applied for different applications because of their unique morphology, chemical structure, and electronic properties. To meet the ever-increasing demands of clean water, graphene-based nanomaterials are intensively dedicated to the prevention, control, and reduction of water pollution. So, they have been focused on different organic materials, heavy metal ions, and inorganic pollutants in water with the help of strong π-π interactions (Lu and Astruc 2020). Graphene oxides and sheets are emerging as competent agents for the removal of antibiotics, a class of pharmaceutical waste, with the help of adsorption or oxidation processes. A summarized report on graphene oxide and related nanomaterials towards the adsorption of antibiotics like tetracycline, sulfamethoxazole, and ciprofloxacin, analgesics like diclofenac, β-blockers like atenolol and propranolol and pharmaceutically active compounds like carbamazepine has been provided by Khan et al. (2017). Many advanced graphene structures have a removal potential against a series of toxic compounds including dyes, pesticides, phenols, aromatic polar and non-polar compounds, halogenated compounds, and nitro compounds which are present in aqueous solutions (Mandeep and Kakkar 2020). Yuan et al. (2013) reported removal of ions of Cr(VI), Cu(II), Pb(II), Fe(III), and Zn(II) with the help of modified graphene. In a study, Tran et al. (2020) made a successful attempt for the removal of anionic (crystal violet), cationic (congo red), and neutral (methyl red) dyes from the wastewater using composites of graphene. The parameters of graphene sheets like roughness, surface functionalization, charge, orientation, size, and wettability decide the extent of adhesion and interaction of bacteria with these sheets (Szunerits and Boukherroub 2016). The antibacterial nature of graphene oxide sheets towards different strains of bacteria is attributed to the mechanisms like membrane rupture with sharp edges of nanosheets, oxidative stress, and membrane wrapping (Perreault et al. 2015).

5.4.3.4 Fullerenes

Fullerenes are carbon compounds with hydrophobic surfaces, higher electron affinity, and surface-to-volume ratios. The structural deficits on the superficial areas of fullerenes aid them in adsorbing different pollutants from aqueous solutions. They showed an adsorbing tendency towards different organic compounds, but the most fascinating results were obtained for the organometallic class (Panahi et al. 2018). Owing to the internal and external coverage of π electrons, these carbon compounds interact strongly with organic molecules (Lucena et al. 2011). Nanocrystalline aggregates of C60 have antimicrobial behaviour towards bacteria (Azizi-Lalabadi et al. 2020), whereas amino fullerenes show activity against viruses and are reported for degrading pharmaceutical wastes (Elessawy et al. 2020). Fullerenes are always looked upon as an attractive solution for wastewater treatment because of their lower affinity to aggregation and larger surface area. Fullerenes differ from other carbon-related nanomaterials because of the extensive adsorption sites availability. They can adsorb pollutants at the superficial layers or in the channels or between the masses of fullerenes (Bagheri et al. 2020).

5.4.4 Nanocomposites

To overcome the drawbacks associated with each one of the nanomaterials, researchers have come up with the strategy of fabricating nanocomposites for the treatment of contaminated water. Because of the enhanced efficiency and limited or no drawbacks, nanocomposites are emerging in the field of nanoscience as the most active subject of the current times. Bentonite-hydroxyapatite nanocomposite is evaluated for the significant removal rates of Pb(II) and Ni(II) and the eradication of residues of fungicides present in aqueous solutions (El-Nagar et al. 2020). Nanocomposites are capable of removing more than one pollutant at the same time and with much greater efficiency. One such nanocomposite fabricated with the cross-linking of iron oxide and sodium alginate was confirmed as an excellent nanosorbent for the simultaneous removal of different heavy metals like Pb, Cd, and Cu from the wastewater (Mahmoud et al. 2020). Oyewo et al. (2020) summarized the removal and degradation of different heavy metals and dyes discharged in surface water bodies with the help of different nanocomposites designed by combining different metal oxides with cellulose. Magnetic nanocomposite of Fe_3O_4 and Cr_2O_3 is competent of degrading 4-chlorophenol followed by its magnetic separation from the pollutant for the recycling of the nanocomposite (Singh et al. 2017). Biogenic Fe_3O_4/Au nanocomposite performs photocatalytic degradation of drugs like imatinib and imipenem. Additionally, it is successfully evaluated for the removal of three human pathogens, thus expressing its antimicrobial character as well (Mirsadeghi et al. 2020). The functionalized graphene/sericin composites perform exceptionally great rates of phenolic compounds decontamination from water due to the presence of chemically active functional groups on its surface (Abdolmaleki and Mahmoudian 2020).

5.4.5 Nanomembranes

Membrane-based decontamination is a rapidly evolving innovation in nanotechnology wherein nano-sized formulations are explored for the treatment of polluted water. Sizes of the membrane pore and contaminant molecules are the deciding features for the successful removal of different pollutants by providing a physical barrier. The efficacy of the nano-sized membrane is governed by polarity, weight, molecular structure, hydrophobicity, hydrophilicity, and pore size. These membranes prohibit the contaminants of size dimensions less than 1 nm, and the mechanism of charge-carried repulsions is followed for the separation of different ions. These nanomembranes because of their amazing permeation properties and characteristics of being antibacterial, antifouling, and photocatalytic are exploited as ultrafast and selective filtration units for water purification (Ying et al. 2017). Nanomembranes studded with hydrophilic NPs develop resistance towards many pathogens and organic compounds (Gehrke et al. 2015). These membranes can be grouped under two categories, namely, inorganic nanomembranes and organic nanomembranes (Lu and Astruc 2020). The incorporation of organic polymers with photocatalytic NPs exhibits the fabrication of new membranes with enhanced degradation rates for the organic pollutants dispersed in water. Two hydrophilic membranes and one hydrophobic membrane were fabricated using TiO_2NP-impregnated polyethersulfone and polyvinylidene fluoride (PVDF) in the former case and only PVDF in the latter case. Hydrophobic membranes were efficient photocatalysts for methylene blue dye, whereas hydrophilic membranes showed degradation of pharmaceutical wastes such as ibuprofen and diclofenac (Fischer et al. 2015). Successful degradation of organic pollutants such as humic acid and cyclohexanoic acid is performed with the integration of Fe_2O_3NPs and CNTs with PVDF membrane (Alpatova et al. 2015).

5.5 Ecotoxicological Risks Associated with the Nanomaterials

The toxicity issues associated with different nanomaterials rest on the parameters such as origin, life, dose, dimensions, surface chemistry, surface area, surface coating, crystallinity, aspect ratio, and aggregation tendency of these nanomaterials (Saleh 2020). NPs after entering into the bodies of the aquatic organisms via direct or indirect interactions (Fig. 5.3) get accumulated into different tissues and organs and thus express their toxicity in many forms such as apoptosis, ROS production, oxidative stress, and death. The toxic effects of NPs in aquatic ecosystems are highly dependent on their mobility and bioavailability. Many reports have thrown light on the toxic impacts of NPs on different aquatic flora and fauna comprising bacteria, algae, plants, cladocera, oligochaete, cnidarian, and fishes. The ecotoxicological impacts of the deliberate and non-deliberate introduction of NPs in the aquatic environment can be studied under the following sub-sections.

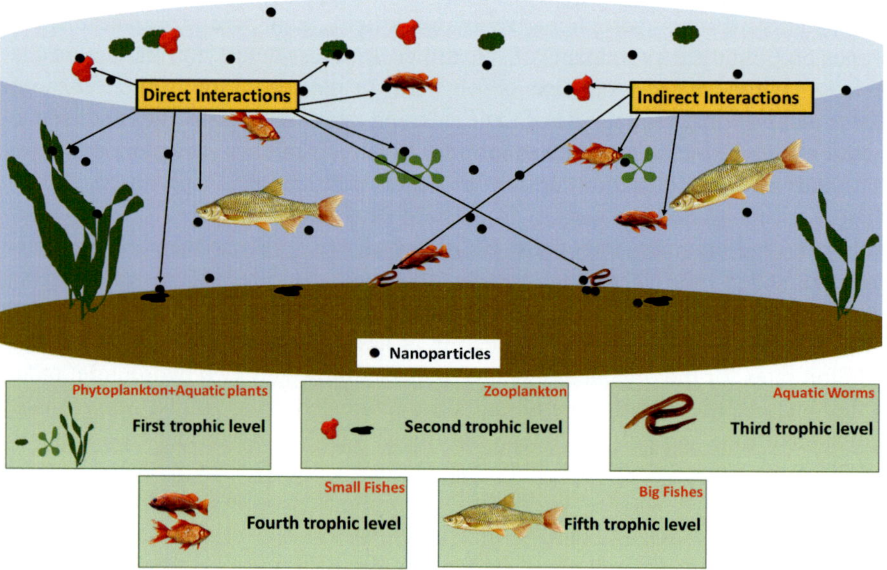

Fig. 5.3 Interaction of NPs with the different trophic levels of aquatic ecosystem

5.5.1 Effects on Aquatic Autotrophs

5.5.1.1 Toxic Impacts on Phytoplankton

Phytoplanktons are the primary producers of the aquatic ecosystems, and their community structure and functions are greatly affected by the presence of nanospecies of the metals. Since marine phytoplanktons occupy the base of a trophic level of food chains and food webs, thus they are more structurally, reproductively, and metabolically susceptible to different nanomaterials. In a study reported by Dauda et al. (2020), significant alterations were obtained in the community structure and dynamics of phytoplankton in the presence of TiO_2NPs. NPs either directly or indirectly impart toxic responses to phytoplankton by the release of ionic species of toxic metals or by generating ROS in the cellular bodies of the organisms (Dauda et al. 2020). The disproportionate levels of ROS attack and mutilate proteins, lipids, and DNA assembly of the cells. These nanomaterials induce mutations or completely degrade the DNA after forming complexes with them. When they get associated with proteins of the cells, serious alterations are noticed in the proteins' structures and functions. Cr_2O_3NPs reported the impairment of the photosystem II in the chlorophyll-containing organelles of *Chlamydomonas reinhardtii* (Dauda et al. 2020). The activity of oxidative stress enzymes is greatly altered by the ZnONPs in the species of *Chlorella vulgaris* (Suman et al. 2015). CuONPs in the presence of UV light express the synergistic response in the form of cell membrane destruction and oxidative stress mismanagement (Cheloni et al. 2016). In a study, the probability of DNA destruction is reflected in the *Dunaliella tertiolecta*, a diatom, when provided with the exposure of TiO_2NPs, SiO_2NPs, and ZnONPs. TiO_2NPs activate the signals needed for cell wall destruction, SiO_2NPs play the role in excessive generation of ROS, whereas ZnONPs induce fragmentation of DNA strands (Cheloni et al. 2016). A significant decline in the photosynthetic efficiency of *Isochrysis galbana*, a prymnesiophyte phytoplankton species, is seen as a physiological endpoint for the concerned aquatic photoautotrophs by the population-level effect of four types of NPs, namely, ZnONPs, AgNPs, CeONPs, and CuONPs (Miller et al. 2017). AgNPs and CeO_2NPs alter the structural, reproductive, and physiological processes of *D. salina* and *C. autotrophica* by causing a decline in chlorophyll content and quantum yield of photosystem II and cell densities and enhancing the cell complexities and generation of ROS in the internal environment of cellular bodies (Sendra et al. 2018). NPs like α-Fe_2O_3NPs and γ-Fe_2O_3NPs have adverse effects on the species of both types of phytoplankton, i.e. freshwater phytoplankton like *Selenastrum capricornutum* and marine phytoplankton like *Nannochloropsis oculata* (Ates et al. 2020).

5.5.1.2 Toxic Impacts on Cyanobacteria

These photosynthesizing single-celled or colony-forming organisms are an integral part of aquatic ecosystems. Generally, they are more prevalent in the form of blooms in freshwater ecosystems but are also found in marine habitats. Investigations are made to establish toxicity concerns of citrate-stabilized AgNPs towards *Prochlorococcus,* a marine dominant cyanobacteria. The detrimental results obtained in the NP-exposed community are largely driven by the generation of superoxides and by the leaching of silver ions from the reduced AgNPs (Dedman et al. 2020). *Synechococcus elongatus*, a species of cyanobacteria, exhibits remarkably high sensitivity towards graphene oxides (Malina et al. 2019). The toxic behaviour of AgNPs towards *Microcystis aeruginosa*, a bloom-forming strain of cyanobacteria, is confirmed by the captured TEM and SEM images of the shrunk and damaged cell walls. The inhibitory effect of around 99% is obtained, resulting in the reduction of cell growth, morphological changes, and changes in cell density (Duong et al. 2016). The toxicity of AgNPs towards cyanobacteria is reflected by their negative impacts on the viability of the member of cyanobacterial community, i.e. *Synechococcus leopoliensis* (Taylor et al. 2016).

5.5.1.3 Toxic Impacts on Aquatic Plants

Aquatic vascular plants comprising of different categories of plants like free-floating, rooted, emergent, and submerged play a key role as producers in ecological energy metabolism. The toxicity of different nanomaterials to aquatic plants like macrophytes has been reported in the literature (Blinova et al. 2020). The toxicity signs are noticed in AgNP-exposed *Halophila stipulacea* (a sea grass) in the form of actin filament disturbances, affected microtubules, undulations in H_2O_2 levels, death of epidermal cells, inhibition in cell elongation, disappearance of grainy structures and massive aggregations in meristematic cells containing endoplasmic reticulum, elevated malondialdehyde content, and enhanced activity of superoxide dismutase and peroxidase, thus initiating oxidative damages and activation of antioxidant defence mechanism (Mylona et al. 2020). AgNPs exhibit toxic impacts on *Lemna gibba* by targeting its physiological state. These NPs cause serious reductions in the aquatic plant's biomass because of drop in photosynthetic activity by inhibiting the transfer of energy to the reaction centres from the light-harvesting complexes, negatively influencing the water splitting system of PS-II, or by the complete inactivation of photosynthetic reaction centres (Dewez et al. 2018). Another plant from the same family, *Lemna minor* when exposed to AgNPs shows increasing rates of chlorosis, decrease in growth rate and fronds per colony, and unregulated enzymatic activity of guaiacol peroxidase and glutathione-S-transferase because of the induction of oxidative stress in the cells of the aquatic plant (Pereira et al. 2018). CuONPs inhibit the PS-II and thus affect the energy fluxes undergoing in photosynthetic membranes of *L. gibba* as a response to the release of Cu ions from the CuONPs (Perreault et al. 2014). A significant reduction in the reducing ratios of chlorophyll a to phaeophytin is observed in *Spirodela polyrhiza* when exposed to ZnONPs (Hu et al. 2013). AgNPs synthesized by chemical routes get accumulated in the tissues of water hyacinth, thus inhibiting its growth (Rani et al. 2016).

5.5.2 *Effects on Aquatic Heterotrophs*

5.5.2.1 Toxic Impacts on Zooplanktons

The shape, size, and chemical properties of NPs affect the structural and functional composition of the zooplanktons. NPs not only carry toxic characters but also influence and increase the toxicity of co-occurring pollutants in the aquatic environment. TiO_2NPs in addition to the toxicity issues of their own also increase the negative impacts of polycyclic aromatic hydrocarbons like phenanthrene and heavy metals like Cd(II). Zooplanktons like *Artemia salina* bioaccumulate large quantities of these pollutants when they come in the vicinity of TiO_2NPs. These TiO_2NPs govern the concentration-dependent mechanisms for the adsorption of hydrocarbons and heavy metals to impart toxic effects towards the zooplanktons (Lu et al. 2018). A high dose of AgNPs for shorter durations reduces abundance, species richness, and biomass of the zooplanktons (Vincent et al. 2017). Studies show the bioaccumulation, biomagnification, and thus elevated toxicity of AgNPs in the water flea (*Moina macrocopa*) (Yoo-iam et al. 2014). Ahmed et al. (2021) provided a detailed insight into toxicity concerns of differently shaped Cu_2ONPs on the multi-generations of *Daphnia magna* in the aquatic ecosystem. Cu_2ONPs impart long-lasting impacts on the body structures of *D. magna* and cause increased Cu accumulation in their offsprings. The negative effects on the body length and reproduction rate of *D. magna* are because of the dissolution of CuONPs in water, resulting in the formation of Cu ions in the exposure medium. ZnONPs are found to be toxic for the marine primary consumers like *A. salina* and *D. magna*. Even the small concentrations of these NPs can exhibit lethal results on the organisms. The mortality rate is parallel with the increasing concentrations and time periods (Danabas et al. 2020). When daphnids are exposed to La_2O_3NPs, serious accumulations are seen in their intestines which induces many toxic impacts and can even lead to mortality within 48 h of exposure (Balusamy et al. 2015). Graphene oxide leads to a shift in the energy paradigm of *Ceriodaphnia dubia* from feeding and reproductive incentives to self-maintenance. Prolonged doses of these nanomaterials cause a noteworthy change in the feeding patterns, reduction in neonates, and ROS generation (Souza et al. 2018).

5.5.2.2 Toxic Impacts on Invertebrates

Caixeta et al. (2020) summarized the toxicity details of different nanomaterials on the diverse species of snails which includes the bioaccumulation and biomagnification of NPs, reproductive and transgenerational impacts, genotoxicity, and embryotoxicity. Short-term exposures of graphene oxides are capable of inducing significant changes in the eastern oyster like reduction in total protein contents, increased lipid peroxidation, epithelial inflammation, and oxidative stress (Khan et al. 2019a). The presence of graphene oxide in the *Crassostrea virginica* causes alterations in the detoxifying enzyme 'glutathione-s-transferase' present in the digestive glands and gills of the animal. It leads to the generation of stress signals which in turn start the uncontrolled generation of ROS, thus causing oxidative damage in the system of the concerned invertebrate. Further, the resulting elevated levels of lipid peroxidase cause cellular damage in the organisms (Khan et al. 2019b). Nano oxides of the graphene induce destructive changes in the physiological feature of regeneration in *Diopatra neapolitana.* Its exposure results in delayed regenerative capacity, decreased metabolism, and significant cellular damage (Marchi et al. 2017). Graphene oxide present in water bodies aggravates the toxicity of co-contaminants like Cd and Zn and thus compromises the routine metabolism of the shrimp '*Palaemon pandaliformis*' (Melo et al. 2019). Fullerenes in water bodies impair the life cycle parameters like breeding of benthic invertebrate *Chironomus riparius* by delaying the emergence time of females (Waissi et al. 2017).

5.5.2.3 Toxic Impacts on Vertebrates

Numerous reports are depicting the toxicity of ZnONPs towards the vertebrates of the aquatic environment. As per a study conducted on the zebrafish exposed with ZnONPs, inhibiting effects on the hatching, induction of metallothionein, variations in the expressions of target genes, and changes in the transcription of pro-inflammatory cytokines were noticed. These transcriptional alterations caused significant toxic impacts on the embryos of zebrafish (Brun et al. 2014). Exposure of AgNPs on the biological system of *Danio rerio* is studied wherein toxicological end points are attained in the form of abnormal hatching rates and heart rates and mortality (Cunningham et al. 2013). Significant toxic impacts such as DNA damage, erythrocytes compositional and functional disruption, nuclear abnormalities like micronuclei formation, generation of oxidative stress, accumulations in the brain and liver, and reticence of antioxidant system are observed in the juveniles of *Ctenopharyngodon idella* which are fed with the polystyrene NP-contained water. Even the nanogram concentrations of these NPs raise cytotoxic, mutagenic, and genotoxic concerns in these organisms (Guimarães et al. 2021). Carbon nanofibres accumulate in the biomass of *Nile tilapia* via the food chain wherein it imparts severe cytotoxic effects. These effects are in association with occurrence rates of erythrocytes nuclear aberrations like kidney-shaped micro-nucleation, constricted nucleated erythrocytes, small-sized nuclei, and a lower ratio of nuclear/cytoplasmic area. The causative agent of mutagenic and cytotoxic changes is transferred from the lower trophic levels of the food chain (Gomes et al. 2021). Exposure of multi-walled CNTs to sexually mature *Xenopus tropicalis* causes toxicological changes in the growth and reproductive phases of their life cycles. The share of inhaled CNTs accumulates in their lungs to cause lung cannons. Reports suggest significant changes in the structure and diversity of microbial communities residing inside their guts. Apart from these, chronic exposures of CNTs raise the potentiality of reproductive abnormalities such as inhibition to the formation of spermatogonia and oocytes along with the decreased cases of fertilization and survival rates of their embryos (Zhao et al. 2020). The inhalation of ZnONP formulation induces cardio-respiratory and metabolic disorders in the white sucker. Gill tissues of *Catostomus commersonii* provide the shreds of evidence of cellular level and oxidative stress involving elevation in levels of malondialdehyde, increase in activity of caspase 3/7, and expression of heat shock protein. The acute exposure of ZnONPs upsurges the parasympathetic nervous signals which in turns reduces the heart rate by around 35%, thereby decreasing energy demand of the tissues. The resultant enhancement in the permeability of epithelial cells and structural remodelling causes a threefold increase in the activity of gill Na^+/K^+-ATPase. Thus a whole new animal hypoxic response is generated to the cellular stress and tissue degeneration caused by the exposure of ZnONPs (Bessemer et al. 2015). Stability of the genomic template of *Dicentrarchus labrax* is highly hampered in the form of oxidative DNA damage by the presence of TiO_2NPs (Vannuccini et al. 2015), whereas these NPs induce immunotoxic effects in the *Pimephales promelas* by disturbing the bactericidal properties of the neutrophils (Jovanović et al. 2015).

5.6 Conclusion

Water environments served the purpose of an ultimate sink of anthropogenically generated waste products and end products of divergent utilities. Due to the pernicious fate of these pollutants in the water bodies, the recent advancements are emanated in the nano-field. Nanomaterials provide a more effective and on-site detection, monitoring, and removal of these contaminants with remarkable results. But the ongoing investigations revealed the associated toxic temperament of these nanomaterials on the aquatic environment and its inhabitants. The summarized reports in this chapter show both the phases of nanomaterials, i.e. nanomaterials as a tool for combating water pollution and ecotoxicological effects of nanomaterials on the quality and life of different trophic levels of water organisms. So, while keeping in view the negative aspects associated with the use of nanomaterials for environmental amelioration (Fig. 5.4), there is an urgent need for designing nano entities with their complete life cycle analysis beforehand. The engineering of nanomaterials with environmental frames help in better understanding of their future transformations in environmental matrices so that minimum negative impacts are imparted to the living organisms.

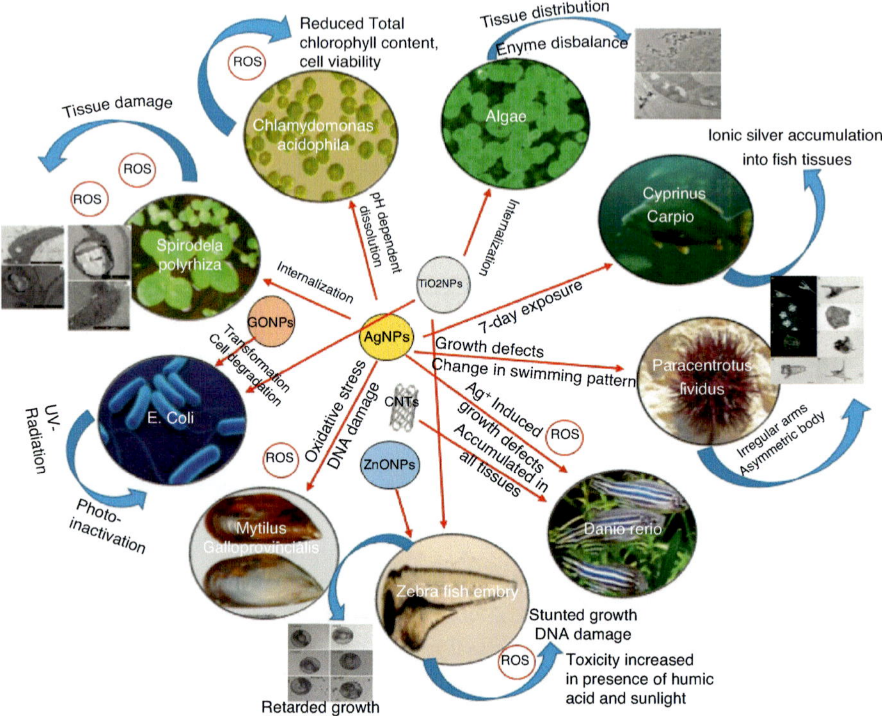

Fig. 5.4 Toxic effects of nanomaterials on different aquatic organisms. (This image is taken from a review by Jahan et al. (2017))

Multiple-Choice Questions

1. Which of the following contaminants can be efficiently treated with nanomaterials?
   (a) Pesticides
   (b) Heavy metals
   (c) Dyes
   (d) All of these
2. Nanomaterials should have 1nm to 100 nm size range in:
   (a) 1 dimension
   (b) 2 dimensions
   (c) 3 dimensions
   (d) All of these

3. 'Nanomaterials are the materials with any external nanoscale dimension or having internal nanoscale structure' is the definition of nanomaterials as per:

   (a) International Organization for Standardization
   (b) European Union
   (c) Both a and b
   (d) None of these

4. Which properties of the nanomaterials are governed by the effects of quantum confinement?

   (a) Optical
   (b) Structural
   (c) Electronic
   (d) All of these

5. Nanomaterials proved as a failure for which of the following solutions?

   (a) Desalination of seawater
   (b) Reduction of toxicants
   (c) Improvement of water quality parameters
   (d) None of these

6. Which of the following is a property of nanoparticles?

   (a) Less surface area
   (b) Tendency of electron transfer
   (c) Non-catalytic nature
   (d) Chemical instability

7. What is the value of standard redox potential for the elemental form of nano zero-valent iron?

   (a) 0.440 V
   (b) 1 V
   (c) 0 V
   (d) 0.34 V

8. nZVI are found to be successful for treating which of the following contaminants from groundwater pollution?

   (a) Chlorinated hydrocarbons
   (b) Halogenated organic pollutants
   (c) Perchlorates
   (d) All of these

Answers

1. (d)
2. (d)
3. (a)
4. (d)
5. (d)
6. (b)
7. (a)
8. (d)

Short Questions

1. How does nZVI interact with groundwater pollutants?

   nZVI interacts with groundwater pollutant either by directly being injected in stabilized form in the underwater aquifers wherein they get adsorbed on the surfaces of aquifer solids after making a zone of iron particles or by the formation of reactive iron particles plume when mobile nZVI is incorporated in the groundwater. In the latter case, the formed plume has the tendency of destroying dissolved organic pollutants, which came into existence from the liquid source of dense non-aqueous phase present in the aquifer.

2. Write a short note on carbon nanotubes.

   Carbon nanotubes are the folded structures of carbon in the shape of a tube with size dimensions ranging in nanoscale. The inimitable behaviour of these nano-sized carbon tubes is ascribed to a list of properties like accessibility of larger number of adsorption sites, large surface area, high mechanical strength, remarkable electrical and thermal conductivities, and flexible surface chemistry. These nanotubes can be further categorized into two types based on structural forms: single-walled carbon nanotubes and multi-walled carbon nanotubes. The former type is comprised of only a single layer of graphene folded in a cylindrical shape, whereas in the latter case, there are more than one layer with length ranging from nano to few microns.

3. What are carbon quantum dots (CQDs)?

   CQDs are emerging as a promising alternative approach to semiconductor quantum dots for water remediation because of their excellent optical and electrical properties. These carbon-based nanomaterials are comprised of carbon nanoparticles with size range less than 10 nm. CQDs are considered at par or superior to other available nano-remediation approaches because of their properties like cost-effectiveness, higher physical and chemical stability, stronger photoluminescence, good thermal and electrical conductivity, appreciable solubility in water, and lower toxicity levels.

4. How do nanoparticles impart toxicity to phytoplanktons?

   Nanoparticles either directly or indirectly impart toxic responses to phytoplanktons by the release of ionic species of toxic metals or by generating reactive oxygen species (ROS) in the cellular bodies of the organisms. The disproportionate levels of ROS attack and mutilate proteins, lipids and DNA assembly of the cells. These nanomaterials induce mutations or completely degrade the DNA after forming complexes with them. When they get associated with proteins of the cells, serious alterations are noticed in the protein structures and functions.

5. What are the toxicity signs induced by the silver nanoparticles on the *Halophila stipulacea*?

   The toxicity signs are noticed in silver nanoparticle-exposed *Halophila stipulacea* in the form of actin filament disturbances, effected microtubules, undulations in H_2O_2 levels, death of epidermal cells, inhibition in cell elongation, disappearance of grainy structures and massive aggregations in meristematic cells containing endoplasmic reticulum, elevated malondialdehyde content, and enhanced activity of superoxide dismutase and peroxidase, thus initiating oxidative damages and activation of antioxidant defence mechanism.

References

Abdolmaleki A, Mahmoudian M (2020) Use of biomass sericin as matrices in functionalized graphene/sericin nanocomposites for the removal of phenolic compounds. Heliyon 6:e04955

Ahmed I, Zhang B, Muneeb-Ur-Rehman M, Fan W (2021) Effect of different shapes of Nano-Cu_2O and humic acid on two-generations of *Daphnia magna*. Ecotoxicol Environ Saf 207:111274

Alghuthaymi M, Asran-Amal, Mostafa M, Abd-Elsalam KA (2020) Carbon nanotubes: an efficient sorbent for herbicide sensing and remediation. In: Carbon nanomaterials for agri-food and environmental applications. Elsevier, Amsterdam, pp. 429–457

Alpatova A, Meshref M, McPhedran KN, Gamal El-Din M (2015) Composite polyvinylidene fluoride (PVDF) membrane impregnated with Fe_2O_3 nanoparticles and multiwalled carbon nanotubes for catalytic degradation of organic contaminants. J Memb Sci 490:227–235

Archer E, Petrie B, Kasprzyk-Hordern B, Wolfaardt GM (2017) The fate of pharmaceuticals and personal care products (PPCPs), endocrine disrupting contaminants (EDCs), metabolites and illicit drugs in a WWTW and environmental waters. Chemosphere 174:437–446

Aredes S, Klein B, Pawlik M (2012) The removal of arsenic from water using natural iron oxide minerals. J Clean Prod 29–30:208–213

Ates M, Cimen ICC, Unal I, Kutlu B, Ertit Tastan B, Danabas D, Aksu O, Arslan Z (2020) Assessment of impact of α-Fe_2O_3 and γ-Fe_2O_3 nanoparticles on phytoplankton species *Selenastrum capricornutum* and *Nannochloropsis oculata*. Environ Toxicol 35:385–394

Azizi-Lalabadi M, Hashemi H, Feng J, Jafari SM (2020) Carbon nanomaterials against pathogens; the antimicrobial activity of carbon nanotubes, graphene/graphene oxide, fullerenes, and their nanocomposites. Adv Colloid Interf Sci 284:102250

Bagheri H, Fakhri H, Ghahremani R, Karimi M, Madrakian T, Afkhami A (2020) Nanomaterial-based adsorbents for wastewater treatment. In: Smart nanocontainers. Elsevier, pp. 467–485

Balusamy B, Taştan BE, Ergen SF, Uyar T, Tekinay T (2015) Toxicity of lanthanum oxide (La_2O_3) nanoparticles in aquatic environments. Environ Sci Process Impacts 17:1265–1270

Bernd N (2010) Pollution prevention and treatment using nanotechnology. In: Nanotechnology. Wiley-VCH Verlag GmbH & Co. KGaA,Weinheim. https://doi.org/10.1002/9783527628155.nanotech010

Bessemer RA, Butler KMA, Tunnah L, Callaghan NI, Rundle A, Currie S, Dieni CA, MacCormack TJ (2015) Cardiorespiratory toxicity of environmentally relevant zinc oxide nanoparticles in the freshwater fish *Catostomus commersonii*. Nanotoxicology 9:861–870

Bharathi SKV, Murugesan P, Moses JA, Anandharamakrishnan C (2021) Recent trends in nanocomposite packaging materials. In: Innovative food processing technologies. Elsevier, Amsterdam, pp 731–755

Bhuiyan MSH, Miah MY, Paul SC, Das AT, Saha O, Rahaman MM, Sharif MJI, Habiba O, Ashaduzzaman M (2020) Green synthesis of iron oxide nanoparticle using *Carica papaya* leaf extract: application for photocatalytic degradation of remazol yellow RR dye and antibacterial activity. Heliyon 6:e04603

Bishop EJ, Fowler DE, Skluzacek JM, Seibel E, Mallouk TE (2010) Anionic homopolymers efficiently target zerovalent iron particles to hydrophobic contaminants in sand columns. Environ Sci Technol 44:9069–9074

Blinova I, Muna M, Heinlaan M, Lukjanova A, Kahru A (2020) Potential Hazard of lanthanides and lanthanide-based nanoparticles to aquatic ecosystems: data gaps, challenges and future research needs derived from bibliometric analysis. Nano 10:328

Bokare V, Jung J, Chang Y-Y, Chang Y-S (2013) Reductive dechlorination of octachlorodibenzo-p-dioxin by nanosized zero-valent zinc: modeling of rate kinetics and congener profile. J Hazard Mater 250–251:397–402

Borysiewicz MA (2019) ZnO as a functional material, a review. Crystals 9:505

Brun NR, Lenz M, Wehrli B, Fent K (2014) Comparative effects of zinc oxide nanoparticles and dissolved zinc on zebrafish embryos and eleuthero-embryos: importance of zinc ions. Sci Total Environ 476–477:657–666

Budimir M, Szunerits S, Markovic Z, Boukherroub R (2020) Nanoscale materials for the treatment of water contaminated by bacteria and viruses. In: Nanomaterials for sustainable energy and environmental remediation. Elsevier, Amsterdam, pp 261–305

Burakov AE, Galunin EV, Burakova IV, Kucherova AE, Agarwal S, Tkachev AG, Gupta VK (2018) Adsorption of heavy metals on conventional and nanostructured materials for wastewater treatment purposes: a review. Ecotoxicol Environ Saf 148:702–712

Cai Y, Li C, Wu D, Wang W, Tan F, Wang X, Wong PK, Qiao X (2017) Highly active MgO nanoparticles for simultaneous bacterial inactivation and heavy metal removal from aqueous solution. Chem Eng J 312:158–166

Caixeta MB, Araújo PS, Gonçalves BB, Silva LD, Grano-Maldonado MI, Rocha TL (2020) Toxicity of engineered nanomaterials to aquatic and land snails: a scientometric and systematic review. Chemosphere 260:127654

Calvache-Muñoz J, Rodríguez-Páez JE (2020) Removal of Rhodamine 6G in the absence of UV radiation using ceria nanoparticles (CeO_2-NPs). J Environ Chem Eng 8:103518

Cheloni G, Marti E, Slaveykova VI (2016) Interactive effects of copper oxide nanoparticles and light to green alga *Chlamydomonas reinhardtii*. Aquat Toxicol 170:120–128

Choudhary BC, Paul D, Gupta T, Tetgure SR, Garole VJ, Borse AU, Garole DJ (2017) Photocatalytic reduction of organic pollutant under visible light by green route synthesized gold nanoparticles. J Environ Sci 55:236–246

Cunningham S, Brennan-Fournet ME, Ledwith D, Byrnes L, Joshi L (2013) Effect of nanoparticle stabilization and physicochemical properties on exposure outcome: acute toxicity of silver nanoparticle preparations in zebrafish (*Danio rerio*). Environ Sci Technol 47:3883–3892

Danabas D, Ates M, Ertit Tastan B, Cicek Cimen IC, Unal I, Aksu O, Kutlu B (2020) Effects of Zn and ZnO nanoparticles on *Artemia salina* and *Daphnia magna* organisms: toxicity, accumulation and elimination. Sci Total Environ 711:134869

Das SK, Das AR, Guha AK (2009) Gold nanoparticles: microbial synthesis and application in water hygiene management. Langmuir 25:8192–8199

Dauda S, Gabriel AM, Idris OF, Chia MA (2020) Combined nanoTiO_2 and nitrogen effects on phytoplankton: a mesocosm approach. J Appl Phycol 32:3123–3132

Dedman CJ, Newson GC, Davies G-L, Christie-Oleza JA (2020) Mechanisms of silver nanoparticle toxicity on the marine cyanobacterium *Prochlorococcus* under environmentally-relevant conditions. Sci Total Environ 747:141229

Dewez D, Goltsev V, Kalaji HM, Oukarroum A (2018) Inhibitory effects of silver nanoparticles on photosystem II performance in *Lemna gibba* probed by chlorophyll fluorescence. Curr Plant Biol 16:15–21

Duong TT, Le TS, Tran TTH, Nguyen TK, Ho CT, Dao TH, Le TPQ, Nguyen HC, Dang DK, Le TTH, Ha PT (2016) Inhibition effect of engineered silver nanoparticles to bloom forming cyanobacteria. Adv Nat Sci Nanosci Nanotechnol 7:035018

Durán-Álvarez JC, Avella E, Ramírez-Zamora RM, Zanella R (2016) Photocatalytic degradation of ciprofloxacin using mono- (Au, Ag and Cu) and bi- (Au–Ag and Au–Cu) metallic nanoparticles supported on TiO_2 under UV-C and simulated sunlight. Catal Today 266:175–187

Elessawy NA, Elnouby M, Gouda MH, Hamad HA, Taha NA, Gouda M, Mohy Eldin MS (2020) Ciprofloxacin removal using magnetic fullerene nanocomposite obtained from sustainable PET bottle wastes: adsorption process optimization, kinetics, isotherm, regeneration and recycling studies. Chemosphere 239:124728

El-Nagar DA, Massoud SA, Ismail SH (2020) Removal of some heavy metals and fungicides from aqueous solutions using nano-hydroxyapatite, nano-bentonite and nanocomposite. Arab J Chem 13:7695–7706. https://doi.org/10.1016/j.arabjc.2020.09.005

Fakhri H, Farzadkia M, Boukherroub R, Srivastava V, Sillanpää M (2020) Design and preparation of core-shell structured magnetic graphene oxide@MIL-101(Fe): Photocatalysis under shell to remove diazinon and atrazine pesticides. Sol Energy 208:990–1000

Feng Q, Zhang Z, Ma Y, He X, Zhao Y, Chai Z (2012) Adsorption and desorption characteristics of arsenic onto ceria nanoparticles. Nanoscale Res Lett 7:84

Fischer K, Grimm M, Meyers J, Dietrich C, Gläser R, Schulze A (2015) Photoactive microfiltration membranes via directed synthesis of TiO_2 nanoparticles on the polymer surface for removal of drugs from water. J Memb Sci 478:49–57

Fu M, Li Y, Wu S, Lu P, Liu J, Dong F (2011) Sol–gel preparation and enhanced photocatalytic performance of Cu-doped ZnO nanoparticles. Appl Surf Sci 258:1587–1591

Gehrke I, Geiser A, Somborn-Schulz A (2015) Innovations in nanotechnology for water treatment. Nanotechnol Sci Appl 8:1–17

Gholami F, Mosmeri H, Shavandi M, Dastgheib SMM, Amoozegar MA (2019) Application of encapsulated magnesium peroxide (MgO_2) nanoparticles in permeable reactive barrier (PRB) for naphthalene and toluene bioremediation from groundwater. Sci Total Environ 655:633–640

Gomes AR, Chagas TQ, Silva AM, de Lima S, Rodrigues A, Marinho da Luz T, de Andrade E, Vieira J, Malafaia G (2021) Trophic transfer of carbon nanofibers among *Eisenia fetida*, *Danio rerio* and *Oreochromis niloticus* and their toxicity at upper trophic level. Chemosphere 263:127657

Goodarzi S, Da Ros T, Conde J, Sefat F, Mozafari M (2017) Fullerene: biomedical engineers get to revisit an old friend. Mater Today 20:460–480

Guesh K, Mayoral Á, Márquez-Álvarez C, Chebude Y, Díaz I (2016) Enhanced photocatalytic activity of TiO_2 supported on zeolites tested in real wastewaters from the textile industry of Ethiopia. Microporous Mesoporous Mater 225:88–97

Guimarães ATB, Estrela FN, Pereira PS, de Andrade Vieira JE, de Lima Rodrigues AS, Silva FG, Malafaia G (2021) Toxicity of polystyrene nanoplastics in *Ctenopharyngodon idella* juveniles: a genotoxic, mutagenic and cytotoxic perspective. Sci Total Environ 752:141937

Gupta VK, Jain R, Nayak A, Agarwal S, Shrivastava M (2011) Removal of the hazardous dye—Tartrazine by photodegradation on titanium dioxide surface. Mater Sci Eng C 31:1062–1067
He F, Zhao D (2007) Manipulating the size and dispersibility of zerovalent iron nanoparticles by use of carboxymethyl cellulose stabilizers. Environ Sci Technol 41:6216–6221
Homhoul P, Pengpanich S, Hunsom M (2011) Treatment of distillery wastewater by the nano-scale zero-valent iron and the supported nano-scale zero-valent iron. Water Environ Res 83:65–74
Hu C, Liu Y, Li X, Li M (2013) Biochemical responses of duckweed (*Spirodela polyrhiza*) to zinc oxide nanoparticles. Arch Environ Contam Toxicol 64:643–651
Hunger K (ed) (2003) Dyes, General Survey. In: Industrial dyes. Wiley-VCH Verlag GmbH & Co. KGaA, Weinheim, pp 1–12
Ismail M, Khan MI, Khan MA, Akhtar K, Asiri AM, Khan SB (2019) Plant-supported silver nanoparticles: efficient, economically viable and easily recoverable catalyst for the reduction of organic pollutants. Appl Organomet Chem 33(1):e4971. https://doi.org/10.1002/aoc.4971
Jaciw-Zurakowsky I, Snowdon MR, Schneider OM, Zhou YN, Liang RL (2020) Advanced oxidation processes using catalytic nanomaterials for air and water remediation. In: Nanomaterials for air remediation. Elsevier, Amsterdam, pp 167–192
Jahan S, Bin YI, Alias YB, Bakar AFBA (2017) Reviews of the toxicity behavior of five potential engineered nanomaterials (ENMs) into the aquatic ecosystem. Toxicol Reports 4:211–220
Jasrotia T, Chaudhary S, Kaushik A, Kumar R, Chaudhary GR (2020) Green chemistry-assisted synthesis of biocompatible Ag, Cu, and Fe_2O_3 nanoparticles. Mater Today Chem 15:100214
Jeevanandam J, Barhoum A, Chan YS, Dufresne A, Danquah MK (2018) Review on nanoparticles and nanostructured materials: history, sources, toxicity and regulations. Beilstein J Nanotechnol 9:1050–1074
Jovanović B, Whitley EM, Kimura K, Crumpton A, Palić D (2015) Titanium dioxide nanoparticles enhance mortality of fish exposed to bacterial pathogens. Environ Pollut 203:153–164
Kagan CR, Murray CB (2015) Charge transport in strongly coupled quantum dot solids. Nat Nanotechnol 10:1013–1026
Kanakaraju D, Glass BD, Oelgemöller M (2014) Titanium dioxide photocatalysis for pharmaceutical wastewater treatment. Environ Chem Lett 12:27–47
Khan A, Wang J, Li J, Wang X, Chen Z, Alsaedi A, Hayat T, Chen Y, Wang X (2017) The role of graphene oxide and graphene oxide-based nanomaterials in the removal of pharmaceuticals from aqueous media: a review. Environ Sci Pollut Res 24:7938–7958
Khan B, Adeleye AS, Burgess RM, Smolowitz R, Russo SM, Ho KT (2019a) A 72-h exposure study with eastern oysters (*Crassostrea virginica*) and the nanomaterial graphene oxide. Environ Toxicol Chem 38:820–830
Khan B, Adeleye AS, Burgess RM, Russo SM, Ho KT (2019b) Effects of graphene oxide nanomaterial exposures on the marine bivalve, *Crassostrea virginica*. Aquat Toxicol 216:105297
Kumar KY, Muralidhara HB, Nayaka YA, Balasubramanyam J, Hanumanthappa H (2013) Low-cost synthesis of metal oxide nanoparticles and their application in adsorption of commercial dye and heavy metal ion in aqueous solution. Powder Technol 246:125–136
Kumar I, Mondal M, Meyappan V, Sakthivel N (2019) Green one-pot synthesis of gold nanoparticles using *Sansevieria roxburghiana* leaf extract for the catalytic degradation of toxic organic pollutants. Mater Res Bull 117:18–27
Latha P, Prakash K, Karuthapandian S (2018) Effective Photodegradation of CR & MO dyes by morphologically controlled cerium oxide nanocubes under visible light illumination. Optik (Stuttg) 154:242–250
Latrous El Atrache L, Hachani M, Kefi BB (2016) Carbon nanotubes as solid-phase extraction sorbents for the extraction of carbamate insecticides from environmental waters. Int J Environ Sci Technol 13:201–208
Lico D, Vuono D, Siciliano C, Nagy JB, De Luca P (2019) Removal of unleaded gasoline from water by multi-walled carbon nanotubes. J Environ Manag 237:636–643
Lim SY, Shen W, Gao Z (2015) Carbon quantum dots and their applications. Chem Soc Rev 44:362–381

Liu H, Liang J, Fu S, Li L, Cui J, Gao P, Zhao F, Zhou J (2020) N doped carbon quantum dots modified defect-rich g-C3N4 for enhanced photocatalytic combined pollutions degradation and hydrogen evolution. Colloids Surfaces A Physicochem Eng Asp 591:124552

Lu F, Astruc D (2020) Nanocatalysts and other nanomaterials for water remediation from organic pollutants. Coord Chem Rev 408:213180

Lu H, Wang J, Stoller M, Wang T, Bao Y, Hao H (2016) An overview of nanomaterials for water and wastewater treatment. Adv Mater Sci Eng 2016:1–10

Lu J, Tian S, Lv X, Chen Z, Chen B, Zhu X, Cai Z (2018) TiO_2 nanoparticles in the marine environment: impact on the toxicity of phenanthrene and Cd^{2+} to marine zooplankton Artemia salina. Sci Total Environ 615:375–380

Lucena R, Simonet BM, Cárdenas S, Valcárcel M (2011) Potential of nanoparticles in sample preparation. J Chromatogr A 1218:620–637

Lugani Y, Sooch BS, Singh P, Kumar S (2021) Nanobiotechnology applications in food sector and future innovations. In: Microbial biotechnology in food and health. Elsevier, Amsterdam, pp 197–225

Mahanty S, Chatterjee S, Ghosh S et al (2020) Synergistic approach towards the sustainable management of heavy metals in wastewater using mycosynthesized iron oxide nanoparticles: biofabrication, adsorptive dynamics and chemometric modeling study. J Water Process Eng 37:101426

Mahmoud WMM, Rastogi T, Kümmerer K (2017) Application of titanium dioxide nanoparticles as a photocatalyst for the removal of micropollutants such as pharmaceuticals from water. Curr Opin Green Sustain Chem 6:1–10

Mahmoud ME, Saleh MM, Zaki MM, Nabil GM (2020) A sustainable nanocomposite for removal of heavy metals from water based on crosslinked sodium alginate with iron oxide waste material from steel industry. J Environ Chem Eng 8:104015

Malina T, Maršálková E, Holá K, Tuček J, Scheibe M, Zbořil R, Maršálek B (2019) Toxicity of graphene oxide against algae and cyanobacteria: Nanoblade-morphology-induced mechanical injury and self-protection mechanism. Carbon N Y 155:386–396

Mandeep GA, Kakkar R (2020) Graphene-based adsorbents for water remediation by removal of organic pollutants: theoretical and experimental insights. Chem Eng Res Des 153:21–36

Marchi LD, Neto V, Pretti C, Figueira E, Brambilla L, Rodriguez-Douton MJ, Rossella F, Tommasini M, Furtado C, Soares MVM, Freitas R (2017) Physiological and biochemical impacts of graphene oxide in polychaetes: the case of *Diopatra neapolitana*. Comp Biochem Physiol Part C Toxicol Pharmacol 193:50–60

Marimuthu S, Antonisamy AJ, Malayandi S, Rajendran K, Tsai P-C, Pugazhendhi A, Ponnusamy VK (2020) Silver nanoparticles in dye effluent treatment: a review on synthesis, treatment methods, mechanisms, photocatalytic degradation, toxic effects and mitigation of toxicity. J Photochem Photobiol B Biol 205:111823

Melo CB, Côa F, Alves OL, Martinez DST, Barbieri E (2019) Co-exposure of graphene oxide with trace elements: effects on acute ecotoxicity and routine metabolism in *Palaemon pandaliformis* (shrimp). Chemosphere 223:157–164

Meziani MJ, Dong X, Zhu L, Jones LP, LeCroy GE, Yang F, Wang S, Wang P, Zhao Y, Yang L, Tripp RA, Sun Y-P (2016) Visible-light-activated bactericidal functions of carbon "quantum" dots. ACS Appl Mater Interfaces 8:10761–10766

Miller RJ, Muller EB, Cole B, Martin T, Nisbet R, Bielmyer-Fraser GK, Jarvis TA, Keller AA, Cherr G, Lenihan HS (2017) Photosynthetic efficiency predicts toxic effects of metal nanomaterials in phytoplankton. Aquat Toxicol 183:85–93

Mirsadeghi S, Zandavar H, Yousefi M, Rajabi HR, Pourmortazavi SM (2020) Green-photodegradation of model pharmaceutical contaminations over biogenic Fe_3O_4/Au nanocomposite and antimicrobial activity. J Environ Manag 270:110831

Mishra J, Pattanayak DS, Das AA, Mishra DK, Rath D, Sahoo NK (2019) Enhanced photocatalytic degradation of cyanide employing Fe-porphyrin sensitizer with hydroxyapatite palladium doped TiO_2 nano-composite system. J Mol Liq 287:110821

Miyashiro CS, Mateus GAP, dos Santos TRT, Paludo MP, Bergamasco R, Fagundes-Klen MR (2021) Synthesis and performance evaluation of a magnetic biocoagulant in the removal of reactive black 5 dye in aqueous medium. Mater Sci Eng C 119:111523

Moghadas MRS, Motamedi E, Nasiri J, Naghavi MR, Sabokdast M (2020) Proficient dye removal from water using biogenic silver nanoparticles prepared through solid-state synthetic route. Heliyon 6:e04730

Mohamed EF (2017) Nanotechnology: future of environmental air pollution control. Environ Manag Sustain Dev 6:429

Mohan S, Singh Y, Verma DK, Hasan SH (2015) Synthesis of CuO nanoparticles through green route using citrus limon juice and its application as nanosorbent for Cr(VI) remediation: process optimization with RSM and ANN-GA based model. Process Saf Environ Prot 96:156–166

Moussavi G, Alahabadi A, Yaghmaeian K, Eskandari M (2013) Preparation, characterization and adsorption potential of the NH4Cl-induced activated carbon for the removal of amoxicillin antibiotic from water. Chem Eng J 217:119–128

Mu B, Wang A (2016) Adsorption of dyes onto palygorskite and its composites: a review. J Environ Chem Eng 4:1274–1294

Mylona Z, Panteris E, Kevrekidis T, Malea P (2020) Silver nanoparticle toxicity effect on the seagrass *Halophila stipulacea*. Ecotoxicol Environ Saf 189:109925

Nasrollahzadeh M, Sajjadi M, Iravani S, Varma RS (2021) Green-synthesized nanocatalysts and nanomaterials for water treatment: current challenges and future perspectives. J Hazard Mater 401:123401

Navalon S, Martin R, Alvaro M, Garcia H (2010) Gold on diamond nanoparticles as a highly efficient Fenton catalyst. Angew Chemie Int Ed 49:8403–8407

Nguyen DTC, Dang HH, Vo D-VN, Bach LG, Nguyen TD, Van Tran T (2020) Biogenic synthesis of MgO nanoparticles from different extracts (flower, bark, leaf) of *Tecoma stans* (L.) and their utilization in selected organic dyes treatment. J Hazard Mater 404:124146

Noreen S, Khalid U, Ibrahim SM, Javed T, Ghani A, Naz S, Iqbal M (2020) ZnO, MgO and FeO adsorption efficiencies for direct sky blue dye: equilibrium, kinetics and thermodynamics studies. J Mater Res Technol 9:5881–5893

Nurhasanah I, Kadarisman GV, Sutanto H (2020) Cerium oxide nanoparticles application for rapid adsorptive removal of tetracycline in water. J Environ Chem Eng 8:103613

Okonji SO, Dominic JA, Pernitsky D, Achari G (2020) Removal and recovery of selenium species from wastewater: adsorption kinetics and co-precipitation mechanisms. J Water Process Eng 38:101666

Oyewo OA, Elemike EE, Onwudiwe DC, Onyango MS (2020) Metal oxide-cellulose nanocomposites for the removal of toxic metals and dyes from wastewater. Int J Biol Macromol 164:2477–2496

Pan H, Feng J, He G-X, Cerniglia CE, Chen H (2012) Evaluation of impact of exposure of Sudan azo dyes and their metabolites on human intestinal bacteria. Anaerobe 18:445–453

Panahi Y, Mellatyar H, Farshbaf M, Sabet Z, Fattahi T, Akbarzadehe A (2018) Biotechnological applications of nanomaterials for air pollution and water/wastewater treatment. Mater Today Proc 5:15550–15558

Pereira SPP, Jesus F, Aguiar S, de Oliveira R, Fernandes M, Ranville J, Nogueira AJA (2018) Phytotoxicity of silver nanoparticles to *Lemna minor*: surface coating and exposure period-related effects. Sci Total Environ 618:1389–1399

Perreault F, Samadani M, Dewez D (2014) Effect of soluble copper released from copper oxide nanoparticles solubilisation on growth and photosynthetic processes of *Lemna gibba* L. Nanotoxicology 8:374–382

Perreault F, de Faria AF, Nejati S, Elimelech M (2015) Antimicrobial properties of graphene oxide nanosheets: why size matters. ACS Nano 9:7226–7236

Piotrowski P, Krogul-Sobczak A, Kaim A (2019) Magnetic iron oxide nanoparticles functionalized with C60 phosphonic acid derivative for catalytic reduction of 4-nitrophenol. J Environ Chem Eng 7:103147

Prairie MR, Evans LR, Stange BM, Martinez SL (1993) An investigation of titanium dioxide photocatalysis for the treatment of water contaminated with metals and organic chemicals. Environ Sci Technol 27:1776–1782
Priyadarshni N, Nath P, Nagahanumaiah CN (2020) Sustainable removal of arsenate, arsenite and bacterial contamination from water using biochar stabilized iron and copper oxide nanoparticles and associated mechanism of the remediation process. J Water Process Eng 37:101495
Qiao K, Tian W, Bai J, Wang L, Zhao J, Du Z, Gong X (2019) Application of magnetic adsorbents based on iron oxide nanoparticles for oil spill remediation: a review. J Taiwan Inst Chem Eng 97:227–236
Qu X, Alvarez PJJ, Li Q (2013) Applications of nanotechnology in water and wastewater treatment. Water Res 47:3931–3946
Rafiq Z, Nazir R, Durr-e-Shahwar SMR, Ali S (2014) Utilization of magnesium and zinc oxide nano-adsorbents as potential materials for treatment of copper electroplating industry wastewater. J Environ Chem Eng 2:642–651
Rafique M, Tahir MB, Irshad M, Nabi G, Gillani SSA, Iqbal T, Mubeen M (2020) Novel *Citrus aurantifolia* leaves based biosynthesis of copper oxide nanoparticles for environmental and wastewater purification as an efficient photocatalyst and antibacterial agent. Optik (Stuttg) 219:165138
Raizada P, Sudhaik A, Patial S, Hasija V, Parwaz Khan AA, Singh P, Gautam S, Kaur M, Nguyen V-H (2020) Engineering nanostructures of CuO-based photocatalysts for water treatment: current progress and future challenges. Arab J Chem 13:8424–8457. https://doi.org/10.1016/j.arabjc.2020.06.031
Rani PU, Yasur J, Loke KS, Dutta D (2016) Effect of synthetic and biosynthesized silver nanoparticles on growth, physiology and oxidative stress of water hyacinth: *Eichhornia crassipes* (Mart) Solms. Acta Physiol Plant 38:58
Rani S, Dilbaghi N, Kumar S, Varma RS, Malhotra R (2020) Rapid redox sensing of p-nitrotoluene in real water samples using silver nanoparticles. Inorg Chem Commun 120:108157
Ray SS, Salehiyan R (2020) Fundamental definition and importance of nanomaterials, nanostructured, and bulk nanostructured materials. In: Nanostructured immiscible polymer blends. Elsevier, Amsterdam, pp 15–28
Sackey J, Bashir AKH, Ameh AE, Nkosi M, Kaonga C, Maaza M (2020) Date pits extracts assisted synthesis of magnesium oxides nanoparticles and its application towards the photocatalytic degradation of methylene blue. J King Saud Univ – Sci 32:2767–2776
Sadegh H, Shahryari-ghoshekandi R, Kazemi M (2014) Study in synthesis and characterization of carbon nanotubes decorated by magnetic iron oxide nanoparticles. Int Nano Lett 4:129–135
Saleh TA (2020) Nanomaterials: classification, properties, and environmental toxicities. Environ Technol Innov 20:101067
Sane PK, Tambat S, Sontakke S, Nemade P (2018) Visible light removal of reactive dyes using CeO_2 synthesized by precipitation. J Environ Chem Eng 6:4476–4489
Sathiyan K, Bar-Ziv R, Mendelson O, Zidki T (2020) Controllable synthesis of TiO_2 nanoparticles and their photocatalytic activity in dye degradation. Mater Res Bull 126:110842
Scher JA, Elward JM, Chakraborty A (2016) Shape matters: effect of 1D, 2D, and 3D isovolumetric quantum confinement in semiconductor nanoparticles. J Phys Chem C 120:24999–25009
Schmidt-Mende L, MacManus-Driscoll JL (2007) ZnO – nanostructures, defects, and devices. Mater Today 10:40–48
Sendra M, Blasco J, Araújo CVM (2018) Is the cell wall of marine phytoplankton a protective barrier or a nanoparticle interaction site? Toxicological responses of *chlorella autotrophica* and *Dunaliella salina* to ag and CeO_2 nanoparticles. Ecol Indic 95:1053–1067
Shaban YA (2019) Solar light-induced photodegradation of chrysene in seawater in the presence of carbon-modified n-TiO_2 nanoparticles. Arab J Chem 12:652–663
Shahwan T, Abu Sirriah S, Nairat M, Boyacı E, Eroğlu AE, Scott TB, Hallam KR (2011) Green synthesis of iron nanoparticles and their application as a Fenton-like catalyst for the degradation of aqueous cationic and anionic dyes. Chem Eng J 172:258–266

Sherin L, Sohail A, Amjad U-S, Mustafa M, Jabeen R, Ul-Hamid A (2020) Facile green synthesis of silver nanoparticles using *Terminalia bellerica* kernel extract for catalytic reduction of anthropogenic water pollutants. Colloid Interface Sci Commun 37:100276

Si Q-S, Guo W-Q, Wang H-Z, Liu B-H, Ren N-Q (2020) Carbon quantum dots-based semiconductor preparation methods, applications and mechanisms in environmental contamination. Chinese Chem Lett 31:2556–2566. https://doi.org/10.1016/j.cclet.2020.08.036

Singh V, Srivastava VC (2020) Self-engineered iron oxide nanoparticle incorporated on mesoporous biochar derived from textile mill sludge for the removal of an emerging pharmaceutical pollutant. Environ Pollut 259:113822

Singh KK, Senapati KK, Borgohain C, Sarma KC (2017) Newly developed Fe_3O_4–Cr_2O_3 magnetic nanocomposite for photocatalytic decomposition of 4-chlorophenol in water. J Environ Sci 52:333–340

Sivaselvam S, Premasudha P, Viswanathan C, Ponpandian N (2020) Enhanced removal of emerging pharmaceutical contaminant ciprofloxacin and pathogen inactivation using morphologically tuned MgO nanostructures. J Environ Chem Eng 8:104256

Souza JP, Venturini FP, Santos F, Zucolotto V (2018) Chronic toxicity in *Ceriodaphnia dubia* induced by graphene oxide. Chemosphere 190:218–224

Sreeju N, Rufus A, Philip D (2017) Studies on catalytic degradation of organic pollutants and antibacterial property using biosynthesized CuO nanostructures. J Mol Liq 242:690–700

Suman TY, Radhika Rajasree SR, Kirubagaran R (2015) Evaluation of zinc oxide nanoparticles toxicity on marine algae *Chlorella vulgaris* through flow cytometric, cytotoxicity and oxidative stress analysis. Ecotoxicol Environ Saf 113:23–30

Sun J, Wang C, Zeng L, Xu P, Yang X, Chen J, Xing X, Jin Q, Yu R (2016) Controllable assembly of CeO_2 micro/nanospheres with adjustable size and their application in Cr(VI) adsorption. Mater Res Bull 75:110–114

Sunitha M, Dinesh Karthik A, Geetha K (2020) Eco synthesis, spectral and antimicrobial studies of copper oxide (CuO) nanoparticles. Mater Today Proc 29:1229–1234

Surendra B, Raju BM, Onesimus KNS, Choudhary GL, Paul PF, Vangalapati M (2020) Synthesis and characterization of Ni doped TiO_2 nanoparticles and its application for the degradation of malathion. Mater Today Proc 26:1091–1095

Szunerits S, Boukherroub R (2016) Antibacterial activity of graphene-based materials. J Mater Chem B 4:6892–6912

Tahir MB, Rafique M, Rafique MS, Nawaz T, Rizwan M, Tanveer M (2020) Photocatalytic nanomaterials for degradation of organic pollutants and heavy metals. In: Nanotechnology and photocatalysis for environmental applications. Elsevier, Amsterdam, pp 119–138

Taylor C, Matzke M, Kroll A, Read DS, Svendsen C, Crossley A (2016) Toxic interactions of different silver forms with freshwater green algae and cyanobacteria and their effects on mechanistic endpoints and the production of extracellular polymeric substances. Environ Sci Nano 3:396–408

Tratnyek PG, Salter AJ, Nurmi JT, Sarathy V (2010) Environmental applications of zerovalent metals: iron vs. zinc. In: Nanoscale materials in chemistry: environmental applications. American Chemical Society, Washington, pp 165–178

Trujillo-Reyes J, Peralta-Videa JR, Gardea-Torresdey JL (2014) Supported and unsupported nanomaterials for water and soil remediation: are they a useful solution for worldwide pollution? J Hazard Mater 280:487–503

Vannuccini ML, Grassi G, Leaver MJ, Corsi I (2015) Combination effects of nano-TiO_2 and 2,3,7,8-tetrachlorodibenzo-p-dioxin (TCDD) on biotransformation gene expression in the liver of European sea bass *Dicentrarchus labrax*. Comp Biochem Physiol Part C Toxicol Pharmacol 176–177:71–78

Verma B, Balomajumder C (2020) Surface modification of one-dimensional carbon nanotubes: a review for the management of heavy metals in wastewater. Environ Technol Innov 17:100596

Verma N, Vaidh S, Vishwakarma GS, Pandya A (2020) Antimicrobial nanomaterials for water disinfection. Nanotoxicity. Elsevier, In, pp 365–383

Vincent JL, Paterson MJ, Norman BC, Gray EP, Ranville JF, Scott AB, Frost PC, Xenopoulos MA (2017) Chronic and pulse exposure effects of silver nanoparticles on natural lake phytoplankton and zooplankton. Ecotoxicology 26:502–515

Wahab R, Khan F, Ahmad N, Alam M, Ahmad J, Al-Khedhairy AA (2020) Rapid sensing response for phenol with CuO nanoparticles. Colloids Surfaces A Physicochem Eng Asp 607:125424

Waissi GC, Väänänen K, Nybom I, Pakarinen K, Akkanen J, Leppänen MT, Kukkonen JVK (2017) The chronic effects of fullerene C60 -associated sediments in the midge *Chironomus riparius* – responses in the first and the second generation. Environ Pollut 229:423–430

Wang B, Liang W, Guo Z, Liu W (2015) Biomimetic super-lyophobic and super-lyophilic materials applied for oil/water separation: a new strategy beyond nature. Chem Soc Rev 44:336–361

Wong K-A, Lam S-M, Sin J-C (2019) Wet chemically synthesized ZnO structures for photodegradation of pre-treated palm oil mill effluent and antibacterial activity. Ceram Int 45:1868–1880

Xu C, Nasrollahzadeh M, Sajjadi M, Maham M, Luque R, Puente-Santiago AR (2019a) Benign-by-design nature-inspired nanosystems in biofuels production and catalytic applications. Renew Sustain Energy Rev 112:195–252

Xu P, Chen M, Lai C, Zeng G, Huang D, Wang H, Gong X, Qin L, Liu Y, Mo D, Wen X, Zhou C, Wang R (2019b) Effects of typical engineered nanomaterials on 4-nonylphenol degradation in river sediment: based on bacterial community and function analysis. Environ Sci Nano 6:2171–2184

Xu Z, Wang Z, Liu M, Yan B, Ren X, Gao Z (2020) Machine learning assisted dual-channel carbon quantum dots-based fluorescence sensor array for detection of tetracyclines. Spectrochim Acta Part A Mol Biomol Spectrosc 232:118147

Yari S, Abbasizadeh S, Mousavi SE, Moghaddam MS, Moghaddam AZ (2015) Adsorption of Pb(II) and Cu(II) ions from aqueous solution by an electrospun CeO2 nanofiber adsorbent functionalized with mercapto groups. Process Saf Environ Prot 94:159–171

Yathisha RO, Nayaka YA (2020) Effect of solvents on structural, optical and electrical properties of ZnO nanoparticles synthesized by microwave heating route. Inorg Chem Commun 115:107877

Ying Y, Ying W, Li Q, Meng D, Ren G, Yan R, Peng X (2017) Recent advances of nanomaterial-based membrane for water purification. Appl Mater Today 7:144–158

Yoo-iam M, Chaichana R, Satapanajaru T (2014) Toxicity, bioaccumulation and biomagnification of silver nanoparticles in green algae (Chlorella sp.), water flea (Moina macrocopa), blood worm (Chironomus spp.) and silver barb (Barbonymus gonionotus). Chem Speciat Bioavailab 26:257–265

Yuan X, Wang Y, Wang J, Zhou C, Tang Q, Rao X (2013) Calcined graphene/MgAl-layered double hydroxides for enhanced Cr(VI) removal. Chem Eng J 221:204–213

Yunus IS, Harwin, Kurniawan A, Adityawarman D, Indarto A (2012) Nanotechnologies in water and air pollution treatment. Environ Technol Rev 1:136–148

Zamri MSFA, Sapawe N (2019) Effect of pH on Phenol Degradation Using Green Synthesized Titanium Dioxide Nanoparticles. Mater Today Proc 19:1321–1326

Zhan J, Kolesnichenko I, Sunkara B, He J, McPherson GL, Piringer G, John VT (2011) Multifunctional Iron−Carbon Nanocomposites through an Aerosol-Based Process for the In Situ Remediation of Chlorinated Hydrocarbons. Environ Sci Technol 45:1949–1954

Zhang D, Gersberg RM, Ng WJ, Tan SK (2014a) Removal of pharmaceuticals and personal care products in aquatic plant-based systems: A review. Environ Pollut 184:620–639

Zhang X, Lin M, Lin X, Zhang C, Wei H, Zhang H, Yang B (2014b) Polypyrrole-Enveloped Pd and Fe3O4 Nanoparticle Binary Hollow and Bowl-Like Superstructures as Recyclable Catalysts for Industrial Wastewater Treatment. ACS Appl Mater Interfaces 6:450–458

Zhang M, Cui J, Lu T, Tang G, Wu S, Ma W, Huang C (2021) Robust, functionalized reduced graphene-based nanofibrous membrane for contaminated water purification. Chem Eng J 404:126347

Zhao J, Luo W, Xu Y, Ling J, Deng L (2020) Potential reproductive toxicity of multi-walled carbon nanotubes and their chronic exposure effects on the growth and development of Xenopus tropicalis. Sci Total Environ 766:142652

Zhou Q, Wang Y, Xiao J, Fan H, Chen C (2019) Preparation and characterization of magnetic nanomaterial and its application for removal of polycyclic aromatic hydrocarbons. J Hazard Mater 371:323–331

Yu F, Ma J, Wang J, Zhang M, Zheng J (2016) Magnetic iron oxide nanoparticles functionalized multi-walled carbon nanotubes for toluene, ethylbenzene and xylene removal from aqueous solution. Chemosphere 146:162–172

Tran T Van, Nguyen H-TT, Dang HH, Nguyen DTC, Nguyen DH, Pham T Van, Tan L Van (2020) Central composite design for optimizing the organic dyes remediation utilizing novel graphene oxide@CoFe2O4 nanocomposite. Surfaces and Interfaces 21:100687

Zhu K, Chen C, Wang H, Xie Y, Wakeel M, Wahid A, Zhang X (2019) Gamma-ferric oxide nanoparticles decoration onto porous layered double oxide belts for efficient removal of uranyl. J Colloid Interface Sci 535:265–275

Chapter 6
Nanotechnology: Emerging Opportunities and Regulatory Aspects in Water Treatment

Yogita Lugani, Venkata Ramana Vemuluri, and Balwinder Singh Sooch

Abstract Nanotechnology is emerging as one of the innovative technologies, which involves the controlled synthesis of structures, materials, and devices in the nano range. Nanomaterials have shown their competence in almost all fields of science due to their entirely different functional properties than bulk counterparts. Environmental nanotechnology is a revolutionary field of science and technology that involves the use of nanomaterials for environmental applications. In this era, the most worrying issue of global concern are water scarcity, which is becoming more intense day by day due to increasing human population, civilization, environmental changes, agricultural activities, and industrialization. The conventional water treatment approaches like coagulation, flocculation, activated carbon adsorption, ozonation, and membrane processing are not competent to remove all the contaminants from wastewater, which necessitate the emergence of developing novel water treatment technologies to overcome this social issue. In view of these, nanotechnology is one of the promising tools to solve the problems of water purification and wastewater treatment. The competency of nanomaterials is due to their high reactivity, large specific surface area, affinity for specific target contaminants, size-dependent properties, and a high degree of functionalization of engineered nanoforms. Different nanomaterials are used in the past for the detection and removal of chemical and biological contaminants. The toxic effect of nanomaterials on ecology and human health is a critical concern in their selection for commercial applications. Hence, regulatory guidelines should be adopted before the marketing of nano-engineered products. This chapter covers diverse applications of nanotechnology in different sectors, the action mechanism of nanomaterials, conventional and advanced tools for wastewater treatment, and the application of nanotechnology in water

Y. Lugani · V. R. Vemuluri (✉)
Microbial Type Culture Collection and Gene Bank (MTCC), CSIR-Institute of Microbial Technology, Chandigarh, India
e-mail: venkat@imtech.res.in

B. S. Sooch (✉)
Department of Biotechnology, Punjabi University, Patiala, Punjab, India

R. Kumar et al. (eds.), *Advanced Functional Nanoparticles "Boon or Bane" for Environment Remediation Applications*, Environmental Contamination Remediation and Management, https://doi.org/10.1007/978-3-031-24416-2_6

purification and wastewater treatment. Further, attention is paid to safety issues with the use of nanomaterials along with regulatory aspects. The commercial processes of nano-based materials related to water application will also be addressed.

Keywords Nanotechnology · Nanomaterials · Water contaminants · Water purification · Waste water treatment · Environment

6.1 Introduction

Nanotechnology is related to the engineering of materials at the atomic or molecular level with sizes ranging from 0.1 to 100 nm, and this process was initiated in 1959 by Richard Feynman (Nikalje 2015). The extremely small size of nanomaterials results in their unique physical, chemical, biological, and electrical properties such as high surface-to-volume ratio, water solubility, physical strength, magnetism, chemical reactivity, optical effects, photocatalytic activity, zeta potential and electrical conductance (Lugani et al. 2018a). The two common methods used for the synthesis of nanoparticles are the top-down approach and the bottom-up approach. The strategies used for the preparation of nanoparticles are self-assembly, layer-by-layer deposition, thermal plasmas, gas phase synthesis and sol-gel processing, crystallization, biogenic strategy, microbial synthesis, biomass reactions, sonochemical processing, cavitation processing, microemulsion processing, and high-energy ball milling (Tiwari et al. 2008). There are numerous benefits of nanotechnology in various industrial sectors like food, pharmaceuticals, medicine, cosmetics, energy, textile and clothing, paint and coating, and catalysis (Huang et al. 2010; Berekaa 2015; Chaturvedi and Dave 2019). Nanoparticles are used in biological research for the detection of biomolecules, sample separation, purification and concentration, substrate coding, and signal transduction and amplification (Liu 2006). Currently, most of the developing and developed countries are facing terrifying challenges to meet the rising demand for fresh water arising due to population growth, extended droughts, enhanced industrialization, stringent health-based regulations, water quality deterioration, and competing demand for a variety of users (EPA 2015; BRSNL 2003). Hence, the major global challenge in the twenty-first century is to provide clean and affordable water to the public for their good health. Contaminated drinking water containing pathogenic bacteria is the major cause of water-borne diseases (approximately 80%) in developing countries like India. According to the guidelines published by World Health Organization (1996), water is considered safe for drinking purposes when its fecal and total coliform count is zero per 100 mL water (WHO 1996). It is well known that India is an agricultural country, and excessive use of pesticides for agricultural activities leads to contamination of groundwater by notorious organochlorine insecticides such as endosulfan, dichlorodiphenyltrichloroethane (DDT), aldrin, hexachlorocyclohexane (HCH), and dioxin. Today, different conventional techniques such as physical and chemical agents, halogens (Cl, Br), sedimentation, boiling, distillation, reverse osmosis, solvent extraction, evaporation, ultraviolet light, low-frequency ultrasonic radiations, ion exchange water softener, neutralization and remineralization, oxidation and reduction, ozonization, water sediment filters (fiber and ceramic), crystallization, pitcher and faucet mount filters, and activated carbon are being used for the water treatment of such contaminated water (Ali 2012; Chaturvedi and Dave 2019). Some enzyme-based methods for the reatment of industrial wastewater were also tested in the past (Sooch et al. 2014, 2016; Sooch and Kauldhar 2015). It is evident that the current methods of water purification and treatment are expensive and create carcinogenic and toxic by-products. Currently, available water treatment technologies are not sufficient to meet human and environmental needs; therefore, the present "need of the day" is to recycle, reuse, and repurpose of fresh water. In view of this, nanotechnology offers leapfrogging opportunities for developing advanced, eco-friendly, and cost-effective water purification and treatment systems. Nanomaterials like metal oxides, zeolites, dendrimers, carbon nanotubes (CNTs), fullerenes, graphene-based nanomaterials, nanosorbents, nanocatalysts, biomimetic membranes, molecularly imprinted polymers (MIP), and zerovalent iron particles are presently used for water treatment applications at pilot scale. The characterization of nanoparticle interactions with bacterial contaminants by atomic force microscopy (AFM), scanning electron microscopy (SEM), transmission electron microscopy (TEM), Raman microscopy (RM), scanning probe microscopy (SPM), and laser confocal microscopy (LCM) showed considerable changes in cell membrane integrity, which result in bacterial cell death. The currently available nanomaterial-based immobilization techniques are leading to a significant loss of treatment efficiency. Keeping this in view, research groups need to focus on developing simple, cost-effective methods for immobilizing nanomaterials without significantly altering their performance (Qu et al. 2013). Although nanomaterials have proved to be potential for their use in various industrial applications, there are some issues associated with the commercialization of nano-based materials due to their harmful effects on human health and the environment (Bumbudsanpharoke and Ko 2015). There is little knowledge about the hazards and exposure risks of nanoparticles to the environment, and hence, risk assessment and management is a great challenge, which needs to be resolved before the use of such nanoparticles in larger-scale water purification systems (Gardner and Dhai 2014). This chapter will briefly update on the action mechanism of nanomaterials, conventional and advanced tools for wastewater treatment, the application of nanotechnology in water purification and wastewater treatment, and safety and regulatory aspects of nanomaterials for their safe use. This review also provides an overview of the availability of commercial systems having nano-based materials in the market for water treatment applications.

6.2 Manufacturing Approaches

There are two major approaches for the synthesis of nanoparticles, which are the top-down approach (reduction of larger structures to nanoscale level by maintaining their original properties) and the bottom-up approach (engineering of atomic or molecular level components by self-assembly) (Fig. 6.1). Bottom-up approach is also called as molecular manufacturing or molecular nanotechnology, and this method is used in energy, healthcare, national security, biotechnology, and information technology (Sanchez and Sobolev 2010). The strategies used for the preparation of nanoparticles are self-assembly, layer-by-layer deposition, thermal plasmas, gas phase synthesis and sol-gel processing, crystallization, biogenic strategy, microbial synthesis, biomass reactions, sonochemical processing, cavitation processing,

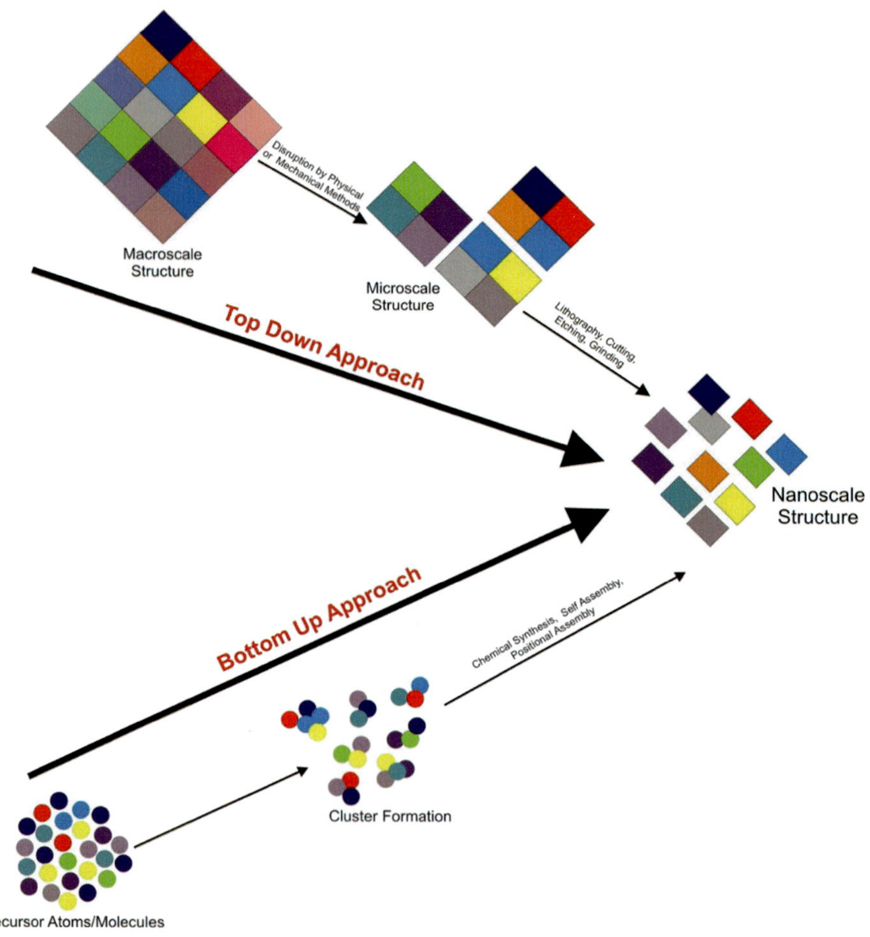

Fig. 6.1 Major approaches used for synthesis of nanoparticles

microemulsion processing, and high-energy ball milling (Tiwari et al. 2008). Self-assembly of molecules is induced by manipulation of physical (temperature, pH) and chemical (solute concentration) conditions to form fibrous nanostructures (Graveland and Kruif 2006). Vesicles/polymersomes are produced by self-assembly through the slow evaporation of organic solvents, and they are used for encapsulation (Lorenceau et al. 2005). The Layer-by-layer deposition is used for the synthesis of bilayer membranes using layering of sodium silicate and poly(allylamine hydrochloride) on gold, which was followed by calcination in the furnace, and this strategy is used for the preparation of lipid bilayers, which are used for the detection of specific proteins (Phillips et al. 2006). Silicide and rare-earth boride are used for preparation of functional nanoparticles by induction of thermal plasmas, and the resulting nanoparticles are used in electromagnetic shielding and solar control windows (Tiwari et al. 2008). Gas phase synthesis and sol-gel processing are used for the generation of nanoparticles (1–10 nm) of the high degree of monodispersity with consistent crystal structure and surface derivatization (Siegel 1994). The crystallization process uses the solution containing amino acids for the synthesis of hydroxyapatite-aspartic acid/glutamic acid (Boanini et al. 2006). There is the self-assembly of organic materials using the "nested level of structural hierarchy" for the preparation of templates/scaffolds for inorganic components (Aksay 1992).

Production of gold and silver nanoparticles by a number of bacterial and fungal species is one of the examples of microbial synthesis of nanoparticles. Gold nanorods and nanoparticles are produced by incubating the dead oat stalks with an aqueous solution of gold ions (Bhattacharya and Gupta 2005). Low-temperature nanoparticle preparation methods are used for the preparation of silver halide particles (Bishop 1990). Sonochemical processing leads to the generation of nanoparticles by the formation of the transient-localized hot zone through the acoustic cavitation process in the presence of extremely high-temperature gradient and pressure, and the materials produced by this process are utilized for various industrial applications (Suslick et al. 1996). Nanoparticles are generated by hydrodynamic cavitation by the creation and release of gas bubbles inside a sol-gel solution. Pressure and solution retention time in a cavitation chamber are two major factors affecting the size of nanoparticles in hydrodynamic cavitation (Sunstrom et al. 1996).

6.3 Applications of Nanotechnology

Nanotechnology has shown potential applications in various areas such as food and agriculture, medicine and drugs, cosmetics, transportation, energy, paper and pulp, defense and security, bioremediation, wastewater treatment, paints and coatings, optical engineering, bioengineering, electronics, and space exploration (Mihindukulasuriya and Lim 2014; Pradhan et al. 2015; Skalickova et al. 2017; Hamad et al. 2018; Chaturvedi and Dave 2019; Lugani et al. 2020; Rattu et al. 2020). Nanomaterials are used in the food sector for food packaging, food preservation (Beyth et al. 2015), nutritional drinks, and nutritional supplements (Skalickova et al. 2017) because they play an important role in enhancing the quality and shelf life of food products; improving the chemical, mechanical, thermal, barrier, and antimicrobial properties; detection of pathogens; and decreasing depletion of nutrients from food products. They are used in nano-fertilizers, nano-sensors, nano-pesticides, waste management, animal husbandry, post-harvest technology, and agricultural engineering aspects (Meetoo 2011; Bharathi et al. 2016; Abobatta 2018; Hamad et al. 2018; Krishnan et al. 2018). Nanoparticles are used in the biomedical sector for screening and early detection of diseases, tissue-specific drug delivery, tissue engineering, organ transplantation, and treatment of many diseases like acute ischemic stroke, brain tumors, Alzheimer's disease, and Parkinson's disease (Lugani et al. 2018b). The antibacterial activity of ciprofloxacin can be improved by supplementing ZnO nanoparticles, which interfere with the antibiotic resistance mechanisms (Banoee et al. 2010). Nanotechnology is used in forensic science in DNA fingerprinting for parental testing and solving criminal cases. The nanoparticles commonly used for DNA fingerprinting are Ag, Au, TiO, ZnS, and CdSe (Harvey 2015; Lee et al. 2016; Kaushik et al. 2017). The nanoparticles that have shown potential use for fingerprint analysis are hybrid cadmium sulfide quantum dots (QDs) nanocomposites assembled in porous phosphate heterostructure (PPH), and functionalized with propionitrile (PPH-CN) and mercaptopropyl (PPH-SH) (Algarra et al. 2011). The nanomaterials that are used nowadays in medicine, environment engineering, consumer products, catalysis, quantum computer, and communication are peptide nanotubes, carbon nanotubes, and QDs (Verma et al. 2011). Nano-engineering is used in the cement industry for the development of new-generation multifunctional, tailored, and cement composites with high durability and superior mechanical performance. Nano-engineered concrete has various novel properties like self-cleaning and self-sensing capability, self-control of cracks, low electrical resistivity, and high ductility (Sanchez and Sobolev 2010). Different nanomaterials (TiO_2, ZnO, dendrimers, polymers, nano-emulsions, nano-crystals, and solid lipid nanoparticles) are used in various hair care products, moisturizers, sunscreen, makeup, and other cosmetic products. Nanoparticles are also used for developing modified medicated textiles, which possess improved UV-protection, antimicrobial activity, anti-odor, easy clean, stain, and water repellent properties (Lugani et al. 2018a, b). Nanotechnology is used in the papermaking process due to the great potential of celluloses and lignocelluloses as nanomaterials, and hence nanomaterials like fillers, sizing and coating minerals, and nanofibers are used to improve the quality of produced paper and for producing new types of paper (Mohieldin et al. 2011). In biological research, nanoparticles are used for the detection of biomolecules, sample separation, purification and concentration, substrate coding, and signal transduction, and amplification (Liu 2006). Nanoparticles are used as fluorophores in fluorescence in situ hybridization (FISH), and quantum dots (QDs) attached with specific oligonucleotide probes or immunoglobulin G (IgG) has been used successfully for the detection of the human Y chromosome (Pathak et al. 2001) and in cellular imaging for the location of cancer markers (Wu et al. 2002). The sensitivity and yield of polymerase chain reaction (PCR) amplification were improved by the addition of gold (Au) nanoparticles (Li et al. 2005). Global climate change due to increasing population growth, industrialization, and improved living standards is one of the major issues, which lead to steady alteration in environment. Though several government- and private-funded projects are in action all over the world to overcome this problem, these problems have not been completely resolved to date. Environmental nanotechnology is a revolutionary field of science and technology, which is gaining interest in the scientific community for the past few decades. Nanomaterials have been explored for the generation of renewable energy in batteries, fuel cells, and supercapacitors (Yu et al. 2014). Nanoparticles are used for environmental remediation, water purification, and wastewater treatment (Carpenter et al. 2015; Hussain et al. 2016; Chaturvedi and Dave 2019).

6.4 Synoptic Applications of Nanotechnology in Water Treatment

One of the major problems in the world is water contamination, which affects human health and the environment and ultimately has negative impacts on social costs and economics. The most basic human goal in the twenty-first century is reliable access to affordable and clean water. Approximately, 780 million people lack access to improved drinking water sources (WHO 2012). Industrial wastewater generated across India is around 13,468 million liters per day (MLD); however, the sewage treatment capacity of our nation is only 8080.8 MLD. The major cities of India generated a total of 38.354 million liters of sewage and wastewater per day; however, their sewage treatment capacity is only 11,786 MLD (Kaur et al. 2012). A large number of pathogenic organisms such as bacteria, viruses, protozoans, and helminths are present in wastewater, which can cause severe diseases like leptospirosis, hepatitis A, campylobacteriosis, etc. (Singh et al. 2016; Kaur et al. 2018). Antibiotic-resistant pathogenic microbes, which are very difficult to remove from wastewater, can be eliminated by bioactive nanoparticles that have emerged as an alternative for new chlorine-free biocides (Prachi et al. 2013). Nonbiodegradable materials are the emerging pollutants in wastewater streams, which pose a risk of causing harmful effects on the environment and human health due to their persistence in the environment and bioaccumulation through the food web (Sidhu et al. 2019). The major contaminants of toxic water are heavy metal ions (Ag(I), Pb(II), Co(II), Cd(II), Hg(II), Cu(II), Ni(II), Cr(III), As(III), and Cr(VI)), which are creating severe public health and environmental problems (Patil 2015). The currently available water treatment technologies such as activated carbon adsorption, ozonation, and advanced oxidation processes, coagulation-flocculation, membrane processes, halogens (Cl, Br), sedimentation, boiling, distillation, reverse osmosis, solvent extraction, evaporation, ultraviolet light, low-frequency ultrasonic radiations, ion exchange water softener, and neutralization and remineralization are competent to remove some specific particulates; however, these approaches are not effective for the removal of endocrine-disruptor chemicals (Arbabi et al. 2015;

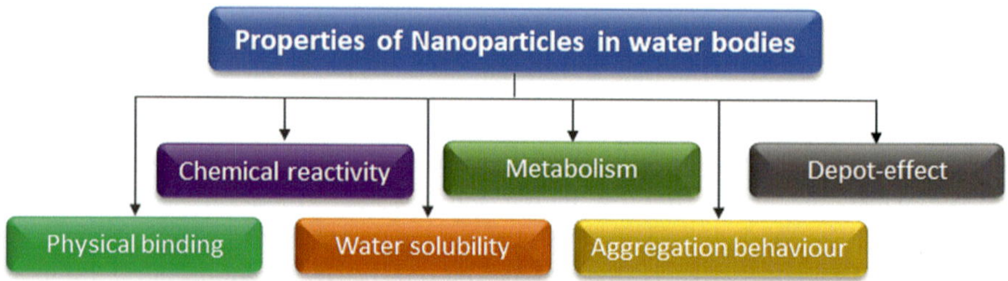

Fig. 6.2 Properties of nanoparticles in water bodies

Bonvin et al. 2016). Therefore, wastewater treatment has become one of the most worrying topics all over the world, and there is a dire need for tailoring treatment methods and the generation of eco-friendly collaborative approaches with efficient wastewater management systems. Nanotechnology is one of the finest and most advanced techniques for wastewater treatment. Nanoparticles have gained special attention in the last few decades for industrial and municipal wastewater treatment due to their enhanced absorbing, interacting, and reacting capabilities (Kamaly et al. 2016). The properties of nanoparticles in water bodies are listed in Fig. 6.2. Nanoparticles having applicability in wastewater ecosystems like nanocomposites, MNPs, CNTs, quantum dots (QDs), and zerovalent metal nanoparticles are promising tools for wastewater treatment due to their ability to remove various contaminants like bacteria, emerging pollutants, inorganic anions, and organic pollutants (Mendez et al. 2017; Prasad and Thirugnanasanbandham 2019). The effective degradation and dechlorination of toxic contaminants have been achieved by immobilization of metallic nanoparticles on polyvinylidene fluoride (PVDF), cellulose acetate, polysulfone, and chitosan membranes, which offer many advantages such as organic partitioning, high reactivity, lack of agglomeration, and reduction of surface passivation (Tian et al. 2011).

6.4.1 Nanomaterials Used for Environmental Applications

For the removal of charged particles, sediments, heavy metals, chemical effluents, microbial pathogens, different nano-structured materials such as carbon nanotubes, magnetic nanoparticles (MNPs), carbon-iron nanoparticles, nanofiltration membranes, photocatalytic titania nanoparticles, silver-impregnated cyclodextrin nanocomposites, nanocatalysts, functionalized silica nanoparticles, quantum dots (QDs), and iron zeolites are employed. The applications of nanomaterials in water purification and wastewater treatment are summarized in Table 6.1.

Based on the type of material, nanoparticles are classified into the metallic, semiconductor, and polymeric nanoparticles (Liu 2006). The nanomaterials used in water purification system are carbon-based nano-adsorbents, metal-based nano-adsorbents, polymeric nano-adsorbents, nanofiber membranes, nanocomposite

Table 6.1 Applications of nanomaterials in water purification and wastewater treatment

S.no.	Application	Type of nanomaterial used	Properties	Technologies adapted for water purification/wastewater treatment	Limitations
1.	Membranes and membrane process	Nano-zeolites, Nano-Ag, CNTs, aquaporin, Nano-TiO_2, Nano-magnetite	High reliable and automated process, molecular sieve, hydrophilicity, small diameter, tunable opening, atomic smoothness of inner surface	Anti-biofouling membranes, aquaporin membranes, reactive membranes, high performance thin film nanocomposite membranes, forward osmosis, sea water desalination	Requires high-energy source
2.	Nanoadsorbent	CNTs, nanoscale metal oxide, nanofibers with core-shell structure	High specific surface area, good adsorption capacity	Adsorption media filters, slurry reactors, reactive nano-adsorbents, contaminant preconcentration/ detection	High production cost
3.	Nanometals and metal oxides	Noble metal NPs, magnetic NPs, CNTs	High specific surface area, short intraparticle diffusion distance, compressible without change in surface area, abrasion-resistant, magnetic and photocatalytic nature	Sample preconcentration and purification groundwater remediation	Less reusable
4.	Photocatalysis	Nano-TiO_2, fullerene derivatives	High stability, low cost, low human toxicity	Photocatalytic reactors, solar disinfection systems	Reaction selectivity
5.	Disinfection and microbial control	Nano-Ag, Nano-TiO_2, CNTs	Ease of use, low human toxicity, strong and wide spectrum antimicrobial activity	Anti-biofouling surface, disinfection and decontamination	Lack of disinfection residue
6.	Sensing and monitoring	Quantum dots, dye-doped silica NPs, CNTs, magnetic NPs	High sensitivity and stability, broad absorption spectrum, superparamagnetism, high mechanical strength and chemical stability, excellent electronic properties, enhanced localized surface plasmon resonance, tunable surface chemistry	Optical detection, electrochemical detection	Less limit of detection, high cost of method implementation

Qu et al. (2013) and Gehrke et al. (2015)

membranes, thin-film nanocomposite (TFN) membranes, and biologically inspired membranes (Zhao et al. 2011a; Qu et al. 2013). Nanomaterials often possess novel properties such as high specific surface area, strong and wide-spectrum antimicrobial activity, high chemical stability, low cost, photocatalytic activity, ease of use, high conductivity, high mechanical strength, tunable surface chemistry, and superparamagnetism, which have been explored by researchers for their applications in water purification and wastewater treatment. Some nanomaterials like nano-TiO_2, nano-Ag, nano-Ce_2O_4, nano-ZnO, fullerenes, and CNTs exhibit antimicrobial properties and lead to less formation of toxic disinfection by-products (DBPs) (Qu et al. 2013). MNPs are becoming very popular for the removal of organic pollutants, natural organic matter, biological contaminants, and toxic heavy metals or elements like cations, arsenic, nitrates, and fluorides due to their large surface area and superparamagnetic nature. They can be produced from materials that are strongly attracted by magnets or be magnetized, and hence these MNPs can be used to capture different agents and biomolecules before their detection (Ito et al. 2005). MNPs are available in different phases and compositions, including pure metals like Co and Fe, alloys like FePt and $CoPt_3$, iron oxides like Fe_2O_3 and g-Fe_2O_3, and spinel-type ferromagnets like $CoFe_2O_4$, $MgFe_2O_4$, and $MnFe_2O_4$ (Patil 2015). Magnetic nanosorbents are interesting tools for the removal of organic contaminants by magnetic forces; they also act as ion exchangers, cleaning agents, and cost-effective regenerated nanosorbents and hence promoted for commercialization (Campos et al. 2011). Dendrimers are tailored polymeric nano-adsorbents in which exterior branches are tailored at the terminals with $-NH_2$ and $-OH$ group for the adsorption of heavy metals, and the interior shell is hydrophobic for sorption of organic compounds and hence used for the removal of both heavy metals and organics (Tian et al. 2011; Qu et al. 2013). Table 6.2 gives an overview of different adsorbing nanomaterials used for the removal of heavy metals, dyes, and pesticides.

Nanostructured catalytic membranes (NCMs) offer many advantages such as the easy capability of optimization, high uniformity of catalytic sites, allowing sequential reactions, limited contact time of catalysts, and ease in industrial scale-up, which allows their use for water contamination treatment (Prachi et al. 2013). Nanocatalysts possess high catalytic activity due to their large surface area and shape-dependent properties, which enhances their reactivity and degradation of various environmental pollutants like azo dyes, organochlorine pesticides, nitro aromatics, polychlorinated biphenyls (PCBs), halogenated herbicides, and halogenated aliphatics (Wang and Gu 2015). QDs refer to the quantum confinement of electrons and hole carriers at dimensions smaller than Bohr radius and QD nanocrystals composed of group II, III, V, and VI atoms such as CdS, CdSe, and CdTe at their core. They are compatible to use with biological molecules, which are present in aqueous solution due to their high stability against photobleaching and insolubility in water with size-dependent tunable emission characteristics (Gao et al. 2002). Membrane processes are used for the removal of undesired constituents from water, and membranes provide physical barriers in water treatment by providing a high degree of automation and less chemical and land use by providing flexible design. The major challenges associated with currently available membrane processes are high energy

Table 6.2 Nanoadsorbents used for removal of different contaminants with their adsorption capacity and rate constant

S.no.	Category of contaminant	Contaminant type	Adsorbing nanomaterial used	Adsorption capacity (mg/g)	Rate constant (k_1, h^{-1})	Reference
1.	Heavy metal	As(V)	Akaganeite nanocrystals	120	–	Deliyanni et al. (2003)
		Co(II)	Magnetic chitosan nanoparticles	27.5	–	Chang et al. (2006)
		As(V)	Aluminosilicate treated with Fe(II) nanoparticles	22.5 for zeolite, <18 for clinoptilolite, 10 for metakaoline	–	Hristovski et al. (2007)
		Zn(II)	Akaganeite nanocrystals	27.61	–	Deliyanni et al. (2007)
		Cd(II)	Graphene	106.3	–	Zhao et al. (2011b)
		Co(II)	Graphene	68.2	–	Zhao et al. (2011b)
		Pb(II)	Diethylenetriamine-modified multiwalled carbon nanotube	58.26	–	Vukovic et al. (2011)
		Zn(II)	Polyaniline/rice husk nanocomposite	24.3	–	Ghorbani et al. (2012)
		Cr(VI)	Multiwalled carbon nanotubes	2.35	0.42	Dehghani et al. (2015)
2.	Dyes	Methylene blue	Co_3O_4/SiO_2	53.87	–	Kannan and Sundaram (2001)
		Reactive blue 19	MgO	166.7	–	Legrouri et al. (2005)
		Nitrofurazone	CNT	59.9	–	Moussavi and Mahmoudi (2009)
		Chlorpyrifos	Graphene	1200	–	Visa et al. (2009)
		Reactive red	Activated carbon	10	–	Rafatullah et al. (2010)

(continued)

Table 6.2 (continued)

S.no.	Category of contaminant	Contaminant type	Adsorbing nanomaterial used	Adsorption capacity (mg/g)	Rate constant (k_1, h^{-1})	Reference
		Reactive red M-2BE	Multiwalled carbon nanotubes	335.7	2.860	Machado et al. (2011)
		Malachite green	$CuFe_2O_4$	200	–	Feng et al. (2012)
		Bromothymol blue	Oxidized multiwalled carbon nanotubes	55	0.042	Ghaedi et al. (2012)
		Rhodamine B	Polyacrylamide/$Ni_{0.02}Zn_{0.98}O$ nanocomposites	–	8.88	Kumar et al. (2014)
		Congo red	Multiwalled carbon nanotubes	352.1	3.18	Zare et al. (2015)
		Bromothymol blue	Polyvinyl alcohol	276.2	4.266	Agarwal et al. (2016)
		Remazole-red (RR-133)	$CoFe_2O_4$-humic acid(HA)	277.77	–	Ibrahim et al. (2019)
3.	Pesticides	Endosulfan	Reduced graphene oxide nanosheets	1100	–	Maliyekkal et al. (2013)
		Malathion	Reduced graphene oxide nanosheets	800	–	Maliyekkal et al. (2013)
		Chlorpyrifos	Reduced graphene oxide nanosheets	1200	–	Maliyekkal et al. (2013)
		Aldrin	Fe_3O_4	24.7	–	Lan et al. (2014)
		Lindane	Fe_3O_4	10.2	–	Lan et al. (2014)
		Dieldrin	Fe_3O_4	21.3	–	Lan et al. (2014)
		Endrin	Fe_3O_4	33.5	–	Lan et al. (2014)
4.	Others	Trichloroethylene	Al_2O_3/multiwalled carbon nanotube	19.84	1.1048	Liang et al. (2015)
		Nitrofurazone	Powder-activated carbon	50.8	0.1129	Wu and Xiong (2016)

consumption, membrane fouling, and complexity of process design and operation. In the past few years, functional nanomaterials are incorporated into membranes to improve fouling resistance, membrane permeability, thermal and mechanical stability, and self-cleaning. The distinguished specialty of nanofiltration membranes is their high rejection rate to monovalent and multivalent ions (Kanchi 2014). The nano-membranes used for water treatment are nanofibers, nanocomposites, thin-film nanocomposites (TFN), and biologically inspired membranes (Qu et al. 2013). The nanomaterials used for constructing nanocomposite membranes are photocatalytic nanomaterials (TiO_2), metal-oxide NPs (TiO_2, Al_2O_3, zeolite), and antimicrobial NPs (CNTs, nano-Ag) (Pendergast et al. 2010).

6.4.2 Action Mechanism of Nanomaterials

Nanomaterials possess the properties of oxidation, reduction, precipitation, and adsorption, and hence they are used worldwide for the degradation of many pollutants like inorganic anions, phosphates, nitrates, phenols, chlorinated and halogenated organic compounds, radio elements, nitroaromatic compounds, and organic dyes (Lu et al. 2016). The different antimicrobial action mechanisms of nanomaterials are discussed in Fig. 6.3.

Silver nanoparticles (Ag NPs) possess good antimicrobial activity, and they have been used extensively for wastewater treatment against a wide range of bacteria, viruses, and fungi. Ag NPs increase the permeability of cell membrane by the generation of free radicals, which finally results in cell apoptosis (Le et al. 2012). Ag

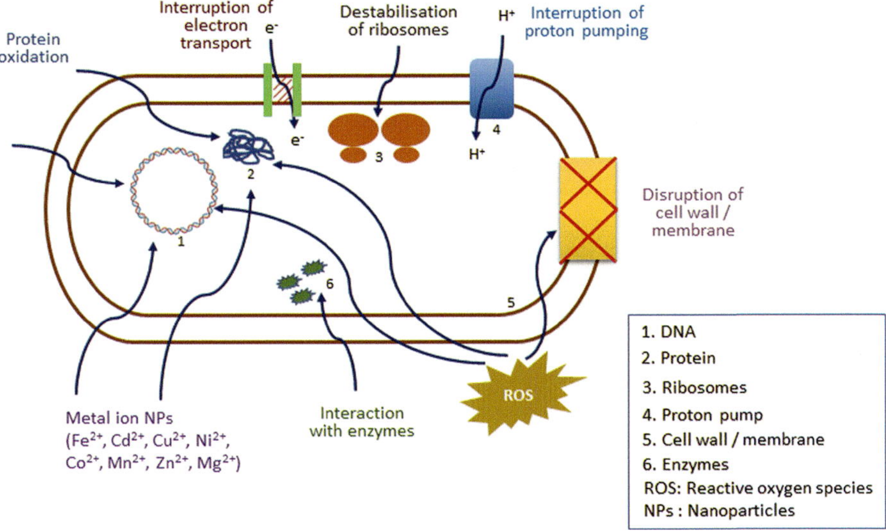

Fig. 6.3 Action mechanisms of different nanomaterials

NPs may also damage DNA by acting on its sulfur (S) and phosphorous (P) elements, or they may alter normal life functioning by inactivating metabolically active enzymes (Aziz et al. 2019). Some research groups have studied the action mechanism of Ag NPs and observed that these NPs can kill pathogenic bacteria by inducing physical perturbation with oxidative stress through the disruption of specific microbial processes by oxidation or disturbing vital cellular components or cell membrane structure (Seo et al. 2014; Prasad et al. 2018). The antagonistic effect of TiO_2 NPs is linked to the synthesis of reactive oxygen species, ROS (H_2O_2 and OH^-) by ultraviolet (UV)-A irradiation through the oxidative and reductive pathways (Li et al. 2014). The action mechanism for the antimicrobial behavior of zinc oxide (ZnO) NPs is a photocatalytic generation of ROS, disorganization of cellular membrane, and penetration of cellular envelope (Bhuyan et al. 2015). Different mechanisms associated with iron oxide (FeO) NPs under the adsorption of contaminants from wastewater are magnetic selective adsorption, surface binding, electrostatic interactions, and ligand combinations (Yang and Yin 2017). Most of the previously reported studies on wastewater treatment using Zn NPs and Fe NPs are based on dehalogenation reactions (Mahgoub and Samaras 2014; Anjum et al. 2016; Ghosh et al. 2017). Fe NPs are oxidized into Fe(II) and H_2 by molecules of water or protons under anaerobic conditions, and the resultant Fe(II) is oxidized to Fe(III) to form $Fe(OH)_3$, which acts as an effective flocculent for the removal of various inorganic and organic contaminants from the ecosystem (Klacanova et al. 2013). The oxidized CNTs possess a major adsorption sites for metal ions by chemical bonding and electrostatic interactions through surface functional groups like –OH, and –COOH. CNTs have proved to be better adsorbents with high adsorption kinetics for heavy metals (Pb^{2+}, Cd^{2+}, Cu^{2+}, and Zn^{2+}) and short intraparticle diffusion distance and faster kinetics and possess highly accessible adsorption sites (Lu et al. 2006; Qu et al. 2013). Alumina nano-adsorbents have a high surface area with good thermal stability, and they can be prepared at a low cost. These nano-adsorbents have been used in wastewater treatment for the removal of Cd, Cr, Hg, and Pb metal ions (Pacheco and Rodriguez 2001). The graphene-based nanomaterials show their antimicrobial action by oxidative stress and microbial membrane damage (Qu et al. 2013). The metal-based nano-adsorbents are involved in adsorption of heavy metals and radionucleotides in a two-step process (Trivedi and Axe 2000), and the sorption process is controlled by complex formation between metals and oxygen present in metal oxides (Koeppenkastrop and Decarlo 1993). The sorption of heavy metals and organic compounds on dendrimers is achieved by hydrogen bonding, complexation, hydrophobic effect, and electrostatic interactions (Crooks et al. 2001). The nanomaterials used as photocatalysts for photocatalytic water treatment are TiO_2, WO_3, and some fullerene derivatives (Kominami et al. 2001). The photocatalyst-based (TiO_2, CeO_2, and CNTs) reaction pathways (radical mediated and nonradical mediated) for the degradation of organic pollutants have also been suggested in the literature (Nawrocki and Hordern 2010).

6.4.3 Conventional and Modern Techniques Used for Water Treatment

There are various conventional methods that have been utilized currently for the treatment of wastewater such as screening, sedimentation, evaporation, precipitation, floatation, filtration, coagulation, gravity separation, distillation, setting-out, oxidation, solvent extraction, adsorption, reverse osmosis (RO), microfiltration and ultrafiltration, electrodialysis, ion exchange, electrolysis, fluidization, centrifugal and membrane separation, oxidation and reduction, neutralization, and remineralization (Westerhoff et al. 2009; Ali 2012). The major limitations of conventional methods are high energy input and generation of toxic chemicals, which are not recommended, and such methods are also time-consuming. Microfiltration systems are not efficient to remove sodium, fluorides, nitrates, metals, volatile organics, color, and viruses, which leads to a common issue of membrane fouling. The conventional pathogen indicator systems such as the detection of coliform bacteria areis very slow and time-consuming. Moreover, this method is inefficient to monitor the presence of some emerging pathogens such as bacteria (*Legionella, Helicobacter*), viruses (coxsackieviruses, adenoviruses, hepatitis A and Echoviruses, Norwalk viruses), and protozoans (*Giardia, Cryptosporidium*) (Theron et al. 2010). The existing water treatment technologies are not 100% effective against notorious anthropogenic pollutants, which pose serious challenges from purifying contaminated water to supply potable water. Therefore, new and improved water purification technologies are extremely in demand. Nanotechnology is one of the modern techniques that has been utilized for the last few years for water purification and wastewater treatment.

6.4.4 Nanotechnology in Water Purification

Nanotechnology can be used for economic purification and utilization of different water resources. Some of the distinctive features of nanomaterials like high aspect ratio, hydrophobic, hydrophilic and electrostatic interactions, reactivity, and tunable pore volume are helpful in sensors, adsorption, and catalysis (Qu et al. 2013). An nanostructured membrane formed by N-doped "nut-like" ZnO nanostructured material was proved efficient in a research study for removal of water contaminants by increasing its photodegradation activity under irradiation of visible light due to its antibacterial activity (Bai et al. 2012). One of the commonly employed techniques for removal of organic and inorganic contaminants from wastewater is adsorption. In this regard, nano-adsorbents offer a significant role in the adsorption process due to their short intraparticle diffusion distance, high specific surface area, tunable pore size, and surface chemistry (Qu et al. 2013). CNTs belong to the carbon family, and based on the layering system of nanotubes, they are differentiated into single-walled and multi-walled CNTs (Sears et al. 2010). CNTs have drawn attention for water treatment applications due to their high chemical stability (Okpalugo et al. 2005), high thermal stability (Kundu et al. 2008), and excellent mechanical and chemical properties (Esawi et al. 2010). CNTs are used as adsorbents for the removal of organic, inorganic, and biological pollutants from wastewater due to their layered structure (Pande et al. 2009), high specific adsorption surface area (Peigney et al. 2001), and hollowness (Zhang and Zhu 2006). The abovementioned properties of CNTs make them a perfect adsorbent for various pollutants, which promise their scaffold functions in wastewater purification systems. The fabricated features of nano-adsorbents are high surface area, small size, a large number of active sites for interaction with contaminants, high catalytic potential, and ease of separation, and these properties make them an ideal adsorbent material for water purification and treatment (Ali 2012). The nanomaterials currently used in nano-adsorbent technologies are magnetic or nonmagnetic oxide composites, metal-based nano-adsorbents, carbon-based nano-adsorbents, zeolites, etc. (Rafiq et al. 2014; Gehrke et al. 2015). The functionalization of CNTs is done by noncovalent wrapping, endohedral filling, and treatment with covalent chemical agents (Spitalsky et al. 2010). CNTs are used for the adsorption of Cd, Pb, and 1,2-dichlorobenzene (DCB) through hydrophobic interactions, hydrogen bonding, covalent bonding, p-p electron coupling, mesopore filling, and electrostatic interactions. CNT bundles contain a significant number of micropores, which are inaccessible to pharmaceuticals and antibiotics. Organic compounds having –OH, –COOH, and $–NH_2$ functional groups can form a hydrogen bond with graphitic CNT surface, and electrostatic interactions are involved in the adsorption of antibiotics to CNT surface at suitable pH (Ji et al. 2009). The literature reports have already proved the significant importance of CNTs for industrial applications due to its regeneration capacity, and the metal recovery rate is approximately 90%, which reaches to 100% at pH <2 (Lu et al. 2006). Previously, multiple regeneration and reuse cycles have been detected with CNTs for Zn^{2+} using statistical analysis based on best-fit regression and found that the nano-adsorbent can be regenerated and reused several times along with maintaining its adsorption capacity (Lu et al. 2007). Metal oxide nanocrystals can be compressed into porous pellets, and they can be used industrially in different forms, that is, porous pellets and fine powders (Lucas et al. 2001). Metal-based nanomaterials have great potential to outcompete activated carbon for the removal of a variety of heavy metals including arsenic (As), mercury (Hg), lead (Pb), nickel (Ni), cadmium (Cd), copper (Cu), and chromium (Cr) (Sharma et al. 2009) due to their low cost, high adsorption capacity, and easy separation and regeneration, which make them more advantageous over conventional adsorbents. Previously, metal hydroxide nanoparticles (MH NPs) were used in point-of-use (POU) applications for the simultaneous removal of As and co-contaminants by impregnating them into the skeleton of activated carbon or other porous materials (Hristovski et al. 2009). Nanocomposites are made up of ordered mesoporous carbons, and they have gained significant attention in water purification for the adsorption of many contaminants like heavy metal ions, pesticides, and dyes due to their large surface area (Amin et al. 2014). The antifouling property of nanocomposites can be improved by incorporation of different photoactive or antimicrobial nanomaterials (Rodrigues et al. 2017). Some of the nanomaterials used in nanocomposite membranes are antimicrobial nanoparticles (nano-Ag, CNTs), (photo)catalytic nanomaterials (TiO_2, bimetallic NPs), and hydrophilic metal oxide NPs (TiO_2, Al_2O_3, zeolites) (Qu et al. 2013). The introduction of carboxyl-functionalized MWCNTs into TFN membranes has improved both chlorine resistance and antifouling property due to improved surface hydrophilicity and greater negative surface charge (Zhao et al. 2014). Significant removal of pathogens and contaminants has been achieved with ceramic and polymeric membranes containing TiO_2 under UV irradiation, and these biomembranes are less vulnerable to organic and biological fouling (Berekaa 2016). Previously, researchers have conducted some studies for developing efficient water purification systems by embedment of nanocomposite membranes with different functional nanoparticles like photocatalytic, antimicrobial, and catalytic NPs (Maximino et al. 2018; Guerra et al. 2018). Nanofiltration membranes are used in industrial water treatment, brackish water treatment, and desalination as two-pass nanofiltration systems (Kanchi 2014). Dendrimers also called dendritic polymers are symmetrical, highly branched, monodispersed, spherical macromolecules with controlled composition, and it includes random hyperbranched polymers, dendrigraft polymers, dendrons, and dendrimers. It consists of three components, that is, core, interior branch cells, and terminal branch cells (Frechet and Tomalia 2001). For the removal of metal ions like Cu^{2+}, a dendrimer-ultrafiltration system was developed, and recovery and regeneration of such system were achieved by decreasing the pH to 4.0 (Diallo et al. 2005). Nanofiber membranes are used for the removal of organic contaminants and heavy metals during filtration at a high rejection rate without fouling problems, and they have been reported to be used for pre-treatment before the reverse osmosis (RO) or ultrafiltration (Ramakrishna et al. 2006). Nanofibrous alumina filters are more efficient for the removal of negatively charged contaminants like bacteria, viruses, and organic and inorganic colloids at a faster rate as compared to conventional filters (Kanchi 2014). The use of polyvinyl-N-carbazole-SWNT composites leads to high bacterial inactivation (>90%) (Ahmed et al. 2012), and the incorporation of photocatalytic NPs into membranes improved both physical separation function and reactivity of catalyst toward degradation of contaminants (Choi et al. 2006). TFN membranes and nanocomposites can improve energy efficiency of water treatment with low osmosis pressure, which are known to have good scalability due to their fabrication, which enhanced their use in industrial manufacturing processes (Elimelech and Phillip 2011). The higher permeability and better salt rejection (>99.4%) have been achieved by using TFN membranes doped with nano-zeolites (250 nm) at 0.2% wt compared to commercial RO membranes (Lind et al. 2010).

6.4.5 *Nanotechnology in Wastewater Treatment*

Nanotechnology-based processes are highly modular, efficient, and multifunctional in nature with high-performance and wastewater treatment solutions, and hence this revolutionary technology has overcome major challenges of existing methods. The major benefits of wastewater treatment using nanofiltration are low energy requirement and operation cost, removal of color, tannins, turbidity, pesticides, volatile organic chemicals (VOCs), sulfates, nitrates and total dissolved solids (TDS), and softening of hard water (Patil 2015). Metal oxide nanoparticles (MOPs) have proved technologically and economically advantageous for wastewater technologies due to their low cost, short intraparticle diffusion distance, high specific surface area, more adsorption sites, easy to reuse, superparamagnetic, and environmentally friendly nature (Corsi et al. 2018). There are several previous studies that have been documented with MOPs for the removal of heavy metals (chromium (Cr), copper (Cu), lead (Pb), mercury (Hg), arsenic (Ar), and nickel (Ni)) from wastewater (Gunatilake 2015). TiO_2 NPs have been proven in the complete mineralization of organic contaminants by ozonation (Panahi et al. 2018), and they are shown to be extensively used as photocatalyst NPs in wastewater treatment due to their cost-effectiveness, high chemical stability, and photocatalytic generation of reactive oxygen species (ROS) (Qu et al. 2013). TiO_2 NPs are also effective for the removal of a large number of viruses like hepatitis B virus (Zan et al. 2007), herpes simplex virus (Hajkova et al. 2007), poliovirus (Liga et al. 2011), and MS2 bacteriophages (Cho et al. 2011). The Ag NPs have proved to show biocidal effects against *Escherichia coli, Klebsiella pneumonia, Pseudomonas aeruginosa*, and *Staphylococcus aureus* (Jain and Pradeep 2005). During in situ remediation experiments, Ag NPs immobilized on cellulose fibers have shown antibacterial activity against *Escherichia coli* and *Enterococcus faecalis* (Xu et al. 2009; Chitra and Annadurai 2014; Peiris et al. 2017). Nowadays, high porosity filters, built with sawdust and clay, have been used to improve the efficiency of Ag NPs for the removal of *E. coli* and other pathogens from wastewater (Amin et al. 2014). Zinc nanoparticles (Zn NPs) are an alternative to iron nanoparticles (Fe NPs), and Zn is a stronger reductant than Fe with a high negative reduction potential and high rate of contaminants degradation, and it has been used extensively in wastewater treatment (Fu et al. 2014). There are many previously reported studies on the potent role of Zn NPs for wastewater treatment in conventional activated sludge (CAS) (Puay et al. 2015; Zheng et al. 2011). The efficient degradation of octachlorodibenzo-p-dioxin into substituent chlorinated congeners has been achieved by using Zn NPs under ambient conditions (Bokare et al. 2013). ZnO NPs are efficient candidates for reducing biological nitrogen and pathogenic bacteria from wastewater due to their strong oxidation ability, wide bandgap in UV spectral region, and better photocatalytic properties (Zheng et al. 2011; Rana et al. 2018). The effectiveness of ZnO NPs can be improved by doping and metal dopants, which have been used are cationic dopants, anionic dopants, codopants, and rare earth dopants (Lee et al. 2016; Badreddine et al. 2018). Zerovalent metal nanoparticles are elemental metallic iron with zero charge-bearing Fe atoms, and they are used for the remediation of contaminated sediments, soils, and wastewater bodies due to their distinctive mechanical, electrical, optical, magnetic, and catalytic properties (Nam and Lead 2016; Prasad et al. 2017). FeO NPs possess novel properties and functions such as ease of synthesis, modifications and coating, high surface volume, nanoscale size, low toxicity, chemical inertness, biocompatibility, and superparamagnetic, which result in their marvelous applications in amalgamation with biotechnology (Dinali et al. 2017). The applications of FeO NPs in the treatment of wastewater are based on photocatalytic technologies or adsorptive technologies. FeO NPs are used as semiconductor photocatalysts in photocatalytic technologies for converting contaminants into less toxic compounds and in adsorptive technologies. FeO NPs are used as immobilization carriers/nano sorbent for the efficient removal of pollutants (Saharan et al. 2014). In 2003, Fe NPs were proven to be effective for the transformation and detoxification of a large number of contaminants such as polychlorinated biphenyls (PCB), chlorinated organic solvents, and organochlorine pesticides from groundwater and contaminated soil (Doyle 2006). Zerovalent iron (ZVI) nanoparticles have shown positive results for developing on-site wastewater remediation technologies for detoxifying chlorinated organic compounds (COCs), hexachlorocyclohexane (lindane), trichloroethane, vinyl chloride, and carbon tetrachloride (Crane and Scott 2012). ZVI NPs are very efficient for removing Ni and Pb from wastewater, and they are 25–30 times faster with higher sorption capacity compared with granular Fe (Liu et al. 2006). The other metals used for zerovalent metal nanoparticles are zinc and silver (Sharma et al. 2015). The limitations associated with the use of ZVI NPs are aggregation, oxidation, and separation difficulty from contaminant degradation system (Guan et al. 2015); therefore, new approaches (surface coating, doping, and emulsification with metal ions or conjugation with supports) have been utilized initially to improve the reactivity, efficiency, and dispersibility of ZVI NPs (Dutta et al. 2015). QD NPs are used for target-based detection and removal of contaminants such as polycyclic aromatic hydrocarbons, organic and inorganic pollutants, and heavy metals from surface water and groundwater, and this detection is done by detecting changes in optical emission signals (Dutta et al. 2015; Zhang et al. 2017). Their functionality is better than dye-based sensors due to their longer emissions and compatibility with molecularly imprinted polymers (MIPs) (Foguel et al. 2017). Previously, the detection of 4-nitrophenol was done using Mn-doped ZnS QDs anchored on MIP, and MIP-QD nanospheres have fast adsorption and desorption kinetics with high recognition selectivity in aqueous media. MIPs are advantageous over other sorbents due to the selective extraction of molecules, and the electrospinning method is commonly used to encapsulate MIP NPs in nanofibers for controlling water pollution (Kiparissides 2010). Elmizadeh et al. (2018) have developed a sensitive nanosensor based on synthetic ligand-coated CdTe QDs for rapid detection of Cr(III) ions in wastewater samples. ZnO NPs having Pd nanocatalysts possess high photocatalytic activity for the removal of *E. coli* from wastewater (Khalil et al. 2011). The adsorption of As(V) in the form of arsenate salts has been done by paramagnetic ferrite-based nanoparticles ($CuFe_2O_4$) from contaminated groundwater (Tu et al. 2012). The halogenated organic compounds (HOCs) are biodegraded using Pd nanocatalysts by initial conversion of HOCs into organic compounds using nano-sized Pd catalysts followed by their degradation in the treatment plant. The Pd catalysts are easily recycled and reused due to their ferromagnetic property, which also indicates their further use for industrial applications (Zhang et al. 2012). Photocatalysis is one of the advanced environmentally friendly, low-cost oxidation process for the removal of microbial pathogens, trace contaminants, and hazardous and nonbiodegradable contaminants. The slow kinetics of photocatalysts due to their limited photocatalytic activity and light fluorescence are the major obstacles to industrial applications of this process. The other technical challenges associated with large-scale applications of this process in wastewater treatment are requirement of efficient photocatalytic reactor design, catalyst recovery/immobilization techniques, catalyst optimization to improve quantum yield, and better reaction selectivity. The efficiency of photocatalytic water treatment depends on the operation parameters and configuration of the photoreactor. Nanomaterials are used in different water treatment-related surfaces to prevent biofilm formation, microbial influenced corrosion, and control of pathogen contamination in distribution pipes and storage tanks. The monitoring of water quality is one of the major challenges for wastewater treatment due to extremely low concentrations of certain contaminants and the lack of a fast and efficient pathogen detection systems with high sensitivity and selectivity. The commonly used nanomaterials used in pathogen detection are QDs, MNPs, CNTs, and dye-doped NPs. CNTs and MNPs have been used extensively for sample concentration, and purification (Qu et al. 2013). QDs are semiconductor materials (e.g., CdSe), which have broad absorption spectra for multiplex detection, and the emission spectra of QDs are 10–20 times brighter than organic fluorophores (Yan et al. 2007). These are 1000 times more stable than conventional dyes (Sukhanova et al. 2004; Zare et al. 2015). Ag NPs doped with organic or inorganic luminescent dyes have been developed to construct ultrasensitive sensors, and their outstanding photostability makes them advantageous for the detection of biomolecules, which require a high intensity of prolonged excitations (Yan et al. 2007). It has been reported in a previous study that bisphenol A could be detected in water at a very low concentration (10 nm) within 5 s using a nanosensor based on CoTe QDs immobilized on a glassy carbon electrode surface (Yin et al. 2010). Multifunctional devices are another advancement of functional nanomaterials, which are used in water treatment devices for performing multiple tasks in one device to enhance overall performance, miniaturizing the footprint, and avoiding excessive redundancy. The small footprint, high performance, and modular design of nanotechnology-based devices also make their potential use for commercial applications (Qu et al. 2013).

6.5 Toxicity Facts and Safety Issues Associated with Nanomaterials

Nanotechnology is better than other techniques used in wastewater treatment; however, currently, the knowledge about the transport, environmental fate, and toxicity of nanomaterials is still in its infancy. There is a poor understanding of the fate and behavior of nanomaterials in the environment and human toxicity. Therefore, the major concern related to nanotechnology is the hazardousness of nanoparticles due to their leakage, spillage, circulation and concentration, and risks associated with their exposure (Gardner and Dhai 2014). The NPs enter the bloodstream through inhalation, ingestion, and skin, from where they can travel to different body parts like the brain, nervous system, bone marrow, spleen, kidney, liver, and heart (Ali 2012). Due to small the size of NPs, their chemical reactivity is very high, which leads to the generation of reactive oxygen species (ROS), which may cause inflammation and oxidative stress resulting in damage to DNA and membrane proteins. Some NPs can alter the mechanism of enzymes and different proteins by absorbing through the skin surface (Hubbs et al. 2011). Most of the previous toxicological studies have been conducted with mammalian cells and carried out in cell culture medium containing biological compounds. However, the toxicological results from mammalian studies cannot be directly correlated to environmental conditions. The toxicity effects of NPs in the environment are due to their agglomeration nature. The use of CNTs is not expected at an industrial scale in wastewater treatment plants due to their high production cost and coagulation phenomenon with some organic contaminants and because of several reports of health-related concerns of NPs (Ali 2012; Volder et al. 2013). There are some players that are highly concerned regarding the problems of toxicity and the safety of nanotechnology for remediation. DuPont is the major pioneer, which the adopted use of ZVI NPs for site remediation until problems regarding their fate and transport, post-remediation persistence, and potential human exposure to NPs at these places have been thoroughly studied (Chaturvedi and Dave 2019). The opportunities and challenges for the role of nanomaterials in water purification and treatment of contaminated water are a matter of continuing concern. Hence, there are various challenges for large-scale industrial applications of nanotechnology in water purification and wastewater treatment, and the behavior of NPs inside the body is still a major question, which needs to be addressed before marketing nano-based products. Further, a thorough understanding of the mobility, toxicity, bioavailability, and persistence of NPs is needed to assess their risks.

6.6 Regulatory Aspects of Nanomaterials

Nanotechnology is one of the most challenging technologies of the current century, and billions of dollars have been spent annually for research and marketing of these processes. Nanomaterials that have been used for water purification and wastewater treatment must be nontoxic and environmentally friendly. Approximately, 1.5 billion USD has been allocated to National Nanotechnology Initiative (Arlington, VA, USA) through the federal budget for nanotechnology-related activities in 20 departments and independent agencies (Nano.gov 2014). In 2020, European Union (EU) provided 110 billion USD for research and innovative projects and 92 million USD for water innovations (Horizon 2020). Several initiatives have been taken by applying laws and regulations at the international level to prevent environmental damage including plants, human beings, and aquatic organisms by nanomaterials. The United States and EU deal with implementing novel regulations related to nanomaterials and their exposure in water bodies (Gehrke et al. 2015). Nanomaterials are regulated by the Registration, Evaluation, Authorization and Restriction of Chemicals (REACH) in Europe, and this program regulates the production and use of chemical substances along with their impact on the environment and human health. There are many non-government organizations that have claimed the integration of current regulations with EU definition of nanomaterials (EC 2011). A framework has been established by European Water Framework Directive for the evaluation of priority substances by REACH and for protection of different water bodies, that is, groundwater, transitional water, coastal water, and inland surface water. Any amendment of REACH related to nanomaterials is implemented automatically to the water directive (EPC 2000). In the United States, the use of nanomaterials in water medium has not been mentioned clearly. European Protection Agency (EPA) has allowed the manufacture and use of nanoscale chemical materials under the Toxic Substances Control Act (EPA 1999, 2015). Although there are several regulatory guidelines that have been implemented by different countries for the safe use of nanomaterials, there are no strict regulations with reliable databases on an international scale. Therefore, their toxicity performance tests must be strictly included in standard operating procedures (SOPs) with proper documentation, before introducing nanomaterials to the industry.

6.7 Commercial Processes of Nano-Based Materials Related to Water Application

The first complete dendrimer family, which was synthesized, characterized, and commercialized in 1990, was PAMAM (poly(amidoamine)) dendrimer (Tomalia 2004). A team of US researchers has developed a sponge that can absorb oil from water, which is made up of pure CNTs with a dash of boron. The oil can be retrieved or burned off, and the sponge can be reused, and the research team is also planning to weld the sheets for oil remediation (Hashim et al. 2012). CNT-based systems are used commercially for the mitigation of different contaminants from wastewater matrices (Qu et al. 2013) and desalination of brackish water and seawater (Das et al. 2014a, b). Two cost-competitive dendrimer-based commercial systems for drinking water treatment systems available in the market are Arsen^{np} and ADSORBSIA™ (Aragon et al. 2007; Qu et al. 2013). The commercially available TFN membrane is QuantumFlux, a seawater TFN RO membrane (http://www.nanoH$_2$O.com). The commercial devices available in the market utilizing nano-Ag are MARATHON® and Aquapure® systems. In some developing countries, nano-Ag has been incorporated into ceramic microfilters to be used as a barrier against pathogens (Varbanets et al. 2009). NanoCeram (Argonide Corporation, Sanford, FL, USA) is a marketed nanofiber filter with a large surface area (300–600 m^2/g) and small diameter, which is used in ultrafiltration for the removal of bacteria, viruses, and proteins through

Coulombic interactions (Karim et al. 2009). A firm named Purific Water (Holiday, FL, USA) has developed a filtration assembly by combining water pre-treatment process with photocatalysis and ceramic filtration membrane with a capacity of >4 million cubic meters/day, and this system has been successfully used for degradation of 1,4-dioxane (http://www.purifics.com/; http://www.nanowerk.com/spotlight/spotid=4662.php). Dynabead® is a commercial nanocomposite available in the market for developing pathogen detection kits (Qu et al. 2013). TFN membranes have been used commercially by LG NanoH_2O Inc., and this technology has twice the flux of polyamide membranes with >99.7% salt rejection (Lau et al. 2015). The first commercial membrane with embedded aquaporins is Aquaporin Inside (Aquaporin A/S, Copenhagen, Denmark). This membrane is used in desalination applications, and it can withstand up to 10 bar and a water flux rate > 100 L/(hm^2) (Gehrke et al. 2015).

6.8 Concluding Remarks and Future Outlook

Nanotechnology-based strategies are gaining momentum globally for water purification and wastewater treatment solutions. This technology is more rapid, reliable, economic, and durable for the removal of a specific types of contaminants from wastewater. There are various reports on the application of nanomaterials for the removal of toxic pollutants like polycyclic aromatic hydrocarbons, volatile organic compounds, pharmaceutical, and personal care products, polychlorinated biphenyls, agrochemicals and pesticides, inorganic pollutants, polycyclic aromatic hydrocarbons, furans, phthalates, bacteria, and viruses. The convergence of existing water treatment technologies with nanotechnology provides great opportunities to revolutionize water purification and wastewater treatment. To provide sustainable water supplies, the nano-based applications have been utilized for water treatment, water filtration, and desalination and as sensors and catalysts. Based on the cost of nanomaterials involved, stages in research and development, and commercial availability, many nano-based systems for water purification and wastewater treatment are still in the research phase. However, it is expected that some of the strategies based on nano-photocatalysts, nano-adsorbents, and nano-membranes will become functional in the upcoming years. Zn NPs are not efficient for the degradation of halogenated and chlorinated organic compounds; therefore, studies should be conducted for the treatment of such compounds using other types of nanomaterials under ambient conditions to achieve maximum degradation of such kinds of pollutants from wastewater. Pilot scale studies should also be carried out at polluted field sites to investigate real-time applications of NPs to achieve better results. In developing countries, pure and safe drinking water is provided to people by developing solar disinfection (SoDis) approach using nano-TiO_2. The pilot scale applications of CNTs are very limited; hence, further improvement in existing characteristics like high water permeability, scalability, robustness, compatibility with industrial settings, energy savings, antifouling, and desalination are highly encouraged.

MIP-QDs has been proven as an efficient system for the removal of contaminants from wastewater, and this technology can be used in the future for revolutionary environment management. Although most of the nanosensors possess excellent sensitivity and photostability, nonspecific binding is still one of the major challenges for wastewater treatment applications. Therefore, there is a dire need of developing strategies for reducing nonspecific binding and preventing undesired aggregation of nanoparticles. Further research is required to develop simple low-cost nanotechnology methods and nano-based immobilizing systems with significantly enhanced performance. Many nano-based systems for water purification and water treatment applications are at the pilot or experimental phase due to various technical hurdles, cost-effectiveness, and potential environmental and human risks. Hence, research efforts addressing the treatment of wastewater under more realistic conditions are needed to assess the efficiency and applicability of nanomaterials and nanotechnologies in water and wastewater treatment. More collaborative efforts of government, regulatory bodies, industries, research institutes, and other stakeholders are required to provide fast, economical, and robust technologies to resolve the global issue of water contamination.

Questions and Answers: A Solution Manual

Objective Questions

Q1. Which technique is used for characterization of nanoparticle interaction with microbial contaminants?

(a) Scanning Probe Microscopy (SPM)
(b) Transmission Electron Microscopy (TEM)
(c) Both of the above
(d) None of the above

Ans. (c).
Q2. The sewage wastewater treatment capacity of India is

(a) 8080.8 MLD
(b) 8008.2 MLD
(c) 8642.0 MLD
(d) 9080.9 MLD

Ans. (a).
Q3. Vesicles are used for

(a) Calcination
(b) Encapsulation
(c) Crystallization
(d) All of the above

Ans. (b).

Q4. Nanoparticles of which of the following metals is used for efficient degradation of octachlorodibenzo-p-dioxin into substituent chlorinated congeners under ambient conditions

(a) Ti
(b) Ni
(c) Mn
(d) Zn

Ans. (d).

Q5. Which of the following groups of atoms, Quantum Dots are composed of?

(a) I
(b) II
(c) Both of the above
(d) None of the above

Ans. (b).

Fill in the Blanks

1. The process of nanotechnology was initiated by ________ in the year _______.

   Answer: Richard Feynman, 1959

2. Nanotechnology is used in forensic science in DNA fingerprinting for __________ and ____________.

   Answer: Parental testing, Solving criminal cases

3. The emerging pollutants in wastewater are nonbiodegradable materials due to their _______ in food web.

   Answer: Bioaccumulation

4. Dendrimers are used as nano-adsorbents for removal of heavy metals by tailoring of external branches with ____ and _____ groups.

   Answer: $-NH_2$ and $-OH$

5. Based on layering system of nanotubes, CNTs can be _____ and ______.

   Answer: Single-walled, multiwalled

Short Answer Questions

1. Which techniques are commonly used for the synthesis of nanoparticles?

   Answer: Top-down approach and Bottom-up approach.

2. What kind of nanomaterials are used for wastewater treatment in the current scenario?

   Answer: Metal oxides, zeolites, dendrimers, carbon nanotubes (CNTs), fullerenes, graphene-based nanomaterials, nanosorbents, nanocatalysts, biomimetic membrane, molecularly imprinted polymers (MIP), and zerovalent iron particles.

3. What are the applications of nanotechnology in environment sector?

   Answer: Environmental nanotechnology is a revolutionary field of science and technology, which is gaining interest of scientific community from past few decades. Nanomaterials have been explored for generation of renewable energy in batteries, fuel cells, and supercapacitors. Nanoparticles are used for environmental remediation, water purification, and wastewater treatment.

4. Which techniques are currently used for wastewater treatment?

   Answer: Activated carbon adsorption, ozonation and advanced oxidation processes, coagulation-flocculation, membrane processes, halogens (Cl, Br), sedimentation, boiling, distillation, reverse osmosis, solvent extraction, evaporation, ultraviolet light, low frequency ultrasonic radiations, ion exchange water softener, and neutralization and remineralization.

5. Which properties of nanoparticles make them suitable for application in wastewater purification and treatment?

   Answer: High specific surface area, strong and wide-spectrum antimicrobial activity, high chemical stability, low cost, photocatalytic activity, ease of use, high conductivity, high mechanical strength, tunable surface chemistry, and superparamagnetism.

6. What are the issues for not adopting the use of CNTs at industrial level?

   Answer: The use of CNTs is not expected at industrial scale in wastewater treatment plants due to their high production cost, coagulation phenomenon with some organic contaminants, and because of several reports of health-related concerns of NPs.

Long Answer Question

1. What are the different action mechanisms of nanomaterials used for wastewater purification and treatment?

   Answer: Nanomaterials are used worldwide for the degradation of many pollutants like inorganic anions, phosphates, nitrates, phenols, chlorinated and halogenated organic compounds, radio elements, nitroaromatic compounds, and organic dyes by different mechanisms such as oxidation, reduction, precipitation, and adsorption. Silver nanoparticles (Ag NPs) possess good antimicrobial activity, and they have been used extensively for wastewater treatment against wide range of bacteria, viruses, and fungi. Ag NPs increase permeability of cell membrane by generation of free radicals, which finally resulting in cell apoptosis (Le et al. 2012). Some research groups have studied the action mechanism of Ag NPs and observed that these NPs can kill pathogenic bacteria by inducing physical perturbation with oxidative stress through disruption of specific microbial process by oxidation or disturbing vital cellular components or cell membrane structure. The antagonistic effect of TiO_2 and ZnO NPs is linked to synthesis of reactive oxygen species, ROS (H_2O_2 and OH^-) by ultraviolet (UV)-A irradiation through oxidative and reductive pathway. Different mechanisms associated with iron oxide (FeO) NPs under adsorption of contaminants from wastewater are magnetic selective adsorption, surface binding, electrostatic interactions, and ligand combinations. Most of the previously reported studies on wastewater treatment using Zn NPs, and Fe NPs are based on dehalogenation reaction. The oxidized CNTs possess major adsorption site for metal ions by chemical bonding and electrostatic interactions through surface functional groups like –OH and –COOH. CNTs have proved to be better adsorbents with high adsorption kinetics for heavy metals (Pb^{2+}, Cd^{2+}, Cu^{2+}, and Zn^{2+}) and short intraparticle diffusion distance, faster kinetics, and highly accessible adsorption sites. Alumina nanoadsorbents have high surface area with good thermal stability, and they can be prepared at low cost. These nanoadsorbents have been used in wastewater treatment for removal of Cd, Cr, Hg, and Pb metal ions. The graphene-based nanomaterials show their antimicrobial action by oxidative stress and microbial membrane damage. The sorption of heavy metals and organic compounds on dendrimers is achieved by hydrogen bonding, complexation, hydrophobic effect, and electrostatic interactions. The photocatalyst-based (TiO_2, CeO_2, and CNTs) reaction pathways (radical mediated and nonradical mediated) for degradation of organic pollutants have also been suggested in literature.

2. What are the available commercial systems of nanomaterials for wastewater treatment?

   Answer: The first complete dendrimer family which was synthesized, characterized, and commercialized in 1990 was PAMAM (poly (amidoamine)) dendrimer. A team of US researchers have developed sponge that can absorb oil from water, which is made up of pure CNTs with a dash of boron. The oil can be retrieved or burned off, and sponge can be reused, and the research team is also planning to weld the sheets for oil remediation. CNT-based systems are used commercially for mitigation of different contaminants from wastewater matrices and desalination of brackish water and seawater. Two cost-competitive dendrimer-based commercial systems for drinking water treatment systems available in the market are Arsen^{np} and ADSORBSIA™. The commercially available TFN membrane is QuantumFlux, a seawater TFN RO membrane. The commercial devices available in the market utilizing nano-Ag are MARATHON® and Aquapure® systems. In some developing countries, nano-Ag has been incorporated into ceramic microfilters to be used as barrier for pathogens. NanoCeram (Argonide Corporation, Sanford, FL, USA) is a marketed nanofiber filter with large surface area (300–600 m^2/g) and small diameter, which is used in ultrafiltration for removal of bacteria, viruses, and proteins through Columbic interactions. Purific Water (Holiday, FL, USA) has developed a filtration assembly by combining water pretreatment process with photocatalysis and ceramic filtration membrane with capacity of >4 million cubic meters/day, and this system has been successfully used for degradation of 1,4-dioxane. Dynabead® is a commercial nanocomposite available in the market for developing pathogen detection kits. TFN membranes have been used commercially by LG NanoH_2O Inc., and this technology has twice the flux of polyamide membrane with >99.7% salt rejection. The first commercial membrane with embedded aquaporins is Aquaporin Inside (Aquaporin A/S, Copenhagen, Denmark). This membrane is used in desalination applications, and it can withstand up to 10 bar and water flux rate > 100 L/(hm^2).

References

Abobatta WF (2018) Nanotechnology applications in agriculture. Acta Sci Agric 2(6):99–102

Agarwal S, Sadegh H, Monajjemi M, Hamdy AS, Ali GA, Memar AO, Ghoshekandi RS, Tyagi I, Gupta VK (2016) Efficient removal of toxic bromothymol blue and methylene blue from wastewater by polyvinyl alcohol. J Mol Liq 218:191–197

Ahmed F, Santos CM, Vergara R, Tria MCR, Advincula R, Rodrigues DF (2012) Antimicrobial applications of electroactive PVK-SWNT nanocomposites. Environ Sci Technol 46(3):1804–1810

Aksay IA (1992) Hierarchically structures materials. MRS Proceeding, p 255

Algarra M, Jimenez JJ, Tost RM, Campos BB, Silva JCGE (2011) CdS nanocomposites assembled in porous phosphate heterostructures for fingerprint detection. Opt Mater 33(6):893–898

Ali I (2012) New generation adsorbents for water treatment. Chem Rev 112(10):5073–5091

Amin MT, Alazba AA, Manzoor U (2014) A review of removal of pollutants from water/wastewater using different types of nanomaterials. Adv Mater Sci Eng 2014:1–24

Anjum M, Miandad R, Waqas M, Gehany F, Barakat MA (2016) Remediation of wastewater using various nano-materials. Arab J Chem 12(8):4897–4919

Aragon M, Kottenstette R, Dwyer B, Aragon A, Everett R, Holub W, Siegel M, Wright J (2007) Arsenic pilot plant operation and results. Sandia National Laboratories, Anthony

Arbabi M, Hemati S, Amiri M (2015) Removal of lead ions from industrial wastewater: a review of removal methods. Int J Epidemiol Res 2(2):105–109

Aziz N, Faraz M, Sherwani MA, Fatrna T, Prasad R (2019) Illuminating the anticancerous efficiency of a new fungal chassis for silver nanoparticle synthesis. Front Chem 7(65):1–11

Badreddine K, Kazah I, Rekaby M, Award R (2018) Structural, morphological, optical, and room temperature magnetic characterization on pure and Mn-doped ZnO nanoparticles. J Nanomater 2018:1–11

Bai H, Liu Z, Sun DD (2012) Hierarchical ZnO nanostructured membrane for multifunctional environmental applications. Colloid Surf A Physicochem Eng Asp 410(20):11–17

Banoee M, Seif S, Nazari ZE, Fesharaki PJ, Shahverdi HR, Moballegh A, Moghaddam KM, Shahverdi AR (2010) ZnO nanoparticles enhanced antibacterial activity of ciprofloxacin against *Staphylococcus aureus* and *Escherichia coli*. J Biomed Mater Res B Appl Biomater 93(2):557–561

Berekaa MM (2015) Nanotechnology in food industry: advances in food processing, packaging and food safety. Int J Curr Microbiol Appl Sci 4(5):345–357

Berekaa MM (2016) Nanotechnology in wastewater treatment: influence of nanomaterials on microbial systems. Int J Curr Microbiol App Sci 5(1):713–726

Beyth N, Haddad YH, Domb A, Khan W, Hazan R (2015) Alternative antimicrobial approach: nano-antimicrobial materials. Evid Based Complement Alternat Med 2015:1–16

Bharathi P, Balasubramanian N, Anitha S, Vijayabharathi V, Bhuvenswari (2016) Improvement of membrane system for water treatment by synthesized gold nanoparticles. J Environ Biol 37:1407–1414

Bhattacharya D, Gupta RK (2005) Nanotechnology and potential of microorganisms. Crit Rev Biotechnol 25(4):199–204

Bhuyan T, Mishra K, Khanuja M, Prasad R, Varma A (2015) Biosynthesis of zinc oxide nanoparticles from *Azadirachta indica* for antibacterial and photocatalytic applications. Mater Sci Semicond Process 32:55–61

Bishop J (1990) Surface modified drug nanoparticles. US Patent application. Docket 61894 Filed 9/17/90

Boanini E, Torricelli P, Gazzano M, Giardino R, Bigi A (2006) Nanocomposites of hydroxyapatite with aspartic acid and glutamic acid and their interaction with osteoblast-like cells. Biomaterials 27(25):4428–4433

Bokare V, Jung JL, Chang YY, Chang YS (2013) Reductive dechlorination of octachlorodibenzo-p-dioxin by nanosized zero-valent zinc: modeling and rate kinetics and congener profile. J Hazard Mater 250–251:397–402

Bonvin F, Jost L, Randin L, Bonvin E, Kohn T (2016) Super-fine powdered activated carbon (SPAC) for efficient removal of micropollutants from wastewater treatment plant effluent. Water Res 90:90–99

BRSNL (Bureau of Reclamation ad Sandia National Laboratories), US (2003) Desalination and water purification technology roadmap a report of the executive committee. Directorate for Information Operations and Reports, Arlington VA, USA

Bumbudsanpharoke N, Ko S (2015) Nano-food packaging: an overview of market, migration research, and safety regulations. J Food Sci 80(5):910–922

Campos AFC, Aquino R, Cotta TAPG, Tourinho FA, Depeyrot J (2011) Using speciation diagrams to improve synthesis of magnetic nanosorbents for environmental applications. Bull Mater Sci 34(7):1357–1361

Carpenter AW, Lannoy CF, Wiesner MR (2015) Cellulose materials in water treatment technologies. Environ Sci Technol 49(9):5277–5287

Chang YC, Chang SW, Chen DH (2006) Magnetic chitosan nanoparticles: studies on chitosan binding and adsorption of Co(II)ions. React Funct Polym 66(3):335–341

Chaturvedi S, Dave PN (2019) Water purification using nanotechnology an emerging opportunity. Chem Methodol 3:115–144

Chitra K, Annadurai G (2014) Antibacterial activity of pH-dependent biosynthesized silver nanoparticles against clinical pathogen. Biomed Res Int 2014:1–6

Cho M, Cates EL, Kim J (2011) Inactivation and surface interactions of MS 2 bacteriophage in a TiO_2 photoelectrolytic reactor. Water Res 45(5):2104–2110

Choi H, Stathatos E, Dionysiou DD (2006) Sol-gel preparation of mesoporous photocatalytic TiO_2 films and TiO_2/Al_2O_3 composite membranes for environmental applications. Appl Catal B-Environ 63(1–2):60–67

Corsi I, Nielse MW, Sethi R, Punta C, Torre CD, Libralato G, Cinuzzi F (2018) Ecofriendly nanotechnologies and nanomaterials for environmental applications: key issue and consensus recommendations for sustainable and ecosafe nanoremediation. Ecotoxicol Environ Saf 154:237–244

Crane RA, Scott TB (2012) Nanoscale zero-valent iron: future prospects for an emerging water treatment technology. J Hazard Mater 211–212:112–125

Crooks RM, Zhao MQ, Sun L, Chechik V, Yeung LK (2001) Dendrimer-encapsulated metal nanoparticles: synthesis, characterization, and applications to catalysis. Acc Chem Res 34(3):181–190

Das R, Ali ME, Hamid SBA, Ramakrishna S, Chowdhury ZZ (2014a) Carbon nanotube membranes for water purification: a bright future in water desalination. Desalination 336:97–109

Das R, Hamid SBA, Ali ME, Ismail AF, Annuar MSM, Ramakrishna S (2014b) Multifunctional carbon nanotubes in water treatment: the present, past and future. Desalination 354:160–179

Dehghani MH, Taher MM, Bajpai AK, Heibati B, Tyagi I, Asif M, Agarwal S, Gupta VK (2015) Removal of noxious Cr(VI) ions using single walled carbon nanotubes and multi walled carbon nanotubes. Chem Eng J 279:344–352

Deliyanni EA, Bakoyannakis DN, Zouboulis AI, Matis KA (2003) Sorption of As(V) ions by akaganeite type nanocrystals. Chemosphere 50(1):155–163

Deliyanni EA, Peleka EN, Matis KA (2007) Removal of zinc ion from water by sorption on iron-based nanoadsorbent. J Hazard Mater 141(1):176–184

Diallo MS, Christie S, Swaminathan P, Johnson JH, Goddard WA (2005) Dendrimer enhanced ultrafiltration, recovery of Cu(II) from aqueous solutions using PAMAM dendrimers with ethylene diamine core and terminal NH_2 groups. Environ Sci Technol 39(5):1366–1377

Dinali R, Ebrahiminezhad A, Harris MM, Ghasemi Y, Berenjian A (2017) Iron oxide nanoparticles in modern microbiology and biotechnology. Crit Rev Microbiol 43(4):493–507

Doyle ME (2006) Nanotechnology: a brief literature review. Food Research Institute, University of Wisconsin, Madison, pp 1–10

Dutta D, Thakur D, Bahadur D (2015) SnO_2 quantum dots decorated silica nanoparticles for fast removal of cationic dye (methylene blue) from wastewater. Chem Eng J 281:482–490

EC (European Commission) (2011) Commission recommendation of 18 October 2011 on the definition of nanomaterial (2011/696/EU). Available from: http://eur-lex.europa.eu/LexUriServ/LexUriServ.do?uri=OJ:L:2011:275:0038:0040:EN:PDF. Accessed on 20 Jan 2020

Elimelech M, Phillip WA (2011) The future of seawater desalination: energy, technology, and the environment. Science 333(6043):712–717

Elmizadeh H, Soleimani M, Faridbod F, Bardajee GR (2018) A sensitive nano-sensor based on synthetic ligand-coated CdTe quantum dots for rapid detection of Cr(III) ions in water and wastewater samples. Colloid Polym Sci 296(9):1581–1590

EPA (Environmental Protection Agency), US (1999) Alternative disinfectants and oxidants guidance manual. EPA Office of Water Report 815-R-99-014. Directorate for Information Operations and Reports, Arlington VA, USA

EPA (Environmental Protection Agency), US (2015) Control of nanoscale materials under the toxic substances control. Available from: https://www.epa.gov/reviewing-new-chemicals-under-toxic-substances-control-act-tsca/control-nanoscale-materials-under. Accessed on 24 Nov 2019

EPC (European Parliament and Council) (2000) Directive 2000/60/EC of the European Parliament and of the Council of 23 October 2000 establishing a framework for community action in the field of water policy. Available from: https://eur-lex.europa.eu/legal-content/EN/TXT/?uri=celex%3A32000L0060. Accessed on 03 March 2023

Esawi AMK, Morsi K, Sayed A, Taher M, Lanka S (2010) Effect of carbon nanotube (CNT) content on the mechanical properties of CNT-reinforced aluminum composites. Compos Sci Technol 70(16):2237–2241

Feng L, Cao M, Ma X, Zhu Y, Hu C (2012) Superparamagnetic high-surface area Fe_3O_4 nanoparticles as adsorbents for arsenic removal. J Hazard Mater 217–218:439–446

Foguel MV, Pedro NTB, Zanoni MVB, Sotomayor MDPT (2017) Molecularly imprinted polymer (MIP): a promising recognition system for development of optical sensor for textile dyes. Procedia Tech 27:299–300

Frechet JMJ, Tomalia DA (2001) Dendrimers and other dendritic polymers. Wiley, New York

Fu F, Dionysiou DD, Liu H (2014) The use of zero-valent iron for groundwater remediation and wastewater treatment: a review. J Hazard Mater 267:194–205

Gao XH, Chan WCW, Nie SM (2002) Quantum-dot nanocrystals for ultrasensitive biological labeling and multi-color optical encoding. J Biomed Opt 7(4):532–537

Gardner J, Dhai A (2014) Nanotechnology and water: ethical and regulatory considerations. In: Mishra AK (ed) Application of nanotechnology in water research. Wiley, Hoboken, pp 1–20

Gehrke I, Geiser A, Schulz AS (2015) Innovations in nanotechnology for water treatment. Nanotech Sci Appl 8:1–17

Ghaedi M, Khajehsharifi H, Yadkuri AH, Roosta M, Asghari A (2012) Oxidized multiwalled carbon nanotubes as efficient adsorbent for bromothymol blue. Toxicol Environ Chem 94(5):873–883

Ghorbani M, Eisazadeh H, Ghoreyshi AA (2012) Removal of zinc ions from aqueous solution using polyaniline nanocomposite coated on rice husk. Iran J Energy Environ 3(1):83–88

Ghosh A, Nayak AK, Pal A (2017) Nano-particle mediated wastewater treatment: a review. Curr Pollut Rep 3(1):17–30

Graveland BJF, Kruif CG (2006) Unique milk protein based nanotubes: food and nanotechnology meet. Trends Food Sci Technol 17(5):196–203

Guan X, Sun Y, Qin H, Li J, Lo IM, He D, Dong H (2015) The limitations of applying zero-valent iron technology in contaminants sequestration and the corresponding counter measures: the development in zero-valent iron technology the last two decades (1994–2014). Water Res 75:224–248

Guerra F, Attia M, Whitehead D, Alexis F (2018) Nanotechnology for environmental remediation: materials and applications. Molecules 23(7):1–23

Gunatilake SK (2015) Methods of removing heavy metals from industrial wastewater. Methods 1(1):12–18

Hajkova P, Spatenka P, Horsky J, Horska I, Kolouch A (2007) Photocatalytic effect of TiO_2 films on viruses and bacteria. Plasma Process Polym 4(S1):S397–S401

Hamad AF, Han JH, Kim BC, Rather IA (2018) The interwine of nanotechnology with food industry. Saudi J Biol Sci 25:27–30

Harvey JMH (2015) Nanoforensics: forensic application of nanotechnology in illicit drug detection. J Forensic Res 6(5):106

Hashim DP, Narayanan NT, Herrera JMR, Cullen DA, Hahm MG, Lezzi P, Suttle JR, Kelkhoff D, Sandoval EM, Ganguli S, Roy AK, Smith DJ, Vajtai R, Sumpter BG, Meunier V, Terrones H, Terrones M, Ajayan PM (2012) Covalently bonded three-dimensional carbon nanotube solids via boron induced nanoconjunctions. Sci Rep 2:363

Horizon (2020) Work programme 2014–2015. 5. Leadership in enabling and industrial technologies. II. Nanotechnologies, Advanced materials, biotechnology and advanced manufacturing and processing. Available from: http://ec.europa.eu/research/participants/data/ref/h2020/wp/2014_2015/main/h2020-wp1415-leit-nmp_en.pdf. Accessed on 20 Jan 2020

Hristovski K, Baumgardener A, Westerhoff P (2007) Selecting metal oxide nanoparticles for arsenic removal in fixed bed columns: from nanoparticles to aggregated nanoparticles media. J Hazard Mater 147(1–2):265–274

Hristovski KD, Nguyen H, Westerhoff PK (2009) Removal of arsenite and 17-ethinyl estradiol (EE2) by iron (hydr)oxide modified activated carbon fibers. J Environ Sci Health A Tox Hazard Subst Environ Eng 44(4):354–361

http://www.nanoH$_2$O.com. Accessed on 15 Jan 2020

http://www.nanowerk.com/spotlight/spotid=4662.php. Accessed on 15 Jan 2020

http://www.purifics.com/. Accessed on 15 Jan 2020

Huang Q, Yu H, Ru Q (2010) Bioavailability and delivery of nutraceuticals using nanotechnology. J Food Sci 75(1):R50–R57

Hubbs AF, Mercer RR, Benkovic SA, Harkema J, Sriram K, Berry DS, Goravanahally MP, Nurkiewicz TR, Castranova V, Sargent LM (2011) Nanotoxicology-pathologist's perspective. Toxicol Pathol 39(2):301–324

Hussain I, Singh NB, Singh A, Singh H, Singh SC (2016) Green synthesis of nanoparticles and its potential application. Biotechnol Lett 38(4):545–560

Ibrahim SM, Badawy AA, Essawy HA (2019) Improvement of dyes removal from aqueous solution by nanosized cobalt ferrite treated with humic acid during coprecipitation. J Nanostructure Chem 9:281–298

Ito A, Shinkai M, Honda H, Kobayashi T (2005) Medical application of functionalized magnetic nanoparticles. J Biosci Bioeng 100(1):1–11

Jain P, Pradeep T (2005) Potential of silver nanoparticle-coated polyurethane foam as an antibacterial water filter. Biotechnol Bioeng 90(1):59–63

Ji LL, Chen W, Duan L, Zhu DQ (2009) Mechanisms for strong adsorption of tetracycline to carbon nanotubes: a comparative study using activated carbon and graphite as adsorbents. Environ Sci Technol 43(7):2322–2327

Kamaly N, Yameen B, Wu J, Farokhzad OC (2016) Degradable controlled-release polymers and polymeric nanoparticles: mechanisms of controlling drug release. Chem Rev 116(4):2602–2663

Kanchi S (2014) Nanotechnology for water treatment. J Environ Anal Chem 1(2):1–3

Kannan N, Sundaram MM (2001) Kinetics and mechanism of removal of methylene blue by adsorption on various carbons- a comparative study. Dyes Pigments 51(1):25–40

Karim MR, Rhodes ER, Brinkman N, Wymer L, Fout GS (2009) New electropositive filter for concentrating enteroviruses and noroviruses from large volumes of water. Appl Environ Microbiol 75(8):2393–2399

Kaur R, Wani SP, Singh AK, Lal K (2012) Wastewater production, treatment and use in India. In: National report presented at the 2nd regional workshop on safe use of wastewater in agriculture, New Delhi, 16–18 May 2012, pp 1–13

Kaur P, Singh S, Kumar V, Singh N, Singh J (2018) Effect of rhizobacteria on arsenic uptake by macrophyte *Eichhornia crassipes* (Mart.) Solms. Int J Phytoremediation 20(2):114–120

Kaushik M, Mahendru S, Chaudhary S, Kukreti S (2017) DNA fingerprints: advances in their forensic analysis using nanotechnology. J Forensic Biomech 8(1):1–4

Khalil A, Gondal MA, Dastageer MA (2011) Augmented photocatalytic activity of palladium incorporated ZnO nanoparticles in the disinfection of *Escherichia coli* microorganism from water. Appl Catal A Gen 402(1–2):162–167

Kiparissides C (2010) Nanotechnology meets water treatment. Dissemination workshop of the Nano4water cluster

Klacanova K, Fodran P, Simon P, Rapta P, Boca R, Jorik V, Miglicrini M, Kolek E, Caplovic L (2013) Formation of Fe(0)-nanoparticles via reduction of Fe(II) compounds by amino acids and their subsequent oxidation to iron oxides. J Chem 2013:1–10

Koeppenkastrop D, Decarlo EH (1993) Uptake of rare-earth elements from solution by metal-oxides. Environ Sci Technol 27(9):1796–1802

Kominami H, Yabutani K, Yamamoto T, Kara Y, Ohtani B (2001) Synthesis of highly active tungsten(VI) oxide photocatalysts for oxygen evolution by hydrothermal treatment of aqueous tungstic acid solutions. J Mater Chem 11(12):3222–3227

Krishnan U, Barbara S, Arifullah M, Nabi NNA, Antonella P (2018) Relevance of nanotechnology in food processing industries. Int J Agric Sci 10(7):5730–5733

Kumar V, Upadhyay N, Kumar V, Kaur S, Singh J, Singh S, Datta S (2014) Environmental exposure and health risks of the insecticide monocrotrophos- a review. J Biodivers Environ Sci 5:111–120

Kundu S, Wang Y, Xia W, Muhler M (2008) Thermal stability and reducibility of oxygen-containing functional groups on multiwalled carbon nanotube surfaces: a quantitative high-resolution XPS and TPD/TPR study. J Phys Chem C 112(43):16869–16878

Lan Y, Cheng Y, Zhao Z (2014) Effective organochlorine pesticides removal from aqueous systems by magnetic nanospheres coated with polystyrene. J Wuhan Univ Technol Mater Sci Ed 29(1):168–173

Lau WJ, Gray S, Matsuura T, Emadzadeh D, Chen JP, Ismail AF (2015) A review on polyamide thin film nanocomposite (TFN) membranes: history, applications, challenges and approaches. Water Res 80:306–324

Le AT, Le TT, Tran HH, Dang DA, Tran QH, Vu DL (2012) Powerful colloidal silver nanoparticles for the prevention of gastrointestinal bacterial infections. Adv Nat Sci Nanosci Nanotechnol 3:1–10

Lee KM, Lai CW, Ngai KS, Juan JC (2016) Recent developments of zinc oxide based photocatalyst in water treatment technology: a review. Water Res 88:428–448

Legrouri K, Khouyab E, Ezzineas M, Hannachea H, Denoyele R, Pallierd R, Naslaind R (2005) Production of activated carbon from a new precursor molasses with sulphuric acid. J Hazard Mater 118(1–3):259–263

Li M, Lin YC, Wu CC, Liu HS (2005) Enhancing the efficiency of PCR using gold nanoparticles. Nucleic Acids Res 33(21):1–10

Li M, Yin JJ, Warmer WG, Lo YM (2014) Mechanistic characterization of titanium dioxide nanoparticle-induced toxicity using electron spin resonance. J Food Drug Anal 22(1):76–85

Liang J, Liu J, Yuan X, Dong H, Zeng G, Wu H, Wang H, Liu J, Hua S, Zhang S, Yu Z (2015) Facile synthesis of alumina-decorated multi-walled carbon nanotubes for simultaneous adsorption of cadmium ion and trichloroethylene. Chem Eng J 273:101–110

Liga MV, Bryant EL, Colvin VL, Li Q (2011) Virus inactivation by silver doped titanium dioxide nanoparticles for drinking water treatment. Water Res 45(2):535–544

Lind ML, Suk DE, Nguyen TV, Hoek EMV (2010) Tailoring the structure of thin film nanocomposite membranes to achieve seawater RD membrane performance. Environ Sci Technol 44(21):8230–8235

Liu WT (2006) Nanoparticles and their biological and environmental applications. J Biosci Bioeng 102(1):1–7

Liu Y, Li J, Qiu X, Burda C (2006) Novel TiO_2 nanocatalysts for wastewater purification: tapping energy from the sun. Water Sci Technol 54(8):47–54

Lorenceau E, Utada AS, Link DR, Joanicot CM, Weitz DA (2005) Generation of polymerosomes from double-emulsions. Langmuir 21(20):9183–9186

Lu CS, Chiu H, Liu CT (2006) Removal of zinc (II) from aqueous solution by purified carbon nanotubes: kinetics and equilibrium studies. Ind Eng Chem Res 45(8):2850–2855

Lu C, Chiu H, Bai H (2007) Comparisons of adsorbent cost for the removal of zinc (II) from aqueous solution by carbon nanotubes and activated carbon. J Nanosci Nanotechnol 7(4–5):1647–1652

Lu H, Wang J, Stoller J, Wang T, Bao Y, Hao H (2016) An overview of nanomaterials for water and wastewater treatment. Adv Mater Sci Eng 2016:1–10

Lucas E, Decker S, Khaleel A, Seitz A, Fultz S, Ponce A, Li WF, Carnes C, Klabunde KJ (2001) Nanocrystalline metal oxides as unique chemical reagents/sorbents. Chem Eur J 7(12):2505–2510

Lugani Y, Kaur G, Oberoi S, Sooch BS (2018a) Nanotechnology: current applications and future prospects. World J Adv Healthc Res 2(5):137–139

Lugani Y, Oberoi S, Rai SK, Sooch BS (2018b) Nanomedicine: the imminent gauntlet of medical science. World J Pharm Pharm Sci 7(9):422–429

Lugani Y, Oberoi S, Rattu G (2020) Nanotechnology in food industry- applications, safety regulations, and upcomings. In: Maurya VK (ed) Micro and nano engineering in food science, vol 1. Springer Nature (in press)

Machado FM, Bergmann CP, Fernandes TH, Lima EC, Royer B, Calvete T, Fagan SB (2011) Adsorption of reactive Red M-2BE dye from water solutions by multi-walled carbon nanotubes and activated carbon. J Hazard Mater 192(3):1122–1131

Mahgoub S, Samaras P (2014) Nanoparticles from biowaste and microbes: focus on role in water purification and food preservation. In: Proceedings of the 2nd international conference on sustainable solid waste management, Athens, 12–14 June 2014, pp 1–39

Maliyekkal SM, Sreeprasad TS, Krishnan D, Kouser S, Mishra AK, Waghmare UV, Pradeep T (2013) Graphene: a reusable substrate for unprecedented adsorption of pesticides. Small 9(2):273–283

Maximino NJ, Alvarez MP, Avila RS, Orta CAA, Regalado EJ, Bello AM, Morones PG, Pliego GC (2018) Oxidation of copper nanoparticles protected with different coatings and stored under ambient conditions. J Nanomater 2018:1–8

Meetoo DD (2011) Nanotechnology and the food sector: from the farm to the table. Emirates J Food Agric 23(5):387–407

Mendez E, Fuentes MAG, Perez GR, Albores AM, Torres E (2017) Emerging pollutant treatments in wastewater: cases of antibiotics and hormones. J Environ Sci Health A 52(3):235–253

Mihindukulasuriya SDF, Lim LT (2014) Nanotechnology development in food packaging: a review. Trends Food Sci Technol 40(2):149–167

Mohieldin SD, Zainudin ES, Paridah MT, Ainun ZM (2011) Nanotechnology in pulp and paper industries: a review. Key Eng Mater 471–472:251–256

Moussavi G, Mahmoudi M (2009) Removal of azo and anthraquinone reactive dyes from industrial wastewaters using MgO nanoparticles. J Hazard Mater 168(2–3):806–812

Nam YJ, Lead J (2016) Properties, sources, pathways and fate of nanoparticles in the environment. In: Xing B, Vecitis CD, Senesi N (eds) Engineered nanoparticles and the environment: biophysicochemical processes and toxicity. Chapter 6, vol 4. Wiley, Hoboken, pp 95–117

Nano.gov What is NNi? (2014). Available from: http://www.nano.gov/about-nni/what. Accessed on 20 Jan 2020

Nawrocki J, Hordern BK (2010) The efficiency and mechanisms of catalytic ozonation. Appl Catal B Environ 99(1–2):27–42

Nikalje AP (2015) Nanotechnology and its applications in medicine. Med Chem 5(2):81–89

Okpalugo TIT, Papakonstantinou P, Murphy H, Laughlin JM, Brown NMD (2005) High resolution XPS characterization of chemical functionalized MWCNTs and SWCNTs. Carbon 43(1):153–161

Pacheco S, Rodriguez R (2001) Adsorption properties of metal ions using alumina nano-particles in aqueous and alcoholic solutions. J Sol-Gel Sci Technol 20(3):263–273

Panahi Y, Mellatyar H, Farshbaf M, Sabet Z, Fattahi T, Akbarzadehe A (2018) Biotechnological applications of nanomaterials for air pollution and water/wastewater treatment. Mater Today Proc 7(3):15550–15558

Pande S, Singh BP, Mathur RB, Dhami TL, Saini P, Dhawan SK (2009) Improved electromagnetic interference shielding properties of MWCNT-PMMA composites using layered structures. Nanoscale Res Lett 4:327–334

Pathak S, Choi SK, Arnheim N, Thompson ME (2001) Hydroxylated quantum dots as luminescent probes for in situ hybridization. J Am Chem Soc 123(17):4103–4104

Patil BBT (2015) Wastewater treatment using nanoparticles. J Adv Chem Eng 5(3):1–2

Peigney A, Laurent C, Flahaut E, Bacsa RR, Rousset A (2001) Specific surface area of carbon nanotubes and bundles of carbon nanotubes. Carbon 39(4):507–514

Peiris MK, Gunasekara CP, Jayaweera PM, Arachchi ND, Fernando N (2017) Biosynthesized silver nanoparticles: are they effective antimicrobials? Mem Inst Oswaldo Cruz 112(8):537–543

Pendergast MTM, Nygaard JM, Ghosh AK, Hoek EMV (2010) Using nanocomposite materials technology to understand and control reverse osmosis membrane compaction. Desalination 261(3):255–263

Phillips KS, Han JH, Martinez M, Wang ZZ, Carter D, Cheng Q (2006) Nanoscale glassification of gold substrates for surface plasmon resonance analysis of protein toxins with supported lipid membranes. Anal Chem 78(2):596–603

Prachi GP, Madathil D, Nair ANB (2013) Nanotechnology in wastewater treatment: a review. Int J ChemTech Res 5(5):2303–2308

Pradhan N, Singh S, Ojha N, Shrivastava A, Barla A, Rai V, Bose S (2015) Facets of nanotechnology as seen in food processing, packaging, and preservation industry. Biomed Res Int 2015:1–17

Prasad R, Thirugnanasanbandham K (2019) Advances research on nanotechnology for water technology. Springer http://www.springer.com/us/book/9783030023805

Prasad R, Bhattacharyya A, Nguyen QD (2017) Nanotechnology in sustainable agriculture: recent developments, challenges, and perspectives. Front Microbiol 8:1–13

Prasad R, Jha A, Prasad K (2018) Exploring the realms of nature for nanosynthesis. Springer. ISBN 978-3-319-99570-0. http://www.springer.com/978-3-319-99570-0

Puay NQ, Qiu G, Ting YP (2015) Effect of zinc oxide nanoparticles on biological wastewater treatment in sequencing batch reactor. J Clean Prod 88:139–145

Qu X, Alvarez PJ, Li Q (2013) Applications of nanotechnology in water and wastewater treatment. Water Res 47(12):3931–3946

Rafatullah M, Sulaiman O, Hashim R, Ahmad A (2010) Adsorption of methylene blue on low-cost adsorbents: a review. J Hazard Mater 117:70–80

Rafiq Z, Nazir R, Shah MR, Ali S (2014) Utilization of magnesium and zinc oxide nano-adsorbents as potential materials for treatment of copper electroplating industry wastewater. J Environ Chem Eng 2(1):642–651

Ramakrishna S, Fujihara K, Teo WE, Yong T, Ma ZW, Ramaseshan R (2006) Electrospun nanofibers: solving global issues. Mater Today 9(3):40–50

Rana N, Ghosh KS, Chand S, Gathania AK (2018) Investigation of ZnO nanoparticles for their applications in wastewater treatment and antimicrobial activity. Indian J Pure Appl Phys 56(1):19–25

Rattu G, Khansili N, Lugani Y, Krishna PM (2020) Benefits and drawbacks of nanomaterials in the food industry. In: Maurya VK (ed) Micro and nano engineering in food science, vol 2. Springer Nature (in press)

Rodrigues SM, Demokritou P, Dokoozlian N, Hendren CO, Karn B, Mauter MS, Sadik OA, Safarpour M, Unrine JM, Viers J, Welle P (2017) Nanotechnology for sustainable food production: promising opportunities and scientific challenges. Environ Sci Nano 4:767–781

Saharan P, Chaudhary GM, Mehta SK, Umar A (2014) Removal of water contaminants by iron oxide nanomaterials. J Nanosci Nanotechnol 14(1):627–643

Sanchez F, Sobolev K (2010) Nanotechnology in concrete- a review. Constr Build Mater 24(11):2060–2071

Sears K, Dumee L, Schutz J, She M, Huynh C, Hawkins S, Duke M, Gray S (2010) Recent developments in carbon nanotube membranes for water purification and gas separation. Materials 3(1):127–149

Seo Y, Hwang J, Jeong Y, Hwang MP, Choi J (2014) Antibacterial activity and cytotoxicity of multi-walled carbon nanotubes decorated with silver nanoparticles. Int J Nanomedicine 9:4621–4629

Sharma YC, Srivastava V, Singh VK, Kaul SN, Weng CH (2009) Nano-adsorbents for the removal of metallic pollutants from water and wastewater. Environ Technol 30(6):583–609

Sharma VK, Filip J, Zboril R, Varma RS (2015) Natural inorganic nanoparticles-formation, fate, and toxicity in the environment. Chem Soc Rev 44(23):8410–8423

Sidhu GK, Singh S, Kumar V, Datta S, Singh D, Singh J (2019) Toxicity, monitoring and biodegradation of organophosphate pesticides: a review. Crit Rev Environ Sci Technol 49(13):1135–1187

Siegel RW (1994) Physics of new materials. In: Fujita FE (ed) Springer series in materials science, vol 27. Springer, Berlin

Singh S, Singh N, Kumar V, Datta S, Wani AB, Singh D, Singh J (2016) Toxicity, monitoring and biodegradation of fungicide carbendazim. Environ Chem Lett 14(3):317–329

Skalickova S, Milosavljevic V, Cihalova K, Horky P, Richtera L, Adam V (2017) Selenium nanoparticles as a nutritional supplement. Nutrition 33:83–90

Sooch BS, Kauldhar BS (2015) Development of an eco-friendly whole cell based continuous system for the degradation of hydrogen peroxide. J Bioprocess Biotech 5:1000233/1–1000233/5

Sooch BS, Kauldhar BS, Puri M (2014) Recent insights into microbial catalases: isolation, production and purification. Biotechnol Adv 32(8):1429–1447

Sooch BS, Kauldhar BS, Puri M (2016) Isolation and polyphasic characterization of a novel hyper catalase producing thermophilic bacterium for the degradation of hydrogen peroxide. Bioprocess Biosyst Eng 39(11):1759–1773
Spitalsky Z, Tasis D, Papagelis K, Galiotis C (2010) Carbon nanotube- polymer composites: chemistry, processing, mechanical and electrical properties. Prog Polym Sci 35(3):357–401
Sukhanova A, Devy M, Venteo L, Kaplan H, Artemyev M, Oleinikov V, Klinov D, Pluot M, Cohen JHM, Nabiev O (2004) Biocompatible fluorescent nanocrystals for immunolabeling of membrane proteins and cells. Anal Biochem 324(1):60–67
Sunstrom JE, Moser WR, Guerts BM (1996) General route of nanocrystalline oxides by hydrodynamic cavitation. Chem Mater 8(8):2061–2067
Suslick KS, Hyeon T, Fang F (1996) Sonochemical synthesis of iron colloids. J Am Chem Soc 118(47):11960–11961
Theron J, Cloete TE, Kwaadsteniet M (2010) Current molecular and emerging nanobiotechnology approaches for the detection of microbial pathogens. Crit Rev Microbiol 36(4):318–339
Tian Y, Wu M, Liu R, Li Y, Wang D, Tan J, Wu R, Huang Y (2011) Electrospun membrane of cellulose acetate for heavy metal ion adsorption in water treatment. Carbohydr Polym 83(2):743–748
Tiwari DK, Behari J, Sen P (2008) Applications of nanoparticles in waste water treatment. World Appl Sci J 3(3):417–433
Tomalia DA (2004) Birth of a new macromolecular architecture: dendrimers as quantized building blocks for nanoscale synthetic organic chemistry. Aldrichimica Acta 37(2):39–57
Trivedi P, Axe L (2000) Modeling Cd and Zn sorption to hydrous metal oxides. Environ Sci Technol 34(11):2215–2223
Tu YJ, You CF, Chang CK, Wang SL, Chan TS (2012) Arsenate adsorption from water using a novel fabricated copper ferrite. Chem Eng J 198–199:440–448
Varbanets MP, Zurbrugg C, Swartz C, Pronk W (2009) Decentralized systems for potable water and the potential of membrane technology. Water Res 43(2):245–265
Verma S, Domb AJ, Kumar N (2011) Nanomaterials for regenerative medicine. Nanomedicine (Lond) 6(1):157–181
Visa M, Carcel RA, Andronic L, Duta A (2009) Advanced treatment of wastewater with methyl orange and heavy metals on TiO_2, fly ash and their mixtures. Catal Today 144(1–2):137–142
Volder MF, Tawfick SH, Baughman RH, Hart J (2013) Carbon nanotubes: present and future commercial applications. Science 339(6119):535–539
Vukovic GD, Marinkovic ADM, Colic M, Ristic MD, Aleksic R, Grujic AAP, Uskokovic (2011) Removal of cadmium from aqueous solutions by oxidized and ethylenediamine-functionalized multi-walled carbon nanotubes. Chem Eng J 157(1):238–248
Wang J, Gu H (2015) Novel metal nanomaterials and their catalytic applications. Molecules 20:17070–17092
Westerhoff P, Moon H, Minakata D, Crittenden J (2009) Oxidation of organics in retentates from reverse osmosis wastewater reuse facilities. Water Res 43(16):3992–3998
WHO (World Health Organization) (1996) Guidelines for drinking water quality, vol 2. WHO, Geneva
WHO (world Health Organization) (2012) Meeting the MDG drinking water and sanitation the urban and rural challenge of the decade
Wu YY, Xiong ZH (2016) Multi-walled carbon nanotubes and powder-activated carbon adsorbents for the removal of nitrofurazone from aqueous solution. J Dispers Sci Technol 37(5):613–624
Wu X, Liu J, Haley KN, Treadway JA, Larson JP, Ge N, Peale F, Bruchez MP (2002) Immunofluorescent labeling of cancer marker Her2 and other cellular targets with semiconductor quantum dots. Nat Biotechnol 21:41–46
Xu J, Bachas L, Bhattacharyya D (2009) Synthesis of nanostructured bimetallic particles in poly ligand functionalized membranes for remediation applications. In: Sustich A (ed) Nanotechnology applications for clean water, chapter-22. William Andrew Publishing, Boston, pp 311–335

Yan JL, Estevez MC, Smith JE, Wang KM, He XX, Wang L, Tan WH (2007) Dye-doped nanoparticles for bioanalysis. Nano Today 2(3):44–50

Yang JC, Yin XB (2017) $CoFe_2O_4$@MIL-100 (Fe) hybrid magnetic nanoparticles exhibit fast and selective adsorption of arsenic with high adsorption capacity. Sci Rep 7:1–15

Yin HS, Zhou YL, Ai SY, Chen QP, Zhu XB, Liu XG, Zhu LS (2010) Sensitivity and selectivity determination of BPA in real water samples using PAMAM dendrimers and CoTe quantum dots modified glassy carbon electrode. J Hazard Mat 174(1–3):236–243

Yu W, Lin W, Shao X, Hu Z, Li R, Yuan D (2014) High performance supercapacitor based on Ni_3S_3/carbon nanofibers and carbon nanofibers electrodes derived from bacterial cellulose. J Power Sources 272:137–143

Zan L, Fa W, Peng T, Gong ZK (2007) Photocatalytic effect of nanometer TiO_2 and TiO_2-coated ceramic plate on Hepatitis B virus. J Photochem Photobiol B 86(2):165–169

Zare K, Sadegh H, Ghoshekandi RS, Maazinejad B, Ali V, Tyagi I, Agarwal S, Gupta VK (2015) Enhanced removal of toxic Congo red dye using multi walled carbon nanotubes: kinetic, equilibrium studies and its comparison with other adsorbents. J Mol Liq 212:266–271

Zhang XX, Zhu CC (2006) Field-emission lighting tube with CNT film cathode. Microrelectron J 37(11):1358–1360

Zhang QK, Kemp KC, Chandra V (2012) Homogeneous anchoring of TiO_2 nanoparticles on graphene sheets for wastewater treatment. Mater Lett 81:127–130

Zhang J, Wang H, Xiao Y, Tang J, Liang C, Li F, Dong H, Xu W (2017) A simple approach for synthesizing of fluorescent carbon quantum dots from tofu wastewater. Nanoscale Res Lett 12(1):611

Zhao X, Lv L, Pan B, Zhang W, Zhang S, Zhang Q (2011a) Polymer-supported nanocomposites for environmental applications: a review. Chem Eng J 170(2–3):381–394

Zhao G, Li J, Ren X, Chen C, Wang X (2011b) Few-layered graphene oxide nanosheets as superior sorbents for heavy metal ion pollution management. Environ Sci Technol 45:10454–10462

Zhao H, Qiu S, Wu L, Zhang L, Chen H, Gao C (2014) Improving the performance of polyamide reverse osmosis membrane by incorporation of modified multi-walled carbon nanotubes. J Membr Sci 450:249–256

Zheng X, Wu R, Chen Y (2011) Effects of ZnO nanoparticles on wastewater biological nitrogen and phosphorous removal. Environ Sci Technol 45(7):2826–2832

Chapter 7
Nanoparticles in Dye Degradation: Achievement and Confronts

Rekha Dhull, Kavita Rathee, and Vikas Dhull

Abstract At present, the textile industries are growing very fast to meet the demand of exponentially growing population. They discharge effluent in an open environment, which is responsible for causing serious health concerns to the life forms and polluting the environment due to the presence of dye. So, it is necessary to degrade the harmful dyes from the effluent before their discharge in the surroundings. Several conventional methods are in use for dye removal such as activated carbon adsorption, ozonation, electrochemical oxidation, forward osmosis, biological degradation, coagulation, and flocculation, but these methods are inefficient in successful degradation of dye from the effluent and are also not environmentally friendly. Nowadays, nanomaterials have found a wide range of applications in different fields such as analytical, cosmetics, agriculture, electronics, and medical applications. This is due to their unique properties like small in size, large surface area, highly electrocatalytic, biocompatible, antimicrobial properties, and so on. These unique properties have attracted different researchers to use them for degradation of dyes from the effluent of various industries. This review highlights the toxicity caused by the dye-containing effluent and the mechanism of degradation of dye using nanomaterials. The chapter also emphasizes on the use of nanomaterials (nanoparticles, carbon nanotubes, nanorods, graphene sheets, and fullerene structure) for dye removal.

Keywords Dye · Nanomaterial · Nanoparticles · Graphene · Nanotube

R. Dhull
Department of Science, DPG Degree College, Gurugram, Haryana, India

K. Rathee
Department of Biochemistry, PDM University, Bahadurgarh, Haryana, India

V. Dhull (✉)
Department of Biotechnology Engineering, University Institute of Engineering & Technology, Maharshi Dayanand University, Rohtak, Haryana, India
e-mail: vikasbtdhull.gf.uiet@mdurohtak.ac.in

R. Kumar et al. (eds.), *Advanced Functional Nanoparticles "Boon or Bane" for Environment Remediation Applications*, Environmental Contamination Remediation and Management, https://doi.org/10.1007/978-3-031-24416-2_7

7.1 Introduction

A dye is a colored compound that possesses affinity to bind at specific substrate and imparts color to that substrate. The dyes have been categorized into ionic and non-ionic dyes. The ionic dyes consist of cationic and anionic dyes. These anionic dyes are further subcategorized into acidic, reactive, and direct dyes. On the other hand, nonionic dyes have been categorized into vat and disperse dyes, as shown in Fig. 7.1 (Tan et al. 2015). Presently, dyes have found a wide range of applications in paper-making, food, pharmaceuticals, cosmetics, paint, textile, and leather industries (Nautiyal et al. 2016). To fulfill the demand of dyes in various industries, around 1.6 million tons of dyes have been produced every year. Around 10–15% of this total produced dyes are being discharged unused as effluent in an open environment (Hunger et al. 2003). These untreated and unused dyes present in effluent are composed of harmful chemicals, resulting in environment contamination. The toxic dyes percolate to groundwater via soil finally leading to groundwater contamination. As a result of this, these toxic dyes enter into the food chain causing serious health concerns in humans as well as in animals. These dyes are responsible for causing mutagenic, carcinogenic, and teratogenic effects in living beings (Alves de Lima et al. 2007). So, it is necessary to treat the dyes from the effluent before their discharge into the environment.

Several methods have been reported to treat these harmful dyes such as adsorption using activated carbon (Ruhl et al. 2014), ozonation of dyes (Punzi et al. 2015), photocatalytic degradation of azo dyes hydrothermally (Saleh 2019), electrochemical oxidation (Gao et al. 2019), biological treatment of textile dyes (Paz et al. 2017), forward osmosis (Korenak et al. 2019), coagulation-flocculation-based treatment of dye (GilPavas et al. 2017) using nanofiltration membranes (Wang et al. 2018), and

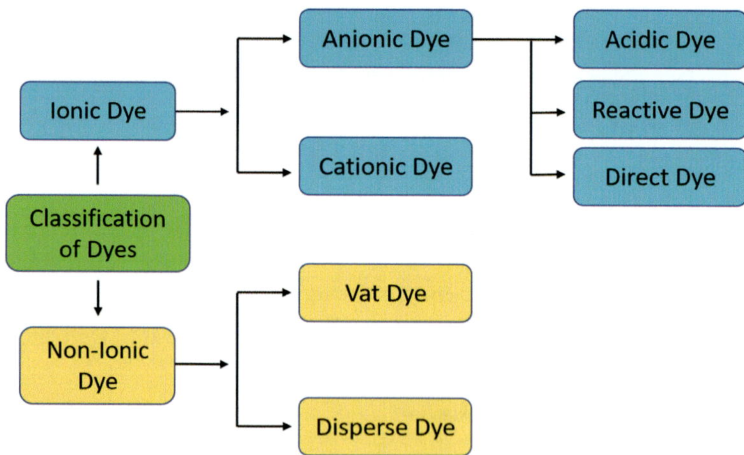

Fig. 7.1 Schematic diagram of classification of dyes

many more. Among all methods reported above, the activated carbon has been extensively used for the successful removal of dyes from effluent (Carrott et al. 1991). But the drawback of using activated carbon for dye removal is that it is expensive as its production and reactivation require steam, which is maintained at high pressure, and it further increases the dye removal cost (Marsh and Rodríguez-Reinoso 2006). To make the process cost-effective, investigators have reported the use of adsorbents for dye removal, which include leaf-based adsorbents (Bulgariu et al. 2019), adsorption using palm oil (Hameed et al. 2009), calcium biochar adsorbent derived from crab shell (Dai et al. 2018), geopolymer paste adsorbent (Maleki et al. 2018), and palm oil-derived microcrystalline cellulose (Tan et al. 2018). The above adsorbents have been successfully used for dye removal from the effluent. As adsorption depends solely on the surface area, it is necessary to enhance surface area of the adsorbents.

This requirement can be easily fulfilled using nanomaterials, especially nanoparticles. Metallic nanoparticles being small in size range from 1 to 100 nm and offer high surface area. Nowadays, nanomaterials are considered as an important candidate for adsorption of dyes due to their high surface area, presence of short intraparticle distance with respect to diffusion, pore size tunability, possess large surface area, high mechanical strength, presence of active sites, and low mass (Sweet et al. 2012; Mallakpour and Rashidimoghadam 2019). This review focuses on dye removal using nanoparticles, carbon nanotubes, nanorods, graphene, and fullerene with the amount of adsorbent used and the efficiency of dye removal.

7.2 Toxic Effect of Dye

In the textile industry, synthetic dyes have been extensively used for developing colorfast and bright hues. But the toxic nature of these synthetic dyes resulted in carcinogenic effects on humans and animals and also affects the environment. The discharge from these industries comprises of dyes along with sulfur, soaps, dye fixing agents, and other nonbiodegradable chemicals resulting in generation of toxic effluent. When this untreated effluent is discharged in open environment, it leads to clogging of soil pores, which results in decline of soil productivity, and also affects the quality of drinking water (Kant 2012).

Another source of toxic effluent is the paper and pulp industry, which contain dyes and lignocellulose material. Their discharge imparts dark color, increase in chemical oxygen demand, and imbalances pH of water (Pokhrel and Viraraghavan 2004). This is mainly due to the presence of dyes and other organic ligands from wood and the tannins, lignin, etc. (Lacorte 2003). The untreated effluent is also responsible for the reduction in transparency of water affecting photosynthetic activity of aquatic plants and animals, resulting in death (Meriläinen and Oikari 2008).

Also, effluent from the cosmetic industry is nonbiodegradable due to the presence of organic dyes, which are polar in nature (Chen et al. 2007). So, its untreated discharge results in increase of chemical oxygen demand. Effluent-containing dyes from other industries such as paint and pigmentation also lead to serious health concerns to all life forms and the environment.

7.3 Mechanism of Dye Removal Using Nanomaterials

The dyes have been successfully removed using a variety of nanomaterials via photodegradation. The process of photodegradation takes place on surface of nanomaterial in the presence of ultraviolet light, which excites electrons from valence band to conduction band leaving behind holes. The potential at valence band (h^+) is positive, which can easily produce hydroxyl radicals (OH) on the surface of the nanomaterial. On the other hand, potential at conduction band (e^-) will be negative, which is helpful in reduction of oxygen. The oxidizing nature of OH radical will degrade dye present in the vicinity of nanomaterial surface as illustrated in Fig. 7.2 (Khataee and Kasiri 2010). Photocatalysis of dye was also reported with SnO_2 nanotube. The mechanism is nearly the same as discussed above in the case of nanoparticle. When nanotubes are exposed to light, photon is absorbed, and an electron is ejected from valence band of the nanotube, which moves toward conduction band leaving behind a hole in valence band. During this, holes start migrating toward conduction band, and electrons start moving to valence band. This movement will increase charge transfer leading to oxidation and reduction of oxygen and hydroxyl molecule, respectively. When the surface of nanotube is exposed to light, oxygen at the surface yield superoxide radical ($°O_2^-$). As a result of this, new energy levels in bandgaps are created, which helped in the degradation of dye molecule to carbon dioxide and water as illustrated in Fig. 7.3 (Sadeghzadeh-Attar 2018). The photocatalytic mechanisms of dye removal using nanoparticles and nanotubes are discussed above, which showed light-dependent degradation of dye. The degradation

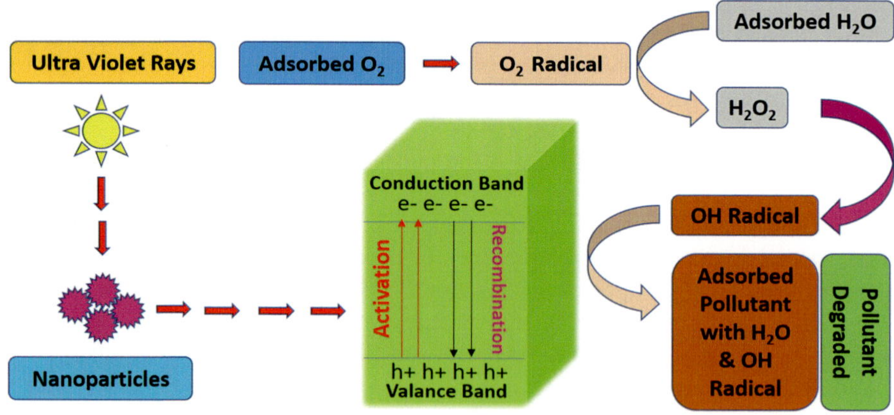

Fig. 7.2 General mechanism of photodegradation of dye using nanoparticle

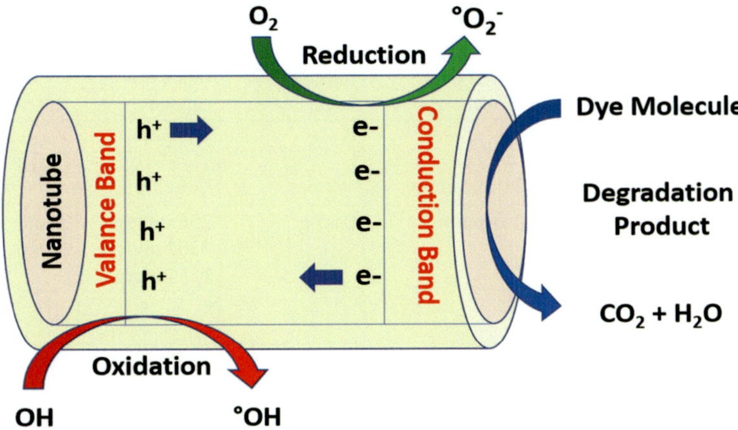

Fig. 7.3 General mechanism of photodegradation of dye using nanotube

process involves excitation of electrons from valence band of nanomaterial leading to the formation of holes in valence band and their successful migration to conduction band. This helps in the enhancement of charge transfer process during the oxidation and reduction resulting in degradation of the dye molecule on nanomaterial surface.

7.4 Different Nanomaterials for Dye Removal

7.4.1 Nanoparticles

A variety of nanoparticles have been used as adsorbent for dye removal from industrial waste as summarized in Table 7.1. Being small in size, nanoparticles provide a large surface area, and they are highly pure with narrow size distribution and reproducible (Verma et al. 2019). In one of the studies, zinc oxide (ZnO) nanoparticles (NPs) were used as adsorbent to remove azo dye such as methyl orange (MO) and amaranth (AM) (Zafar et al. 2018). In another study, ZnO-NPs were used to adsorb three dyes: malachite green (MG), acid fuchsin (AF), and Congo red (CR) (Zhang et al. 2016) and ZnO-NPs using alginate to adsorb methylene blue (MB) (Tamer et al. 2018). Acid Black 210 (AB 210) and Reactive Blue 19 (RB 19) dyes were adsorbed on 0.2 g of ZnO-NPs with adsorption capacities of 34.13 mg/g and 38.02 mg/g, respectively (Monsef Khoshhesab and Souhani 2018). Reactive Blue 21 (RB 21) dye was also degraded via photodegradation using 50 mg of green-synthesized ZnO-NPs. In this study, it was also observed that RB 21 was fully decolorized in 270 min time. This shows that ZnO NPs are used as broad range nanoparticles for removal of a variety of dyes through adsorption. Other NPs such as copper nanoparticles (Davar et al. 2015) (CuNPs) have been used for catalytic

Table 7.1 Summarizes the nanoparticles used in dye removal

Nanoparticle	Dyes removed	Method of removal	Amount of nanoparticle used	Adsorption/ reduction capacity	Reference
Zinc oxide nanoparticle	Methyl orange Amaranth	Adsorption	0.3 g	40 ppm	Zafar et al. (2018)
	Malachite green Acid Fuchsin Congo red	Adsorption	0.05 g	2963 mg/g 3307 mg/g 1554 mg/g	Zhang et al. (2016)
	Methylene blue	Adsorption	0.025–0.3 g	33.87–45.38%	Tamer et al. (2018)
	Acid black 210 Reactive blue 19	Adsorption	0.2 g	34.13 mg/g 38.02 mg/g	Monsef Khoshhesab and Souhani (2018)
	Reactive blue 21	Photodegradation	50 mg	NR	Davar et al. (2015)
Copper nanoparticles	Methylene blue Methyl red Congo red	Photocatalytic Degradation	100 μM	91.53% 73.89% 84.89%	Fathima et al. (2018)
Gold nanoparticles	Rhodamine B Methyl orange	Photocatalytic Degradation	5 mg	87.64% 83.25%	Baruah et al. (2018)
	Congo red Methylene blue	Catalytic degradation	50 μg/ml	98% 88%	Nadaf and Kanase (2016)
Silica oxide nanoparticles	Methylene blue	Photocatalytic degradation	10 g/lt	85%	Aly and Abd-Elhamid (2018)
	Methylene blue	Adsorption	0.2 g	80.8 mg/g	Dhmees et al. (2018)
	Methylene blue	Adsorption	1000 mg/lt	679.9 mg/g	Peres et al. (2018)

NR Not reported

degradation of MB, methyl red (MR), and CR. 100 μM of CuNPs were used for catalytic degradation of these three dyes (Fathima et al. 2018). The photocatalytic degradation of Rhodamine B (RB) and MO had been achieved using gold nanoparticles (AuNPs). In this, 5 mg of each dye has been used for the process, which resulted in adsorption of 87.64% of RB and 83.25% of MO using AuNPs (Baruah et al. 2018). AuNPs were also biologically synthesized from *Bacillus marisflavi* and used for catalytic degradation of CR and MB. Ninety-eight percent of CR was degraded in 20 min, and 88% of MB was catalytically degraded in 10 min using

AuNPs (Nadaf and Kanase 2016). The AuNPs are efficient in dye removal, but their synthesis is expensive, which finally adds to the cost of dye removal. Silica oxide nanoparticles (SiO_2-NPs) were used to degrade dye via photodegradation. This is possibly due to the presence of silanols, which can interact with dyes and make them chemically stable that can successfully degrade dye on silica surface. In a study, 10 g/l of the adsorbent is used, which decolorized MB to 85% within 90 s (Aly and Abd-Elhamid 2018). Adsorption of MB was also reported using silica nanoparticles (SiNPs) obtained from blast furnace in which 0.2 g of SiNPs as adsorbent showed maximum sorption capacity of 80.8 mg/g (Dhmees et al. 2018). SiO_2-NPs obtained from rice husk used for degradation of MB dye showed adsorption capacity of 679.9 mg/g with 80% dye removal using 1 g/l of the adsorbent (Peres et al. 2018). Among all reported nanoparticles, ZnO NPs are best for the purpose of dye removal as they are easy to synthesize and are less costly as compared to the other nanoparticles. They also possess the inherent capacity to remove a broad range of toxic dyes from effluent.

7.4.2 Carbon Nanotubes

Carbon nanotubes (CNT) are allotropes of carbon, which achieve cylindrical shape during synthesis and are used for a variety of purposes. They are used for applications in optics and electronics and are one of the most important nanomaterials, which can be used to adsorb dyes present in wastewater effluent from various industries as summarized in Table 7.2. Here, we discuss some reported methods, which successfully used CNTs for dye removal. One of the methods exploits amorphous CNT for removal of two textile dyes, MO and RB. For this, two dye degradation methods were used: one was adsorption based, and the other was UV-based catalysis. In this method, concentration of dyes used was 4.79 mg/L and 3.27 mg/L for RB and MO, respectively. Though the concentration of adsorbent used was not reported, but it was stated that adsorbent took lesser time of 30 min to degrade MO as compared to 45 min for the RB dye. For both dyes, adsorbent had 90% removal efficiency (Dutta et al. 2018). In some reported studies, composites were also used for dye removal process. A composite mixture of modified CNT has been reported for removal of cationic dyes (MB, MG, RB) and anionic dye (MO). For achieving dye removal, multiwalled carbon nanotubes (MWCNT) were prepared and functionalized by acid treatment (ACNT), amine treatment (NH_2CNT), and finally with the heat treatment (HCNT). It was clear from the study that cationic dyes were adsorbed with high efficiency using ACNT and MWCNT, whereas the anionic dyes were decolorized using NH_2CNT. So, a composite of ACNT/NH_2CNT and MWCNT/NH2CNT was finally used for dye removal (Dutta et al. 2017). A nanohybrid of microporous carbon xerogels and MWCNT (CX/MWCNT) was successfully used for removal of RB dye in which concentration of adsorbent varied from 1 to 4 g/L. It was documented that nanohybrid prepared possessed dye removal efficiency ranging from 154 to 256 mg/g (Shouman and Fathy 2018). MWCNTs were

Table 7.2 Use of carbon nanotubes in dye removal

Carbon nanotubes	Dyes removed	Method of removal	Amount of nanomaterial used	Adsorption/ reduction capacity	Reference
Amorphous CNT	Methyl orange Rhodamine B	Adsorption and UV-assisted catalysis	NR	More than 90%	Dutta et al. (2018)
ACNT/ NH2CNT MWCNT/ NH2CNT	Methylene blue Malachite green Rhodamine B Methyl orange	Adsorption	5 mg ACNT/ MWCNT 5 mg NH2CNT Mixture	93% 91%	Dutta et al. (2017)
CX/ MWCNT	Rhodamine B	Adsorption	1 to 4 g/L	154–256 mg/g	Shouman and Fathy (2018)
MWCNTs/ Gly/β-CD	Methylene blue Acid blue 113 Methyl orange Disperse red 1	Adsorption	0.01 g	90.90 mg/g 172.41 mg/g 96.15 mg/g 500 mg/g	Mohammadi and Veisi (2018)
ZnO/ MWCNT	Congo red	Adsorption	9 mg	99.8%	Seyed Arabi et al. (2019)

NR Not reported

functionalized using β-cyclodextrin and glycine (Gly) for the adsorption of organic dyes such as MB, Acid Blue 113 (AB113), MO, and disperse red 1 (DR1). The 0.01 g of adsorbent (MWCNT/Gly/β-CD) was used for the dye removal (Mohammadi and Veisi 2018). One work revealed that when ZnO-NPs were loaded on MWCNT, they successfully removed CR dye in the aqueous medium (Seyed Arabi et al. 2019). So, it is clear from the above reported methods that CNTs work better when they are used as composite mixtures for dye removal.

7.4.3 Nanorods

Nanorods are one of the morphological structures of nanomaterials with size ranging from 1 to 100 nm and standard aspect ratio of 3–5. They can be easily synthesized from metal oxides and other semiconducting material through chemical synthesis. They have found a wide range applications in analytical methods and also used for adsorption of unused dyes. Different earlier reported methods have been documented in Table 7.3 given below. Among one of these methods, manganite

Table 7.3 Nanorods used for dye degradation

Nanorods	Dyes removed	Method of removal	Amount of nanorod used	Adsorption/ reduction capacity/% dye removal	Reference
γ-MnOOH	Methylene blue	Adsorption	8 mg/L	89%	Varghese et al. (2017)
AgCl-NR-AC	Methylene blue	Adsorption	15 mg	96%	Nekouei et al. (2016)
$ZnCo_2O_4$	Methylene blue	Adsorption	5 mg	2400 mg/g	Lin et al. (2018)
ZnO-NR-AC	Bromocresol green Eosin Y	Ultrasonic assisted adsorption	0.01 to 0.03 g	57.80 mg/g 61.73 mg/g	Ansari et al. (2016)
SBP	Methylene blue	Adsorption	0.05 g	1691.8 mg/g	Zhang et al. (2015)
ZnS/ SnS/A-FA	Congo red	Photocatalytic degradation	10 mg	NR	Kalpana and Selvaraj (2016)
WO_3	Methylene blue	Adsorption	5.0 mg	57.6 mg/g	Park and Nam (2017)
Cu-doped-ZnO	Methyl Orange	Solar-assisted photodegradation	0.3 g/L	99%	Perillo and Atia (2018)
SnS nanorods	Tryphan blue	Photocatalytic degradation	0.9 mg/ml	More than 95%	Das and Dutta (2015)

NR Not reported

(γ-MnOOH) nanorods were used as adsorbent for adsorption of MB dye. The concentration of adsorbent used was 8 mg/L, which successfully decolorized 89% of MB dye in aqueous solution (Varghese et al. 2017). In one other method, AgCl nanorods were modified on the activated carbon (AC) and AgCl-NR-AC composite when used in concentration of 15 mg successfully removed MB in aqueous solution and observed dye removal efficiency of about 96% in 16 min (Nekouei et al. 2016). $ZnCo_2O_4$ nanorods have also been used, which were synthesized through hydrothermal method and applied for dye removal. It was observed that 5 mg of nanorods when used as adsorbent removed MB dye with adsorption capacity of 2400 mg/g (Lin et al. 2018). An ultrasonic assisted adsorption of the Bromocresol Green (BCG) and Eosin Y (EY) dyes was also reported using ZnO nanorods and AC complex. The adsorption capacities recorded were 57.80 mg/g and 61.73 mg/g for BCG and EY, respectively, when 0.01–0.03 g of adsorbent was used (Ansari et al. 2016). In this, 0.05 g of strontium phosphate and barium phosphate (SBP) composite of nanorod had been used for MB dye removal with adsorption capacity of 1691.8 mg/g (Zhang et al. 2015). Zinc sulfide (ZnS) in combination with tin sulfide (SnS) was used to synthesize ZnS/SnS/A-FA nanorods for removal of CR dye from wastewater. The nanorods were in concentration of 10 mg for removal of dye in the wastewater. The dye was completely photodegraded in 150 min (Kalpana and Selvaraj 2016). Other methods for dye removal using nanorods are also available using tungsten trioxide (WO_3) (Park and Nam 2017), copper (Cu)-doped-ZnO (Perillo and Atia 2018), SnS nanorods (Das and Dutta 2015), etc. All the above-reported methods demonstrated that nanorods are more efficient in dye removal when they are used in combination with materials, which enhances their dye removal capacity.

7.4.4 Graphene

Graphene is an allotropic form of carbon in which carbon atoms are arranged in single-layer hexagonal lattice. Its basic structure reveals that there is a small overlap between valence band and conduction band, which makes it an important nanomaterial for dye removal. Many researchers have used graphene for dye removal as summarized in Table 7.4. A composite of 3D graphene and calcium carbonate ($CaCO_3$) was synthesized in the presence of calcium carbonate and iron oxide nanoparticles. Further, this nanocomposite of graphene was used for the removal of Acid Red 88 (AR88) dye. The amount of adsorbent used was 87 mg to remove 100% of AR88 in 18 min (Arsalani et al. 2018). Three-dimensional graphene was functionalized with magnetic citric acid to prepare a nanocomposite (MCF3DG) for removal of crystal violet (CV) in aqueous medium (Nasiri and Arsalani 2018). Reduced graphene oxide (rGO) coupled with bismuth vanadate ($BiVO_4$) was used to remove MG and RB dye. It was reported that the adsorbent used possessed low catalytic and dye removal efficiency. MG was degraded to 99.5% in two hours, and RB was decolorized 99.84% in four hours (Zhang et al. 2018). Several other methods are also available, which include liquid laser-treated rGO (Russo et al. 2015), polyvinyl alcohol (PVA) hydrogel in combination with GO (Li et al. 2014), magnetic sulfonic and graphene nanocomposite G-SO_3H/Fe_3O_4 (Wang et al. 2013), GO nanosheet functionalized using dithiocarbamate (GO-DTC) (Mahmoodi et al. 2017), GO and cellulose nanowhisker hydrogel nanocomposite removed MO and RB dye in 20 min (Soleimani et al. 2018), rGO and TiO_2 nanocomposite (TiO_2@rGO) (Ali et al. 2018), GO and chitosan aerogel composite (GOCA) (Lai et al. 2019), magnetite nanoparticles and GO nanocomposite (Fe_3O_4@GO) (Mishra 2018), graphene-tannic acid hydrogel (GT hydrogel) (Tang et al. 2018), sulfonated GO (SGO) (Wei et al. 2018), and GO and magnetic iron oxide NPs (GO-MNP) (Othman et al. 2018).

7.4.5 Fullerene

Fullerene is also an allotrope of carbon, which is hollow sphere in shape. They are also known as buckminsterfullerene or the bulkyball. Fullerene is also used for dye removal, and the reported methods are summarized in Table 7.5. Fullerene (C_{60}) has been used as composite with TiO_2 for removal of CV dye. The dye-removing capacity of 82% was achieved when TiO_2 and C_{60} fullerene nanocomposites were used in

Table 7.4 Graphene sheets used in dye removal

Graphene	Dyes removed	Method of removal	Amount of graphene used	Adsorption/ reduction capacity/% dye removal	Reference
3D graphene aerogel/$CaCO_3$	Acid red 88	Adsorption	87 mg	100%	Arsalani et al. (2018)
MCF3DG	Crystal violet	Adsorption	28 mg	100%	Nasiri and Arsalani (2018)
$BiVO_4$-rGO	Malachite green Rhodamine B	Adsorption enhanced with visible light irradiation	50 mg	99.5% 99.84%	Zhang et al. (2018)
rGO	Methylene blue	Adsorption	0.17 mg/ml	746 mg/g	Russo et al. (2015)
PVA/GO hydrogels	Methylene blue	Adsorption	1 g	NR	Li et al. (2014)
G-SO_3H/Fe_3O_4	Safranine T Neutral red Victoria blue	Adsorption	10 mg	199.3 mg/g 216.8 mg/g 200.6 mg/g	Wang et al. (2013)
GO-DTC	Basic blue 41 Basic red 46	Adsorption	0.01–0.04 g	128.5 mg/g 111 mg/g	Mahmoodi et al. (2017)
GO-cellulose nanowhisker hydrogel	Methylene blue Rhodamine B	Adsorption	0.025 g	100% 90%	Soleimani et al. (2018)
TiO_2@rGO	Rhodamine B	Photodegradation	0.1–0.5 g	97%	Ali et al. (2018)
GOCA	Metanil yellow	Adsorption	8 mg	430.99 mg/g	Lai et al. (2019)
Fe_3O_4@GO	Rhodamine 6G	Adsorption	0.005–0.02 g	68–89%	Mishra (2018)
GT hydrogel	Methylene blue	Adsorption	NR	714 mg/g	Tang et al. (2018)
SGO	Methylene blue	Adsorption	1 mg/25 ml	2530 mg/g	Wei et al. (2018)
GO-MNP	Methylene blue	Adsorption	10–15 mg	98%	Othman et al. (2018)

NR Not reported

10 mg concentration (Panahian et al. 2018). Fullerene-modified SiO_2 material was used for efficient removal of MB dye. The percentage of dye removed via photodegradation was 45% and 90% under VIS and UVC light, respectively (Rogozea et al. 2015). One of the studies was also reported in which Fe_2O_3 was doped on C_{60} to achieve a composite of C_{60}-Fe_2O_3. The composite was further used for

Table 7.5 Fullerene structures for dye removal

Fullerene	Dyes removed	Method of removal	Amount of fullerene used	Adsorption/ reduction capacity/% dye removal	Reference
F-TiO_2(B)/ fullerene	Crystal violet	Photocatalysis	10 mg	82%	Panahian et al. (2018)
Fullerene-modified silica	Methylene blue	Photodegradation	20 mg	45% via VIS 90% via UVC	Rogozea et al. (2015)
C_{60}-Fe_2O_3	Methylene blue Rhodamine B Methyl orange	Photodegradation	50 mg/L	98.9%	Zou et al. (2018)

NR Not reported

photocatalytic degradation of three dyes MB, RhB, and MO in the presence of hydrogen peroxide. 98.9% dye decolorization was achieved in 80 min (Zou et al. 2018). The fullerene structures are less exploited as compared to other nanomaterials. This may be due to complexity in their structure,and they are not easy to synthesize.

7.5 Conclusion and Future Prospects

Untreated industrial discharge contains toxic dye, and they are being released in open environment, which is extensively harmful to living beings and environment. Variety of conventional methods is available for effluent treatment, but these methods suffer from limitations of less efficiency, costly, labor-intensive, and time-consuming. Therefore, the use of nanomaterials for dye degradation is considered as novel and eco-friendly approach. Many types of nanomaterials are being reported for dye removal such as nanoparticles, nanotubes, nanorods, graphene, and fullerene. Nanomaterials possess unique properties like highly electroactive, small size, high surface area, easy to synthesize in laboratory, nontoxic nature, biodegradability, and biocompatibility, which make them suitable for dye removal. Due to these properties, nanomaterials can be used with other composites as reported in this review. It is suggested that photocatalytic degradation of dyes is based on light irradiation, which helps in excitation of electrons and causes movement of electron to form electron hole pair, and formation of radicals takes place, which caused degradation of dyes. Out of nanomaterials, which are discussed here, nanoparticles are more exploited. Although the researcher has used a variety of nanomaterials, but fullerene is less exploited. Fullerene should be used as nonabsorbent in combination with the other nanomaterials. Quantum dots are also to be explored for their ability to adsorb dye from the effluent. Besides this toxicity of nanomaterials used as nanosorbents in dye removal must be examined. There are also some challenges in using nanomaterials for dye degradation such as photocatalyst loading, dye concentration, pH of the medium, intensity of light, temperature, photocatalyst morphology, wavelength of light used in photodegradation, effect of oxidizing species, and so on. These above parameters should be optimized carefully before using any nanomaterial for photodegradation of harmful dye.

Acknowledgments Author sincerely acknowledges University Institute of Engineering & Technology, Maharshi Dayanand University, Rohtak, India, for providing necessary facilities.

Questions

1. Write a note on classification of dyes.
2. What are the toxic effects of dyes on the environment and living beings?
3. Discuss in detail conventional methods used to remove dyes.
4. Write down different types of nanomaterials along with their properties.
5. How nanomaterials can be used for removal of dyes and what are their inherent properties, which make them an excellent candidate for dye removal?
6. Discuss the mechanism dye removal via photodegradation.

References

Ali MHH, Al-Afify AD, Goher ME (2018) Preparation and characterization of graphene – TiO2 nanocomposite for enhanced photodegradation of Rhodamine-B dye. Egypt J Aquat Res 44:263–270. https://doi.org/10.1016/J.EJAR.2018.11.009

Alves de Lima RO, Bazo AP, Salvadori DMF et al (2007) Mutagenic and carcinogenic potential of a textile azo dye processing plant effluent that impacts a drinking water source. Mutat Res Toxicol Environ Mutagen 626:53–60. https://doi.org/10.1016/J.MRGENTOX.2006.08.002

Aly HF, Abd-Elhamid AI (2018) Photocatalytic degradation of methylene blue dye using silica oxide nanoparticles as a catalyst. Water Environ Res 90:807–817. https://doi.org/10.2175/106143017X15131012187953

Ansari F, Ghaedi M, Taghdiri M, Asfaram A (2016) Application of ZnO nanorods loaded on activated carbon for ultrasonic assisted dyes removal: experimental design and derivative spectrophotometry method. Ultrason Sonochem 33:197–209. https://doi.org/10.1016/J.ULTSONCH.2016.05.004

Arsalani N, Nasiri R, Zarei M (2018) Synthesis of magnetic 3D graphene decorated with CaCO3 for anionic azo dye removal from aqueous solution: kinetic and RSM modeling approach. Chem Eng Res Des 136:795–805. https://doi.org/10.1016/J.CHERD.2018.06.024

Baruah D, Goswami M, Yadav RNS et al (2018) Biogenic synthesis of gold nanoparticles and their application in photocatalytic degradation of toxic dyes. J Photochem Photobiol B Biol 186:51–58. https://doi.org/10.1016/J.JPHOTOBIOL.2018.07.002

Bulgariu L, Escudero LB, Bello OS et al (2019) The utilization of leaf-based adsorbents for dyes removal: a review. J Mol Liq 276:728–747. https://doi.org/10.1016/J.MOLLIQ.2018.12.001

Carrott PJM, Carrott MMLR, Roberts RA (1991) Physical adsorption of gases by microporous carbons. Colloids Surf 58:385–400. https://doi.org/10.1016/0166-6622(91)80217-C

Chen D, Zeng X, Sheng Y et al (2007) The concentrations and distribution of polycyclic musks in a typical cosmetic plant. Chemosphere 66:252–258. https://doi.org/10.1016/j.chemosphere.2006.05.024

Dai L, Zhu W, He L et al (2018) Calcium-rich biochar from crab shell: an unexpected super adsorbent for dye removal. Bioresour Technol 267:510–516. https://doi.org/10.1016/J.BIORTECH.2018.07.090

Das D, Dutta RK (2015) A novel method of synthesis of small band gap SnS nanorods and its efficient photocatalytic dye degradation. J Colloid Interface Sci 457:339–344. https://doi.org/10.1016/J.JCIS.2015.07.002

Davar F, Majedi A, Mirzaei A (2015) Green synthesis of ZnO nanoparticles and its application in the degradation of some dyes. J Am Ceram Soc 98:1739–1746. https://doi.org/10.1111/jace.13467

Dhmees AS, Khaleel NM, Mahmoud SA (2018) Synthesis of silica nanoparticles from blast furnace slag as cost-effective adsorbent for efficient azo-dye removal. Egypt J Pet 27:1113. https://doi.org/10.1016/J.EJPE.2018.03.012

Dutta DP, Venugopalan R, Chopade S (2017) Manipulating carbon nanotubes for efficient removal of both cationic and anionic dyes from wastewater. ChemistrySelect 2:3878–3888. https://doi.org/10.1002/slct.201700135

Dutta AK, Ghorai UK, Chattopadhyay KK, Banerjee D (2018) Removal of textile dyes by carbon nanotubes: a comparison between adsorption and UV assisted photocatalysis. Phys E Low-Dimens Syst Nanostructures 99:6–15. https://doi.org/10.1016/J.PHYSE.2018.01.008

Fathima JB, Pugazhendhi A, Oves M, Venis R (2018) Synthesis of eco-friendly copper nanoparticles for augmentation of catalytic degradation of organic dyes. J Mol Liq 260:1–8. https://doi.org/10.1016/J.MOLLIQ.2018.03.033

Gao X, Li W, Mei R et al (2019) Effect of the B2H6/CH4/H2 ratios on the structure and electrochemical properties of boron-doped diamond electrode in the electrochemical oxidation process of azo dye. J Electroanal Chem 832:247–253. https://doi.org/10.1016/J.JELECHEM.2018.11.009

GilPavas E, Dobrosz-Gómez I, Gómez-García MÁ (2017) Coagulation-flocculation sequential with Fenton or photo-Fenton processes as an alternative for the industrial textile wastewater treatment. J Environ Manag 191:189–197. https://doi.org/10.1016/J.JENVMAN.2017.01.015

Hameed BH, Ahmad AA, Aziz N (2009) Adsorption of reactive dye on palm-oil industry waste: equilibrium, kinetic and thermodynamic studies. Desalination 247:551–560. https://doi.org/10.1016/J.DESAL.2008.08.005

Hunger K, Wiley J, Sons. (2003) Industrial dyes: chemistry, properties, applications. Wiley-VCH

Kalpana K, Selvaraj V (2016) Development of ZnS/SnS/A-FA nanorods at ambient temperature: binary catalyst for the removal of Congo red dye and pathogenic bacteria from wastewater. J Ind Eng Chem 41:105–113. https://doi.org/10.1016/J.JIEC.2016.07.016

Kant R (2012) Textile dyeing industry an environmental hazard. Nat Sci 04:22–26. https://doi.org/10.4236/ns.2012.41004

Khataee AR, Kasiri MB (2010) Photocatalytic degradation of organic dyes in the presence of nanostructured titanium dioxide: influence of the chemical structure of dyes. J Mol Catal A Chem 328:8–26. https://doi.org/10.1016/J.MOLCATA.2010.05.023

Korenak J, Hélix-Nielsen C, Bukšek H, Petrinić I (2019) Efficiency and economic feasibility of forward osmosis in textile wastewater treatment. J Clean Prod 210:1483–1495. https://doi.org/10.1016/J.JCLEPRO.2018.11.130

Lacorte S (2003) Organic compounds in paper-mill process waters and effluents. TrAC Trends Anal Chem 22:725–737. https://doi.org/10.1016/S0165-9936(03)01009-4

Lai KC, Hiew BYZ, Lee LY et al (2019) Ice-templated graphene oxide/chitosan aerogel as an effective adsorbent for sequestration of metanil yellow dye. Bioresour Technol 274:134–144. https://doi.org/10.1016/J.BIORTECH.2018.11.048

Li C, She M, She X et al (2014) Functionalization of polyvinyl alcohol hydrogels with graphene oxide for potential dye removal. J Appl Polym Sci 131:39–87. https://doi.org/10.1002/app.39872

Lin K, Qin M, Geng X et al (2018) ZnCo2O4 nanorods as a novel class of high-performance adsorbent for removal of methyl blue. Adv Powder Technol 29:1933–1939. https://doi.org/10.1016/J.APT.2018.05.006

Mahmoodi NM, Ghezelbash M, Shabanian M et al (2017) Efficient removal of cationic dyes from colored wastewaters by dithiocarbamate-functionalized graphene oxide nanosheets: from synthesis to detailed kinetics studies. J Taiwan Inst Chem Eng 81:239–246. https://doi.org/10.1016/J.JTICE.2017.10.011

Maleki A, Mohammad M, Emdadi Z et al (2018) Adsorbent materials based on a geopolymer paste for dye removal from aqueous solutions. Arab J Chem 13:3017. https://doi.org/10.1016/J.ARABJC.2018.08.011

Mallakpour S, Rashidimoghadam S (2019) Carbon nanotubes for dyes removal. Compos Nanoadsorbents:211–243. https://doi.org/10.1016/B978-0-12-814132-8.00010-1

Marsh H, Rodríguez-Reinoso F (2006) Activated carbon. Elsevier

Meriläinen P, Oikari A (2008) Exposure assessment of fishes to a modern pulp and paper mill effluents after a black liquor spill. Environ Monit Assess 144:419–435. https://doi.org/10.1007/s10661-007-0004-9

Mishra A (2018) Study of organic pollutant removal capacity for magnetite@ graphene oxide nanocomposites. Vacuum 157:524–529. https://doi.org/10.1016/J.VACUUM.2018.08.034

Mohammadi A, Veisi P (2018) High adsorption performance of β-cyclodextrin-functionalized multi-walled carbon nanotubes for the removal of organic dyes from water and industrial wastewater. J Environ Chem Eng 6:4634–4643. https://doi.org/10.1016/J.JECE.2018.07.002

Monsef Khoshhesab Z, Souhani S (2018) Adsorptive removal of reactive dyes from aqueous solutions using zinc oxide nanoparticles. J Chin Chem Soc 65:1482–1490. https://doi.org/10.1002/jccs.201700477

Nadaf NY, Kanase SS (2016) Biosynthesis of gold nanoparticles by Bacillus marisflavi and its potential in catalytic dye degradation. Arab J Chem 12:4806. https://doi.org/10.1016/J.ARABJC.2016.09.020

Nasiri R, Arsalani N (2018) Synthesis and application of 3D graphene nanocomposite for the removal of cationic dyes from aqueous solutions: response surface methodology design. J Clean Prod 190:63–71. https://doi.org/10.1016/J.JCLEPRO.2018.04.143

Nautiyal P, Subramanian KA, Dastidar MG (2016) Adsorptive removal of dye using biochar derived from residual algae after in-situ transesterification: alternate use of waste of biodiesel industry. J Environ Manag 182:187–197. https://doi.org/10.1016/J.JENVMAN.2016.07.063

Nekouei F, Kargarzadeh H, Nekouei S et al (2016) Novel, facile, and fast technique for synthesis of AgCl nanorods loaded on activated carbon for removal of methylene blue dye. Process Saf Environ Prot 103:212–226. https://doi.org/10.1016/J.PSEP.2016.07.010

Othman NH, Alias NH, Shahruddin MZ et al (2018) Adsorption kinetics of methylene blue dyes onto magnetic graphene oxide. J Environ Chem Eng 6:2803–2811. https://doi.org/10.1016/J.JECE.2018.04.024

Panahian Y, Arsalani N, Nasiri R (2018) Enhanced photo and sono-photo degradation of crystal violet dye in aqueous solution by 3D flower like F-TiO2(B)/fullerene under visible light. J Photochem Photobiol A Chem 365:45–51. https://doi.org/10.1016/J.JPHOTOCHEM.2018.07.035

Park S-M, Nam C (2017) Dye-adsorption properties of WO3 nanorods grown by citric acid assisted hydrothermal methods. Ceram Int 43:17022–17025. https://doi.org/10.1016/J.CERAMINT.2017.09.111

Paz A, Carballo J, Pérez MJ, Domínguez JM (2017) Biological treatment of model dyes and textile wastewaters. Chemosphere 181:168–177. https://doi.org/10.1016/J.CHEMOSPHERE.2017.04.046

Peres EC, Slaviero JC, Cunha AM et al (2018) Microwave synthesis of silica nanoparticles and its application for methylene blue adsorption. J Environ Chem Eng 6:649–659. https://doi.org/10.1016/J.JECE.2017.12.062

Perillo PM, Atia MN (2018) Solar-assisted photodegradation of methyl orange using Cu-doped ZnO nanorods. Mater Today Commun 17:252–258. https://doi.org/10.1016/J.MTCOMM.2018.09.010

Pokhrel D, Viraraghavan T (2004) Treatment of pulp and paper mill wastewater—a review. Sci Total Environ 333:37–58. https://doi.org/10.1016/j.scitotenv.2004.05.017

Punzi M, Nilsson F, Anbalagan A et al (2015) Combined anaerobic–ozonation process for treatment of textile wastewater: removal of acute toxicity and mutagenicity. J Hazard Mater 292:52–60. https://doi.org/10.1016/J.JHAZMAT.2015.03.018

Rogozea EA, Meghea A, Olteanu NL et al (2015) Fullerene-modified silica materials designed for highly efficient dyes photodegradation. Mater Lett 151:119–121. https://doi.org/10.1016/J.MATLET.2015.03.069

Ruhl AS, Zietzschmann F, Hilbrandt I et al (2014) Targeted testing of activated carbons for advanced wastewater treatment. Chem Eng J 257:184–190. https://doi.org/10.1016/J.CEJ.2014.07.069

Russo P, D'Urso L, Hu A et al (2015) In liquid laser treated graphene oxide for dye removal. Appl Surf Sci 348:85–91. https://doi.org/10.1016/J.APSUSC.2014.12.014

Sadeghzadeh-Attar A (2018) Efficient photocatalytic degradation of methylene blue dye by SnO2 nanotubes synthesized at different calcination temperatures. Sol Energy Mater Sol Cells 183:16–24. https://doi.org/10.1016/J.SOLMAT.2018.03.046

Saleh SM (2019) ZnO nanospheres based simple hydrothermal route for photocatalytic degradation of azo dye. Spectrochim Acta Part A Mol Biomol Spectrosc 211:141–147. https://doi.org/10.1016/J.SAA.2018.11.065

Seyed Arabi SM, Lalehloo RS, Olyai MRTB et al (2019) Removal of Congo red azo dye from aqueous solution by ZnO nanoparticles loaded on multiwall carbon nanotubes. Phys E Low-Dimens Syst Nanostructures 106:150–155. https://doi.org/10.1016/J.PHYSE.2018.10.030

Shouman MA, Fathy NA (2018) Microporous nanohybrids of carbon xerogels and multi-walled carbon nanotubes for removal of rhodamine B dye. J Water Process Eng 23:165–173. https://doi.org/10.1016/J.JWPE.2018.03.014

Soleimani K, Tehrani AD, Adeli M (2018) Bioconjugated graphene oxide hydrogel as an effective adsorbent for cationic dyes removal. Ecotoxicol Environ Saf 147:34–42. https://doi.org/10.1016/J.ECOENV.2017.08.021

Sweet MJ, Chessher A, Singleton I (2012) Review: metal-based nanoparticles; size, function, and areas for advancement in applied microbiology. Adv Appl Microbiol 80:113–142. https://doi.org/10.1016/B978-0-12-394381-1.00005-2

Tamer TM, Abou-Taleb WM, Roston GD et al (2018) Formation of zinc oxide nanoparticles using alginate as a template for purification of wastewater. Environ Nanotechnol Monit Manag 10:112–121. https://doi.org/10.1016/J.ENMM.2018.04.006

Tan KB, Vakili M, Horri BA et al (2015) Adsorption of dyes by nanomaterials: recent developments and adsorption mechanisms. Sep Purif Technol 150:229–242. https://doi.org/10.1016/J.SEPPUR.2015.07.009

Tan CHC, Sabar S, Hussin MH (2018) Development of immobilized microcrystalline cellulose as an effective adsorbent for methylene blue dye removal. South African J Chem Eng 26:11–24. https://doi.org/10.1016/J.SAJCE.2018.08.001

Tang C-Y, Yu P, Tang L-S et al (2018) Tannic acid functionalized graphene hydrogel for organic dye adsorption. Ecotoxicol Environ Saf 165:299–306. https://doi.org/10.1016/J.ECOENV.2018.09.009

Varghese SP, Babu AT, Babu B, Antony R (2017) γ-MnOOH nanorods: efficient adsorbent for removal of methylene blue from aqueous solutions. J Water Process Eng 19:1–7. https://doi.org/10.1016/J.JWPE.2017.06.001

Verma M, Kumar A, Singh KP (2019) Reactive magnetron sputtering based synthesis of WO3 nanoparticles and their use for the photocatalytic degradation of dyes. Solid State Sci 99:105847. https://doi.org/10.1016/J.SOLIDSTATESCIENCES.2019.02.008

Wang S, Wei J, Lv S et al (2013) Removal of organic dyes in environmental water onto magnetic-sulfonic graphene nanocomposite. Clean (Weinh) 41:992–1001. https://doi.org/10.1002/clen.201200460

Wang T, He X, Li Y, Li J (2018) Novel poly(piperazine-amide) (PA) nanofiltration membrane based poly(m-phenylene isophthalamide) (PMIA) hollow fiber substrate for treatment of dye solutions. Chem Eng J 351:1013–1026. https://doi.org/10.1016/J.CEJ.2018.06.165

Wei M, Chai H, Cao Y, Jia D (2018) Sulfonated graphene oxide as an adsorbent for removal of Pb2+ and methylene blue. J Colloid Interface Sci 524:297–305. https://doi.org/10.1016/J.JCIS.2018.03.094

Zafar MN, Dar Q, Nawaz F et al (2018) Effective adsorptive removal of azo dyes over spherical ZnO nanoparticles. J Mater Res Technol 8:713. https://doi.org/10.1016/j.jmrt.2018.06.002

Zhang F, Song W, Lan J (2015) Effective removal of methyl blue by fine-structured strontium and barium phosphate nanorods. Appl Surf Sci 326:195–203. https://doi.org/10.1016/J.APSUSC.2014.11.143

Zhang F, Chen X, Wu F, Ji Y (2016) High adsorption capability and selectivity of ZnO nanoparticles for dye removal. Colloids Surf A Physicochem Eng Asp 509:474–483. https://doi.org/10.1016/J.COLSURFA.2016.09.059

Zhang M, Gong J, Zeng G et al (2018) Enhanced degradation performance of organic dyes removal by bismuth vanadate-reduced graphene oxide composites under visible light radiation. Colloids Surf A Physicochem Eng Asp 559:169–183. https://doi.org/10.1016/J.COLSURFA.2018.09.049

Zou C, Meng Z, Ji W et al (2018) Preparation of a fullerene[60]-iron oxide complex for the photo-Fenton degradation of organic contaminants under visible-light irradiation. Chinese J Catal 39:1051–1059. https://doi.org/10.1016/S1872-2067(18)63067-0

Chapter 8
Safe Appraisal of Carbon Nanoparticles in Pollutant Sensing

Manisha Kumari, G. R. Chaudhary, and Savita Chaudhary

Abstract The quantification of different ranges of pollutants and contaminants in the environment has earned enormous progress in the past few years owing to its extensive necessity in the domain of public health, clinical research, occupational hygiene, and mankind's welfare. The release of carcinogenic pollutants from industrial, agriculture, and various other sectors is one of the serious threats faced by human beings. For the safety of living beings, there is an urgent need for rapid, selective, and sensitive detection strategies for the screening of different types of pollutants. Out of numerous advanced functional nanoparticles, the fluorescent carbon-based nanoparticles have initiated a new approach for establishing easy, selective, and expeditious real-time analysis of pollutants. The purpose of this chapter is to examine the unique properties of carbon dots, which make it useful in the area of sensing. The profiling of multifunctional luminescent CDsat nanometric range has unlocked new horizons of development in materials science and nanoremediation. In particular, we review the chemical and physical properties of carbon dots and their extensive potential in the fate of perilous pollutants detection for a better and cleaner environment.

Keywords Environment · Carbon dots · Sensing · Quantitation · Pollutants

8.1 Introduction

The "emerging pollutants" are the synthetic or naturally arising chemicals that are not usually existing in the environment but which have the capability to intrude into the environment (Geissen et al. 2015) and affect living beings. The term "emerging"

M. Kumari · G. R. Chaudhary · S. Chaudhary (✉)
Department of Chemistry and Centre of Advanced Studies in Chemistry, Panjab University, Chandigarh, India
e-mail: schaudhary@pu.ac.in

R. Kumar et al. (eds.), *Advanced Functional Nanoparticles "Boon or Bane" for Environment Remediation Applications*, Environmental Contamination Remediation and Management, https://doi.org/10.1007/978-3-031-24416-2_8

is used as a demonstration that the occurrence of pollutants in the ecological system. It has attracted significant attentiveness across scientific communities for the safety of human beings (Li and Migliaccio et al. 2010). Apart from their evolution in social economy, the different forms of environmental pollutants have emerged as a serious threat to mankind. A broad range of pollutants that include highly toxic substances gives rise to major ill effects on human health even if their small concentrations remain in the environment (Liua et al. 2019; Jiang et al. 2015; Arisido et al. 2015). For instance, phenols, dyestuffs, aromatic nitro compounds, heavy metals, polycyclic aromatic hydrocarbons (PACs), pesticides, perfluorinated compounds (PFCs), and antibiotics have adverse side effects on the ecosphere and health of living beings (Pérez-Estrada et al. 2015; Ren et al. 2018). Due to global industrialization, the continuous increase in ecological imbalance and environmental pollution has made the detection and separation of pollutants a vast topic of research hotspot. For example, rhodamine B (RhB) is extensively used in the textile industry as an organic dye is a vital organic pollutant in the environment (Fig. 8.1) (Jiao et al. 2018). Phenol and its compounds have low biodegradation in water with high toxicity and stability.

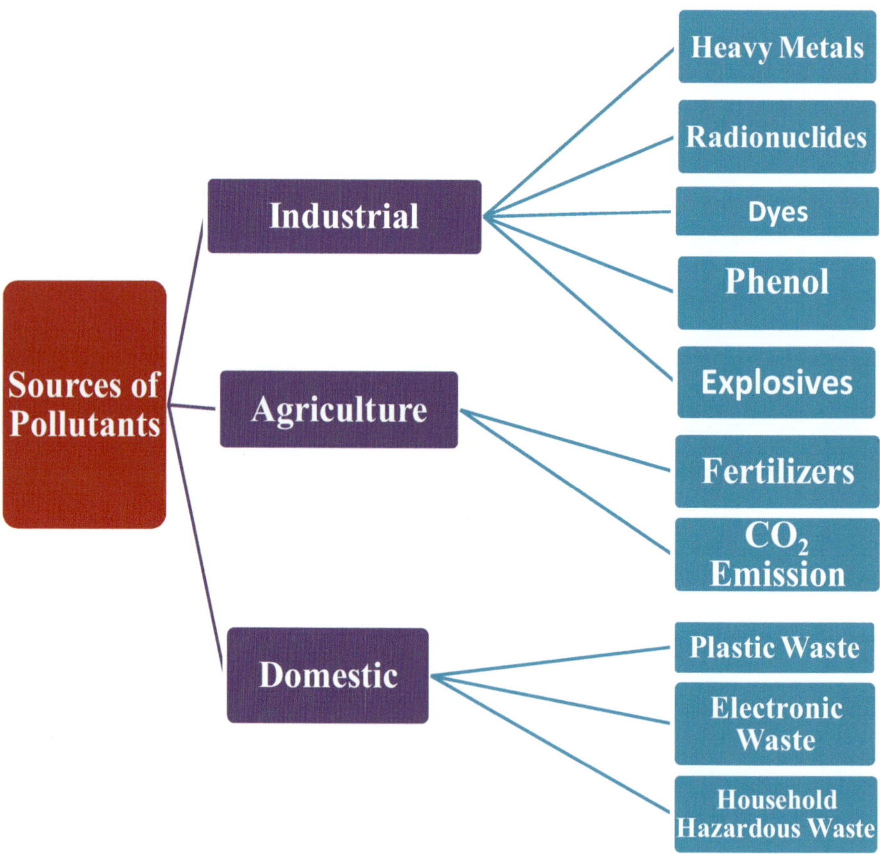

Fig. 8.1 A diagrammatic representation of diverse sources of pollutants in the environment

It possesses hazardous effects on the health of human beings. Among several phenolic compounds, the most potential peril has been posed by phenol because it is more vexatious to degrade as compared to chlorinated and nitro-substituted phenol (Darabdhara et al. 2016; Mohapatra et al. 2018; Du et al. 2006). Commonly used antibiotics such as tetracycline in human and animal medicine has been excreted quite largely into the ecosystem due to its poor absorbance in the digestive tract of animals (about 50–80%) (Gothwal and Shashidar et al. 2015). As a consequence, the dealing with phenol, rhodamine B, tetracycline, and various other pollutants in our environment has become an area of concern for the safety of human beings.

An extensive increase in the production of more fibers, grains, medicines, and erection of abounding buildings has been compelled due to huge increase in the population. Our environment has been under the burden of tremendous load due to the industrial revolution accompanied with information technology. The production of enough food to feed the current population has been surpassed by the demand. The high yield of food obtained from the huge utilization of pesticides, insecticides, herbicides, and other plant growth promoters and protecting agents have a deleterious effect on the healthy environment. Different kinds of processes such as industrial, domestic, and agriculture processes arbitrarily enumerate toxic and carcinogenic pollutants to our environment, such as fertilizers, biocides, pesticides, chlorinated solvents, chlorinated phenol, heavy metals, polyaromatic hydrocarbon, dyes, and plastic (Damlas and Eleftherohorinos et al. 2011; Wan et al. 2019; Lew et al. 2013; Nicolopoulou-Stamati et al. 2016; Wasim et al. 2009). The origin of pollutants from different sources are broadly classified into the following categories (Fig. 8.1).

Heavy metal ions such as lead, mercury, cadmium, and arsenic can cause various carcinogenic effects on human health. Different kinds of diseases can be posed by these toxic heavy metals, when ingested or inhaled above the bio-recommended levels (Martin and Griswold et al. 2009). Mercury (Hg^{2+}) can pervade respiratory, gastrointestinal, and skin tissues, which can further damage the deoxyribonucleic acid (DNA) and influence the central nervous system peremptory (Weng et al. 2011; Lu et al. 2012). Long exposure to toxic heavy metals can prove to be very lethal and malignant. In order to eliminate the noxious effects of these pollutants on mankind, animals, and other living beings due to their slow degradation and recalcitrant nature, it needed to remove them from the environment (Kumar and Bharadvaja 2019). For this purpose, it is urgent to quantify and identify deleterious pollutants in our environmental media. Over the last decade, significant interest has been developed among researchers in nanoparticles, due to the invention in the changes of chemical properties of materials upon their conversion in nanoparticles form. The dimensions of nanoparticles ultimately determine the physicochemical properties of materials. The influence of size-dependent effect on optical properties of nanoparticles, prominently photoluminescence properties, has further been used for the sensing of harmful pollutants (Yang et al. 2013; Prashanth et al. 2015).

In the past years, semiconductor nanomaterials such as quantum dots (e.g., cadmium sulfide, lead selenide) have gained colossal attention due to their eminent photoluminescence properties. Unfortunately, these inorganic quantum dots incorporate toxic elements, which have carcinogenic and adverse effect on living things in the environment. The excessive use of semiconductor quantum dots in several applications like sensors and in devices becomes limited due to the fact that their eventual discharge is known to be detrimental. Accordingly, it is the need of the hour to develop organic photoluminescent nanomaterials, which are biocompatible in nature and can be fabricated via effective, simple, green, cost-effective synthesis processes (Wang et al. 2013).

In this regard, carbon dots (CDs), a fascinating young material of nano-family, are a kind of fluorescent nanoparticles with extraordinary potential in pollutant sensing. CDs have exceptional and remarkable properties such as high specific surface area, nontoxicity, good chemical stability, biocompatibility, aqueous dispersibility, and high quantum yield. Historically, CDs were serendipitously invented during the separation and purification process of single-walled carbon nanotubes using arc-discharge method in 2004 (Xu et al. 2004).

8.2 Carbon Dots Structure

The quantum-sized carbon nanoparticles have excellent photoluminescence, high water solubility, and biocompatibility. The existence of numerous carboxylic and hydroxyl groups on the exterior facade has made them soluble in aqueous media. The existence of exceptionally well-coordinated structure has further contributed to the novel and peculiar properties of CDs (Lim et al. 2015; Baker and Baker et al. 2010; Weng et al. 2011; Tao et al. 2016; Namdari et al. 2017). Usually, all nanosized carbon materials are known as nanocarbons, if at least one dimension is less than 10 nm in size. However, in the case of CDs, all three dimensions are below 10 nm. CDs have sp^2/sp^3 spatial hybridization featuring luminescence as their innate characteristic property (Baker and Baker et al. 2010; Yan et al. 2010). Based on the synthesis process, CDs have exhibited distinct chemical structure and photoluminescence properties. The facade of CDs can comprise or modify diverse chemical groups, for instance, oxygen-containing group, nitrogen-containing groups, polymer chains, nitrogen-based amino-containing groups, etc. (Baker and Baker et al. 2010). The novel and tunable properties of CDs can be enhanced by employing surface passivation and chemical modification with various biological, polymeric, inorganic, and organic materials. Moreover, carbon is present in their basic skeleton; due to this, CDs have nontoxicity, good conductivity, and superior optical and electronic properties (Yang et al. 2013; Tetsuka et al. 2012).

8.3 Carbon Dots Synthesis

A diverse range of synthetic methods have been reported in the literature to produce CDs. Broadly, it can be classified into two main approaches: one is bottom-up, and another one is top-down approach (Fig. 8.2). Top-down method includes synthesis from larger carbon precursors such as graphite sheets, carbon nanotubes, etc. to

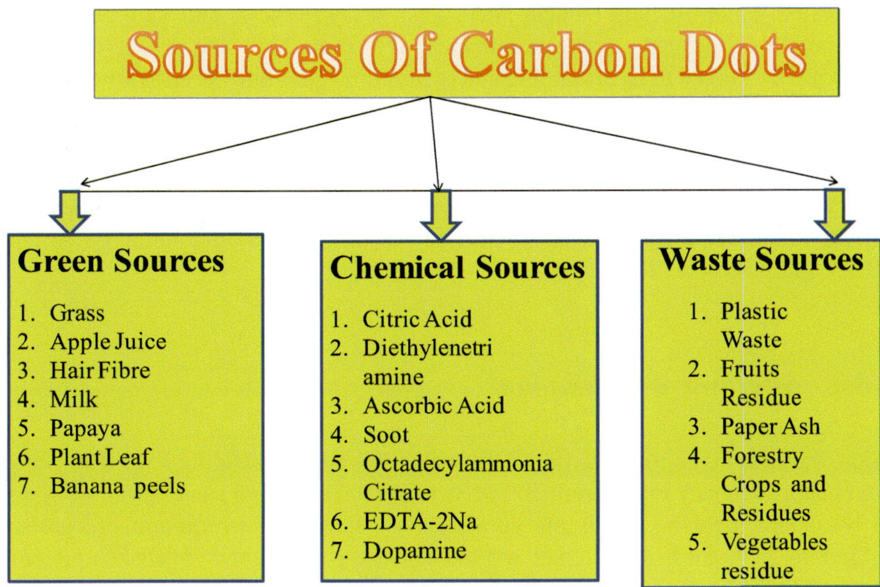

Fig. 8.2 A variety of precursors used for the synthesis of CDs

produce CDs. Bottom-up methods include formation of CDs from small carbon compounds, for instance, carbohydrates, ascorbic acid, ethylene diamine, etc. (Robertson and O'Reilly et al. 1987; Chen et al. 2016). Posttreatment methods can be utilized for further possible modification.

During the synthesis of CDs, three major concerns are involved: (1) centrifugation, dialysis, and electrophoresis are posttreatment techniques applied to control size and uniformity. (2) Pyrolysis and solvent chemotherapy are the electrochemical synthesis used for carbon capture during carbonization. (3) Preparation and post-preparation steps optimize the functional groups and surface properties (Cao et al. 2007; Baker and Baker et al. 2010; Briscoe et al. 2015). The superior and intriguing properties of the CDs can be achieved by other methods, which involve hydrothermal carbonization in a microwave hydrothermal, plasma hydrothermal, and microreactor method (He et al. 2011; Zong et al. 2011; Tang et al. 2012). The properties of CDs can be influenced by the methods and techniques described above. In this section, the prominent methods are illustrated.

8.3.1 Hydrothermal/Solvothermal Synthesis Method

The hydrothermal method is a common, low-cost, eco-friendly, and nontoxic pathway to synthesize carbon-based nanomaterials. In this method, a chemical reaction takes place in a solvent media inside a closed system between the precursors under high boiling point and high-pressure condition. CDs have been generated by several precursors, which include banana juice, glucose, citric acid, chitosan, dopamine, diammonium hydrogen citrate, proteins, etc. (He et al. 2011; Qu et al. 2013). Two types of CDs known as hydrophobic and hydrophilic CDs have been synthesized by carbonization of carbohydrates within a size range of less than 10 nm (Bhunia et al. 2013). For example, synthesis of hydrophilic CDs produced by applying a broad range of pH to the reaction media. On the other hand, preparation of hydrophobic CDs done via combining octadecylamine and octadecene with different amounts of carbohydrates for a time period of 30 minutes at a temperature range of 70 and 300 °C (Bhunia et al. 2013).

8.3.2 Microwave Irradiation

Microwaves are electromagnetic waves with a wavelength ranging from 100 cm to 1 mm and frequency range from 300 MHz to 300 GHz, which can distribute energy to break chemical bonds of the compounds. Microwave irradiation of carbon-containing material is an efficient and inexpensive pathway to synthesize uniform-sized CDs by breaking bonds of the substance. In this synthesis procedure, the possibility to modify the compound structure, morphology, and combination can be produced by using various temperatures ranges (Jaiswal et al. 2012). For example, green luminescent CDs were fabricated using diethylene glycol (DEG) as the carbon source by microwave irradiation process (Jaiswal et al. 2012). The produced diethylene glycol-CDs (DEG-CQDs) can be dispersed in water with a clear appearance and eventually absorbed by glioma C6 cells and become ideal for bioimaging applications (Liu et al. 2014).

8.3.3 Carbonization

Recent reports have confirmed that carbonization of carbon-containing materials can be achieved by providing external heat and develop them into CDs. This pathway of synthesizing CDs has provided significant eminence such as nonsolvent media, facile synthesis procedure, short reaction time, a wide range of carbon sources, and economic viable approach for large-scale production. As reported earlier, pristine blue-emitting CDs were made by thermal decomposition of citric acid, which is an inexpensive and efficient bottom-up method to produce homogenous distribution in an aqueous media (Dong et al. 2012). Thermal decomposition of citric acid was done by heating it for 30 minutes at 200 °C. On the other hand, by heating the citric acid for a long period of time, which is 120 minutes, we can obtain graphene oxide (GO) particles with controllable size and shape. In graphene oxide, the carbon content that is shown by elemental analysis is high as compared to CDs and citric acid, and hence the process is called "carbonization" (Dong et al. 2012). Since then, a wide range of temperatures and time durations were examined for the fabrication of CDs (Dong et al. 2012; Wu et al. 2013; Amjadi et al. 2014; Bagheri et al. 2017). However, chemical thermal decomposition is not the only pathway to carbonize carbon-containing material. Another powerful technique to make CDs is known as "electrochemical carbonization" using a broad range of carbon-containing compounds as precursors. For instance, electrochemical carbonization can convert alcohol into CDs (Deng et al. 2014). It has been found that the enhancement in the applied potential has increased the size of CDs to a greater extent (Wan et al. 2019; Deng et al. 2014).

8.3.4 Chemical Ablation

Small-sized organic molecules can be carbonized by strong oxidizing acids into carbonic materials via controlled oxidation process, which can further be cut down into sheets (Baker and Baker et al. 2010; Dong et al. 2010; Qiao et al. 2010; Luo et al. 2013; Zhu et al. 2013). For instance, an easy and efficient method for fabricating fluorescent CDs is that the carbonic structure of carbohydrates, which is dehydrated in an aqueous medium with concentrated sulfuric acid, was disintegrated into CDs by using nitric acid (Peng and Travas-Sejdic 2009). The respective photoluminescence properties can be enhanced by surface passivation of CDs in posttreatment of CDs by employing amine-terminated compounds. The respective emission properties of CDs have further been controlled by varying the amount of the precursors and reaction conditions.

8.4 Photoluminescence Mechanism in CDs

Photoluminescence (PL) is the phenomenon of the emission of light from any matter when an electromagnetic wave in the visible light wavelength range is absorbed by that particular material. Photo excitation initiates the process of photoluminescence (light emission). The time lag may alter up to minutes or hours ranging from short femtoseconds to retard emission under possible conditions. The origin of the detected optoelectronic properties of CDs has grabbed attention among researchers (Baker and Baker 2010). In the literature, various reasons such as surface states, distinct functional groups, defects, and quantum confinement effect, which include the association of electron hole pairs in small sp^2 carbon clusters or an sp^3 matrix, are responsible for the photoluminescence properties of CDs (Baker and Baker 2010; Wang and Hu 2014).

8.4.1 Quantum Confinement-Derived PL

The bandgap transition via conjugated π-domains is associated with a quantum confinement-based photoluminescence mechanism in carbon materials. For optoelectronic materials, the expansion of the new π-conjugated molecules in CDs made it effective material in fabricating organic semiconducting, light-emitting, and photovoltaic materials in optoelectronic devices. Extensive delocalization of electrons is caused by the electronic structure of π-systems present in CDs. The conjugated π-domains have further attributed to the wide range of ultraviolet-infrared (UV-IR) absorption, and peaks were assigned to the n-π^* transition in C=O bonds and π-π^* transition in C-C bonds (Bailey et al. 2005). π-π * electronic levels in the sp^2 positions are strongly localized. However, the photoluminescence of the CDs leads to the renewal of electron-hole pairs in carbon nanoparticles. Intense visible emission has further been attributed to the bandgap of the ε and ε^* in sp^3 and sp^2 matrix (Robertson and O'Reilly 1987). In conclusion, the enhancement of chemical structures can adjust the highest occupied molecular orbital/lowest occupied molecular orbital (HOMO/LUMO) energies, as to be highest occupied and the lowest unoccupied molecular orbitals, and their intermolecular interplays of π-conjugated molecules (Shen et al. 2014). The optical absorption and fluorescence emission properties have further related to the π-plasmon and radiative recombination of the surface-limited electron and holes in the system rather than the bandgap, which is opposite from semiconductor metallic quantum dots (Hu et al. 2015). Size-tunable pathways for the synthesis of CDs are because of the unique size-related photoluminescence properties of CDs, which are further affected by altering the reaction factors like temperature, reaction time period, and precursors proportion/ratio (Robertson and O'Reilly 1987). The photoexcited form of CDs has exhibited electron-accepting and electron-donating abilities due to the existence of radiative recombination and PL-quenching abilities of CDs (Wang et al. 2009). A productive technique to prepare tunable size CDs by altering the composition of the reactants has already been demonstrated by Zhang et al. in detail (Zhang et al. 2015). The surface states are mainly determined by the existence of different types of hybridized groups and energy traps in the backbone structure of CDs. The energy traps have further been affected by the electronic transitions in the CDs. Although, the size range of CDs has also affected the PL spectrum. However, the surface states are believed to play a dominant role in the PL properties of CDs. Therefore, by engineering these trapping states, the luminescence properties of CDs can be controlled and optimized for better outcomes.

8.4.2 Surface State-Derived PL

CD facade forms energy traps due to the existence of surface state having various types of hybridized groups in the carbon backbone. It has been manifested that PL mechanism can be controlled by energy traps in CDs. PL center in CDs emerged from the surface state in most of the cases. In addition, size can also affect the PL spectrum. Therefore, surface states played a prominent role in PL transitions in CDs (Hilderbrand et al. 2009). PL behavior can be restrained by organizing the trapping states in CDs (Wang et al. 2010; Zhu et al. 2015). Quantum yield is the measure of the efficiency of the photon emission of a quantum dot and is defined by the number of emitted photon to the number of photons absorbed. The chosen synthetic procedure and surface functionality can remarkably increase the quantum yield of CDs (Shi et al. 2016). In the past, inadequate knowledge about the surface chemistry of CDs has mainly contributed to the low quantum yield of prepared CDs. The quantum yield of CDs has further been reduced due to the formation of energy traps on the surface in a nonradiative way. Surface passivation can intensify the photoluminescence of CDs by expediting the radiative recombination of electron-hole pairs (Sun et al. 2006). The stabilization can also be achieved by surface passivation of CDs. Various kinds of passivating agents such as polymers, natural proteins, and other organic molecules are efficient for modifying the surface properties of CDs. Therefore, it is not wrong to say that different pathways to synthesize CDs are a key factor to improve the PL mechanism in nano dots (Zheng et al. 2011). In addition, the quantum yield of formed CDs has further been modified by passivating the surface of CDs with the heteroatoms as dopant. The most commonly used dopant atoms are oxygen, nitrogen, sulfur, and phosphorus (Zheng et al. 2011; Chandra et al. 2013). Co-doping is also possible to utilize the synergic effect of different atoms (Lim et al. 2015). For instance, synthesis of CDs by adopting amine molecule as a precursor can be termed as both nitrogen-doping agent (N-doping agent) and surface-modifying agent (Wang and Hu 2014). In conclusion, an enhancement in the physicochemical properties of CDs and their utilities in different bio-applications can be optimized by surface engineering and modification of CDs (Namdari et al. 2017).

8.5 Biocompatibility and Cytotoxicity

CDs have attracted considerable interest among researchers due to their unique and exceptionally high valued characteristic properties that proved to be playing a critical role in materials science, biomedicine, and biochemistry. They have also been utilized in various biological and medical applications. Biocompatibility and cytotoxicity analysis have also been employed for the safer use of CDs in the biomedical field. Biocompatibility is a term that generally refers to the ability of a material or to probe the compatibility of a material within a tissue, cells, and living organisms. The biocompatible substances do not produce any kind of noxious or immunological reaction within the body system. On the other hand, cytotoxicity is attributed to the fatal effects generated to the cells. In the case of CDs, many studies have been observed to ensure the biocompatibility of these nano-materials. For instance, the polyethylene glycol (PEG)-functionalized CQDs have possessed no cytotoxic effect on living cells. Also, the presence of CDs has not produced any kind of toxic impact on the mice after vaccinating with CDs. In conclusion, it was clear from this study that CDs were appropriate for in vivo optical imaging in the biomedical field. Surface passivation was also studied on CDs via employing poly-propionylethylenimine-co-ethylenimine (PPEI-EI) agent. These surface-passivated CDs have not possessed any kind of cytotoxic impact on the living system (Cao et al. 2007). The bare CDs did not exhibit toxicity up to 0.4 mg/ml high dose concentration for cells. Furthermore, cytotoxicity was also evaluated in human kidney cells in the presence of CDs synthesized by using electrochemical method (Zhao et al. 2008). Another preparation of CDs is obtained through nitric acid oxidation. The toxic effect of N-doped CDs appraised by using MTT and trypan blue assay has shown higher cell survival rate in the presence of synthesized CDs and supported nontoxic nature of prepared particles. 0.1–1 mg/ml concentrations of CDs were employed on cells for 24 h. The amount of dose required was 100–1000 times more than the amount needed for bioimaging. The cell viability was around 90–100% for concentration less than 0.5 mg/ml, and there is a certain rise on the mortality rate of cells at higher concentrations. The results have indicated that bioimaging and biomedical applications can be utilized from these particles (Ray et al. 2009). Recent researches have shown that PEGylated CDs can be congenial at the higher concentration for bioimaging applications without observing any cytotoxic effect. For instance, mice were injected with PEG1500N-passivated CDs for 28 days and have shown nontoxic effect on living cells (Yang et al. 2009).

8.6 Bioimaging

Imaging is a term that can be perceived in many ways. Most of the people understood imaging as a sort of photography, but in a scientific way, the significance of imaging is far beyond this. Different techniques such as computed tomography (CT) imaging (Bar-Shalom et al. 2003), Raman spectroscopy, radio-imaging by relevant nuclides (Love et al. 2003), near-infrared imaging (Giavalisco et al. 2004; Luo et al. 2014), positron emission tomography (Bailey et al. 2005), and even higher complex scanning methods such as laser ablation (Geohegan et al. 1998) have been used for imaging the in vivo samples. Out of all these methods, most of the imaging techniques are lethal and malignant or entailed magnified synthetic protocols for samples. Thus, it is clear that in living arrangements or intact tissues system, fluorescence-based imaging is highly recommended methodology. Recently, in biomedical sciences, fluorescence bioimaging is the most wide-ranging method for internal organ imaging and is considered better than magnetic resonance imaging (MRI) and conventional light microscopic methods (Wolfbeis et al. 2015). For example, fluorescence does not entail an exclusive spectroscopic method, but it includes a vast diversity in its approach. Through the fluorescence technique, various effects caused by quenching can be easily studied (Song et al. 2014). Resonance energy transfer (Stanisavljevic et al. 2015), photo-induced electron transfer, and images can be attained by analyzing the decay time (Smith and Ghiggino et al.

2015), intensity (Das et al. 2014), and polarization (Camacho et al. 2015). The higher selectivity and sensitivity, with acceptable spatial resolution, and absence of radioactivity risk have further enhanced the scope of fluorescence imaging (Zhao et al. 2015a, b). It has become an easy tool to verify the biological processes inside living cells. In the last decade, imaging resolution confined up to the range of nanometers has been growing tremendously due to the application of CDs. There are two diverse methods available for fluorescence imaging. The first technique incorporates inherently fluorescent chemicals and biochemical species, for instance, nicotinamide adenine dinucleotide (NADH) in tissues or chlorophyll in most kinds of plants (Sarder et al. 2015). The second technique involves an imaging procedure that utilizes chemical synthetic fluorescent probes, nanosensors, labels, or nanoparticles to display fluorescence (Walker et al. 2015). Fluorescent CDs have several benefits, such as comparable optical properties and ample chemical and photochemical stability (Lim et al. 2015), which help in making effective bioimaging agents. More significantly, CDs are mainly biocompatible, nontoxic, and eco-friendly (Namdari et al. 2017). As earlier reported, CDs are distinguished from other fluorescent-labeling agents and probes on the basis of its tendency to reveal multicolor emission. This authorizes researchers to select and control the excitation and emission wavelengths in CDs (Ding et al. 2015). These characteristics constrain CDs as a substitute to heavy metal quantum dots to anticipate biological systems, operation, and mechanism for in vivo and in vitro bioimaging (Luo et al. 2013).

8.7 Biosensing

Fluorescence properties of CDs make them a useful applicant in biosensing. As mentioned earlier, this property of CDs resulted from three major factors, which involve surface states, quantum confinement effect, and molecular state. These physicochemical states determine theoretically that the resultant CDs are fluorescent or not. Fluorescence is a procedure in which external radiation, such as ultraviolet (UV) radiation absorbed by substance, causes electron to excite, and ultimately, photon emission is produced. The interaction between electrons in functional groups of CDs with surrounding molecules of analytes leads to two strategies for fluorescent CD-based sensors, that is: (1) strong fluorescence is attained when the analyte attaches robustly to the CDs, which is known as fluorescence quenching, and analyte concentration can be measured quantitatively; (2) if an intermediate substance is bind to the CDs, which is quenched already, and intermediate attaches to analyte due to its greater binding affinity toward it as compared to CDs. This binding to the intermediate results in the recovery of intrinsic fluorescence of CDs, which can easily detect the analytes. CD-based sensors can be beneficial, due to the presence of hydrophilic functional groups such as –OH and –COOH on the facade of CDs, which further enhance their water solubility and biocompatibility (Mazrad et al. 2018). For instance, CDs have been used to detect Cu^{2+} in a selective and sensitive manner by employing luminescence properties of CDs. These fluorescent CDs were successful for Cu^{2+} sensing and imaging due to their enormous biocompatibility and probing impact on the cells (Salinas-Castillo et al. 2013). In addition, sensing of silver nanoparticles (Ag NPs) in cosmetics has also been achieved by using fluorescent CDs. Amine-derivatized CDs have exhibited an emission band at 440 nm. The existence of Ag NPs has induced significant quenching in the fluorescence of CDs. The interaction takes place between Ag NPs and free amine groups on the facade of CDs, which causes decrease in the fluorescence emission due to the occurrence of redshift in the absorption wavelength of CDs. This technique provides high selectivity and sensitivity and inexpensive nature as compared to other sensing methods (Cayuela et al. 2015). Cholesterol is another molecule that is significant for cancer detection. Cholesterol is a substance that is interconnected to several metabolites, which further behaves as a steroid precursor for biologically synthesized organic compound and essential structural component in the body. Its presence in high amounts leads to cardiovascular diseases, and low amount of it causes cancer and cerebral hemorrhage. The main techniques that are employed for sensing require a long duration of time and are expensive in nature with the utilization of enzymes. But in recent times, scientists have opted to take low-cost, easily synthesize materials for detection of cholesterol such as fluorescent CDs. Cholesterol molecules were detected from the use of a carbon quantum dot (CQD)/hemoglobin complex. The decrease in fluorescence emission of CQD was determined due to the presence of noncovalent interactions between hemoglobin and CQD. It is well known that cholesterol has the compatibility to associate with the hemoglobin. Therefore, the fluorescence of CQD restores back with the accession of cholesterol to the CQD/hemoglobin complex, and ultimately, cholesterol sensing is performed (Bui and Park et al. 2016). Li et al. have detected morin by using the CQD in aqueous media. Morin is an important compound that connects to enzymes, nucleic acids, and proteins (Li et al. 2017). In diabetes and some cancers, it is essential to detect glucose for diagnosis and medical purposes due to its involvement in these two diseases. The detection of glucose was achieved because of the reversible covalent interactions of boronic acid in saccharides by binding strongly to cis-1,2 or 1,3- diols. Saccharides especially glucose become effective for sensing by using CDs (Kang et al. 2017). Recently, Fe^{3+} (iron) sensing in body samples was performed by nitrogen and phosphorus co-doped CQDs. The high selectivity and sensitivity of CQDs were attained in this method. This was mainly due to the existence of many ions in ethylenediaminetetraacetic acid (EDTA), and thus, CQDs can easily detect Fe^{3+} ion (Shangguan et al. 2017).

8.8 Drug Delivery

Drug delivery can be defined as a system that transports or carries therapeutic agents with more drug solubility and stability to the targeted parts of the body with fewer side effects. Different methods and materials have been used in the past for effective drug delivery. A broad range of materials have been reported and studied as a possible carrier such as quantum dots, polymers, magnetic nanoparticles, and carbon-based nanoparticles in drug delivery system. Among these various materials, excellent properties of CDs like great biocompatibility, excellent PL, and high surface-to-volume ratio make them as a perfect choice for drug delivery and monitoring carrier simultaneously. In addition to the transport and carrying of drugs into cells, CDs can also be employed as a bioimaging agent because of its intrinsic fluorescence properties, which have huge capability in the tracking of the drugs/genes. CDs can be implemented to release different agents, for instance, gene and neuro-disease drugs and chemotherapeutic drugs, and can liberate the cargo by internal or external stimuli.

Various studies of CQD-based drug/gene delivery experiments have further attributed to cancer treatment (Peng et al. 2017; Hassan et al. 2018). In the past, a facile and cost-effective method was utilized to synthesize CDs using glycerol as a precursor by the pyrolysis method. The synthesis of CDs was done in the existence of mesoporous silica nanoparticles ($mSiO_2$). The encapsulation with PEG has further induced the stabilization and biocompatibility in these CDs. After that, conjugate formation of CQDs with SiO_2-PEG complex and doxorubicin (DOX), an anticancer drug, has further fabricated DOX@CQDs@$mSiO_2$-PEG nanocomposite. This nanocomposite has further behaved as in situ marker for probing the potency of anticancer drug release by shifting the blue PL emission versus the red emission at the cytoplasm and the nucleus by DOX (Lai et al. 2012). Active therapeutic function can also be performed by using CDs. For instance, an inhibitory effect of the drug on cancer cells has also been achieved in the presence of CDs. A study on CDs produced from green tea has exhibited an inhibiting effect on cancer cells by apoptosis. Inhibition of cells such as MCF-7 and MDA-MB-231 performed precisely by CDs as compared to catechin. In addition to better inhibition, biocompatibility and safety have also been high in the presence of CDs. High qualities of CDs such as intrinsic PL, nontoxicity, and cancer prohibition properties make them useful for in situ tracking of CDs in mouse bodies and monitoring the cell secretion and inhibition in the growth of cancer cells (Hsu et al. 2013). The gene therapy was reported by Kim et al. by using fluorescent CDs. CDs combined with gold nanoparticles (Au NPs) have posed transfection of pDNA. Firstly, a cationic polymer was formed when CDs and Au NPs were electrostatically dissolved in polyethylenimine (PEI). Afterward, non-labeled plasmid DNA was computed to the solution, and a nano-assembly of CQDs-PEI/Au-PEI/pDNA was attained as a delivery agent. High salt concentrations have further dissociate the nano-assembly complex. At the time of transfection at the post-endosomal step, when the distance between the carbon and Au NPs increased, the photoluminescence emission was revived, determined, and imaged at the intracellular level (Kim et al. 2013).

8.9 Quantitative Multiassay Approach

8.9.1 Antibacterial Activity

The objective of the respective activity is to find out the minimal inhibitory concentration (MIC) of particles applied for testing the antibacterial activity of bacterium by examining in suitable test conditions. MIC values can figure out the vulnerability of bacteria to investigated substance. This is an efficient and rapid method for analyzing the antibacterial activity of the CDs by avoiding the usage of any complex instrumentation technique. In this method, a sample solution of known concentration is explicitly encapsulated within agar system. After the solidification of agar solution in the disk, the bacterial inoculum made to spread on the facade of the agar and leave in the incubator for 24 h at 37 °C. Experiments can be again set to replicate and can analyze a range of concentrations of the sample solution by bisecting the surface of the agar into wedges or squares on the same plate. Various kinds of bacterial species can also be encapsulated on the same plate to examine the sample. Through this technique, in a single experiment, only a number of bacteria can be screened down. Eventually, growth of bacteria involving the test sample is measured either as present/absent on the plate or percentage of control, like 0%, 25%, 50%, 75%, and 100%. Yang et al. synthesized CDs by using lauryl betaine and tested their antimicrobial activity against gram-positive bacteria (Fig. 8.3). The particles are found to be selective due to the existence of hydrophobic hydrocarbon chains and quaternary ammonium groups over the surface of CDs with a minimum inhibitory concentration of 8 μg/mL for *Staphylococcus aureus* (Yang et al. 2016).

8.9.2 Antioxidant Activity Analysis

Free radical method can determine the antioxidant activity of CDs by evaluating the reduction of diphenyl picrylhydrazyl (DPPH) in methanol solution (Zhao et al. 2015a, b). In the analyzing process, different concentrations of CDs were used and a specific amount of DPPH solution was mixed, then the system was diluted to 1.00 ml. After that, the solution was added to quartz cuvette. The absorption at a particular value was recorded, and the ability of antioxidant activity toward DPPH was measured according to the following equation:

$$\text{Inhibition}(\%) = (A_C - A_s) / A_c \times 100\%$$

where *A* refers to the absorbance of the DPPH, *C* refers to the absorbance in the absence of CDs, and *S* refers to absorbance in the presence of CDs.

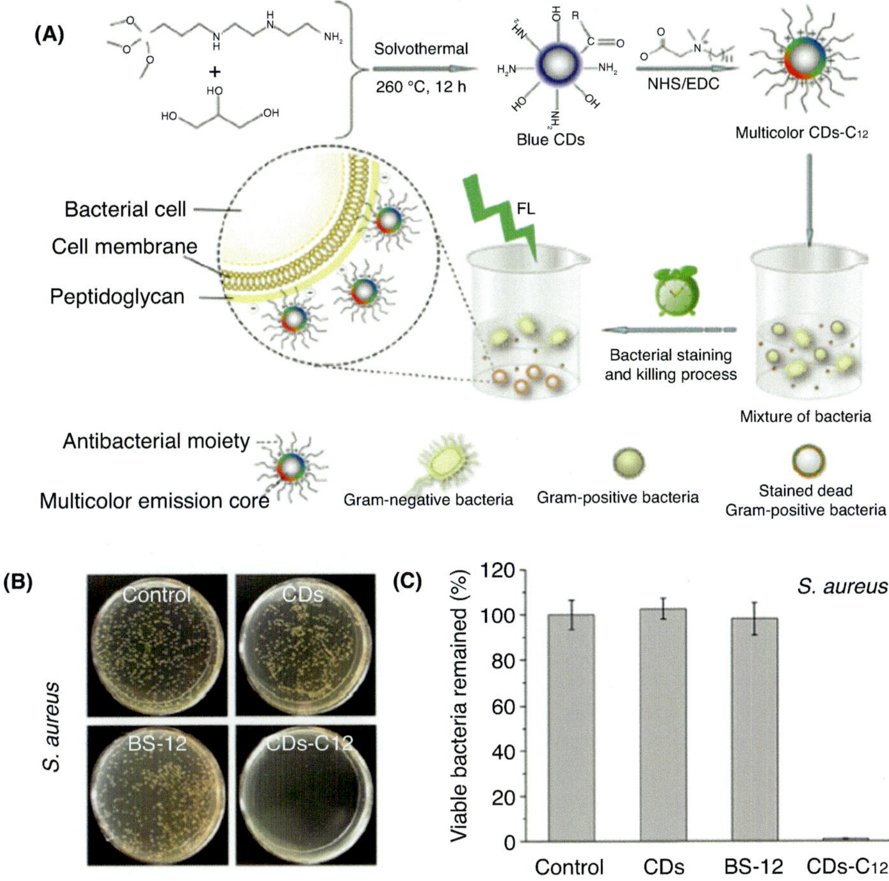

Fig. 8.3 (**a**) Diagrammatic illustration of synthesis of multicolor CDs-C_{12} and their selective gram-positive bacteria imaging. (**b**) *Staphylococcus aureus* colonies photographs. (**c**) Statistical histograms after conducting with control, CDs, BS-12, CDs-C_{12} at 10 μg/mL concentration. (Adapted figure from Yang et al. with permission from copyright (2016), American Chemical Society (Washington, DC, USA))

8.9.3 Seed Germination Assay

Seed germination assay is another method to determine the phytotoxic behavior of the samples. Root length is the most affected region due to the fact that all kinds of nanoparticles have aptness to gather over the root surfaces. The toxicity level can be ascertained by evaluating the root length difference in the surface functionalized CDs at different concentrations. In this method, different types of seeds can be utilized for the estimation of toxic level of CDs using the germination bioassay. The seeds can be sterilized properly for excavating all infectious agent and other impurities from the exterior surface of the seed surface. After that, the seeds can be soaked in deionized water at 37 °C overnight. The solutions of different concentrations of CDs can be used for batches of seeds and placed in dark and warm conditions for germination. After the time period of 72 h, measurement of root length and number of rootlets is done. The experiment can also be executed in triplicates.

8.9.4 Allium cepa *Chromosome Aberration Assay*

The genotoxicity assay by using *Allium cepa* (*A. cepa*) can be employed to evaluate and review the effect of carbon-based nanoparticles on genetic level and to determine the mechanism of their reaction with cells. In this method, root primordial can be exposed after scraping 3–4 g weighing bulbs of *A. cepa* and placed in solution containing CDs for 96 h or until the root primordial evolve around 2–3 cm. A solution containing water in exposure media of onions can be used as the control. The above experiment can be replicated to explore the corresponding effects of different qualities of onions by electing different groups of *A. cepa* with the same concentration of as-prepared CDs. A properly sterilized scissors can be utilized to cut down the rootlets of around 1 cm. The following root tip can be cleaned with distilled water and pigmented with acetocarmine dye. This stained root tip can be prepared by placing it over a glass slide. In order to disperse the cells over the glass slide, coverslip can be gently positioned over the glass slide. Each slide underwent for microscopic investigation to obtain the data of the number of cells sustaining the mitotic stage in about 1000 cells in as processed root tips (Chaudhary et al. 2016).

8.10 Environment Pollutant Sensing by Carbon Nanoparticles

In the past years, the vast development in industrialization, agriculture, and global economy growth has also brought up a broad range of pollutants and contamination in our environment. In this regard, carbon-based nanoparticles have gained huge attention worldwide in the field of sensing due to their unique and tunable chemical and physical properties. The presence of great surface-to-volume ratio and outstanding optical properties has made CDs as a potential candidate to test the analytes. The carbon nanoparticles were utilized in the field of sensing and applied for the detection of pollutants existed in our environment. The standard detection techniques include atomic absorption spectroscopy, anodic stripping voltammetry, inductively coupled plasma-mass spectrometry, ultraviolet-visible spectroscopy, inductively coupled plasma atomic emission spectroscopy, and instrumental neutron activation analysis (INAA) (Bings et al. 2010; Lorber et al. 1986; Kunkel and Manahan et al. 1973; López-Artíguez et al. 1993). These techniques are known to be very selective and peculiar for detection methods. However, these techniques also entail various drawbacks like high-cost apparatus, complex sample preparation, critical experimental conditions, and technically trained operators for equipment handling and analysis. Therefore, these disadvantages restrain the feasibility for smooth field detection (Fen and Yunus 2013; Kim et al. 2001; Li et al. 2013a, b). Consequently, in order to serve this purpose, a simpler qualitative methodology is desired to be developed for authentic and sensitive sensing. Currently, various spectroscopy techniques have emerged as a useful and practical method for selective detection of toxic pollutants (Oehme and Wolfbeis et al. 1997). The fluorescence technique for CDs emerged as a great deal in the detection of pollutants due to simplification of design and effectiveness of selectivity and specificity toward analytes. In this section, we will present carbon nanoparticle-based sensors to identify various pollutants and contaminations exist in our environment.

8.10.1 Lead

Lead (Pb) exhibits highly noxious and carcinogenic adverse effects on human health. It has the capability to bind with bioligands containing nitrogen, oxygen, and sulfur group in the form of complexes. Various adverse effects are produced from these interactions such as it can damage activities of nucleic acids, break hydrogen bonds, deteriorate molecular structure of protein, or retard functions of enzymes in living beings (Gemma et al. 2011). Also, these kinds of heavy metal ions are considered to be included in the category of highest toxic pollutants. Lead is not biodegradable, and due to this, it can easily aggregate in the ecological system and food chain (Zhou et al. 2016). A series of neurological activities can be impaired by an excess amount of Pb^{2+} exposure to a human being. CDs have gained remarkable attention in past years as a fluorescent probe in sensing heavy metal ions. CDs are synthesized via an easy, low-cost, and eco-friendly method, that is, solvothermal method from bamboo leaves. The prepared CDs can work as nano-sensors by attaining a low detection limit of 0.14 nM for Pb^{2+} ions. Hence, this proves that the synthesized CDs work as the most selective and sensitive sensing system for the detection of Pb^{2+} ion.

8.10.2 Mercury

Mercury (Hg), one of the most toxic compounds on earth, ranked third in toxicity right after arsenic and lead by the Agency for Toxic Substances and Disease Registry. It can be accumulated by humans through three main pathways: pigments applied in plastics and paints, advanced industrial procedures, and waste incinerated for medical purposes (Omichinski et al. 2007). Various complicated diagnoses generated from exposure to mercury can end up in greater than 250 unique symptoms in living beings. CDs have been exceedingly studied and explored as fluorescent probes for sensing of mercury metal ions. Zhang and Chen have synthesized nitrogen-doped (N-doped) CQDs, which further show quenching in the presence of Hg^{2+}. This is mainly assigned because Hg^{2+} induced change to the surface states of N-CQDs. The selectivity of as-prepared CQD for Hg^{2+} in the presence of other metal ions in high concentrations was evidenced with the data (Zhang and Chen 2014). Nitrogen and chloride co-doped CDs (N/Cl-CDs) were prepared by Gao et al. with high quantum yield and convenient one-step hydrothermal technique. These nitrogen and chloride co-doped CDs were utilized for the sensing of Hg^{2+} with a linear range of 0.2–40 μM and a detection limit of 0.05 μM (Gao et al. 2019). Mondal et al. synthesized amino-functionalized CDs from microwave method for visual sensing of Hg^{2+}. In the ubiquity of a very less amount of Hg^{2+}, a color change has been examined from blue to green under a UV lamp. The sensitivity and selectivity of these CDs toward Hg^{2+} have been detected in the presence of other metal ions. Electronic band diagram backed by zeta potential has been interpreted by photoluminescence quenching mechanism (Mondal et al. 2018). Another sensing system exhibited by Lu et al. for the detection of Hg^{2+} was established based on the fluorescence quenching of carbon nanoparticles with a lower detection limit of 0.23 nM (Lu et al. 2012). The photoemission spectra of these carbon nanoparticles along with the quenching that occurred in the presence of Hg^{2+} is given in Fig. 8.4.

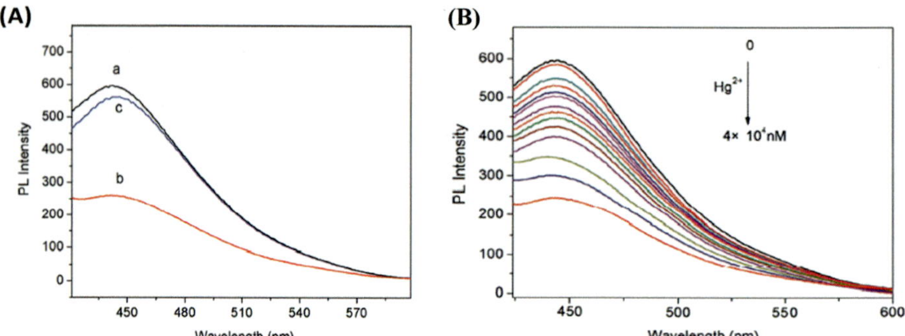

Fig. 8.4 (**a**) (Curve a) Fluorescence emission spectra of carbon nanoparticles solution, (Curve b) carbon nanoparticles –Hg^{2+} solution where quenching occurred, (Curve c) carbon nanoparticles–Hg^{2+}-Cys mixture. (**b**) Luminescence spectra of carbon nanoparticles taking different concentrations of Hg^{2+}. (Adapted figure from (Lu et al.) with permission from copyright (2012), American Chemical Society (Washington, DC, USA))

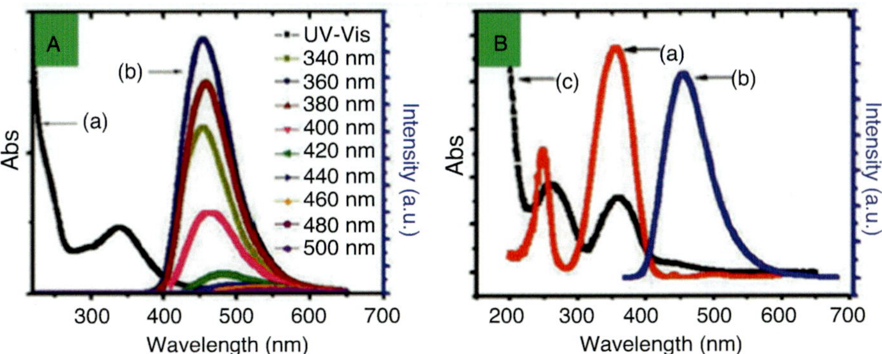

Fig. 8.5 (**a**) UV-vis absorption spectra and (**b**) photoluminescent absorption spectra of CDs in the presence of Cr (VI). (Adapted figure from (Athika et al.) with permission from copyright (2019), American Chemical Society (Washington, DC, USA))

8.10.3 Chromium

Chromium (Cr) is also one of those heavy metal ions that have been constantly released into the environment by industrial and other processes. Chromium ion has greater detrimental effects on human health due to its highly toxic and hazardous properties. The trivalent and hexavalent oxidation states of chromium are genotoxic and most dangerous to human and environment health. Therefore, detection of chromium through highly sensitive and selective sensors is required. Zheng et al. synthesized an on-off fluorescent CDs probe for detecting Cr (VI) by using the inner filter effect (IFE). The high sensitivity and selectivity have been achieved by using the CD-based nano-sensor for Cr (VI) ions (Zheng et al. 2013). Athika et al. applied a sustainable way to produce CDs from denatured milk via a simple and convenient hydrothermal technique. The excellent photophysical property of CDs has demonstrated higher potential applicability in selective sensing of chromium ion (Cr^{6+}) in control and environmental water samples with a low detection limit of 14 nM (Fig. 8.5) (Athika et al. 2019).

8.10.4 Cadmium

Cadmium (Cd) is highly poisonous and known as lethal to the human body and environment. It is chiefly applied in electroplating of steel, agriculture, field of metallurgy, and pigments for plastic industry. It can enter in food chain easily due to its nonbiodegradable nature, and hence, it can cause various health-related hazards to human beings. It can also accumulate in the human body through smoking and breathing in an area in which it is presented in excess amount. The high intake of cadmium can harm kidneys, bones, and liver and can cause diseases like cancer, diabetes, etc. Gunnlaugsson et al. developed two fluorometric chemosensors (Cd-7 and Cd-8) for selective detection of Cd^{2+} by using CDs, which include photoinduced electron transfer (PET) sensor mechanism. One or two iminodiacetates are applied as a receptor with methylene spacer linking anthracene as a fluorophore. The obtained results revealed both chemosensors have great selectivity and sensitivity for Cd^{2+} (Gunnlaugsson et al. 2003).

8.10.5 Nickel

Heavy metals are among the main pollutants that are mostly scrutinized due to the exceeding environmental pollution. Nickel (Ni), as being one of the heavy metals, is known as a contaminant in the ecosystem. Therefore, the qualitative and quantitative detection of Ni is highly required for environmental remediation purpose. The presence of nickel element is expansive in the environment, and most of the detection techniques available still have various drawbacks like arid sample preparation, high-cost equipment, intricate operation plan, complex optical devices, and lower limit of detection. In this regard, a facile method to detect nickel ions (Ni^{2+}) from imidazole-modified CDs was demonstrated by Gong et al. on the basis of fluorescence quenching technique. Spectroscopic analyses were studied by fluorescence and ultraviolet absorption spectrum. The as-synthesized CDs have displayed robust selectivity for Ni^{2+} among other metal ions with a detection limit of 0.93 mM. These results showed that imidazole-modified CDs can be solicited to detect Ni^{2+} (Gong and Liang 2019).

8.10.6 Pesticides

Organophosphorus pesticides broadly used in agriculture is the most common and crucial agrochemicals in controlling or eliminating insects and pests. As a result, it increases the production of vegetables and crops (Yan et al. 2015; Liu et al. 2012; Zhang et al. 2010). Consequently, due to its widespread use and improper disposal, the adverse effects on the ecosystem and food chain have brought up people's great concerns. Pesticides have possessed a serious threat to the sensitive ecosystem cycle. Lia et al. produced acetylcholinesterase (AChE)-controlled quenching of fluorescent CDs for sensitive sensing of organophosphorus pesticides (OPs) through fluorometric and colorimetric methods. The reaction of acetylthiocholine and acetylcholinesterase (AChE) yields thiocholine, which decomposes the 5,5-dithiobis(2-nitrobenzoic acid) to fabricate yellow-colored 5-thio-2-nitrobenzoic acid (TNBA). TNBA being a strong absorber with a positive charge performs quenching in the fluorescence emission of CDs. The enzyme functioning of acetylcholinesterase (AChE) was hindered with the presence of organophosphorus pesticides resulting in the recovery of fluorescence signal and declining the absorbance intensity with color change. This method imparts efficient sensitivity for fast detection of paraoxon, which is an organophosphorus pesticide with a detection limit of 0.4 ng mL^{-1}. Thus, it can be concluded that dual-readout method can be a practicable and highly performing candidate to detect organophosphorus pesticides by taking the benefit of attractive optical property of CDs and specificity of enzyme (Lia et al. 2018). Also, Wang et al. studied CDs synthesized through a one-step refluxing method, which is utilized to detect fenitrothion, a widely used organophosphorus pesticide, which is toxic in nature (Fig. 8.6) (Wang et al. 2017).

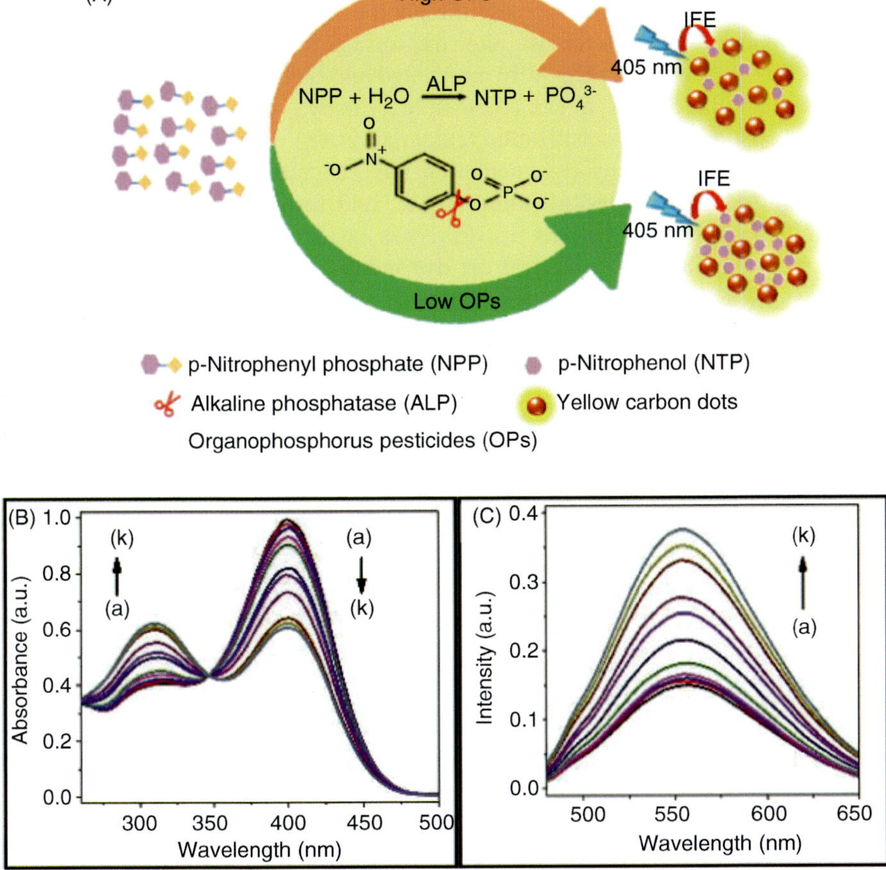

Fig. 8.6 (**a**) Schematic representation of fenitrothion detection with inner filter effect (IFE) by using CDs. (**b**) UV-vis absorption spectrum of CDs. (**c**) Photoluminescence spectrum of CDs after addition of different concentration of fenitrothion. (Adapted figure from Wang et al. with permission from copyright (2017), American Chemical Society (Washington, DC, USA))

8.10.7 Degradation of Organic Dyes by Using CDs

During the last few decades, dye manufacturing and textile industries have liberated a large scale of organic dyes into our water system. Dyes such as methyl orange (MO), methylene blue (MB), and rhodamine B (RhB) can be lethal and produced malignant effects on living beings. Different kinds of harmful effects can be generated on human health from organic dyes due to difficulty in their biodegradation. Over the years, various studies on the photocatalytic degradation of organic dyes by utilizing carbon nanoparticles have gained the attention of researchers. The modification of CDs with semiconductor and appropriate metal materials usually enhanced the photocatalytic efficiency. The CDs with diverse functional groups on their surface improve the migration rate of carriers (electron and hole) and promoted the photocatalytic redox rate and increase the photocatalytic activity of CDs. Various studies reveal that doping can also alter the surface functional groups and photocatalytic activity can be conclusively modulated and increased (Cheng et al. 2019). The photodegradation of dyes can be possible by three possible ways: photocatalysis, photolysis, and photosensitization (Liu et al. 2018). In photosensitization process, excitation of dyes by light radiation has generated photoelectrons, which transmit to conduction band of photocatalyst, and formation of $\cdot O_2$ occurs after reaction with oxygen. In the photolysis process, oxidative degradation of dye persuaded from the electron impel and the dye interact with oxygen to form singlet oxygen atom. Carbon nanomaterial-modified photocatalysts displayed increased photocatalytic activity upon the decay of organic dyes, which caused the introduction of π-system or the generation of heterojunctions (He et al. 2018). For instance, organic dye, C_{60}-modified $Bi_2TiO_4F_2$ microspheres, revealed great photocatalytic performance for the degradation of RhB dye. Nanocomposite of one-dimensional TiO_2@ MWCNTs (multiwalled carbon nanotubes) displayed strong degradation effect on methylene blue (MB) and RhB (Zhang et al. 2017). Two-dimensional carbon nanomaterials such as graphene oxide and reduced graphene oxide showed better reaction with photocatalysts as compared to zero-dimensional and one-dimensional carbon nanomaterials. Degradation of RhB from reduced GO/bismuth tungstate (BiWO) composite that is synthesized with graphene oxide presented great photocatalytic activity (Gao et al. 2011). Herein, various carbon-based nanoparticles utilized in sensing application are listed below in Table 8.1.

8.11 Summary

This chapter gives insight into carbon-based nanoparticles and their application in eradicating the pollutants from the environment. It also intended and focused on the important aspects of CDs such as their synthesis, structure, and their excellent

Table 8.1 Performance of diverse sensors based on carbon-based nanoparticles

Synthetic sources	Detection method	Detector	LOD	Reference
Papaya powder	Fluorescence	Fe^{3+}	0.48 $\mu molL^{-1}$	Wang et al. (2016)
Citric acid and triethylamine	Fluorescence	Hg^{2+}	2.8 nM	Wang et al. (2019)
Water hyacinth waste	Fluorescence	Pretilachlor (herbicide)	2.9 μM	Deka et al. (2019)
Chocolate	Fluorescence	Pb^{2+}	12.7 nM	Liu et al. (2016)
Lotus root	Fluorescence	Hg^{2+}	18.7 nM	Gu et al. (2016)
1,2 ethylenediamine, phthalic acid	Fluorescence	Mn (VII)	48.3 nM	Du et al. (2018)
Groundnut	Fluorescence	Cr (VI)	0.1 mg/L	Roshni V et al. (2019)
Eleusine coracana	Fluorescence	Cu^{2+}	10 nM	Murugan et al. (2019)
Kelp	Fluorescence	Co^{2+}	0.39 μmol/L	Zhao et al. (2019)
Glutathione	Fluorescence	Fe^{3+} and Zn^{2+}	0.8 μmol/L and 1.2 μmol/L	Wang et al. (2019)
Burning ash of waste paper	Fluorescence	Chlorpyrifos (pesticides)	3 ng/ml	Lin et al. (2018)
Quinoline derivative	Fluorescence	Zn^{2+}	6.4 nM	Zhang et al. (2014)
Commercial chlorophyll	Fluorescence	Paraoxon (pesticides)	0.05 $\mu g\ L^{-1}$	Wu et al. (2017)
Orthophenylene diamine	Fluorescence	Atrazine (herbicide)	3 pM	Sharma et al. (2019)
Mesophorus silica	Fluorescence	Cu^{2+} and L-cysteine	2.3×10^{-8} M and 3.4×10^{-10} M	Zong et al. (2014)

applications in diverse fields so that students can acquire a brief overview of these fascinating nanoparticles. We also emphasize on sensing-based applications of CDs for detection of various carcinogenic pollutants. The sensing ability of CDs has been extensively explored since new sensing approaches have been proclaimed in an ever-expanding rate. Indeed, this information will be a beneficial aid for the researchers also. Succinct analysis on representative CD-based sensors is reviewed in this chapter, and for deep-seated studies, the reader is introduced to several profound reviews cited in the entire section.

Appendix

Multiple Choice Questions (MCQs)

Q-1. Which of the following properties determine the dimensions of nanoparticles?

(i) Size distribution
(ii) Solubility
(iii) Physicochemical properties
(iv) All of the above

Q-2. What are the dimensions of CDs?

(i) 1–30 nm
(ii) 30–70 nm
(iii) 80–90 nm
(iv) 1–10 nm

Q-3. Which of the following material utilized as a larger carbon precursor in top-down method for the synthesis of CDs?

(i) Ascorbic acid
(ii) Nitric acid
(iii) Graphite sheets
(iv) Ethylene diamine

Q-4. The characterization techniques for the determination of general structure and chemical composition of CDs include which of the following methods?

(i) Fourier transform infrared (FTIR)
(ii) High-resolution TEM (HRTEM)
(iii) X-ray photoelectron spectroscopy
(iv) All of the above

Q-5. Which method is not used to control the size and uniformity of CDs?

(i) Pyrolysis
(ii) Centrifugation
(iii) Dialysis
(iv) Electrophoresis

Q-6. Which one of the following compounds is a major contributor to pollution caused by organic dyes?

(i) Indigo dye
(ii) Methylene blue
(iii) Madder
(iv) Saffron yellow

Q-7. From which of the following methods, toxicity of carbon-based nanoparticles on the genetic level can be probed?

(i) *Allium cepa* chromosome aberration assay
(ii) Antioxidant activity
(iii) Antibacterial assay
(iv) Seed germination assay

Q-8. How to confirm the existence of carboxylic (-COOH) group on the facade of carbon-based nanoparticles?

(i) Raman spectroscopy
(ii) X-ray diffraction (XRD)
(iii) Fourier transform infrared spectroscopy (FTIR)
(iv) Nuclear magnetic resonance (NMR)

Q-9. Which of the following heavy metal ion can cause various carcinogenic effects on the ecosystem and human health?

(i) Mercury
(ii) Tetracycline
(iii) Rhodamine B
(iv) Nitrophenol

Q-10. What is the principal factor of fluorescence in carbon-based nanoparticles?

(i) Emission
(ii) Absorption
(iii) Quantum yield
(iv) All of the above

Answers of Multiple Choice Questions (MCQs)

Q-1. Answer: All of the above
Q-2. Answer: 1–10 nm
Q-3. Answer: Graphite sheets
Q-4. Answer: All of the above
Q-5. Answer: Pyrolysis
Q-6. Answer: Methylene blue
Q-7. Answer: *Allium cepa* chromosome aberration assay
Q-8. Answer: Fourier transform infrared spectroscopy (FTIR)
Q-9. Answer: Mercury
Q-10. Answer: All of the above

Short Questions

Q-1. What is the meaning of "emerging pollutants?"
Q-2. Give a short definition of CDs?
Q-3. Describe the photoluminescence mechanism in CDs?
Q-4. Give the detailed information about "drug delivery" process by using the application of CDs?
Q-5. Can pesticides impose a serious threat to environment?

Answers of Short Questions

Q-1. Answer: The term "emerging pollutants" is the synthetic or naturally arising chemicals that do not usually exist in the environment but that have the capability to intrude into the environment. A broad range of pollutants that include highly toxic substances gives rise to a major menace to human health even if their small concentrations remain in the environment.

Q-2. Answer: CDs, a fascinating young material of nano-family with extraordinary potentials, are a kind of fluorescent nanoparticles. CDs have exceptional and remarkable properties such as high specific surface area, nontoxicity, good chemical stability, high biocompatibility, aqueous dispersibility, high quantum yield, cost-effectiveness, etc.

Q-3. Answer: Photo excitation initiates the photoluminescence, and it is a form of luminescence (light emission). In the literature, various reasons are elaborated for photoluminescence in CDs like effect of the surface states, which is governed by distinct functional groups on their surface, defects, and quantum confinement effect, which include the association of electron hole pairs in small sp^2 carbon clusters or sp^3 matrix.

Q-4. Answer: Drug delivery can be defined as a system which transports or carries therapeutic agents with more drug solubility and stability to the targeted parts of the body with fewer side effects. Different methods and materials are there to perform drug delivery. A broad range of materials have been reported and studied as possible carriers, such as quantum dots, polymers, magnetic nanoparticles, carbon-based nanoparticles, etc., in drug delivery systems.

Q-5. Answer: Pesticides broadly used in agriculture are the most common and crucial agrochemicals due to their outstanding potential in excluding, controlling, or eliminating insects and pests; as a result, it increases the production of vegetables and crops. Consequently, due to its widespread use and improper disposal, the adverse effects on ecosystem and food chain have brought up people's great concerns, and also, they pose a serious threat to sensitive ecosystem cycle.

References

Amjadi M, Manzoori JL, Hallaj T (2014) Chemiluminescence of graphene quantum dots and its application to the determination of uric acid. J Lumin 153:73–78

Arisido WM (2015) Functional data analysis for environmental pollutants and health. Cancer Res 75:3203–3208

Athika M, Prasath A, Duraisamy E, Devi VS, Sharma AS, Elumalai P (2019) Carbon-quantum dots derived from denatured milk for efficient chromium-ion sensing and supercapacitor applications. Mater Lett 241:156–159

Bagheri Z, Ehtesabi H, Rahmandoust M, Ahadian MM, Hallaji Z, Eskandari F, Jokar E (2017) New insight into the concept of carbonization degree in synthesis of CDs to achieve facile smartphone based sensing platform. Sci Rep 7(1):11013

Bailey DL, Townsend DW, Valk PE, Maisey MN (2005) Positron emission tomography. Springer

Baker SN, Baker GA (2010) Luminescent carbon nanodots: emergent nanolights. Angew Chem Int Ed 49(38):6726–6744

Bar-Shalom R, Yefremov N, Guralnik L, Gaitini D, Frenkel A, Kuten A, Altman H, Keidar Z, Israel O (2003) Clinical performance of PET/CT in evaluation of cancer: additional value for diagnostic imaging and patient management. J Nucl Med 44(8):1200–1209

Bhunia SK, Saha A, Maity AR, Ray SC, Jana NR (2013) Carbon nanoparticle-based fluorescent bioimaging probes. Sci Rep 3:1473

Bings NH, Bogaerts A, Broekaert JA (2010) Atomic spectroscopy: a review. Anal Chem 82:4653–4681

Briscoe J, Marinovic A, Sevilla M, Dunn S, Titirici M (2015) Biomass-derived carbon quantum dot sensitizers for solid-state nanostructured solar cells. Angew Chem Int Ed 54(15):4463–4468

Bui TT, Park SY (2016) A CDs–hemoglobin complex-based biosensor for cholesterol detection. Green Chem 18(15):4245–4253

Camacho R, Tubasum S, Southall J, Cogdell RJ, Sforazzini G, Anderson HL, Pullerits T, Scheblykin IG (2015) Fluorescence polarization measures energy funneling in single light-harvesting antennas—LH2 vs conjugated polymers. Sci Rep 5:15080

Cao L, Wang X, Meziani MJ, Lu F, Wang H, Luo PG, Lin Y, Harruff BA, Veca LM, Murray D (2007) Carbon dots for multiphoton bioimaging. J Am Chem Soc 129(37):11318–11319

Cayuela A, Soriano ML, Valcárcel M (2015) Reusable sensor based on functionalized carbon dots for the detection of silver nanoparticles in cosmetics via inner filter effect. Anal Chim Acta 872:70–76

Chandra S, Patra P, Pathan SH, Roy S, Mitra S, Layek A, Bhar R, Pramanik P, Goswami A (2013) Luminescent S-doped carbon dots: an emergent architecture for multimodal applications. J Mater Chem B 1(18):2375–2382

Chaudhary S, Sharma P, Renu R, Kumar R (2016) Hydroxyapatite doped CeO_2 nanoparticles: impact on biocompatibility and dye adsorption properties. RSC Adv 6:62797–62809

Chen YC, Nien CY, Albert K, Wen CC, Hsieh YZ, Hsu HY (2016) Pseudo-multicolor carbon dots emission and the dilution-induced reversible fluorescence shift. RSC Adv 6(50):44024–44028

Cheng Y, Bai M, Su J, Fang C, Li H, Chen J, Jiao J (2019) Synthesis of fluorescent carbon quantum dots from aqua mesophase pitch and their photocatalytic degradation activity of organic dyes. J Mater Sci Technol 35:1515–1522

Damlas CA, Eleftherohorinos IG (2011) Pesticide exposure, safety issues, and risk assessment indicators. Int J of Environ Res Public Health 8:1402–1419

Darabdhara G, Boruah PK, Borthakur P, Hussain N, Das MR, Ahamad T, Alshehri SM, Malgras V, Wu KCW, Yamauchid Y (2016) Reduced graphene oxide nanosheets decorated with Au–Pd bimetallic alloy nanoparticles towards efficient photocatalytic degradation of phenolic compounds in water. Nanoscale 8:8276–8287

Das SK, Liu Y, Yeom S, Kim DY, Richards CI (2014) Single-particle fluorescence intensity fluctuations of carbon nanodots. Nano Lett 14(2):620–625

Deka MJ, Dutta P, Sarma S, Medhi OK, Talukdar NC, Chowdhury D (2019) Carbon dots derived from water hyacinth and their application as a sensor for pretilachlor. Heliyon 5:1985–1993
Deng J, LuQ MN, Li H, Liu M, Xu M, Tan L, Xie Q, Zhang Y, Yao S (2014) Electrochemical synthesis of carbon nanodots directly from alcohols. Chem Eur J 20(17):4993–4999
Ding H, Yu SB, Wei JS, Xiong HM (2015) Full-color light-emitting carbon dots with a surface state- controlled luminescence mechanism. ACS Nano 10(1):484–491
Dong Y, Zhou N, Lin X, Lin J, Chi Y, Chen G (2010) Extraction of electro-chemiluminescent oxidized carbon quantum dots from activated carbon. Chem Mater 22(21):5895–5899
Dong Y, Shao J, Chen C, Li H, Wang R, Chi Y, Lin X, Chen G (2012) Blue luminescent graphene quantum dots and graphene oxide prepared by tuning the carbonization degree of citric acid. Carbon 50(12):4738–4743
Du Y, Zhou M, Lei L (2006) Role of the intermediates in the degradation of phenolic compounds by Fenton-like process. J Hazard Mater B136:859–865
Du F, Li G, Gong X, Zhonghui G, Shuang S, Xian M, Dong C (2018) Facile, rapid synthesis of N,P-dual-doped carbon dots as a label-free multifunctional nanosensor for Mn(VII) detection, temperature sensing and cellular imaging. Sensors & Actuators B Chem 277:492–501
Fen YW, Yunus WMM (2013) Surface plasmon resonance spectroscopy as an alternative for sensing heavy metal ions: a review. Sens Rev 33:305–314
Gao E, Wang W, Shang M, Xu J (2011) Synthesis and enhanced photocatalytic performance of graphene-Bi_2WO_6 composite. Phys Chem Chem Phys 13:2887–2893
Gao Z, Li X, Shi L, Yang Y (2019) Deep eutectic solvents-derived carbon dots for detection of mercury (II), photocatalytic antifungal activity and fluorescent labeling for C. albicans. Spectrochim Acta, Part A 220:117080
Geissen V, Mol H, Klumpp E, Umlauf G, Nadal M, Ploeg MVD, Zee SEVD, Ritsema CJ (2015) Emerging pollutants in the environment: a challenge for water resource management. Int Soil Water Cons Res 3:57–65
Gemma A, Pons PJ, Arben M (2011) Recent trends in macro-, micro-, and nanomaterials based tools and strategies for heavy-metal detection. Chem Rev 111:3433–3458
Geohegan DB, Puretzky AA, Duscher G, Pennycook SJ (1998) Time-resolved imaging of gas phase nanoparticle synthesis by laser ablation. Appl Phys Lett 72(23):2987–2989
Giavalisco M, Ferguson H, Koekemoer A, Dickinson M, Alexander D, Bauer F, Bergeron J, Biagetti C, Brandt W, Casertano S (2004) The great observatories origins deep survey: initial results from optical and near-infrared imaging. Astrophys J Lett 600(2):L93
Gong Y, Liang H (2019) Nickel ion detection by imidazole modified carbon dots. Spectrochim. Acta, Part A 211:342–347
Gothwal R, Shashidhar T (2015) Antibiotic pollution in the environment: a review. Clean-Soil Air Water 43:479–489
Gu D, Shang S, Yu Q, Shen J (2016) Green synthesis of nitrogen-doped carbon dots from lotus root for Hg(II) ions detection and cell imaging. Appl Surf Sci 390:38–42
Gunnlaugsson T, Lee TC, Parkesh R (2003) Cd (II) sensing in water using novel aromatic iminodiacetate based fluorescent Chemosensors. Org Lett 5:4065
Hassan M, Gomes VG, Dehghani A, Ardekani SM (2018) Engineering carbon quantum dots for photomediated theranostics. Nano Res 11(1):1–41
He X, Li H, Liu Y, Huang H, Kang Z, Lee ST (2011) Water soluble carbon nanoparticles: hydrothermal synthesis and excellent photoluminescence properties. Colloids Surf B Biointerfaces 87(2):326–332
He H, Huang L, Zhong Z, Tan S (2018) Constructing three-dimensional porous graphene- carbon quantum dots/g-C_3N_4 nanosheet aerogel metal-free photocatalyst with enhanced photocatalytic activity. Appl Surf Sci 441:285–294
Hilderbrand SA, Shao F, Salthouse C, Mahmood U, Weissleder R (2009) Upconverting luminescent nanomaterials: application to in vivo bioimaging. Chem Commun 28:4188–4190
Hsu PC, Chen PC, Ou CM, Chang HY, Chang HT (2013) Extremely high inhibition activity of photoluminescent carbon nanodots toward cancer cells. J Mater Chem B 1(13):1774–1781

Hu S, Trinchi A, Atkin P, Cole I (2015) Tunable photoluminescence across the entire visible spectrum from carbon dots excited by white light. Angew Chem Int Ed 54(10):2970–2974
Jaiswal A, Ghosh SS, Chattopadhyay A (2012) One step synthesis of C-dots by microwave mediated caramelization of poly(ethylene glycol). Chem Commun 48(3):407–409
Jiang K, Sun S, Zhang L, Lu Y, Wu A, Cai C, Lin H (2015) Red, green, and blue luminescence by carbon dots: full-color emission tuning and multicolor cellular imaging. Angew Chem Int Ed 54(18):5360–5363
Jiao Y, Wan C, Bao W, Gao H, Liang D, Li J (2018) Facile hydrothermal synthesis of Fe_3O_4@ cellulose aerogel nanocomposite and its application in Fenton-like degradation of Rhodamine B. Carbohydr Polym 189:371–378
Kang EB, Choi CA, Mazrad ZAI, Kim SH, In I, Park SY (2017) Determination of cancer cell based ph-sensitive fluorescent carbon nanoparticles of cross-linked polydopamine by fluorescence sensing of alkaline phosphatase activity on coated surfaces and aqueous solution. Anal Chem 89(24):13508–13517
Kim JRC, Hupp JT (2001) Gold nanoparticle-based sensing of spectroscopically silent heavy metal ions. Nano Lett 1:165–167
Kim J, Park J, Kim H, Singha K, Kim WJ (2013) Transfection and intracellular trafficking properties of CDs-gold nanoparticle molecular assembly conjugated with PEI-pDNA. Biomaterials 34(29):7168–7180
Kumar L, Bharadvaja N (2019) Enzymatic bioremediation a smart tool to fight environmental pollutants. Smart Biorem Tech Microb Enz 99:118
Kunkel R, Manahan SE (1973) Atomic absorption analysis of strong heavy metal chelating agents in water and waste water. Anal Chem 45:1465–1468
Lai CW, Hsiao YH, Peng YK, Chou PT (2012) Facile synthesis of highly emissive carbon dots from pyrolysis of glycerol; gram scale production of CDs/mSiO_2 for cell imaging and drug release. J Mater Chem 22(29):14403–14409
Lew S, Lew M, Biedunkiewicz M, Szarek J (2013) Impact of pesticide contamination on aquatic microorganism populations in the Littoral zone. Arch Environ Contam Toxicol 64:399–409
Li Y, Migliaccio K (2010) Water quality concepts, sampling, and analyses. CRC Press
Li M, Gou H, Al-Ogaidi I, Wu N (2013a) Nanostructured sensors for detection of heavy metals: a review. ACS Sustain Chem Eng 1:713–723
Li G, Jiang B, Li X, Lian Z, Xiao S, Zhu J, Zhang D, Li H (2013b) $C_{60}/Bi_2TiO_4F_2$ heterojunction photocatalysts with enhanced visible-light activity for environmental remediation. ACS Appl Mater Inter 5:7190–7197
Li JY, Liu Y, Shu QW, Liang JM, Zhang F, Chen XP, Deng XY, Swihart MT, Tan KJ (2017) One-pot hydrothermal synthesis of carbon dots with efficient up-and down-converted photoluminescence for the sensitive detection of Morin in a dual-readout assay. Langmuir 33(4):1043–1050
Lia H, Yana X, Lua G, Su X (2018) CDs-based bioplatform for dual colorimetric and fluorometric sensing of organophosphate pesticides. Sensors Actuators B 260:563–570
Lim SY, Shen W, Gao Z (2015) Carbon quantum dots and their applications. Chem Soc Rev 44(1):362–381
Lin B, Yan Y, Guo M, Cao Y, Yu Y, Zhang T, Huang Y, Wu D (2018) Modification-free carbon dots as turn-on fluorescence probe for detection of organophosphorus pesticides. Food Chem 245:1176–1182
Liu DB, Chen WW, Wei JH, Li XB, Wang Z, Jiang XY (2012) A highly sensitive, dual readout assay based on gold nanoparticles for organophosphorus and carbamate pesticides. Anal Chem 84:4185–4191
Liu Y, Xiao N, Gong N, Wang H, Shi X, Gu W, Ye L (2014) One-step microwave-assisted polyol synthesis of green luminescent carbon dots as optical nanoprobes. Carbon 68:258–264
Liu Y, Zhou Q, Li J, Lei M, Yan X (2016) Selective and sensitive chemosensor for lead ions using fluorescent carbon dots prepared from chocolate by one-step hydrothermal method. Sensors Actuators B Chem 237:597–604

Liu FY, Jiang YR, Chen CC, Lee WW (2018) Novel synthesis of $PbBiO_2Cl/BiOCl$ nanocomposite with enhanced visible-driven-light photocatalytic activity. Catal Today 300:112–123
Liua Z, Jina W, Wanga F, Lia T, Niea J, Xiaoa W, Zhangb Q, Liua YZZ, Jina W, Wanga F, Lia T, Niea J, Xiaoa W, Zhangb Q, Zhanga Y (2019) Ratiometric fluorescent sensing of Pb^{2+} and Hg^{2+} with two types of CDs nanohybrids synthesized from the same biomass. Sens Actuators, B Chemical 296:126698
López-Artíguez M, Cameán A, Repetto M (1993) Preconcentration of heavy metals in urine and quantification by inductively coupled plasma atomic emission spectrometry. J Anal Toxicol 17:18–22
Lorber KE (1986) Monitoring of heavy metals by energy dispersive X-ray fluorescence spectrometry. Waste Manag Res 4:3–13
Love C, Din AS, Tomas MB, Kalapparambath TP, Palestro CJ (2003) Radionuclide bone imaging: an illustrative review. Radiographics 23(2):341–358
Lu W, Qin X, LiuS CG, Zhang Y, Luo Y (2012) Economical, green synthesis of fluorescent carbon nanoparticles and their use as probes for sensitive and selective detection of mercury (II) ions. Anal Chem 84:5351–5357
Luo PG, Sahu S, Yang ST, Sonkar SK, Wang J, Wang H, LeCroy GE, Cao L, Sun YP (2013) Carbon "quantum" dots for optical bioimaging. J Mater Chem B 1(16):2116–2127
Luo PG, Yang F, Yang ST, Sonkar SK, Yang L, Broglie JJ, Liu Y, Sun YP (2014) Carbon based quantum dots for fluorescence imaging of cells and tissues. RSC Adv 4(21):10791–10807
Martin S, Griswold W (2009) Human health effects of heavy metals. Environmental Science and Technology Briefs for Citizens 15:1–6
Mazrad ZAI, Lee K, Chae A, In I, Lee H, Park SY (2018) Progress in internal/external stimuli responsive fluorescent carbon nanoparticles for theranostic and sensing applications. J Mater Chem B 6(8):1149–1178
Mohapatra S, Bera MK, Das RK (2018) Rapid "turn-on" detection of atrazine using highly luminescent N-doped carbon quantum dot. Sensors Actuators B Chem 263:459–468
Mondal TK, Ghorai UK, Saha SK (2018) Dual-emissive carbon quantum dot-Tb nanocomposite as a fluorescent indicator for a highly selective visual detection of Hg(II) in water. ACS Omega 3:11439–11446
Murugan N, Prakash M, Jayakumar M, Sundaramurthy A, Sundramoorthy AK (2019) Green synthesis of fluorescent carbon quantum dots from Eleusine coracana and their application as a fluorescence 'turn-off' sensor probe for selective detection of Cu^{2+}. Appl Surf Sci 476:468–480
Namdari P, Negahdari B, Eatemadi A (2017) Synthesis, properties and biomedical applications of carbon-based quantum dots: an updated review. Biomed Pharmacother 87:209–222
Nicolopoulou-Stamati P, Maipas S, Kotampasi C, Stamatis P, Hens L (2016) Chemical pesticides and human health: the urgent need for a new concept in agriculture. Front Public Health 4:148
Oehme I, Wolfbeis OS (1997) Optical sensors for determination of heavy metal ions. Mikrochim Acta 126:177–192
Omichinski JG (2007) Toward methylmercury bioremediation. Science 317(5835):205–206
Peng H, Travas-Sejdic J (2009) Simple aqueous solution route to luminescent carbogenic dots from carbohydrates. Chem Mater 21(23):5563–5565
Peng Z, Han X, Li S, Al-Youbi AO, Bashammakh AS, El-Shahawi MS, Leblanc RM (2017) Carbon dots: biomacromolecule interaction, bioimaging and nanomedicine. Coord Chem Rev 343:256–277
Pérez-Estrada LA, Malato S, Gernjak W, Agüera A, Thurman EM, Ferrer I, Fernández-Alba AR (2015) Photo-Fenton degradation of diclofenac identification of main intermediates and degradation pathway. Environ Sci Technol 39:8300–8306
Prashanth PA, Raveendra RS, Krishna RH, Ananda S, Bhagya NP, Nagabhushana BM, Lingaraju K, Naika HR (2015) Synthesis, characterizations, antibacterial and photoluminescence studies of solution combustion-derived α-Al2O3 nanoparticles. J Asian Ceramic Soc 3:345–351
Qiao ZA, Wang Y, Gao Y, Li H, Dai T, Liu Y, Huo Q (2010) Commercially activated carbon as the source for producing multicolor photoluminescent carbon dots by chemical oxidation. Chem Commun 46(46):8812–8814

Qu K, Wang J, Ren J, Qu X (2013) Carbon dots prepared by hydrothermal treatment of dopamine as an effective fluorescent sensing platform for the label-free detection of iron (III) ions and dopamine. Chem Eur J 19:7243–7249

Ray S, Saha A, Jana NR, Sarkar R (2009) Fluorescent carbon nanoparticles: synthesis, characterization, and bioimaging application. J Phys Chem C 113(43):18546–18551

Ren W, Gao J, Lei C, Xie Y, Cai Y, Ni Q, Yao J (2018) Effects of sodium montmorillonite on the preparation and properties of cellulose aerogels. Chem Eng J 349:766–774

Robertson J, O'Reilly E (1987) Electronic and atomic structure of amorphous carbon. Phys Rev B 35(6):2946

Roshni V, Misra S, Santra MK, Ottoor D (2019) One pot green synthesis of C-dots from groundnuts and its application as Cr (VI) sensor and in vitro bioimaging agent. J Photochem Photochem Photobiol A 373:28–36

Salinas-Castillo A, Ariza-Avidad M, Pritz C, Camprubí-Robles M, Fernández B, Ruedas-Rama MJ, Megia-Fernández A, Lapresta-Fernández A, Santoyo-Gonzalez F, Schrott-Fischer A (2013) Carbon dots for copper detection with down and upconversion fluorescent properties as excitation sources. Chem Commun 49(11):1103–1105

Sarder P, Maji D, Achilefu S (2015) Molecular probes for fluorescence lifetime imaging. Bioconjug Chem 26(6):963–974

Shangguan J, Huang J, He D, He X, Wang K, Ye R, Yang X, Qing T, Tang J (2017) Highly Fe^{3+}-selective fluorescent nanoprobe based on ultrabright N/P co-doped carbon dots and its application in biological samples. Anal Chem 89(14):7477–7484

Sharma P, Rohilla D, Chaudhary S, Kumar R, Singh AN (2019) Nanosorbent of hydroxyapatite for atrazine: a new approach for combating agricultural runoffs. Sci Total Environ 653:264–273

Shen LM, Chen Q, Sun ZY, Chen XW, Wang JH (2014) Assay of biothiols by regulating the growth of silver nanoparticles with C-dots as reducing agent. Anal Chem 86(10):5002–5008

Shi L, Yang JH, Zeng HB, Chen YM, Yang SC, Wu C, Zeng H, Yoshihito O, Zhang Q (2016) Carbon dots with high fluorescence quantum yield: the fluorescence originates from organic fluorophores. Nanoscale 8(30):14374–14378

Smith TA, Ghiggino KP (2015) A review of the analysis of complex time-resolved fluorescence anisotropy data. Methods Appl Fluoresc 3(2):022001

Song Y, Zhu S, Yang B (2014) Bioimaging based on fluorescent carbon dots. RSC Adv 4(52):27184–27200

Stanisavljevic M, Krizkova S, Vaculovicova M, Kizek R, Adam V (2015) Quantum dots-fluorescence resonance energy transfer-based nanosensors and their application. Biosens Bioelectron 74:562–574

Sun YP, Zhou B, Lin Y, Wang W, Fernando KS, Pathak P, Meziani MJ, Harruff BA, Wang X, Wang H (2006) Quantum-sized carbon dots for bright and colorful photoluminescence. J Am Chem Soc 128(24):7756–7757

Tang L, Ji R, Cao X, Lin J, Jiang H, Li X, Teng KS, Luk CM, Zeng S, Hao J (2012) Deep ultraviolet photoluminescence of water-soluble self-passivated graphene quantum dots. ACS Nano 6(6):5102–5110

Tao Z, Huang Y, Liu X, Chen J, Lei W, Wang X, Pan L, Pan J, Huang Q, Zhang Z (2016) High-performance photo-modulated thin-film transistor based on quantum dots/reduced graphene oxide fragment-decorated ZnO nanowires. Nano Micro Lett. 8(3):247–253

Tetsuka H, Asahi R, Nagoya A, Okamoto K, Tajima I, Ohta R, Okamoto A (2012) Optically tunable amino-functionalized graphene quantum dots. Adv Mater 24(39):5333–5338

Walker CL, Lukyanov KA, Yampolsky IV, Mishin AS, Bommarius AS, Duraj Thatte AM, Azizi B, Tolbert LM, Solntsev KM (2015) Fluorescence imaging using synthetic GFP chromophores. Curr Opin Chem Biol 27:64–74

Wan Y, Lao S, Ding W, Zhang Z, Liu S (2019) A novel ratiometric fluorescent probe for detection of iron ions and zinc ions based on dual-emission carbon dots. Sensors Actuators B Chem 284:186–192

Wang Y, Hu A (2014) Carbon quantum dots: synthesis, properties and applications. J Mater Chem C 2(34):6921–6939

Wang X, Cao L, Lu F, Meziani MJ, Li H, Qi G, Zhou B, Harruff BA, Kermarrec F, Sun YP (2009) Photoinduced electron transfers with carbon dots. Chem Commun 25(3774):3776

Wang X, Cao L, Yang ST, Lu F, Meziani MJ, Tian L, Sun KW, Bloodgood MA, Sun YP (2010) Bandgap-like strong fluorescence in functionalized carbon nanoparticles. Angew Chem Int Ed 122(31):5438–5442

Wang K, Gao Z, Gao G, Wo Y, Wang Y, Shen G, Cui D (2013) Systematic safety evaluation on photoluminescent carbon dots. Nanoscale Res Lett 8:122

Wang N, Wang Y, Guo Y, Yang T, Chenn M, Wang J (2016) Green preparation of carbon dots with papaya as carbon source for effective fluorescent sensing of iron (III) and Escherichia coli. Biosens Bioelectron 85:68–75

Wang TY, Chen CY, Wang CM, Tan YZ, Liao WS (2017) Multicolor functional carbon dots via one-step refluxing synthesis. ACS Sens 2(3):354–363

Wang BB, Jin JC, Xu Z, Jiang Z, Li X, Jiang F, Liu Y (2019) Single-step synthesis of highly photoluminescent carbon dots for rapid detection of Hg^{2+} with excellent sensitivity. J Colloid Interface Sci 551:101–110

Wasim A, Sengupta D, Chowdhury A (2009) Impact of pesticides use in agriculture: their benefits and hazards. Interdiscip Toxicol 2(1):1–12

Weng GE, Ling AK, Lv XQ, Zhang JY, Zhang BP (2011) III-nitride-based quantum dots and their optoelectronic applications. Nano Micro Lett 3(3):200–207

Wolfbeis OS (2015) An overview of nanoparticles commonly used in fluorescent bioimaging. Chem Soc Rev 44(14):4743–4768

Wu X, Tian F, Wang W, Chen J, Wu M, Zhao JX (2013) Fabrication of highly fluorescent graphene quantum dots using l-glutamic acid for in vitro/in vivo imaging and sensing. J Mater Chem C 1(31):4676–4684

Wu X, Song Y, Yan X, Zhu C, Ma Y, Du D, Lin Y (2017) Carbon quantum dots as fluorescence resonance energy transfer sensors for organophosphate pesticides determination. Biosens Bioelectron 94:292–297

Xu X, Ray R, Gu Y, Ploehn HJ, Gearheart L, Raker K, Scrivens WA (2004) Electrophoretic analysis and purification of fluorescent single-walled carbon nanotube fragments. J Am Chem Soc 126:12736–12737

Yan X, Cui X, Li LS (2010) Synthesis of large, stable colloidal graphene quantum dots with tunable size. J Am Chem Soc 132(17):5944–5945

Yan X, Li HX, Han XS, Su XG (2015) A ratiometric fluorescent quantum dots based biosensor for organophosphorus pesticides detection by inner-filter effect. Biosens Bioelectron 74:277–283

Yang ST, Wang X, Wang H, Lu F, Luo PG, Cao L, Meziani MJ, Liu JH, Liu Y, Chen M (2009) Carbon dots as nontoxic and high-performance fluorescence imaging agents. J Phys Chem C 113(42):18110–18114

Yang Z, Li Z, Xu M, Ma Y, Zhang J, Su Y, Gao F, Wei H, Zhang L (2013) Controllable synthesis of fluorescent carbon dots and their detection application as nanoprobes. Nano Micro Lett. 5(4):247–259

Yang J, Zhang X, Ma YH, Gao G, Chen X, Jia HR, Li YH, Chen Z, Wu FG (2016) CDs-based platform for simultaneous bacterial distinguishment and antibacterial applications. ACS Appl Mater Interfaces 8(47):32170–32181

Zhang R, Chen W (2014) Nitrogen-doped carbon quantum dots: facile synthesis and application as a "turn-off" fluorescent probe for detection of Hg^{2+} ions. Biosens Bioelectron 55:83–90

Zhang K, Mei QS, Guan GJ, Liu BH, Wang SH, Zhang ZP (2010) Ligand replacement induced fluorescence switch of quantum dots for ultrasensitive detection of organophosphorothioate pesticides. Anal Chem 82:9579–9586

Zhang Z, Shi Y, Pan Y, Cheng X, Zhang L, Chen J, Li MJ, Yi C (2014) Quinoline derivative-functionalized carbon dots as a fluorescent nanosensor for sensing and intracellular imaging of Zn^{2+}. J Mater Chem B 2:5020–5027

Zhang J, Abbasi F, Claverie J (2015) An efficient templating approach for the synthesis of redispersible size-controllable carbon quantum dots from graphitic polymeric micelles. Chem Eur J 21(43):15142–15147

Zhang X, Cao S, Wu Z, Zhao S, Piao L (2017) Enhanced photocatalytic activity towards degradation and H_2 evolution over one dimensional TiO_2@MWCNTs heterojunction. Appl Surf Sci 402:360–368

Zhao QL, Zhang ZL, Huang BH, Peng J, Zhang M, Pang DW (2008) Facile preparation of low cytotoxicity fluorescent carbon nanocrystals by electrooxidation of graphite. Chem Commun 41:5116–5118

Zhao A, Chen Z, Zhao C, Gao N, Ren J, Qu X (2015a) Recent advances in bioapplications of C-dots. Carbon 85:309–327

Zhao SJ, Lan MH, Zhu XY, Xue HT, Ng TW, Meng XM, Lee CSP, Wang F, Zhang WJ (2015b) Green synthesis of bifunctional fluorescent carbon dots from garlic for cellular imaging and free radical scavenging. ACS Appl Mater Interfaces 7:17054–17060

Zhao C, Li X, Cheng C, Yang Y (2019) Green and microwave-assisted synthesis of carbon dots and application for visual detection of cobalt(II) ions and pH sensing. Microchem J 147:183–190

Zheng H, Wang Q, Long Y, Zhang H, Huang X, Zhu R (2011) Enhancing the luminescence of carbon dots with a reduction pathway. Chem Commun 47(38):10650–10652

Zheng M, Xie Z, Qu D, Li D, Du P, Jing X, Sun Z (2013) On–off–on fluorescent CDs Nanosensor for recognition of chromium(VI) and ascorbic acid based on the inner filter effect. ACS Appl Mater Interfaces 524:13242–13247

Zhou YY, Tang L, Zeng GM, Zhang C, Zhang Y, Xie X (2016) Current progress in biosensors for heavy metal ions based on DNAzymes/DNA molecules functionalized nanostructures: a review. Sensors Actuators B Chem 223:280–294

Zhu B, Sun S, Wang Y, Deng S, Qian G, Wang M, Hu A (2013) Preparation of carbon nanodots from single chain polymeric nanoparticles and theoretical investigation of the photoluminescence mechanism. J Mater Chem C 1(3):580–586

Zhu S, Song Y, Zhao X, Shao J, Zhang J, Yang B (2015) The photoluminescence mechanism in CDs(graphene quantum dots, carbon nanodots, and polymer dots): current state and future perspective. Nano Res 8(2):355–381

Zong J, Zhu Y, Yang X, Shen J, Li C (2011) Synthesis of photoluminescent carbogenic dots using mesoporous silica spheres as nanoreactors. Chem Commun 47(2):764–766

Zong J, Yang X, Trinchi A, Hardin S, Cole I, Zhu Y, Li C, Muster T, Wei G (2014) Carbon dots as fluorescent probes for "off–on" detection of Cu^{2+} and L-cysteine in aqueous solution. Biosens Bioelectron 51:330–335

Chapter 9
Advanced Nanomaterials in Biomedicine: Benefits and Challenges

Avtar Singh and Jaspreet Singh Dhau

Abstract Nanotechnology-based materials have unique physical, chemical, and biological properties for use in biomedicine. In medical research, researchers study human physiology and find ways to prevent and treat diseases for improvement in lives for people and animals. Nanotechnology in the biomedical field promises a solution to the challenges associated with this field. The last three decades have seen a significant interest in nanomaterial-based materials and techniques, which have been revolutionized the field of health and medicine. Biomedical nanotechnology holds great promise for a variety of applications, which include diagnostic techniques, drug delivery, gene delivery, vaccine delivery, nanodrugs, and prostheses and implants. Numerous nanomaterial-based approaches are widely used at the commercial level, which are approved by FDA. In the last two decades, metallic gold nanoparticles (GNPs) have been studied extensively because of their wide applications in the biomedical field owing to their remarkable physicochemical properties. This chapter provides an overview of the benefits associated with gold nanoparticle-based materials for applications in drug delivery systems and in gene delivery systems. Major challenges associated with nanotechnology in biomedicine are also highlighted.

Keywords Nanotechnology · Nanoparticles · Biomedicine · Drug delivery · Gene delivery

A. Singh (✉)
Clean Energy Research Center, University of South Florida, Tampa, Florida, USA

Department of Chemistry, Sri Guru Teg Bahadur Khalsa College, Anandpur Sahib, Punjab, India

Research and Development, Molekule Inc, Tampa, Florida, USA
e-mail: avtarldh007@gmail.com; avtar.singh@molekule.com

J. S. Dhau
Research and Development, Molekule Inc, Tampa, Florida, USA
e-mail: jdhau@molekule.com

R. Kumar et al. (eds.), *Advanced Functional Nanoparticles "Boon or Bane" for Environment Remediation Applications*, Environmental Contamination Remediation and Management, https://doi.org/10.1007/978-3-031-24416-2_9

Abbreviations

CBNs	Carbon-based nanomaterials
CNTs	Carbon nanotubes
DDS	Drug delivery system
DG	2-Deoxyglucose
DOX	Doxorubicin
FDA	Food and Drug Administration
FT-IR	Fourier transform infrared spectroscopy
GDS	Gene delivery system
GLP	Glucagon-like peptide
GN	Graphene
GNPs	Gold nanoparticles
GNS	Gold nanostars
GO	Graphene oxide
GSH	Glutathione
HPLC	High-performance liquid chromatography
LSPR	Localized surface plasmon resonance
MBC	Minimum bactericidal concentration
MBNs	Metal-based nanomaterials
MIC	Minimal inhibitory concentration
MTX	Methotrexate
NGF	Nerve growth factors
NM	Nanomaterials
OBN	Organic-based nanomaterials
PBMCs	Peripheral blood mononuclear cells
PEG	Polyethylene glycol
PGHNs	Protein-gold hybrid nanocubes
TNF	Tumor necrosis factor

9.1 Introduction

Over the past several decades, numerous opportunities have arisen for nanotechnology researchers to develop and apply nanomaterials in diverse fields of science (Bayda et al. 2020; Alivisatos 2008; Sonkaria et al. 2012; Yang et al. 2011; Singh et al. 2022). Advancement of nanotechnology in biomedicine has revolutionized the field of health and medicine (Raliya et al. 2016; Leary et al. 2006; Fries et al. 2020; Shin et al. 2020; Patil et al. 2008; Shrivastava et al. 2009). Nanotechnology-based materials can range from 10^{-7} to 10^{-9} m in size (1–100 nm) and have unique physical, chemical, and biological properties (Yang et al. 2011; Dhand et al. 2015; Ren et al. 2020). In the biomedicine field, nanotechnology holds great promise for a variety of applications, including diagnostic techniques, drug delivery, gene delivery, nanodrugs, and prostheses and implants (Laroui et al. 2013; Onoue et al. 2014; Jahangirian et al. 2017; Gavaskar et al. 2018). These applications are revolutionary in the biomedical field and result in new materials, techniques, and devices with more sensitivity, selectivity, and sophistication, which are required for this field.

For cancer treatment, researchers are continually devoting their efforts to develop a nanomaterial-based platform, which could provide better therapeutic systems from thebench scale to clinical applications. In cancer diagnosis, nanoparticles are being used to detect cancer biomarkers, such as cancer-associated proteins, circulating tumor DNA, circulating tumor cells, and exosomes that are secreted by tumors (Jia et al. 2017; Zhang et al. 2019). Nanotechnology-based approaches for cancer diagnosis are in different phases of clinical trials (Engelberth et al. 2014). Numerous nanoparticles are widely used at the commercial level, which are approved by the FDA. Gold nanoparticles are used as label or probe molecules for detection in lateral flow immunoassay devices (Banerjee and Jaiswal 2018). Nanotechnology-based drug delivery and gene delivery systems are major applications and have been gaining a lot of interest from researchers and industries (Jahangirian et al. 2017; Shi et al. 2010; Farokhzad and Langer 2009).

Carbon nanotubes, dendrimers, and liposomes are important tissue engineering scaffold materials for drug delivery systems and have the ability to control the release of a drug or therapeutic compound to impaired tissues (Mody et al. 2014). The first generation of nanomaterial-based products included lipid systems like liposomes and micelles, which are now FDA-approved (Longo et al. 2018). These liposomes and micelles can contain inorganic nanoparticles like gold or magnetic nanoparticles in order to use as targeted drug delivery systems (Park et al. 2006). These applications have increased the interest of researchers for the use of inorganic nanoparticles.

In the last two decades, metallic gold nanoparticles (GNPs) have been studied extensively because of their attractive qualities in the biomedical field owing to their remarkable physical, chemical, and biological properties, like size, inertness, ease of synthesis, and biocompatibility (Tiwari et al. 2011; Dykman and Khlebtsov 2011). This chapter provides an overview of benefits associated with gold nanoparticle-based materials for applications in drug delivery systems and in gene delivery systems. Major challenges associated with biomedical nanotechnology are also highlighted.

9.2 Nanomaterials and Their Classification

During the annual meeting of the American Physical Society, Nobel laureate Richard P. Feynman was the first who presented the concept of nanotechnology at the California Institute in December 1959 (Feynman 1960). In 1974, Norio Taniguchi, a Professor at Tokyo University of Science, first defined the term nanotechnology as: "Nanotechnology mainly consists of the processing of, separation, deformation, and consolidation of material by one atom or by one molecule"

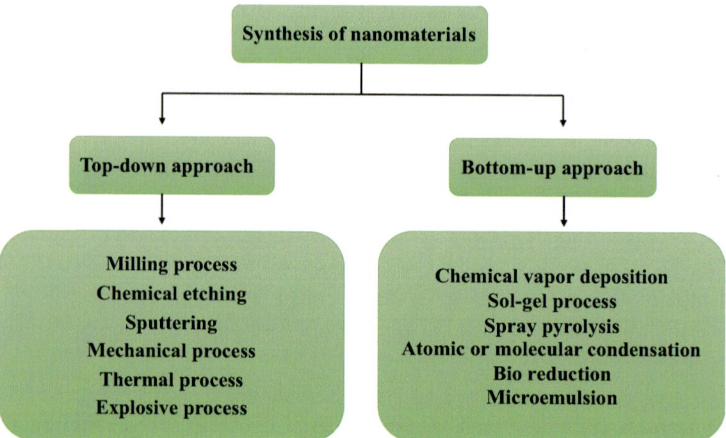

Fig. 9.1 Synthesis of nanomaterials

(Taniguchi et al. 1974). Since then, research in nanotechnology has been catching the interest of scientists for the development and advancement of science.

Two approaches have been developed for the synthesis of nanostructures: first is top-down, and second is bottom-up (Kumar and Kumbhat 2016). The top-down approach involves the breaking down of bulk material to get nano-sized material, which is achieved by sputtering, chemical etching, mechanical process, thermal process, and explosion process. The bottom-up approach refers to the buildup of nanostructures from the bottom means by assembling atoms or molecules together. This approach involves different processes like chemical vapor deposition, sol-gel process, spray pyrolysis, atomic or molecular condensation, bio-reduction, and microemulsion. Methods involved in top-down and bottom-up concepts are summarized in Fig. 9.1.

Nanotechnology deals with synthesis, characterization, application, and resulting devices of nano-scaled (1–100 nm) materials. The unique properties of nanoscale materials have made them attractive for a number of innovative and sustainable processes in different fields of science (Grainger and Castner 2008; Bayda et al. 2020). Nanomaterials can be classified into three major categories: (1) metal-based nanomaterials, (2) carbon-based nanomaterials, and (3) organic-based nanomaterials (Fig. 9.2).

9.2.1 Metal-Based Nanomaterials

Metal-based nanomaterials (MBNs) have a metal core composed of inorganic metal or metal oxide surrounded with organic, inorganic, or hybrid material. These materials have unique properties that differ from their bulk counterparts and are desirable for a large number of advanced functional applications (Sweet et al. 2012). MBNs

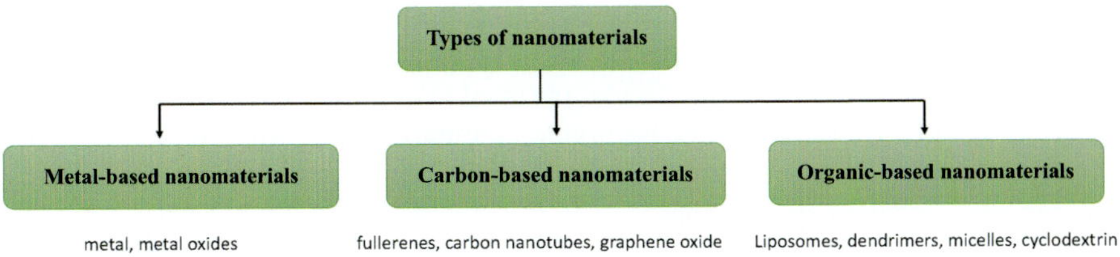

Fig. 9.2 Classification of nanomaterials

such as silver, gold, copper, and metal oxides (zinc oxide, titanium dioxide, iron oxide, etc.) are applied for many uses in biomedical and other fields (Yaqoob et al. 2020). Researchers are continually devoting their efforts to fabricate new metallic nanomaterials and investigate their properties with the ultimate goal of new applications of these materials. Biomedicine is one of the most attractive fields where the use of gold nanoparticles has been studied broadly. In the biomedical field, MBNs have numerous applications in diagnosis, therapeutic, bioimaging, drug carriers, gene carriers, and protein carriers (Raliya et al. 2016).

9.2.2 Carbon-Based Nanomaterials

Due to its unique allotropic characteristics, carbon forms compounds that have different properties depending on the arrangement of the adjacent carbon atoms. Based upon bonding relationships with the neighboring atoms, carbon can exist in three different hybridization sp., sp^2, and sp^3. Carbon-based nanomaterials (CBNs) mainly include fullerenes, carbon nanotubes (CNTs), graphene (GN), graphene oxide (GO), and its derivatives and quantum dots. In applied nanotechnology research, carbon-based nanomaterials have attracted wide attention due to special features of these materials such as large surface area, high mechanical strength, high conductivity, and electrical, thermal, optical, and chemical properties (Li et al. 2020; Mauter and Elimelech 2008; Singh et al. 2021). In the field of biomedical nanotechnology, CBNs have been widely considered as highly appealing biomaterials due to their inimitable features. CBNs have been widely exploited in diverse biomedical applications including biosensing, bioimaging, drug delivery, stem cell engineering, florescence labeling of cells, diagnosis, and cancer therapy (Maiti et al. 2019).

9.2.3 Organic-Based Nanomaterials

Organic-based nanomaterials (OBNs) are mostly based upon liposomes, micelles, lipids, and polymeric nanoparticles (Srinivasan et al. 2015). These nanomaterials have been widely investigated and are employed for imaging or drug and gene delivery techniques. OBNs like micelles, liposomes, and polymers are composed of molecular units with weaker noncovalent intermolecular interactions and can change size and shape into desired structures. Organic nanomaterials like lipid systems, liposomes, and micelles are FDA-approved (Ventola 2017). These nanomaterials can load other molecules like inorganic nanoparticles or drug molecules either by conjugation on the surface or in the core, which makes them appealing delivery systems in biomedical applications.

9.3 Benefits of Nanomaterials in Biomedicine

Advanced nanomaterials have prompted rapid progress in biomedicine to prevent and treat diseases for improvement in the lives of people and animals. Nanomaterials of all dimensions (0D, 1D, 2D, and 3D) are used extensively in various applications due to unique physical, chemical, and biological properties. Considerable number of books, reviews, and research papers on gold nanoparticles (GNPs) and their biomedical applications are proof of growth in this field. In 1994, the first nanomaterial-based product, Doxils, was launched in the market for medical use (Longo et al. 2020). Doxils is a nano-sized phospholipid liposome containing the chemotherapy drug doxorubicin that was developed for cancer therapy. The benefits of nanomaterials to the field of biomedicine are numerous and hold great promises for a variety of applications, which include diagnostic techniques, drug delivery, nanodrugs, and prostheses and implants. This section provides the benefits and contributions of advanced nanomaterials in two major branches of biomedicine: first is in drug delivery systems (DDS), and second is in gene delivery systems (GDS) (Fig. 9.3).

9.3.1 Gold Nanoparticles in Drug Delivery System

A drug delivery system refers to a device that permits the introduction of a drug or pharmaceutical compound to the targeted body site in the human body. Delivering drugs at a slow rate and sustained release in targeted sites of the human body are the two important measures for efficient drug delivery systems (Tiwari et al. 2012). The

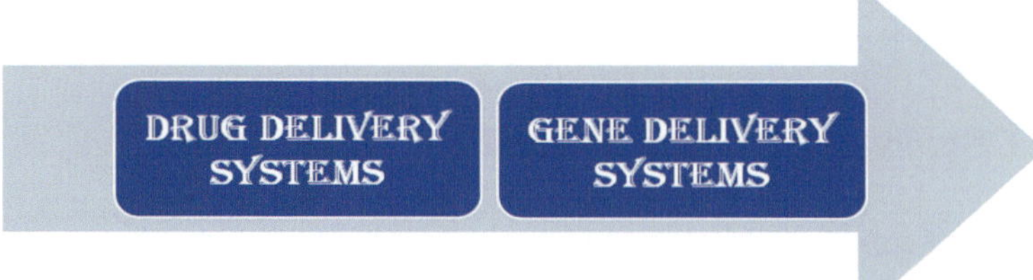

Fig. 9.3 Application of advanced nanomaterials in biomedicine

use of nanomaterials in drug delivery systems has gained increasing attention in the last two decade, and numerous nanomaterials have been developed to use as efficient and reliable drug delivery systems. Metallic nanoparticles are extensively used in biomedical nanotechnology because of their unique properties such as large surface area, high stability, facile chemical modification, biocompatibility, and retention (Raliya et al. 2016). In biomedical nanotechnology, most commonly studied metal-based nanoparticles are of gold, silver, platinum nickel, iron oxides, zinc oxide, and titanium dioxide (Yaqoob et al. 2020). Among these, gold nanoparticles (AuNPs) are most commonly being used for various applications in biomedicine like drug delivery and diagnostic and imaging.

In a drug delivery study, Ding et al. (2015) developed novel protein–gold hybrid nanocubes (PGHNs), including gold nanoclusters, bovine serum albumin, and tryptophan. They used eco- friendly and toxicity-free process for synthesis. In order to check the capability of PGHNs for use as drug delivery vehicle, they used Rhodamine 6G dye, which is highly fluorescent. The results suggested that PGHNs are capable of delivering drugs into eukaryotic cells and unload them at the cytoplasm. The authors found that protein-metal hybrid nanocubes can be used for two purposes: first is a blue-emitting cell marker in bioimaging investigation, and second is a nanocarrier in drug delivery studies.

Paciotti et al. (2004) developed colloidal gold nanoparticle vector for tumor-targeted drug delivery, in which they targeted the delivery of highly toxic anticancer protein biologic and tumor necrosis factor (TNF) to a solid tumor growing in mice. The PT-cAu-TNF drug vector consists of molecules of thiol-derivatized PEG (PT) and recombinant human TNF that are directly bound onto the surface of the gold nanoparticles. PT-cAu-TNF is administrated in C57/BL6 mice implanted with the colon carcinoma tumor cells, MC-38, and results showed little to no accumulation in the livers, spleens, or other healthy organs of the animals.

Hong and co-workers (2006) reported a monolayer-protected gold nanoparticle-based delivery system and release system by utilizing GSH-mediated release. GSH-mediated release is a nonenzymatic approach for the selective intracellular activation of prodrugs. Based upon the large difference in intracellular and extracellular GSH concentration, GSH acts as a reducing agent in biochemical processes and providing a mechanism for selective intracellular release. Using this strategy, Hong et al. (2006) demonstrated a release of payload from nanoparticle surfaces both in vitro and in cell cultures. They used 2 nm core gold nanoparticles (AuNP) with a mixed monolayer composed of a tetra(ethylene glycol)-lyated cationic ligand TTMA and a thiolated BODIPY dye, HSBDP.

Targeting biocompatible drug delivery systems, Monti and co-workers (2019) explored theoretical and experimental approaches to study nanoparticle carriers for in situ controlled antibiotic release. For the drug delivery system, they combined gold nanoparticles with chitosans. Chitosan is a natural biodegradable and biocompatible linear polysaccharide. They loaded chitosan-gold nanoparticles with gentamicin antibiotic to investigate the physicochemical properties of drug delivery system. In the results, they found that conjugated nanoparticle system was very efficient for drug delivery system in which antibiotic activity of gentamicin was preserved, and the drug release from the carrier could be tuned by changing the chitosan/gentamicin weight ratio and the deposition pattern of the adsorbed layers. In another study, Salouti et al. (2016) used conjugation of gold nanorods (GNRs) with gentamicin to use in the delivery of gentamicin to infection foci due to *Staphylococcus aureus*. FT-IR spectroscopy, high-performance liquid chromatography (HPLC), and atomic absorption spectroscopy analyses were used to study interactions and number of gentamicin molecules attached to each gold nanorod. The minimal inhibitory concentration (MIC) and minimum bactericidal concentration (MBC) results showed that gentamicin–GNR conjugate was able to facilitate gentamicin delivery to *Staphylococcal* infection sites and also played role in the enhancement of antibacterial effect of gentamicin.

Rastogi and co-workers (2012) studied gold nanoparticles capped with protein-bovine serum albumin for the effective delivery of amino-glycosidic antibiotics such as streptomycin sulfate, neomycin sulfate, gentamicin sulphate, and kanamycin sulfate. Results showed that conjugated nanoparticles exhibited enhanced antibacterial activity, compared to pure antibiotic.

In another different approach, Pérez-Ortiz et al. (2017) conjugated gold nanoparticles with glucagon-like peptide-1 (GLP-1) for potential application as drug delivery system. They concluded that conjugation to AuNPs may provide an effective delivery system for GLP-1 and potential analogues.

Gold nanoparticles with properties of high surface area, high colloidal stability, small size, and high tissue permeability have gained attention for use in various biomedical uses. Gold nanoparticles are attractive drug delivery systems and have been used in the treatment of various tumors. In comparison to conventional antitumor drug delivery systems, nanotechnology-based materials improve the bioavailability and therapeutic efficiency of antitumor drugs. There are a number of drug delivery systems successfully approved and employed in the biomedical field, and several others are understudy in different clinical trial phases.

Methotrexate has been a widely used medication for treating certain types of cancer of the breast, skin, head, and neck. In order to obtain control release and improved therapeutic efficacy of MTX, researchers studied MTX with conjugation to different systems including protein, polymers, large macroparticles, and nanoparticles. Chen et al. (2007) used new drug formulation by conjugation of MTX with 13 nm gold nanoparticles for delivery of MTX to cancer cells. The spectroscopic characterization of MTX–AuNP complex showed direct bonding gold nanoparticle and carboxyl group of MTX. They studied the cytotoxic effect of MTX–AuNP on eight different cancer cell lines and antitumor effect of MTX–AuNP in the LL2 ascites tumor model. The results showed faster and higher accumulation of MTX in MTX-AuNP-treated tumor cells than that treated with free MTX.

Doxorubicin is an anthracycline antibiotic, and effective anticancer drug has been widely used in chemotherapy for treatment of various cancers. Aryal et al. (2009) used conjugation of thiol-stabilized gold nanoparticles and doxorubicin for drug release. They presented a new technique to covalently conjugate DOX through pH-sensitive bonds onto stabilized Au NPs. The drug release behavior from the

DOX-conjugated Au NPs is very stable at physiological conditions. Results found enhancement of the cytotoxicity of DOX. In the effective chemotherapy of cancer, multidrug resistance is a major impediment, which means resistance of cancer cells to different anticancer drugs. Biomedical nanotechnology holds great promise in order to reverse MDR. Researchers have studied numerous nanomaterial-based approaches for this purpose. Wang and co-workers (2011) developed a drug delivery system through binding doxorubicin onto the surface of gold nanoparticles with a poly(ethylene glycol) as a spacer (DOX-Hyd@AuNPs). They used multidrug-resistant cancer cell line MCF-7/ADR for accumulation and retention of doxorubicin and found better results with DOX-Hyd@AuNPs compared to free doxorubicin.

9.3.2 Gold Nanoparticles in Gene Delivery

Gene delivery is a unique process by which foreign genomic material is transferred to the host cell to prevent or recover any genetic as well as acquired diseases (Nayerossadat et al. 2012). Numerous research groups are working to develop suitable methods for use in gene therapy. Recent studies have suggested that in order to achieve successful gene therapy, development of proper gene delivery systems could be one of the most important factors (Sung and Kim 2019). Various gene delivery systems have been developed in recent years, which mainly include polymer-based system, lipid-based systems, and functionalized inorganic nanoparticle-based systems (Patra et al. 2018). Gene delivery systems mainly include two categories: viral-based and nonviral-based systems (Nayerossadat et al. 2012). Viral-based gene delivery systems utilize the ability of virus to inject their DNA inside host cells. Nonviral gene delivery systems are alternatives to viral gene delivery systems and are categorized based upon preparation and physical and chemical types. Gold nanoparticle-based materials have been widely studied for improving gene delivery due to special features of these materials, such as high specific surface area, small size, nontoxic, and localized surface plasmon resonance (LSPR) (Dreaden et al. 2012).

Kim et al. used gold nanoparticles functionalized with covalently attached oligonucleotides as intracellular gene regulation agents (Kim et al. 2012). They showed that human peripheral blood mononuclear cells (PBMCs) exposed to gold nanoparticle–oligonucleotide complexes lead to widespread transcriptional activation of innate immune responses. For cancer gene therapy, researchers reported a localized gene delivery system, consisting of gold nanorods (Au NRs) loaded with hTERT siRNA attached to the surface of electrospun $ZnGa_2O_4$:Cr ZGOC nanofibers. Under the excitation of LED light, ZGOC nanofiber emission could be absorbed by gold nanorods (AuNRs), which induced a mild photothermal effect that significantly facilitates the escape of siRNA from endosomes and improves the siRNA transfection efficiency. They showed 40–65% enhancement in efficiency of gene silencing. Qin and coworkers anticipated that with this new therapeutic approach, based on

LED-amplified gene silencing, it will lead to a potential new generation of enhanced gene therapy treatments for a range of cancers.

For the first time, Rosi and co-workers demonstrated the relevance of gold nanoparticle-oligonucleotide complexes as intracellular gene regulation agents for the control of protein expression in cells in gene delivery (Rosi et al. 2006). These oligonucleotide-modified nanoparticles have affinity constants for complementary nucleic acids that are higher than their unmodified oligonucleotide counterparts, less susceptible to degradation by nuclease activity, less toxic, and higher cellular uptake. Cellular uptake of nucleotide-grafted AuNPs primarily occurs through endocytosis and is proportional to grafting density, which is controlled by varying the functionality (e.g., -SH) of nucleotide molecules.

Hongyan Zhu and co-workers prepared gold nanostar (GNS) complexes, namely, siCOX-2/AuNS, siCOX-2/9R-AuNS, siCOX-2/9R/DG- AuNS, and siCOX-2/9R/DG-AuNS (hydrazone) (Zhu et al. 2017). Ligands 2-deoxyglucose (DG), polyethylene glycol (PEG), lipoic acid (LA), Lys-9-poly-D, and arginine (9R) (hydrazone) were designed to functionalize GNS and create the nanoparticles named as 9R/DG-GNS (hydrazone). In this study, they used engineering of gold nanostars (GNS) to deliver small interfering siRNA to downregulate cyclooxygenase-2 (COX-2) expression, targeting the human hepatocellular carcinoma cells (HepG2) and SGC7901 cells. COX-2, a key enzyme used as catalyst in the conversion of arachidonic acid to prostaglandin, is involved in tumor invasiveness and angiogenesis. Increased COX-2 expression is one of the signs of early tumorigenesis. Results showed cyclooxygenase-2 siRNA/9-poly-D-arginine/2-deoxyglucose-gold nanostars (hydrazone) have good uptake by tumor cells through its targeting effect. The hydrazone bond was destroyed in the low endosomal pH microenvironment of tumor cells to release DG-PEG from the carrier and return it to the extracellular space through endocytosis, while the siCOX-2/9R-GNS remained in the intracellular space irreversibly. They anticipated that the GNS complex nanocarrier system can be used to deliver siCOX-2 into tumor cells efficiently.

Pancreatic cancer is a disease in which malignant (*cancer*) cells form in the tissues of the *pancreas*. Pancreatic tumors actively assist the growth of neurites and stimulate neurogenesis through nerve growth factors (NGFs). Targeting NGF gene is the most important to tackle pancreatic cancer. Lei et al. prepared gold nanocluster-siRNA having ability for the delivery of siRNA of NGF (GNC–siRNA), which allows efficient NGF gene silencing and pancreatic cancer treatment (Lei et al. 2017). Results suggested a promising therapeutic direction for pancreatic cancer. In one another study, researcher modified siRNA with polyethylene glycol (PEG) through disulfide bonding and then grafted onto amine-functionalized AuNPs. This study demonstrated the application of amine-functionalized gold nanoparticles for intracellular delivery of siRNA. Results showed that cellular uptake and target gene silencing efficacy of PEG-modified siRNA-complexed AuNPs (AuNPs/siRNA-PEG) were higher than regular AuNPs/siRNA complexes in human prostate carcinoma cells as AuNPs/siRNA rapidly get aggregated.

9.4 Challenges for Nanotechnology in Biomedicine

The use of advanced nanomaterials in biomedical field has been rapidly emerging as a field of research. Researchers are continually devoting their efforts to explore new ways and techniques with the ultimate goal of developing effective treatment and cures. Despite various applications and benefits associated with nanomaterials, these materials in biomedicine have lots of challenges before the transformation from the laboratory to the market for actual use (Feng et al. 2019; Longo et al. 2020). As the material at nanoscale behaves differently than in their bulk form, the first challenge is to learn more about materials in details. There is a definite uncertainty over their physicochemical effect on living life. Numerous studies have been published with the concern of toxicity of materials at nanoscale (Ganguly et al. 2018). Researchers across the world are rigorously studying and trying to solve this problem. However, the research community still learning about how quantum mechanics impact different materials at nano-level. Another major challenge is to analyze the effect of nanomaterials on the environment, and it depends upon the complex set of factors like type, size, surface properties, charge, etc. of nanomaterials (Ray et al. 2009). These physicochemical parameters decide the effect of these materials on our environment and living beings. All over the world, extensive research is being conducted on different nanomaterials to design, modify, and alter their behavior, but there is definite lack in understanding the toxicity and the effect on environment (Longo et al. 2020). There is a need to establish a general principle to evaluate all the materials under a common domain. All these challenges are as follow:

- Lack of standardized test assay
- Lack of uniform biological study in vivo
- Lack of uniformity of toxicity
- Lack of efficient analytical tools

Other major challenges that exist in scale-up of these advanced nanomaterials from laboratory to the market for actual use are the need for a well-trained workforce and specific machinery for design, synthesis, characterization, and testing at the pharmaceutical level. All these factors will increase the production cost, which will restrict the accessibility to common people.

9.5 Conclusion

The use of nanotechnology has shown great potential in biomedicine owing to their unique physicochemical properties. Gold nanoparticles with properties of high surface area, high colloidal stability, small size, and high tissue permeability have gained attention for use in various biomedical uses. In biomedicine, gold nanoparticle-based materials are an obvious choice over other nanomaterials due to their amenability of synthesis and functionalization, less toxicity, and ease of detection. In the last two decades, there has been growing interest in gold nanoparticles, which can be used as delivery of drugs, genes, DNA, RNA, proteins, and peptides. The objective of this chapter is to provide researchers the knowledge of application of gold-based nanomaterials in the biomedicine field and biomedical nanotechnology. The applications presented in this chapter strongly emphasized on the current research on advanced gold-based nanomaterials and their benefits in the biomedical field and are divided into two parts: (1) application of gold nanoparticles in drug delivery systems and (2) application of gold nanoparticles in gene delivery systems. In addition, challenges for nanotechnology in biomedicine are also discussed.

Appendix

Multiple Choice Questions

1. Nanotechnology comprises the materials with the size
   (i) 1–10 nm
   (ii) 1–100 nm
   (iii) 1–1000 nm
2. The color of gold in nanometric range
   (i) Red
   (ii) Yellow
   (iii) Orange
3. The highest antimicrobial activity is of
   (i) Ag
   (ii) CNT
   (iii) Graphene
4. Which of these nanomaterials amplify biorecognition signals and improve the analytical performance of biosensors?
   (i) Ag
   (ii) CNT
   (iii) Au
5. SWCNTs have diameters in the range
   (i) 0.4 nm–2 nm
   (ii) 4 nm–20 nm
   (iii) 0.8 nm–2 nm

Answers

1. (ii) 1–100 nm
2. (i) Red
3. (i) Ag
4. (iii) Au
5. (i) 0.4 nm–2 nm.

Short Questions and Answers

1. *What are the main applications of gold nanoparticles?*

   Answer: AuNPs have attracted tremendous interest from different fields of science, because of their particular features: high X-ray coefficient of absorption, ease of synthetic manipulation, strong binding affinity, and unique optical and distinct electronic properties (Dykman and Khlebtsov 2011). Gold nanoparticles are especially utilized in genomics, biosensors, immune analysis, clinical chemistry, detection, and photothermolysis of microorganisms and cancer cells; the targeted delivery of drugs, DNA and antigens; optical bio-imaging; and the monitoring of cells and tissues using modern registration systems.

2. *How Ag NPs are useful in wastewater treatment?*

   Answer: Silver nanoparticles possess antimicrobial properties that may be used to disinfect water. When Ag NPs come in contact with the bacteria, they interfere with the structure of the cell membrane, which causes death of the cell eventually. Another perspective that explains the mechanism of Ag NPs on their contact with bacteria is that the dissolution of Ag NPs will release antimicrobial Ag^+ ions, which can interact with the thiol groups of many vital enzymes, inactivate them, and disrupt normal functions in the cell. Ag NPs have been used along with certain filter membranes due to their cost-effectiveness and high antimicrobial properties.

3. *Write a note on the application of ferrites nanocomposites?*

   Answer: Ferrites have magnetic properties that make them suitable for separation at the end of the process. For example, Congo red (CR) was removed using manganese ferrites, nickel ferrites, and zinc ferrites. They were synthesized using sol-gel method. After various characterization methods and adsorption tests, it was concluded that nickel ferrites showed the best adsorption properties, followed by cobalt ferrites.

References

Alivisatos AP (2008) Birth of a nanoscience building block. ACS Nano 2:1514–1516
Aryal S, Grailer JJ, Pilla S, Steeber DA, Gong S (2009) Doxorubicin conjugated gold nanoparticles as water-soluble and pH-responsive anticancer drug nanocarriers. J Mater Chem 19:7879–7884
Banerjee R, Jaiswal A (2018) Recent advances in nanoparticle-based lateral flow immunoassay as a point-of-care diagnostic tool for infectious agents and diseases. Analyst 143:1970–1996
Bayda S, Adeel M, Tuccinardi T, Cordani M, Rizzolio F (2020) The history of nanoscience and nanotechnology: from chemical–physical applications to nanomedicine. Molecules 25:112
Chen Y-H, Tsai C-Y, Huang P-Y, Chang M-Y, Cheng P-C, Chou C-H, Chen D-H, Wang C-R, Shiau A-L, Wu C-L (2007) Methotrexate conjugated to gold nanoparticles inhibits tumor growth in a syngeneic lung tumor model. Mol Pharm 4:713–722
Dhand C, Dwivedi N, Loh XJ, Ying ANJ, Verma NK, Beuerman RW, Lakshminarayanan R, Ramakrishna S (2015) Methods and strategies for the synthesis of diverse nanoparticles and their applications: a comprehensive overview. RSC Adv 5:105003–105037
Ding H, Yang D, Zhao C, Song Z, Liu P, Wang Y, Chen Z, Shen J (2015) Protein−gold hybrid nanocubes for cell imaging and drug delivery. ACS Appl Mater Interfaces 7:4713–4719
Dreaden EC, Alkilany AM, Huang X, Murphy CJ, El-Sayed MA (2012) The golden age: gold nanoparticles for biomedicine. Chem Soc Rev 41:2740–2779
Dykman LA, Khlebtsov NG (2011) Gold nanoparticles in biology and medicine: recent advances and prospects. Acta Nat 3(2):34–55
Engelberth SA, Hempel N, Bergkvist M (2014) Development of nanoscale approaches for ovarian cancer therapeutics and diagnostics. Crit Rev Oncog 19:281–215
Farokhzad OC, Langer R (2009) Impact of nanotechnology on drug delivery. ACS Nano 3:16–20
Feng J, Markwalter CE, Tian C, Armstrong M, Prud'homme RK (2019) Translational formulation of nanoparticle therapeutics from laboratory discovery to clinical scale. J Transl Med 17:200
Feynman RP (1960) There's plenty of room at the bottom. Eng Sci 23:22–36
Fries CN, Curvino EJ, Chen JL, Permar SR, Fouda GG, Collier JH (2020) Advances in nanomaterial vaccine strategies to address infectious diseases impacting global health. Nat Nanotechnol 16:1–14
Ganguly P, Breen A, Pillai SC (2018) Toxicity of nanomaterials: exposure, pathways, assessment, and recent advances. ACS Biomater Sci Eng 4:2237–2275
Gavaskar A, Rojas D, Videla F (2018) Nanotechnology: the scope and potential applications in orthopedic surgery. Eur J Orthop Surg Traumatol 28:1257–1260
Grainger DW, Castner DG (2008) Nanobiomaterials and nanoanalysis: how to improve the nanoscience for biotechnology. Adv Mater 20:867–877
Hong R, Han G, Fernandez JM, Kim B, Forbes NS, Rotello VM (2006) Glutathione-mediated delivery and release using monolayer protected nanoparticle carriers. J Am Chem Soc 128:1078–1079
Jahangirian H, Lemraski EG, Webster TJ, Rafiee-Moghaddam R, Abdollahi Y (2017) A review of drug delivery systems based on nanotechnology and green chemistry: green nanomedicine. Int J Nanomedicine 12:2957–2978
Jia S, Zhang R, Li Z, Li J (2017) Clinical and biological significance of circulating tumor cells, circulating tumor DNA, and exosomes as biomarkers in colorectal cancer. Oncotarget 8:55632–55645
Kim E-Y, Schulz R, Swantek P, Kunstman K, Malim MH, Wolinsky SM (2012) Gold nanoparticle-mediated gene delivery induces widespread changes in the expression of innate immunity genes. Gene Ther 19(3):347–353. https://doi.org/10.1038/gt.2011.95
Kumar N, Kumbhat S (2016) Essentials in nanoscience and nanotechnology. John Wily & Sons, Inc, pp 31–74
Laroui H, Rakhya P, Xiao B, Viennois E, Merlin D (2013) Nanotechnology in diagnostics and therapeutics for gastrointestinal disorders. Dig Liver Dis 45:995–902

Leary SP, Liu CY, Apuzzo MLJ (2006) Toward the emergence of nanoneurosurgery: part III--nanomedicine: targeted nanotherapy, nanosurgery, and progress toward the realization of nanoneurosurgery. Neurosurgery 58:1009–1026

Lei Y, Tang L, Xie Y, Xianyu Y, Zhang L, Wang P, Hamada Y, Jiang K, Zheng W, Jiang (2017) Gold nanoclusters-assisted delivery of NGF siRNA for effective treatment of pancreatic cancer. Nat Commun 8:15130

Li F, Chen J, Hu X, He F, Bean E, Tsang DCW, Ok YS, Gao B (2020) Applications of carbonaceous adsorbents in the remediation of polycyclic aromatic hydrocarbon-contaminated sediments: a review. J Clean Prod 255:120263

Longo JPF, Lucci CM, Muehlmann LA, Azevedo RB (2018) Nanomedicine for cutaneous tumors–lessons since the successful treatment of the kaposi sarcoma. Nanomedicine 13:2957–2959

Longo JPF, Mussi S, Azevedo RB, Muehlmann LA (2020) Issues affecting nanomedicines on the way from the bench to the market. J Mater Chem B 8:10681–10685

Maiti D, Tong X, Mou X, Yang K (2019) Carbon-based nanomaterials for biomedical applications: a recent study. Front Pharmacol 9:Article 1401

Mauter MS, Elimelech M (2008) Environmental applications of carbon-based nanoaterials. Environ Sci Technol 42:5843–5859

Mody N, Tekade RK, Mehra NK, Chopdey P, Jain NK (2014) Dendrimer, liposomes, carbon nanotubes and PLGA nanoparticles: one platform assessment of drug delivery potential. AAPS Pharm Sci Tech 15:388–399

Monti S, Jose J, Sahajan A, Kalarikkal N, Thomas S (2019) Structure and dynamics of gold nanoparticles decorated with chitosan–gentamicin conjugates: ReaxFF molecular dynamics simulations to disclose drug delivery. Phys Chem Chem Phys 21:13099

Nayerossadat N, Maedeh T, Ali PA (2012) Viral and nonviral delivery systems for gene delivery. Adv Biomed Res 1:27

Onoue S, Yamada S, Chan HK (2014) Nanodrugs: pharmacokinetics and safety. Int J Nanomedicine 9:1025–1037

Paciotti GF, Myer L, Weinreich D, Goia D, Pavel N, McLaughlin RE, Tamarkin L (2004) Colloidal gold: a novel nanoparticle vector for tumor directed drug delivery. Drug Deliv 11:169–183

Park S-H, Oh S-G, Mun J-Y, Han S-S (2006) Loading of gold nanoparticles inside the DPPC bilayers of liposome and their effects on membrane fluidities. Coll Surf B 48:112–118

Patil M, Mehta DS, Guvva S (2008) Future impact of nanotechnology on medicine and dentistry. J Indian Soc Periodontol 12:34–40

Patra JK, Das G, Fraceto LF et al (2018) Nano based drug delivery systems: recent developments and future prospects. J Nanobiotechnol 16:71

Pérez-Ortiz M, Zapata-Urzúa C, Acosta GA, Álvarez-Lueje A, Albericio F, Kogana MJ (2017) Gold nanoparticles as an efficient drug delivery system for GLP-1 peptides. Colloids Surf B: Biointerfaces 158:25–32

Raliya R, Chadha TS, Hadad K, Biswas P (2016) Perspective on nanoparticle technology for biomedical use. Curr Pharm Des 22:2481–2490

Rastogi L, Kora AJ, Arunachalam J (2012) Highly stable, protein capped gold nanoparticles as effective drug delivery vehicles for amino-glycosidic antibiotics. Mater Sci Eng C 32:1571–1577

Ray PC, Yu H, Fu PP (2009) Toxicity and environmental risks of nanomaterials: challenges and future needs. J Environ Sci Health C Environ Carcinog Ecotoxicol Rev 27:1–35

Ren Q, Ga L, Lu Z, Ai J, Wang T (2020) Aptamer-functionalized nanomaterials for biological applications. Mater Chem Front 4:1569–1585

Rosi NL, Giljohann DA, Thaxton CS, Lytton-Jean AKR, Han MS, Mirkin CA (2006) Oligonucleotide-modified gold nanoparticles for intracellular gene regulation. Science 312:1027–1030

Salouti M, Heidari Z, Ahangari A, Zare S (2016) Enhanced delivery of gentamicin to infection foci due to staphylococcus aureus using gold nanorods. Drug Deliv 23:49–54

Shi J, Votruba AR, Farokhzad OC, Langer R (2010) Nanotechnology in drug delivery and tissue engineering: from discovery to applications. Nano Lett 10:3223–3230

Shin MD, Shukla S, Chung YH, Beiss V, Chan SK, Ortega-Rivera OA, Wirth DM, Chen A, Sack M, Pokorski JK, Steinmetz NF (2020) COVID-19 vaccine development and a potential nanomaterial path forward. Nat Nanotechnol 15:646–655

Shrivastava S, Dash D (2009) Applying nanotechnology to human health: revolution in biomedical sciences. J Nanotechnol. Article ID 184702

Singh A, Dhau JS, Kumar R (2021) Application of carbon-based nanomaterials for removal of hydrocarbons. In: Kumar R, Kumar R, Kaur G (eds) New frontiers of nanomaterials in environmental science. Springer, Singapore. https://doi.org/10.1007/978-981-15-9239-3_9

Singh A, Singh P, Kumar R, Kaushik A (2022) Exploring nanoselenium to tackle mutated SARS-CoV-2 for efficient COVID-19 management. Front Nanotechnol 41004729. https://doi.org/10.3389/fnano.2022.1004729

Sonkaria S, Ahn SH, Khare V (2012) Nanotechnology and its impact on food and nutrition: a review. Food Nutr Agric 4:8–18

Srinivasan M, Rajabi M, Mousa SA (2015) Multifunctional nanomaterials and their applications in drug delivery and cancer therapy. Nano 5:1690–1703

Sung Y, Kim S (2019) Recent advances in the development of gene delivery systems. Biomater Res 23:8

Sweet MJ, Chessher A, Singleton I (2012) Metal-based nanoparticles; size, function, and areas for advancement in applied microbiology. Adv Appl Microbiol 80:113–142

Taniguchi N, Arakawa C, Kobayashi T (1974) On the basic concept of nano-technology. In Proceedings of the International Conference on Production Engineering, Tokyo, Japan, 26–29 August 1974

Tiwari PM, Vig K, Dennis VA, Singh SR (2011) Functionalized gold nanoparticles and their biomedical applications. Nanomaterials (Basel) 1:31–63

Tiwari G, Tiwari R, Sriwastawa B, Bhati L, Pandey S, Pandey P, Bannerjee SK (2012) Drug delivery systems: an updated review. Int J Pharm Invest 2:2–11

Ventola CL (2017) Progress in nanomedicine: approved and investigational nanodrugs. P & T: Peer-Rev J Formulary Manag 42:742–755

Wang F, Wang Y-C, Dou S, Xiong M-H, Sun T-M, Wang J (2011) Doxorubicin-tethered responsive gold nanoparticles facilitate intracellular drug delivery for overcoming multidrug resistance in cancer cells. ACS Nano 5:3679–3692

Yang L, Zhang L, Webster TJ (2011) Nanobiomaterials: state of the art and future trends. Adv Eng Mater 6:B197–B217

Yaqoob AA, Ahmad H, Parveen T, Ahmad A, Oves M, Ismail IMI, Qari HA, Umar K, Nasir M, Ibrahim M (2020) Recent advances in metal decorated nanomaterials and their various biological applications: a review. Front Chem 8:Article 341

Zhang Y, Li M, Gao X, Chen Y, Liu T (2019) Nanotechnology in cancer diagnosis: progress, challenges and opportunities. J Hematol Oncol 12:137

Zhu H, Liu W, Cheng Z, Yao K, Yang Y, Xu B, Su G (2017) Targeted delivery of siRNA with pH-responsive hybrid gold nanostars for cancer treatment. Int J Mol Sci 18:2029

Chapter 10
New Perspective Application and Hazards of Nanomaterial in Aquatic Environment

Renuka Choudhary, Sunil Kumar, and Pooja Sethi

Abstract Nanomaterials (NMs) are known to have at least one dimension between about 1 and 100 nm. These nanomaterials can be categorized into nanoparticles (NPs), nanoplates, and nanofibers. In the last few decades, nanomaterial has got attention due to their wide applications in various sectors (such as environmental, pharmaceutical, personal care products, disease diagnosis, and treatment) based on their physicochemical properties. However, there are more concerns among environmentalists and scientists about the release of nanoparticles into the environment through different sources, which can cause environmental problems, human health hazard, and ecotoxicological issues. Aquatic biota is predominantly affected by the pollution caused by NPs. The fate, transport, and toxicity of NPs in aquatic ecosystem are affected by the transformation, which is dependent upon the initial NM properties, and its surrounding chemical and biological environment. Numerous research investigations have explored different types of NPs and their properties, applications, and hazards, but only a few have focused on the effect of NPs on aquatic ecosystem. In this chapter, ecotoxicological effects of NPs on aquatic ecosystem have been discussed based on the physiochemical properties by exploiting the latest research reports and future directions are highlighted.

Keywords Aquatic ecosystem · Ecotoxicological · Nanomaterial · Nanoparticles (NPs) · Physiochemical properties

R. Choudhary (✉) · S. Kumar
Department of Biotechnology, Maharishi Markandeshwar (Deemed to be University), Mullana (Ambala), Haryana, India

P. Sethi
Department of Chemistry, Maharishi Markandeshwar (Deemed to be University), Mullana (Ambala), Haryana, India

R. Kumar et al. (eds.), *Advanced Functional Nanoparticles "Boon or Bane" for Environment Remediation Applications*, Environmental Contamination Remediation and Management, https://doi.org/10.1007/978-3-031-24416-2_10

10.1 Introduction

Nanomaterials (NMs) are ubiquitously utilized in our day-to-day life in the form of professional, recreational, and daily-care items (Monteiro-Riviere et al. 2011). The most commonly used materials at the nanosized level are known as nanoparticles (NPs), which are retrieved from various sources and can be categorized into different types, such as incidental, engineered, and naturally released NPs (Turan et al. 2019). Over the last few decades, applications of nanomaterials in different sectors (industrial, consumer, and medical) have been increased ever than before. The diverse range of products and usage of nanoscale material include food packaging, water treatment, pharmaceutical, fuel cells, electronics, medical devices, and textiles (Aitken et al. 2006; Karnik et al. 2005). Aquatic ecosystem has become the major sufferer of environmental pollution due to the release of NPs (Keller et al. 2013). The fate of NPs entering the aquatic system is dependent upon various physicochemical properties like size, shape, charge, etc. and will be critical in bioavailability controls (Gatoo et al. 2014). The concern about their hazards and toxicity has stimulated studies to predict environmental concentration for their ecotoxicological effects on aquatic biota. The ecotoxicity of NPs is categorized into two main streams based on their distribution pathways in the aquatic ecosystem. First one is primary ecotoxicity, which arises due to the direct release of NPs into the environment without proper treatment. The second one is secondary ecotoxicity, which is an indirect discharge that occurs as a result of usage of NPs as safeguard method. In aquatic environment, NPs undergo transformation in different ways like physical, biological, and chemical and affect the toxicity of NPs (Lowry et al. 2012).The majority of the effects of toxicity were on aquatic plants, phytoplankton, algae, microalgae, cyanobacteria, and other marine organisms. Till now, scanty information has been made available about the toxicity of nanomaterials in aquatic system due to their properties (Holden et al. 2014). In this chapter, focused information about the nanomaterial sources, properties, applications, and hazards/toxicity in aquatic ecosystem have been elaborated.

10.2 Nanoparticles

The branch of science that produces material at the level of nano scale is known as nanotechnology. According to the International Organization for Standardization (Adams-Haduch et al. 2008), nanomaterial can be classified into three main groups: nanoparticles (all three dimensions between 1 and 100 nm), nanoplates (one dimension), and nanofibers (two dimensions). The particles produced by the technology are known as nanoparticles (NPs) and are defined as the wide class of materials that include particulate substances, which have at least one dimension less than 100 nm (Laurent et al. 2010). NPs are composed of three layers including surface layer, shell layer, and the core (Shin et al. 2016). These particles can be of variable size and shapes, which can further influence the physicochemical properties of them.

Owing to exceptional properties, NPs have numerous applications in multidisciplinary fields, such as pharmaceutical industry for drug delivery (Lee et al. 2011), personal, and health care (cosmetics) and also in chemical, biological, and gas sensing (Ullah et al. 2017).

10.3 Sources of NPs

In the environment, NPs get released from various sources, which can be categorized into different types such as incidental, engineered, and naturally released NPs (Turan et al. 2019). Incidental release of nanoparticle can be due to dust storms and cosmic dust, which include carbide, oxide, nitride, silicate, carbon, and organic-based NMs (Jeevanandam et al. 2018); volcanic eruption, which releases about 106 tons of NPs in the form of ash (Taylor 2002); forest fires, which contains micro- and nanosized particles (Sapkota et al. 2005); and evaporation of ocean and sea water, which leads to the formation of sea salt aerosols, which are considered as a type of NPs (Buseck and Posfai 1999). Automobile exhaust acts as a source of engineered NPs. Carbon NPs (20–130 nm and 20–60 nm size) get released from diesel and gasoline engines (Sioutas et al. 2005). In addition to this, cigarette smoke and building demolition also release engineered NPs. Smoke due to cigarette release about 100,0000 NPs in the 10–700 nm range (Ning et al. 2006) and smaller than 10 μm size NPs are released into the atmosphere during building demolition (Stefani et al. 2005). Numerous biomedical and healthcare products (antioxidants, antireflectants, and paints) are also responsible for acting as a source of nanoparticle release (Monteiro-Riviere et al. 2011). NPs are also produced naturally from organisms known as nano-organisms, which cover nanobacteria, viruses, fungi, algae, yeast, and insects (Jeevanandam et al. 2018). The scheme of different sources of NPs entering into the environment is shown in Fig. 10.1.

10.4 Classification of NPs

NPs are broadly classified into different types: size, morphology, physical and chemical properties, etc. Some of the well-known NPs are carbon-based, ceramic, metal, semiconductor, polymeric, and lipid-based nanoparticles (Khan et al. 2019).

10.4.1 Carbon-Based NPs

Carbon-based nanoparticles include carbon nanotubes (CNTs) and fullerenes. CNTs are graphene sheets rolled into a tube. These materials are used for the structural reinforcement because they are 100 times stronger than steel. CNTs are unique

Fig. 10.1 Different sources of NPs entering the environment. (Source: Donia and Carbone 2019)

as they are thermally conductive along the length and nonconductive across the tube. Fullerenes are the allotropes of carbon having a structure of hollow cage of 60 or more carbon atoms. For instance, C-60 is called Buckminsterfullerene. These have commercial applications due to their electrical conductivity, structure, high strength, and electron affinity (Saeed and Khan 2014, 2016).

10.4.2 Ceramic NPs

They are inorganic solids made up of carbides, carbonates, oxides, and phosphates, having high heat resistance and chemical inertness. They are synthesized via heat and successive cooling. They have applications in photocatalysis, photodegradation of dyes, drug delivery, and imaging (Thomas et al. 2015). They perform as a good drug delivery agent by controlling some of the characteristics of ceramic NPs like size, surface area, porosity, surface-to-volume ratio, etc. These NPs have been used effectively as a drug delivery system for a number of diseases like bacterial infections, glaucoma, cancer, etc.

10.4.3 Metal NPs

Metal-based NPs are synthesized from metal precursors, by chemical, electrochemical, or photochemical methods. These NPs have environmental and bioanalytical applications. These can also be used for detection and imaging of biomolecules (Dreaden et al. 2012; Khan et al. 2019).

10.4.4 Semiconductor NPs

These have properties between metals and nonmetals and therefore have numerous applications. They are used in photocatalysis, electronics devices, photo-optics, and water splitting applications (Sun et al. 2000). Examples of semiconductor NPs are GaN, GaP, InP, InAs, ZnO, ZnS, CdS, CdSe, CdTe, and silicon and germanium.

10.4.5 Polymeric NPs

They are organic-based and have shape like nanocapsular or nanospheres depending upon the method of preparation (Mansha et al. 2017). A nanosphere has a matrix-like morphology, whereas the nanocapsular has core-shell structure. In the former, the active compounds and the polymer are uniformly dispersed, whereas in the latter, the active compounds are surrounded by a polymer shell (Rao and Geckeler 2011). Applications of polymeric NPs include controlled release, ability to combine therapy and imaging, specific targeting, protection of drug molecules, and many more. They have applications in drug delivery and diagnostics. The drug deliveries with polymeric NPs are highly biodegradable and biocompatible (Abd Ellah and Abouelmagd 2017; Abouelmagd et al. 2016).

10.4.6 Lipid-Based NPs

These are generally spherical in shape with a diameter ranging from 10 to 100 nm. It consists of a solid core made of lipid and a matrix containing soluble lipophilic molecules. The external core of these NPs is stabilized by surfactants and emulsifiers. These NPs have application in the biomedical field as a drug carrier and delivery (Puri et al. 2009) and RNA release in cancer therapy (Gujrati et al. 2014).

10.5 Properties of NPs

The properties of NPs influencing their toxicity include their size, surface area, charge, shape or structure, solubility, and surface coating. NPs can be used in different sectors like medicine, electronics, and energy production due to their unique physicochemical (magnetic, optical, thermal, mechanical, and electrical) properties. However, toxicity of NPs can be influenced by their properties.

10.5.1 Nanoparticle Size and Surface Area

Size affects the toxicity of NPs as it leads to change in surface area. As the size of the material decreases, the surface area increases exponentially, which enhances the reactivity of nanoparticle surface. It has been reported that various biological mechanisms like endocytosis are dependent upon the size of the material for the modification of macromolecules structure and impacting the biological system. Lankvel et al. (2010) reported Ag NPs' size-specific tissue distribution and toxicity. Another study reported that smallest Ag NPs have the lowest aggregation potential in liver and gills (Scown et al. 2010; Gatoo et al. 2014). Gatoo et al. reported that the smaller the size of NPs, the more it will be able to form reactive oxygen species (ROS), which can affect biological system through DNA damage and inflammatory responses. Oral toxicity also gets enhanced by the size of nanoparticle. For instance, with decrease in size, the oral toxicity increases in copper NPs (Chen et al. 2006). Some studies on rodents also showed the effect of surface area with size in displaying the toxic manifestations (Holgate 2010). With increase in the surface area of NPs as the size decreases, oxidation and DNA damaging abilities also increase in a dose-dependent manner and were found to be much higher in comparison to the same dose of larger particles (Risom et al. 2005; Donaldson and Stone 2003).

10.5.2 Nanoparticle Shape

Numerous NPs such as carbon nanotubes, silica, allotropies, nickel, gold, and titanium have reported the shape-dependent toxicity, which can further influence the wrapping processes especially during phagocytosis and endocytosis (Petersen and Nelson 2010). The shape of the NPs such as spherical and nonspherical has been known to affect the toxicity (Hamilton et al. 2009). It was also reported that spherical NPs are relatively less toxic causing the biological consequences, in contrast to nonspherical shape NPs, which are disposed to flow through capillaries (Champion and Mitragotri 2006). Hsiao and Huang (Hsiao and Huang 2011) reported that TiO_2 spherical entities are less cytotoxic than fibers. Moreover, the uptake of gold spherical nanospheres was faster than nanorods (Chithrani et al. 2006). SWCNT spherical carbon fullerenes slowly block potassium (K^+) ion channels compared to rod shaped, which is two to three times more efficient (Park et al. 2003). Additionally, it was also reported that the length of fibers also influences the toxicity. In mice, TiO_2 fibers with a length of 5 mm are less toxic compared to 15 mm fibers (Hamilton et al. 2009).

10.5.3 Aggregation

Aggregation of NPs also plays a crucial role in influencing the toxicity of the latter. In aquatic environment, discarded NPs enter as aggregates and influence the toxicity and behavior. Aggregation was found to be dependent upon numerous environmental factors like pH, temperature, and ionic strength (Walters et al. 2016). It has been reported that due to elevated temperature, smaller aggregates formed and shows higher toxicity (Walters et al. 2013). It has been observed that carbon nanotubes induce the cytotoxicity in liver, spleen, and lungs due to aggregation for longer time (Yang et al. 2008). It also enhances the pulmonary interstitial fibrosis.

10.5.4 Concentration

Nanoparticles' concentrations inversely affect their toxicity. As concentrations of the NPs decrease, toxicity decreases or vice versa.

10.5.5 Surface Charge

The interaction of NPs with the biological system is dependent upon surface charge. It determines the particle dispersion and affects the ion and biomolecule adsorption (Powers et al. 2007). Studies reported that various biological activities like transmembrane permeability were altered by nanoparticle surface charge (Kohli and Alpar 2004). Goodman et al. reported that toxicity of gold NPs is dependent upon the charge present on it (Goodman et al. 2004). Anionic particles were found to be less toxic compared to cationic particles. Moreover, toxicity was caused to the system, due to hemolysis and platelet aggregation. Badawy et al. reported that the toxicity of silver NPs is dependent upon surface charge (El Badawy et al. 2011). Surface charge also affects the colloidal behavior of the NPs.

10.5.6 Composition

Toxicity of NPs varies with the diversity of NPs. It has been reported that nanosilver and nanocopper caused toxicity but TiO_2 showed zero toxicity in different models like zebrafish, daphnids, and algal species. This indicates that composition of NPs influences the toxicity issues (Griffitt et al. 2008).

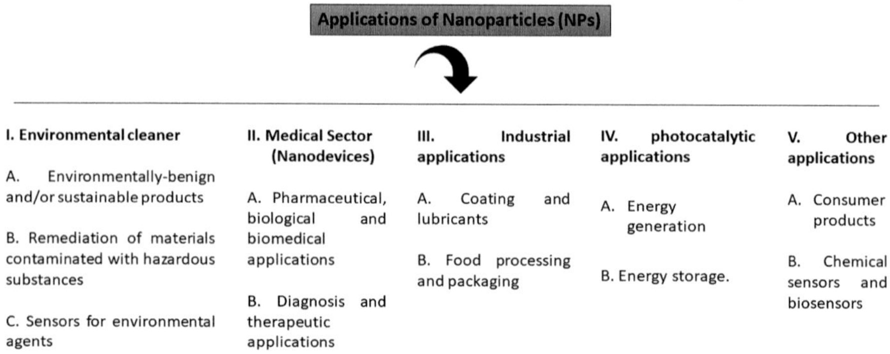

Applications of Nanoparticles (NPs)

I. Environmental cleaner	II. Medical Sector (Nanodevices)	III. Industrial applications	IV. photocatalytic applications	V. Other applications
A. Environmentally-benign and/or sustainable products	A. Pharmaceutical, biological and biomedical applications	A. Coating and lubricants	A. Energy generation	A. Consumer products
B. Remediation of materials contaminated with hazardous substances	B. Diagnosis and therapeutic applications	B. Food processing and packaging	B. Energy storage.	B. Chemical sensors and biosensors
C. Sensors for environmental agents				

Fig. 10.2 Schematic illustration of NP applications

10.6 Applications of NPs

NPs have wide applications in different sectors such as environment, drugs and medications, personal care products, disease diagnosis and treatment, food supplements, clothing sector, manufacturing, electronics sector, for harvesting energy, and in mechanical industries. The scheme of various applications of NPs in different sectors is shown in Fig. 10.2.

10.6.1 NPs Act as an Environmental Cleaner

In environmental sector, NPs get released into the aquatic environment from industrial and household applications. NPs play a crucial role in the water partitioning due to high surface-to-mass ratio. The contaminants can be absorbed to the surface of NPs and get trapped. This mechanism of interaction is dependent upon properties of NPs like size, morphology, and porosity. The majority of environmental applications of NPs fall into following categories: (A) environmentally benign and/or sustainable products (green chemistry or pollution prevention), (B) remediation of materials contaminated with hazardous substances, and (C) sensors for environmental agents (Tratnyek and Johnson 2006). In addition to the above, NPs can also be used for photodegradation due to high surface area and small size. NiO/ZnONPs are utilized for photodegradation as reported by Rogozea (Rogozea et al. 2017).

10.6.2 Medical Applications

NPs can be used in the development of various novel nanodevices. These nanodevices have various applications such as physical, biological, biomedical, and pharmaceutical (Loureiro et al. 2016; Martis et al. 2012; Nikalje 2015). Usage of NPs has increased the efficiency of drugs due to their ability to deliver drugs in optimum dosage range. The most commonly used NPs for biomedical applications are Fe_3O_4 or Fe_2O_3 (Ali et al. 2016). As NPs have some optical properties, they can also be used in photothermal therapeutic applications as well as in biological and cell imaging applications for achieving efficient contrast. Polyethylene oxide (PEO) and polylactic acid (PLA) NPs were found to be excellent systems as drug carriers (Calvo et al. 1997). The in vivo biomedical applications of NPs such as superparamagnetic iron oxide include MRI contrast enhancement, tissue repair and immunoassay, detoxification of biological fluid hyperthermia, drug delivery, and cell separation (Laurent et al. 2010). NPs also have applications in cancer diagnosis and therapy especially the semiconductor and metallic ones. Au NPs can be used for laser photothermal therapy of cancer as they can convert the strong absorbed light into localized heat efficiently. Additionally, NPs can also be used to inhibit the tumor growth (Khan et al. 2019).

10.6.3 Application in Energy Generation and Storage

NPs have photocatalytic applications and are used to generate energy (Avasare et al. 2015; Mueller and Nowack 2008). This energy is generated by photochemical and electrochemical water splitting, and electrochemical CO_2 reduction to fuel precursors, solar cells, and piezoelectric generators (Fang et al. 2013; Lei et al. 2015; Young et al. 2012). Besides energy generation, NPs can also be used to store energy at nanoscale level (Greeley and Markovic 2012; Liu et al. 2015).

10.6.4 Industrial Applications

NPs have applications in industries involving coating and lubricants due to their mechanical properties. Coating with NPs can improve the toughness and wear resistance. Examples are alumina, titania, and carbon-based NPs (Kot et al. 2016; Shao et al. 2012). Lubrication effects can be increased with NPs as they offer good sliding and delamination properties (Guo et al. 2014). In addition to this, NPs also have applications in food processing and packaging industry.

10.6.5 Other Applications

NPs have certain other applications in medical, commercial, and ecological sectors due to their physicochemical characteristics that induce mechanical, optical, and electrical properties (Dong et al. 2014). NPs also have crucial role in consumer products such as health fitness, electronics and computers, and in home and garden category. Some categories of NPs such as noble metals also have application in chemical sensors and biosensors (Unser et al. 2015).

10.7 Aquatic System

Aqua or water is one of the most essential chemical compounds on earth. No life can be assumed without aqua. Life present in water (organisms and plants) is called aquatic life (Field et al. 1998). Examples of aquatic ecosystems include lakes, ponds, bogs, rivers, estuaries, and the open ocean. There are mainly two types of aquatic ecosystems available: (1) marine ecosystems and (2) freshwater ecosystems (Alexander 1999). Ecosystems containing saltwater vary from coastal areas to the barren ocean bottom and cover almost 70% of the earth. In marine locale, the food chain instigates with plankton, microorganisms that require sunlight for energy and growth. In marine ecosystem, animal life ranges from microscopic zooplankton through fishes of all sizes to marine mammals, including seals, whales, and manatees. The organisms living in oceans adapt to the dissolved salts and to the level of salt content. Freshwater ecosystem has little or no salt content and supports its own aquatic ecosystems. It covers nearly about 1% of the earth and is further divided into lotic, lentic, and wetlands (Vaccari 2005). This type of ecosystem includes lakes, ponds, rivers, streams, wetlands, and even groundwater. These habitats are affected by altitude, temperature, and humidity. Freshwater ecosystem includes animal life like insects, amphibians, and fish. In freshwater system, innumerable varieties of plants, mollusks, worms, bacteria, and algae live. Additionally, a massive number of animals and birds use freshwater as food supply. Though aquatic ecosystem plays a pivotal role in human health and survival, it gets affected by human activities. For example, both fresh and marine ecosystem provide food, transportation, and recreation. Besides these, the ecosystems are threatened by pollution caused by urban runoff, agricultural wastes, the introduction (unintended or not) of exotic species to specific locale, overfishing, coastal development, global warming, and also by the release of NPs in the environment.

10.8 Nanoparticles in Aquatic Environment

The accumulation of NPs in aquatic environment begins once they are released into the water bodies (Iavicoli et al. 2014). Aquatic pollution was mainly caused by urban discharge and agricultural runoff, which releases NPs into the environment in different forms. In addition to this, through atmospheric release and water infiltration to the soil, NPs especially engineered are able to enter the aquatic environment (Weinberg et al. 2011). NPs such as personal care products, wastewater, and paints enter intentionally into the aquatic environment and affect the ecosystem

(Fig. 10.3). Entered nanoparticle can impact the aquatic system through numerous routes as explained in Fig. 10.4.

Firstly, through transportation, contaminants including NPs were known to be transported via aquatic ecosystem. For example, filter-feeders may consume NPs attached to algae and transfer to further tropic levels. Secondly, it can be by forming nanoparticle aggregates in the aqueous phase and posing a threat to benthic organisms as these aggregates further gets settled into sediment. These aggregates also reduce the direct uptake of NPs by organisms. Sediments and soil have large surface area, and NPs are generally associated with it (Klaine et al. 2008). Another route of entry into aquatic environment is through direct passage across the gill. It has been

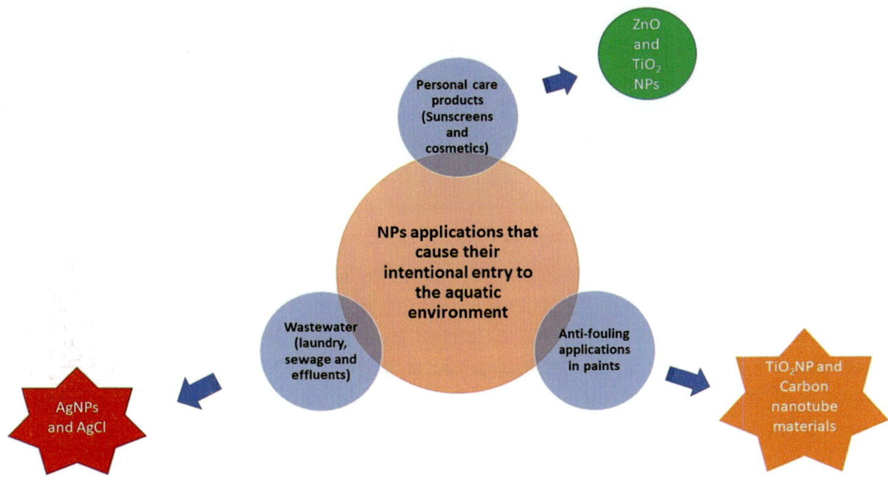

Fig. 10.3 Sources of intentional entry of NPs in aquatic environment

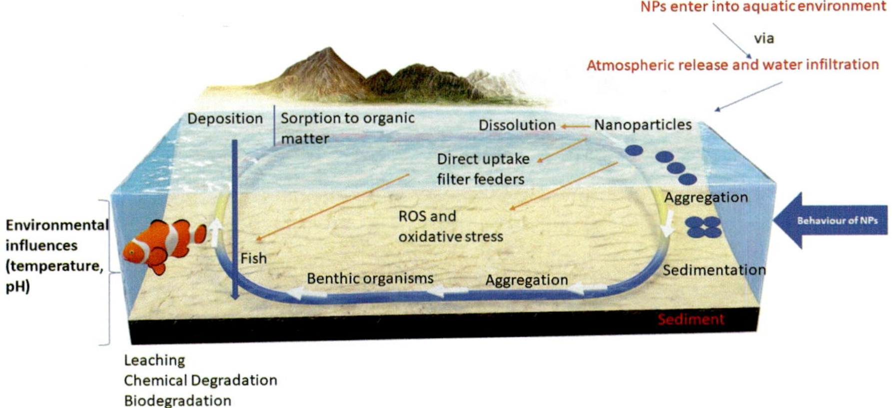

Fig. 10.4 Nanoparticles and aquatic environment

reported that NPs taken via this process accumulate in the hepatopancreas (Lee 2001). Environmental factors such as pH, temperature, and presence of organic matter also affect the release of NPs into the aquatic system. For instance, aggregation is affected by pH (due to change in surface charge) and temperature. At higher range of pH, aggregation and disaggregation with larger aggregate radius was reported (Gilbert et al. 2007). Studies reported that at higher temperature, smaller aggregates were formed indicating higher toxicity (Walters et al. 2013). Dissolution rates of AgNPs get affected by temperature (Liu and Hurt 2010).

10.9 Transformation

NPs undergo transformation once they reach the aquatic environment due to the environmental factors. Therefore, fate of nanomaterial is dependent not only upon their physical and chemical properties but also on environmental factors like temperature, pH, organic matter concentration, and salinity. The NP's persistence, bioavailability, reactivity, and toxicity get influenced by transformation, which can be physical, chemical, and biological. The novel properties exhibited by NPs have raised the concern all around the world due to their uncertain effects. So, citizens have demanded the increase in research activities, which can correlate the nanoparticle properties with living organisms and their behavior in the environment. Lack of information about the types and extent of transformation by the nanoparticle is there, due to which we are unable to understand the impact of it. For this, understanding of toxicological effects from acute and chronic NP exposure and their potential route is required.

In biological and environmental system, redox reactions, aggregation, dissolution, and reactions with biological macromolecules may occur as transformation, which can influence the transportation, fate, and toxicity of NPs (Lowry et al. 2012). Some transformations incline the potential of toxicity, while some decrease the toxicity effects (Fabrega et al. 2009). Studies also reported that some types of transformation have been shown to limit the nanoparticle persistence in the environment (Bian et al. 2011). The type of transformation with its rate and extent depends upon the properties of nanoparticle, its coating, and its surrounding environment (chemical or biological). The chemical, physical, and biological were the different forms of NP transformation occurring in natural systems.

Chemical transformation involves transfer of electrons using naturally occurring coupled processes (oxidation and reduction) to and from chemical moieties. Metal NPs such as silver and iron undergo oxidation and reduction by interacting in aquatic environment. Sulfidation (passivation process) of NPs, which includes Ag, ZnO, and CuO, frequently occurs under several environmental conditions, which leads to inert NP surfaces and makes them reactive and toxic to microorganisms (Kraas et al. 2017). For instance, sulfidation of Ag NPs forms the core–shell AgO–Ag_2S structures (Thalmann et al. 2016). Another process known as dissolution occurs, which is affected by nanoparticle chemistry. The dissolution of nanoparticles is dependent upon particle inherent factors and environmental parameters. Under aerobic conditions, oxide layer can be formed around the particle like in the case of AgNP, Ag_2O layer formed, which further releases Ag^+ ions (Elzey and Grassian 2010).

Physical transformation involves aggregation of NPs, which reduces the surface area to volume effects on nanomaterial reactivity. In aquatic environment, NPs after their release interact with numerous organic and inorganic compounds and affect their aggregation, which is known as colloidal stability (Sani-Kast et al. 2017). This aggregation can be of two types: homo-aggregation and hetero-aggregation. When the interaction between same NP occurs, it is known as homo-aggregation, whereas when the concentration of another particle in the environment increases and it interacts with NPs, the aggregation occurred as hetero-aggregation. As the aggregation decreases the surface area, the reactivity of the particle also gets declined. This implies that the toxicity of the particle also gets decreased with decrease in reactivity, especially in the case of surface-mediated reaction (ROS generation and dissolution).

Biological transformation involves living tissues and environmental media (Lowry et al. 2012). This type of transformation changes the behavior of nanomaterials, which further affects the bioavailability, transport, and toxicity. Nanomaterial behavior includes its surface charge, aggregation state, and reactivity. The OH radicals produced from horseradish peroxidase can lead to the oxidation and carboxylation of carbon nanotubes (Allen et al. 2008). The surface charge gets increased, and hydrophobicity decreases due to this oxidation of carbon nanotubes, which further affected the toxicity of nanotubes (Kang et al. 2008).

Another category of transformation is interaction with macromolecules, which occur when proteins and natural organic matter interact with NPs. The different forms are explained in Fig. 10.5.

10.10 Toxicity of Nanomaterial in Aquatic Environment

The use of nanomaterial in different sectors has resulted in contamination in aquatic environment. The majority of toxicity of NPs was on aquatic plants, phytoplankton, algae, microalgae, cyanobacteria, and other marine organisms.

10.10.1 Effect on Aquatic Plants

Plants play a crucial role in fate and toxicity of NPs and are considered as one of the important components of aquatic ecosystem. Scanty information is available on the effect of NPs on aquatic plants in comparison to toxic effect on microorganisms and animals, which is well documented (Asharani et al. 2008; Kim et al. 2011; Turner et al. 2012). For toxicity testing, Lemnaceae aquatic plants are mostly used a model

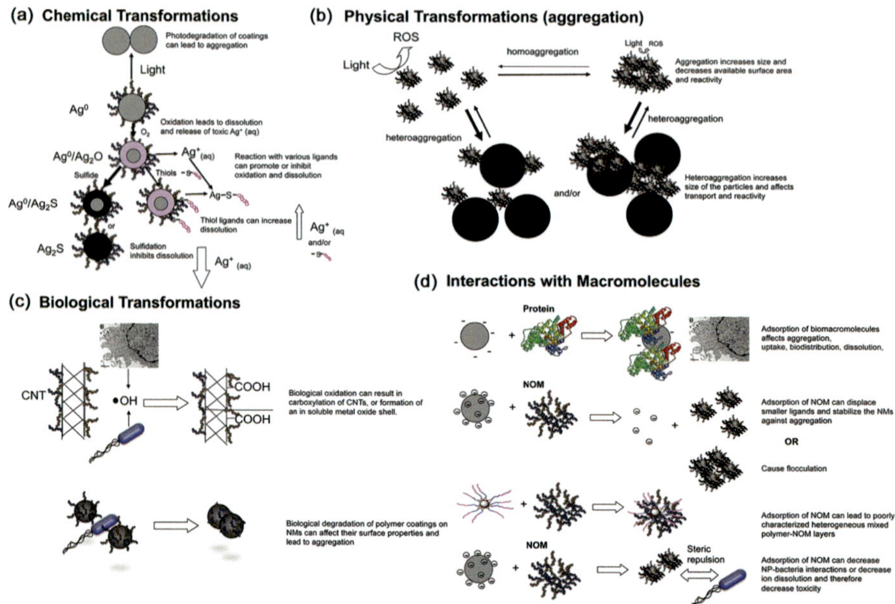

Fig. 10.5 Nanoparticle transformations in aquatic environment in different forms: (**a**) chemical transformations of NPs and the potential impacts on their behavior and effects in the environment. (**b**) Physical transformations (aggregation and hetero aggregation) of NPs. (**c**) Biologically mediated transformations of NPs and their coatings and the subsequent impact on fate, transport, and effects. (**d**) Interactions of NPs with macromolecules such as proteins and natural organic matter. (Source: Lowry et al. 2012)

due to their sensitivity to pollutants and rapid growth. It has been reported that Ag NPs inhibit the growth of *Lemna minor* and *Lemna gibba* due to the uptake of silver ions after the exposure of Ag NP with 10 mg/L concentration (Gubbins et al. 2011). The cell-permeable indicator revealed the high production of ROS (340% compared to the control) when plants are exposed to 10 mg/L of Ag NPs. Jiang et al. also reported the increase in oxidative stress in *Spirodela polyrhiza* plants when tested for the comparative activity of three different types of silver nanoparticles, indicating activation of enzymatic antioxidant system after the exposure of Ag NPs (Jiang et al. 2014). Jiang et al. observed decline in *S. polyrhiza* plant tissue contents due to exposure of $AgNO_3$ (decreases chlorophyll a and phosphate) and Ag NPs (DW decreases) (Jiang et al. 2012). In addition to Ag NPs, titanium dioxide nanoparticles (TiO2-NPs) and nickel oxide nanoparticles (NiO-NPs) were also used for wide range of purposes worldwide. Movafeghi et al. studied the effect of TiO_2-NPs on *S. polyrhiza* aquatic plant and reported the damage on plant defense system due to the ROS generation (Movafeghi et al. 2018). It was also observed that growth parameters decreased as the concentration of TiO_2-NPs increased and antioxidant enzyme activities change. Similarly, Oukarroum et al. reported the cellular oxidative stress induction in aquatic plant *L. gibba* due to the strong increase in reactive oxygen species (ROS) formation by NiO-NPs at high concentration (1000 mg/L) and alteration of the photosynthetic electron transport performance (Oukarroum et al. 2013). Some scientists have examined the toxicity of TiO_2-NPs on algae and reported that retardation of algal growth due to oxidative enzyme production after 72 h (Li et al. 2015). In addition to this, GO-NP toxicity to freshwater algae was also evaluated, and results indicated that GO-NPs lead to membrane damage and nutrient depletion due to the induction of oxidative stress.

10.10.2 Effect on Phytoplankton

Phytoplankton acts as a key component of aquatic webs (oceans, seas, and freshwater ecosystem). These include single-celled biotic microscopic photosynthetic organisms, which create the organic compounds from carbon dioxide present in aquatic environment. About 25,000 species of phytoplankton were known, including eubacterial and eukaryotic species belonging to eight phyla Maranon. These include crustaceans, mollusks, bacteria, algae, coelenterates, and protozoans. With the increase use of NPs, it is crucial to evaluate the effect of NPs on phytoplankton, which is the primary producer of aquatic food web. It has been reported that growth rate of phytoplankton gets suppressed by NPs. They also decline the equilibrium densities of phytoplankton as well as zooplankton, as the contact with them and phytoplankton increases. Studies revealed that Ag-NPs reduce the photosynthetic efficiency of phytoplankton at concentration higher than 500 mg/L (Baptista et al. 2015). TiO_2, which is one of the most commonly used nanomaterial, poses a notable threat to aquatic ecosystem as it may reach to maximum concentration at the surface of water. TiO_2 can act as a photocatalyst by absorbing photons and can directly oxidize the adsorbed species. With similar mechanism, TiO_2 can get absorbed and diffused on the surface of phytoplankton and leads to the oxidation of phytoplankton cell wall. It has been reported that with increase in concentration of TiO_2-NPs, ROS in seawater also increases under low levels of ultraviolet radiations, which further inclines the overall oxidative stress and leads to decreased resiliency of marine ecosystems. Additionally, interactions between TiO_2 and plankton can also attack organic compounds (Miller et al. 2010).

10.10.3 Effect on Aquatic Microbes

Dominant toxicity was observed in aquatic microorganism especially due to engineered nanomaterials such as TiO_2-NPs and Ag-NPs (Jahan et al. 2017). For example, one of the studies reported the Ag-NP toxicity to *E. coli*. The toxicity was due to the pH dependence dissolution of Ag^+, which further binds with the cell cytoplasm and leads to cell death due to oxidative stress (Xiu et al. 2012; Levard et al. 2012). Similarly, TiO_2-NP toxicity to *E. coli* was studied in dark and cellular leakage of K^+, and Mg^{2+} was reported in *E. coli* due to depolarization (Sohm et al. 2015).

Another study reported the fate and behavior of GO-NPs into the aquatic system. The toxicity of GO-NPs to aquatic organisms such as aquatic plants, vertebrates, and bacteria was influenced by the solution chemistry, presence of colloidal, and biocolloidal particles (Zhao et al. 2014). The impact of ZnO-NPs to ammonia-oxidizing bacteria, *Nitrosomonas europaea*, was also evaluated, and it was found that ZnO-NPs exerted dose-dependent impairment to the bacterial cells. For example, the cells were completely restored in a 12 h recovery incubation after the exposure of 50 mg/L ZnO-NPs in terms of cell density, membrane integrity, and the nitrite production rate (Wu et al. 2017).

10.10.4 Effect on Cyanobacteria

In aquatic environments, cyanobacteria act as a crucial part due to their role as primary producers, biofilms, and mats (benthic cyanobacteria) formation. They are also important for ecotoxicological assessment. Higher concern of cyanobacteria was due to the toxins produced by it, which causes health risk (Domingo et al. 2019). Cyanobacteria blooms were found in terrestrial, freshwater, and marine habitats. NPs affect their viability depending upon their concentration. Ag-NPs inhibited the growth of *Cyanobacteria synechococcus* species after 72 hour of exposure at 0.5–5 μM concentration. As the concentration of NPs increased to 10 μM, complete inhibition of growth was observed for it (Burchardt et al. 2012). In another study done on *Microcystis aeruginosa*, damaged and shrunk cell walls were observed using scanning and transmission electron microscopy, as a result of toxicity of Ag-NPs. At high concentration (1 mg/L) of Ag-NPs, the efficiency of growth inhibition was 98.7% for it (Duong et al. 2016). In addition to this, comparison of Ag-NPs and ionic silver in *Synechococcus leopoliensis* revealed that only ionic silver showed negative effects on viability, thereby indicating its higher toxicity (Taylor et al. 2016).

10.10.5 Effect on Algae and Microalgae

Algae play a crucial role in functioning of water ecosystem due to its contribution in global biomass productivity. Approximately, there are 1500 genera and 17,400 species of algae. They are also called as primary producers in aquatic system food chain. The biological activity of aquatic system is dependent upon it as they act as the source of carbon and energy (chemical) for other organisms. They are particularly sensitive to a wide range of pollutants, which can cause alteration in their species composition and densities and may break the balance of the whole ecosystem. Due to this reason, algae can be used as model organisms in ecotoxicological studies (USEPA UEPA 1996; OECD 1984). Studies on NPs evaluating their impact on protozoa, algae, and bacteria revealed that CuO-NP and ZnO-NPs have highest toxicity, and algae was found to be the most sensitive organism (Aruoja et al. 2015). It has also been reported that different algal species showed varied levels of toxicity due to ZnO-NPs indicating that toxic effect of nanoparticle is species dependent. For instance, *Phaeodactylum tricornutum* (a less silicified marine diatom) was found to be less sensitive in contrast to *Thalassiosira pseudonana* (more silicified diatom) to ZnO-NPs (Peng et al. 2011). In addition to it, ZnO-NP toxicity is also dependent upon size, shape, and production of ROS, that is, physicochemical properties (Peng et al. 2011; Wong et al. 2010). Study on *T. pseudonana* revealed that dissolved zinc ions released from ZnO-NPs also have been demonstrated as the reason of toxicity in spite of NPs themselves. Miller et al. in their study on marine algae *T. pseudonana*, *Skeletonema marinoi*, *Dunaliella tertiolecta*, and *Isochrysis galbana* have demonstrated the effects of free zinc ions coupled with NPs (Miller et al. 2010). Studies on the effect of Ag-NPs and ionic silver on biofilms dominated by *Diatomophyceae*, *Chlorophyceae*, and *Cyanophyceae* revealed that these NPs modified the structure of biofilm with decline in biomass and chlorophyll concentration. Dash and his co-workers also reported the decrease in chlorophyll content after the exposure of Ag-NPs in *Pithophora oedogonia* and *Chara vulgaris* (Das et al. 2012). Both Ag and $AgNO_3$-NPs decrease the photosynthetic yield and alter the cell morphology in *E. gracilis*, a green algae without any cell wall (Li et al. 2015). In addition to the effect on photosynthetic activity, Ag-NPs can also affect metabolic activity and photosynthetic gene expression levels (He et al. 2012). Ag-NPs were found to affect diatom *S. costatum* because of ROS formation and photosynthesis inhibition (Huang et al. 2016). These NPs (both Ag NPs and Ag ions) result into decrease of photochemical capacity.

10.10.6 Effect on Other Marine Organisms

NPs like ZnO-NPs and Ag-NPs can affect numerous other marine organisms, which include mollusks, crustaceans, and fish. Benthic invertebrates (filter feeders and suspension feeders) have a high risk of exposure to NPs, which aggregate and deposit on surface of seawater (Keller et al. 2010). Study on exposure of ZnO-NPs (1–10 mg L^{-1}) for 4 days to a coastal marine mussel *Mytilus galloprovincialis* have reported the accumulation of zinc and pseudofeces containing zinc (biotransformed ZnO-NPs), which was further consumed by other marine benthic suspension feeders (clams), deposit feeders (polychaete worms, amphipods), and grazers (sea urchins). In this way, biotransformed NPs could affect their fate and transport in the marine ecosystem (Montes et al. 2012).

Filter feeders, such as copepod, amphipods, and brine shrimp (small crustaceans), were more susceptible to many pollutants compared with other pelagic marine animals. Depending upon the physicochemical nature of the NPs, such as size, dissolution, and aggregation, exposure of zooplanktons to NPs especially (ZnO-NPs) has been associated with numerous harmful effects. For instance, higher toxicity was observed with smaller primary particle size of ZnO-NPs than those with a large particle size in brine shrimp *Artemia salina* (Ates et al. 2013). Effects of NPs on marine crustaceans are also species dependent. Brine shrimps seems to be less sensitive to ZnO-NPs than copepods and amphipods (Fabrega et al. 2012; Larner et al. 2012).

Through ingestion, gut epithelial cells, or via gill, are the major routes of NP uptake by the fish. Toxicity of NPs in freshwater fish was well documented in comparison to marine. In zebrafish adults, different adverse effects of ZnO-NPs toxicity have been observed such as hatching rate of embryos, malformation of larvae, and inhibition of antioxidant activity (Zhu et al. 2009; Xiong et al. 2011).

10.11 Future Recommendations

NPs have become an unavoidable part of our daily life, especially in the last few years. Due to its widespread applications, it can be found in all the sectors related to routine life such as cosmetic, pharma, clothing, food industry, and many more. In this chapter, ecotoxicological effect of NPs with respect to its physiochemical properties has been elaborated. It is recommended that properties of NPs such as composition, size, shape, charge, and aggregation should be taken in consideration in order to optimize the specific characteristic product for application. In addition to this, environmental factors affecting the properties of NPs should also be taken into account, which can enhance the toxicity of NPs in the aquatic system and in future can also affect human beings through food chain.

Questions

1. What type of nanoparticles are 100 times stronger than steel and can be used for the structural reinforcement?
   - A. Carbon-based nanoparticles.
   - B. Semiconductor nanoparticles.
   - C. Ceramic nanoparticles.
   - D. None of the above.
2. How elevated temperature affects the aggregation property in relevance to toxicity of nanoparticles?
3. How does size and surface area affect the toxicity of nanoparticles?
4. What are the different categories of environmental applications of nanoparticles?
5. What are the different sources of nanoparticles release into the environment?
6. What is transformation of nanoparticles? What are the different forms of nanoparticle transformation occurring in natural systems?
7. What are the ecotoxicological effects of NPs on phytoplankton, which are the primary producers of aquatic food web.

Answers

1. (A) Carbon-based nanoparticles
2. With increase in temperature, smaller aggregates will be formed, which further increases toxicity of nanoparticles.
3. Size affects the toxicity of NPs as it leads to change in surface area. As the size of the material decreases, the surface area increases exponentially, which enhances the reactivity of nanoparticle surface. For example, with decrease in size of copper nanoparticles, the oral toxicity also increases.
4. Three categories:

    A. Environmentally benign and/or sustainable products (e.g., green chemistry or pollution prevention).
    B. Remediation of materials contaminated with hazardous substances.
    C. Sensors for environmental agents.

5. There are three main sources, which include: incidental, engineered, and naturally released nanoparticles.
6. The process of converting the nanoparticles into different reactive forms has potential impacts on their behavior, and effects in the environment are called as transformation of nanoparticles. The chemical, physical, and biological were different forms of nanoparticle transformation occurring in natural systems.
7. NPs affect the phytoplankton in numerous ways, such as the growth rate of phytoplankton gets suppressed by NPs. They also decline the equilibrium densities of phytoplankton as well as zooplankton, as the contact with NPs and phytoplankton increases. Photosynthetic efficiency of them was also reduced by the toxicity of NPs.

References

Abd Ellah NH, Abouelmagd SA (2017) Surface functionalization of polymeric nanoparticles for tumor drug delivery: approaches and challenges. Expert Opin Drug Deliv 14(2):201–214. https://doi.org/10.1080/17425247.2016.1213238

Abouelmagd SA, Meng F, Kim BK, Hyun H, Yeo Y (2016) Tannic acid-mediated surface functionalization of polymeric nanoparticles. ACS Biomater Sci Eng 2(12):2294–2303. https://doi.org/10.1021/acsbiomaterials.6b00497

Adams-Haduch JM, Paterson DL, Sidjabat HE, Pasculle AW, Potoski BA, Muto CA et al (2008) Genetic basis of multidrug resistance in Acinetobacter baumannii clinical isolates at a tertiary medical center in Pennsylvania. Antimicrob Agents Chemother 52(11):3837–3843

Aitken RJ, Chaudhry MQ, Boxall AB, Hull M (2006) Manufacture and use of nanomaterials: current status in the UK and global trends. Occup Med (Lond) 56(5):300–306. doi:56/5/300 [pii]10.1093/occmed/kql051

Alexander DE (1999) Encyclopedia of environmental science. Springer. ISBN 0-412-74050-8

Ali A, Zafar H, Zia M, Ul Haq I, Phull AR, Ali JS et al (2016) Synthesis, characterization, applications, and challenges of iron oxide nanoparticles. Nanotechnol Sci Appl 9:49–67

Allen BL, Kichambare PD, Gou P, Vlasova II, Kapralov AA, Konduru N et al (2008) Biodegradation of single-walled carbon nanotubes through enzymatic catalysis 8(11):3899–3903. https://doi.org/10.1021/nl802315h

Aruoja V, Pokhrel S, Sihtma¨e M, Mortimer M, Ma¨dler L, Kahru A (2015) Toxicity of 12 metal-based nanoparticles to algae, bacteria and protozoa. Environ Sci Nano 630(2)

Asharani PV, Lian WY, Gong Z, Valiyaveettil S (2008) Toxicity of silver nanoparticles in zebrafish models. Nanotechnology 19(25):255102. doi:S0957-4484(08)71001-8 [pii]10.1088/0957-4484/19/25/255102

Ates M, Daniels J, Arslan Z, Farah IO, Rivera HF (2013) Comparative evaluation of impact of Zn and ZnO nanoparticles on brine shrimp (Artemia salina) larvae: effects of particle size and solubility on toxicity. Environ Sci Process Impacts 15(1):225–233. https://doi.org/10.1039/c2em30540b

Avasare V, Zhang Z, Avasare D, Khan I, Qurashi A (2015) Room-temperature synthesis of TiO2 nanospheres and their solar driven photoelectrochemical hydrogen production. Int J Energy Res 39:1714–1719

Baptista MS, Miller RJ, Halewood ER, Hanna SK, Almeida CM, Vasconcelos VM et al (2015) Impacts of silver nanoparticles on a natural estuarine plankton community. Environ Sci Technol 49(21):12968–12974. https://doi.org/10.1021/acs.est.5b03285

Bian SW, Mudunkotuwa IA, Rupasinghe T, Grassian VH (2011) Aggregation and dissolution of 4 nm ZnO nanoparticles in aqueousenvironments: influence of ph, ionic strength, size, and adsorption ofhumic acid. Langmuir 27:6059–6068

Burchardt AD, Carvalho RN, Valente A, Nativo P, Gilliland D, Garcia CP et al (2012) Effects of silver nanoparticles in diatom Thalassiosira pseudonana and cyanobacterium Synechococcus sp. Environ Sci Technol 46(20):11336–11344. https://doi.org/10.1021/es300989e

Buseck PR, Posfai M (1999) Airborne minerals and related aerosol particles: effects on climate and the environment. Proc Natl Acad Sci U S A 96(7):3372–3379. https://doi.org/10.1073/pnas.96.7.3372

Calvo P, Remuoon-Lopez C, Vila-Jato JL, Alonso MJ (1997) Novel hydrophilic chitosan-polyethylene oxide nanoparticles as protein carriers. J Appl Polym Sci 63:125–132

Champion JA, Mitragotri S (2006) Role of target geometry in phagocytosis. Proc Natl Acad Sci USA 103(13):4930–4934. doi:0600997103 [pii]10.1073/pnas.0600997103

Chen Z, Meng H, Xing G, Chen C, Zhao Y, Jia G et al (2006) Acute toxicological effects of copper nanoparticles in vivo. Toxicol Lett 163(2):109–120. doi:S0378-4274(05)00317-6 [pii]10.1016/j.toxlet.2005.10.003

Chithrani BD, Ghazani AA, Chan WC (2006) Determining the size and shape dependence of gold nanoparticle uptake into mammalian cells. Nano Lett 6(4):662–668. https://doi.org/10.1021/nl052396o

Das P, Williams CJ, Fulthorpe RR, Hoque ME, Metcalfe CD, Xenopoulos MA (2012) Changes in bacterial community structure after exposure to silver nanoparticles in natural waters. Environ Sci Technol 46(16):9120–9128. https://doi.org/10.1021/es3019918

Domingo G, Bracale M, Vannini C (2019). Phytotoxicity of silver nanoparticles to aquatic plants, algae, and microorganisms

Donaldson K, Stone V (2003) Current hypotheses on the mechanisms of toxicity of ultrafine particles. Ann Ist Super Sanita 39(3):405–410

Dong H, Wen B, Melnik R (2014) Relative importance of grain boundaries and size effects in thermal conductivity of nanocrystalline materials. Sci Rep 4:7037. doi:srep07037 [pii]10.1038/srep07037

Donia DT, Carbone M (2019) Fate of the nanoparticles in environmental cycles. Int J Environ Sci Technol 16:583–600

Dreaden EC, Alkilany AM, Huang X, Murphy CJ, El-Sayed MA (2012) The golden age: gold nanoparticles for biomedicine. Chem Soc Rev 41(7):2740–2779. https://doi.org/10.1039/c1cs15237h

Duong TT, Le TS, Huong Tran TT, Nguyen TK, Ho CT, Dao TH, Dang DK, Ha PT (2016) Inhibition effect of engineered silver nanoparticles to bloom forming cyanobacteria. Adv Nat Sci Nanosci Nanotechnol 7(3)

El Badawy AM, Silva RG, Morris B, Scheckel KG, Suidan MT, Tolaymat TM (2011) Surface charge-dependent toxicity of silver nanoparticles. Environ Sci Technol 45(1):283–287. https://doi.org/10.1021/es1034188

Elzey S, Grassian V (2010) Agglomeration, isolation and dissolution of commercially manufactured silver nanoparticles in aqueous environments. J Nanopart Res 12(5):1945–1958

Fabrega J, Fawcett SR, Renshaw JC, Lead JR (2009) Silver nanoparticle impact on bacterial growth: effect of pH, concentration, and organic matter. Environ Sci Technol 43(19):7285–7290. https://doi.org/10.1021/es803259g

Fabrega J, Tantra R, Amer A, Stolpe B, Tomkins J, Fry T et al (2012) Sequestration of zinc from zinc oxide nanoparticles and life cycle effects in the sediment dweller amphipod Corophium volutator. Environ Sci Technol 46(2):1128–1135. https://doi.org/10.1021/es202570g

Fang XQ, Liu JX, Gupta V (2013) Fundamental formulations and recent achievements in piezoelectric nano-structures: a review. Nanoscale 5(5):1716–1726. https://doi.org/10.1039/c2nr33531j

Field CB, Behrenfeld MJ, Randerson JT, Falkowski P (1998) Primary production of the biosphere: integrating terrestrial and oceanic components. Science 281(5374):237–240. https://doi.org/10.1126/science.281.5374.237

Gatoo MA, Naseem S, Arfat MY, Dar AM, Qasim K, Zubair S (2014) Physicochemical properties of nanomaterials: implication in associated toxic manifestations. Biomed Res Int 2014:498420. https://doi.org/10.1155/2014/498420

Gilbert B, Lu G, Kim CS (2007) Stable cluster formation in aqueous suspensions of iron oxyhydroxide nanoparticles. J Colloid Interface Sci 313(1):152–159. doi:S0021-9797(07)00479-1 [pii]10.1016/j.jcis.2007.04.038

Goodman CM, McCusker CD, Yilmaz T, Rotello VM (2004) Toxicity of gold nanoparticles functionalized with cationic and anionic side chains. Bioconjug Chem 15(4):897–900. https://doi.org/10.1021/bc049951i

Greeley J, Markovic NM (2012) The road from animal electricity to green energy: combining experiment and theory in electrocatalysis. Energy Environ Sci:5

Griffitt RJ, Luo J, Gao J, Bonzongo JC, Barber DS (2008) Effects of particle composition and species on toxicity of metallic nanomaterials in aquatic organisms. Environ Toxicol Chem 27(9):1972–1978. doi:08-002 [pii]10.1897/08-002.1

Gubbins EJ, Batty LC, Lead JR (2011) Phytotoxicity of silver nanoparticles to Lemna minor L. Environ Pollut 159(6):1551–1559. doi:S0269-7491(11)00131-X [pii]10.1016/j.envpol.2011.03.002

Gujrati M, Malamas A, Shin T, Jin E, Sun Y, Lu ZR (2014) Multifunctional cationic lipid-based nanoparticles facilitate endosomal escape and reduction-triggered cytosolic siRNA release. Mol Pharm 11(8):2734–2744. https://doi.org/10.1021/mp400787s

Guo D, Xie G, Luo J (2014) Mechanical properties of nanoparticles: basics and applications. J Phys D Appl Phys 47(13001)

Hamilton RF, Wu N, Porter D, Buford M, Wolfarth M, Holian A (2009) Particle length-dependent titanium dioxide nanomaterials toxicity and bioactivity. Part Fibre Toxicol 6:35. doi:1743-8977-6-35 [pii]10.1186/1743-8977-6-35

He D, Dorantes-Aranda JJ, Waite TD (2012) Silver nanoparticle-algae interactions: oxidative dissolution, reactive oxygen species generation and synergistic toxic effects. Environ Sci Technol 46(16):8731–8738. https://doi.org/10.1021/es300588a

Holden PA, Klaessig F, Turco RF, Priester JH, Rico CM, Avila-Arias H et al (2014) Evaluation of exposure concentrations used in assessing manufactured nanomaterial environmental hazards: are they relevant? Environ Sci Technol 48(18):10541–10551. https://doi.org/10.1021/es502440s

Holgate ST (2010) Exposure, uptake, distribution and toxicity of nanomaterials in humans. J Biomed Nanotechnol 6(1):1–19. https://doi.org/10.1166/jbn.2010.1098

Hsiao IL, Huang YJ (2011) Effects of various physicochemical characteristics on the toxicities of ZnO and TiO nanoparticles toward human lung epithelial cells. Sci Total Environ 409(7):1219–1228. doi:S0048-9697(10)01364-1 [pii]10.1016/j.scitotenv.2010.12.033

Huang J, Cheng J, Yi J (2016) Impact of silver nanoparticles on marine diatom Skeletonema costatum. J Appl Toxicol 36(10):1343–1354. https://doi.org/10.1002/jat.3325

Iavicoli I, Leso V, Ricciardi W, Hodson LL, Hoover MD (2014) Opportunities and challenges of nanotechnology in the green economy. Environ Health 13:78. doi:1476-069X-13-78 [pii]10.1186/1476-069X-13-78

Jahan S, Yusoff IB, Alias YB, Bakar A (2017) Reviews of the toxicity behavior of five potential engineered nanomaterials (ENMs) into the aquatic ecosystem. Toxicol Rep 4:211–220. https://doi.org/10.1016/j.toxrep.2017.04.001S2214-7500(17)30017-3[pii]

Jeevanandam J, Barhoum A, Chan YS, Dufresne A, Danquah MK (2018) Review on nanoparticles and nanostructured materials: history, sources, toxicity and regulations. Beilstein J Nanotechnol 9:1050–1074. https://doi.org/10.3762/bjnano.9.98

Jiang HS, Li M, Chang FY, Li W, Yin LY (2012) Physiological analysis of silver nanoparticles and AgNO3 toxicity to Spirodela polyrhiza. Environ Toxicol Chem 31(8):1880–1886. https://doi.org/10.1002/etc.1899

Jiang HS, Qiu XN, Li GB, Li W, Yin LY (2014) Silver nanoparticles induced accumulation of reactive oxygen species and alteration of antioxidant systems in the aquatic plant Spirodela polyrhiza. Environ Toxicol Chem 33(6):1398–1405. https://doi.org/10.1002/etc.2577

Kang S, Mauter MS, Elimelech M (2008) Physicochemical determinants of multiwalled carbon nanotube bacterial cytotoxicity. Environ Sci Technol 42(19):7528–7534. https://doi.org/10.1021/es8010173

Karnik BS, Davies SH, Baumann MJ, Masten SJ (2005) Fabrication of catalytic membranes for the treatment of drinking water using combined ozonation and ultrafiltration. Environ Sci Technol 39(19):7656–7661. https://doi.org/10.1021/es0503938

Keller AA, Wang H, Zhou D, Lenihan HS, Cherr G, Cardinale BJ et al (2010) Stability and aggregation of metal oxide nanoparticles in natural aqueous matrices. Environ Sci Technol 44(6):1962–1967. https://doi.org/10.1021/es902987d

Keller AA, McFerran S, Lazareva A, Suh S (2013) Global life cycle releases of engineered nanomaterials. J Nanopart Res 15:1–17

Khan I, Saeed K, Khan I (2019) Nanoparticles: properties, applications and toxicities. Arab J Chem 12(7):908–931

Kim J, Kim S, Lee S (2011) Differentiation of the toxicities of silver nanoparticles and silver ions to the Japanese medaka (Oryzias latipes) and the cladoceran Daphnia magna. Nanotoxicology 5(2):208–214. https://doi.org/10.3109/17435390.2010.508137

Klaine SJ, Alvarez PJ, Batley GE, Fernandes TF, Handy RD, Lyon DY et al (2008) Nanomaterials in the environment: behavior, fate, bioavailability, and effects. Environ Toxicol Chem 27(9):1825–1851. https://doi.org/10.1897/08-090.1

Kohli AK, Alpar HO (2004) Potential use of nanoparticles for transcutaneous vaccine delivery: effect of particle size and charge. Int J Pharm 275(1–2):13–17

Kot M, Major Ł, Lackner JM, Chronowska-Przywara K, Janusz M, Rakowski W (2016) Mechanical and tribological properties of carbon-based graded coatings. J Nanomater:1–14

Kraas M, Schlich K, Knopf B, Wege F, Kagi R, Terytze K et al (2017) Long-term effects of sulfidized silver nanoparticles in sewage sludge on soil microflora. Environ Toxicol Chem 36(12):3305–3313. https://doi.org/10.1002/etc.3904

Lankveld DP, Oomen AG, Krystek P, Neigh A, Troost-de Jong A, Noorlander CW et al (2010) The kinetics of the tissue distribution of silver nanoparticles of different sizes. Biomaterials 31(32):8350–8361. doi:S0142-9612(10)00888-4 [pii]10.1016/j.biomaterials.2010.07.045

Larner F, Dogra Y, Dybowska A, Fabrega J, Stolpe B, Bridgestock LJ et al (2012) Tracing bioavailability of ZnO nanoparticles using stable isotope labeling. Environ Sci Technol 46(21):12137–12145. https://doi.org/10.1021/es302602j

Laurent S, Bridot JL, Elst LV, Muller RN (2010) Magnetic iron oxide nanoparticles for biomedical applications. Future Med Chem 2(3):427–449. https://doi.org/10.4155/fmc.09.164
Lee R (2001) Bioavailability, biotransformation and fate of organic contaminants in estuarine animals. CRC Press
Lee JE, Lee N, Kim T, Kim J, Hyeon T (2011) Multifunctional mesoporous silica nanocomposite nanoparticles for theranostic applications. Acc Chem Res 44(10):893–902. https://doi.org/10.1021/ar2000259
Lei YM, Huang WX, Zhao M, Chai YQ, Yuan R, Zhuo Y (2015) Electrochemiluminescence resonance energy transfer system: mechanism and application in Ratiometric Aptasensor for lead ion. Anal Chem 87(15):7787–7794. https://doi.org/10.1021/acs.analchem.5b01445
Levard C, Hotze EM, Lowry GV, Brown GE Jr (2012) Environmental transformations of silver nanoparticles: impact on stability and toxicity. Environ Sci Technol 46(13):6900–6914. https://doi.org/10.1021/es2037405
Li X, Schirmer K, Bernard L, Sigg L, Pillai P, Behra R (2015) Silver nanoparticle toxicity and association with the alga Euglena gracilis. Environ Sci Nano 2(6):594e602
Liu J, Hurt RH (2010) Ion release kinetics and particle persistence in aqueous nano-silver colloids. Environ Sci Technol 44(6):2169–2175. https://doi.org/10.1021/es9035557
Liu D, Li C, Zhou F, Zhang T, Zhang H, Li X et al (2015) Rapid synthesis of monodisperse Au nanospheres through a laser irradiation-induced shape conversion, self-assembly and their electromagnetic coupling SERS enhancement. Sci Rep 5:7686. doi:srep07686 [pii]10.1038/srep07686
Loureiro A, Azoia NG, Gomes AC, Cavaco-Paulo A (2016) Albumin-based Nanodevices as drug carriers. Curr Pharm Des 22(10):1371–1390. doi:CPD-EPUB-73211 [pii]10.2174/1381612822666160125114900
Lowry GV, Gregory KB, Apte SC, Lead JR (2012) Transformations of nanomaterials in the environment. Environ Sci Technol 46(13):6893–6899. https://doi.org/10.1021/es300839e
Mansha M, Khan I, Ullah N, Qurashi A (2017) Synthesis, characterization and visible-light-driven photoelectrochemical hydrogen evolution reaction of carbazole-containing conjugated polymers. Int J Hydrog Energy. https://doi.org/10.1016/j
Martis E, Badve R, Degwekar M (2012) Nanotechnology based devices and applications in medicine: an overview. Chron Young Sci 3(68)
Miller RJ, Lenihan HS, Muller EB, Tseng N, Hanna SK, Keller AA (2010) Impact of metal oxide-nanoparticles on marine phytoplankton. Environ Sci Technol 44(19):7329–7334
Monteiro-Riviere NA, Wiench K, Landsiedel R, Schulte S, Inman AO, Riviere JE (2011) Safety evaluation of sunscreen formulations containing titanium dioxide and zinc oxide nanoparticles in UVB sunburned skin: an in vitro and in vivo study. Toxicol Sci 123(1):264–280. doi:kfr148 [pii]10.1093/toxsci/kfr148
Montes MO, Hanna SK, Lenihan HS, Keller AA (2012) Uptake, accumulation, and biotransformation of metal oxide nanoparticles by a marine suspension-feeder. J Hazard Mater 225-226:139–145. doi:S0304-3894(12)00491-8 [pii]10.1016/j.jhazmat.2012.05.009
Movafeghi A, Khataee A, Abedi M, Tarrahi R, Dadpour M, Vafaei F (2018) Effects of TiO2 nanoparticles on the aquatic plant Spirodela polyrrhiza: evaluation of growth parameters, pigment contents and antioxidant enzyme activities. J Environ Sci (China) 64:130–138. doi:S1001-0742(17)30321-2 [pii]10.1016/j.jes.2016.12.020
Mueller NC, Nowack B (2008) Exposure modeling of engineered nanoparticles in the environment. Environ Sci Technol 42(12):4447–4453. https://doi.org/10.1021/es7029637
Nikalje AP (2015) Nanotechnology and its applications in medicine. Med Chem 5
Ning Z, Cheung CS, Fu J, Liu MA, Schnell MA (2006) Experimental study of environmental tobacco smoke particles under actual indoor environment. Sci Total Environ 367(2–3):822–830. doi:S0048-9697(06)00155-0 [pii]10.1016/j.scitotenv.2006.02.017
OECD OfEC, Development. Guidelines for testing of chemicals, section 2: effects on biotic systems, procedure 201. Algal Growth Inhibition Test 1984; Organisation for Economic Cooperation and Development, Paris

Oukarroum A, Barhoumi L, Pirastru L, Dewez D (2013) Silver nanoparticle toxicity effect on growth and cellular viability of the aquatic plant Lemnagibba. Environ Toxicol Chem 32(902e907)

Park KH, Chhowalla M, Iqbal Z, Sesti F (2003) Single-walled carbon nanotubes are a new class of ion channel blockers. J Biol Chem 278(50):50212–50216. https://doi.org/10.1074/jbc.M310216200M310216200[pii]

Peng X, Palma S, Fisher NS, Wong SS (2011) Effect of morphology of ZnO nanostructures on their toxicity to marine algae. Aquat Toxicol 102(3–4):186–196. doi:S0166-445X(11)00021-X [pii]10.1016/j.aquatox.2011.01.014

Petersen EJ, Nelson BC (2010) Mechanisms and measurements of nanomaterial-induced oxidative damage to DNA. Anal Bioanal Chem 398(2):613–650. https://doi.org/10.1007/s00216-010-3881-7

Powers KW, Palazuelos M, Moudgil BM, Roberts SM (2007) Characterization of the size, shape, and state of dispersion of nanoparticles for toxicological studies. Nanotoxicology 1(1):42–51

Puri A, Loomis K, Smith B, Lee JH, Yavlovich A, Heldman E et al (2009) Lipid-based nanoparticles as pharmaceutical drug carriers: from concepts to clinic. Crit Rev Ther Drug Carrier Syst 26(6):523–580. doi:7a0c3bee026fb778,313cd9ee34e9e406 [pii]10.1615/critrevtherdrugcarriersyst.v26.i6.10

Rao JP, Geckeler KE (2011) Polymer nanoparticles: preparation techniques and size-control parameters. Prog Polym Sci 36:887–913

Risom L, Moller P, Loft S (2005) Oxidative stress-induced DNA damage by particulate air pollution. Mutat Res 592(1–2):119–137. doi:S0027-5107(05)00246-0 [pii]10.1016/j.mrfmmm.2005.06.012

Rogozea EA, Petcu AR, Olteanu NL, Lazar CA, Cadar D, Mihaly M (2017) Tandem adsorption-photodegradation activity induced by light on NiO-ZnO p–n couple modified silica nanomaterials. Mater Sci Semicond Process 57:1–11

Saeed K, Khan I (2014) Preparation and properties of single-walled carbon nanotubes/poly(butylene terephthalate) nanocomposites. Iran Polym J 23:53–58

Saeed K, Khan I (2016) Preparation and characterization of singlewalled carbon nanotube/nylon 6,6 nanocomposites. Instrum Sci Technol 44:435–444

Sani-Kast N, Labille J, Ollivier P, Slomberg D, Hungerbuhler K, Scheringer M (2017) A network perspective reveals decreasing material diversity in studies on nanoparticle interactions with dissolved organic matter. Proc Natl Acad Sci U S A 114(10):E1756–E1E65. doi:1608106114 [pii]10.1073/pnas.1608106114

Sapkota A, Symons JM, Kleissl J, Wang L, Parlange MB, Ondov J et al (2005) Impact of the 2002 Canadian forest fires on particulate matter air quality in Baltimore city. Environ Sci Technol 39(1):24–32. https://doi.org/10.1021/es035311z

Scown TM, Santos EM, Johnston BD, Gaiser B, Baalousha M, Mitov S et al (2010) Effects of aqueous exposure to silver nanoparticles of different sizes in rainbow trout. Toxicol Sci 115(2):521–534. doi:kfq076 [pii]10.1093/toxsci/kfq076

Shao W, Nabb D, Renevier N, Sherrington I, Luo JK (2012) Mechanical and corrosion resistance properties of TiO2 nanoparticles reinforced Ni coating by electrodeposition. IOP Conf Ser Mater Sci Eng 40(12042)

Shin WK, Cho J, Kannan AG, Lee YS, Kim DW (2016) Cross-linked composite gel polymer electrolyte using mesoporous methacrylate-functionalized SiO2 nanoparticles for lithium-ion polymer batteries. Sci Rep 6:26332. doi:srep26332 [pii]10.1038/srep26332

Sioutas C, Delfino RJ, Singh M (2005) Exposure assessment for atmospheric ultrafine particles (UFPs) and implications in epidemiologic research. Environ Health Perspect 113(8):947–955. https://doi.org/10.1289/ehp.7939

Sohm B, Immel F, Bauda P, Pagnout C (2015) Insight into the primary mode of action of TiO2 nanoparticles on Escherichia coli in the dark. Proteomics 15(1):98–113. https://doi.org/10.1002/pmic.201400101

Stefani D, Wardman D, Lambert T (2005) The implosion of the Calgary general hospital: ambient air quality issues. J Air Waste Manag Assoc 55(1):52–59. https://doi.org/10.1080/10473289.2005.10464605

Sun S, Murray CB, Weller D, Folks L, Moser A (2000) Monodisperse FePt nanoparticles and ferromagnetic FePt nanocrystal superlattices. Science 287(5460):1989–1992. doi:8349 [pii]10.1126/science.287.5460.1989

Taylor DA (2002) Dust in the wind. Environ Health Perspect 110(2):A80–A87. doi:sc271_5_1835 [pii]10.1289/ehp.110-a80

Taylor C, Matzke M, Kroll A, Read DS, Svendsen C, Crossley A (2016) Toxic interactions of different silver forms with freshwater green algae and cyanobacteria and their effects on mechanistic endpoints and the production of extracellular polymeric substances. Environ Sci Nano 3(396e408)

Thalmann B, Voegelin A, Morgenroth E, Kaegi R (2016) Effect of humic acid on the kinetics of silver nanoparticle sulfidation. Environ Sci Nano 3(1):203–212

Thomas SC, Harshita, Mishra PK, Talegaonkar S (2015) Ceramic nanoparticles: fabrication methods and applications in drug delivery. Curr Pharm Des 21(42):6165–6188. doi:CPD-EPUB-71344 [pii]10.2174/1381612821666151027153246

Tratnyek PG, Johnson RL (2006) Nanotechnologies for environmental cleanup. Nano Today 1:44–48

Turan NB, Erkan HS, Engin GO, Bilgili MS (2019) Nanoparticles in the aquatic environment: usage, properties,transformation and toxicity—a review. Process Saf Environ Prot 130:238–249

Turner A, Brice D, Brown MT (2012) Interactions of silver nanoparticles with the marine macroalga. Ulva lactuca Ecotoxicology 21(1):148–154. https://doi.org/10.1007/s10646-011-0774-2

Ullah H, Khan I, Yamani ZH, Qurashi A (2017) Sonochemical-driven ultrafast facile synthesis of SnO2 nanoparticles: growth mechanism structural electrical and hydrogen gas sensing properties. Ultrason Sonochem 34:484–490. doi:S1350-4177(16)30223-1 [pii]10.1016/j.ultsonch.2016.06.025

Unser S, Bruzas I, He J, Sagle L (2015) Localized surface Plasmon resonance biosensing: current challenges and approaches. Sensors (Basel) 15(7):15684–15716. doi:s150715684 [pii]10.3390/s150715684

USEPA UEPA (1996) Ecological effect test guidelines. OPPTS 850.5400. Algal Toxicity, Tiers I and II

Vaccari DA (2005) Environmental biology for engineers and scientists. Wiley-Interscience. ISBN 0-471-74178-7

Walters C, Pool E, Somerset V (2013) Aggregation and dissolution of silver nanoparticles in alaboratory-based freshwater microcosm under simulated environmental conditions. Toxicol Environ Chem 95(10):1690–1701

Walters C, Pool E, Somerset V (2016) Nanotoxicology: a review, toxicology – new aspects to this scientific conundrum. Larramendy. IntechOpen. https://doi.org/10.5772/64754

Weinberg H, Galyean A, Leopold M (2011) Evaluating engineered nanoparticlesin natural waters. TrAC Trends Anal Chem 30:72–83

Wong SW, Leung PT, Djurisic AB, Leung KM (2010) Toxicities of nano zinc oxide to five marine organisms: influences of aggregate size and ion solubility. Anal Bioanal Chem 396(2):609–618. https://doi.org/10.1007/s00216-009-3249-z

Wu J, Lu H, Zhu G, Chen L, Chang Y, Yu R (2017) Regulation of membrane fixation and energy production/conversion for adaptation and recovery of ZnO nanoparticle impacted Nitrosomonas europaea. Appl Microbiol Biotechnol 101(7):2953–2965. https://doi.org/10.1007/s00253-017-8092-010.1007/s00253-017-8092-0[pii]

Xiong D, Fang T, Yu L, Sima X, Zhu W (2011) Effects of nano-scale TiO2, ZnO and their bulk counterparts on zebrafish: acute toxicity, oxidative stress and oxidative damage. Sci Total Environ 409(8):1444–1452

Xiu ZM, Zhang QB, Puppala HL, Colvin VL, Alvarez PJ (2012) Negligible particle-specific antibacterial activity of silver nanoparticles. Nano Lett 12(8):4271–4275. https://doi.org/10.1021/nl301934w

Yang ST, Wang X, Jia G, Gu Y, Wang T, Nie H et al (2008) Long-term accumulation and low toxicity of single-walled carbon nanotubes in intravenously exposed mice. Toxicol Lett 181(3):182–189. doi:S0378-4274(08)01199-5 [pii]10.1016/j.toxlet.2008.07.020

Young KJ, Martini LA, Milot RL, Snoeberger RC III, Batista VS, Schmuttenmaer CA et al (2012) Light-driven water oxidation for solar fuels. Coord Chem Rev 256(21–22):2503–2520. https://doi.org/10.1016/j.ccr.2012.03.031

Zhao J, Wang Z, White JC, Xing B (2014) Graphene in the aquatic environment: adsorption, dispersion, toxicity and transformation. Environ Sci Technol 48(17):9995–10009. https://doi.org/10.1021/es5022679

Zhu X, Wang J, Zhang X, Chang Y, Chen Y (2009) The impact of ZnO nanoparticle aggregates on the embryonic development of zebrafish (Danio rerio). Nanotechnology 20(19):195103. doi:S0957-4484(09)03849-5 [pii]10.1088/0957-4484/20/19/195103

Chapter 11
Risk Governance Policies for Sustainable Use of Nanomaterials

Pooja Chauhan, Priyanka Sharma, Savita Chaudhary, and Rajeev Kumar

Abstract The innovational research in the field of nanotechnology is aimed to support and transform every sphere of human lifestyle. However, the environmental impact of engineered nanomaterials remains the topic of debate among researchers. Nano-components, more specifically nanoparticles, may have produced adverse effects on flora and fauna if accidently or deliberately released into the environment. Till date, there are no specific regulatory policies for the safe use of nanomaterials as they are considered under the broad categories of hazardous chemicals and reactive wastes. The regulations of nanomaterials cannot be similar to the bulk materials due to the huge gap in their properties. The rules should be based upon the scientific understanding of the nanomaterials. With the ongoing globalization of nanomaterials, international coordination, and harmonization, there is an urgent need for making the sound regulatory approaches for the safer use of nanomaterials. This chapter reviews and outlines the risk associated with nanomaterials, current regulatory approaches, and efforts in developing the sustainable nanotechnology.

Keywords Nanomaterials · Nanotoxicity · Reusability of nanomaterials · Nano-contamination · Risk governance policies · Sustainable development

P. Chauhan · S. Chaudhary (✉)
Department of Chemistry and Centre of Advanced Studies in Chemistry, Panjab University, Chandigarh, India
e-mail: schaudhary@pu.ac.in

P. Sharma · R. Kumar
Department of Environment Studies, Panjab University, Chandigarh, India
e-mail: rajeev@pu.ac.in

R. Kumar et al. (eds.), *Advanced Functional Nanoparticles "Boon or Bane" for Environment Remediation Applications*, Environmental Contamination Remediation and Management, https://doi.org/10.1007/978-3-031-24416-2_11

11.1 Introduction

Structures or particles in the size range between 1 and 100 nanometers (nm) are referred as nanostructures or nanoparticles. The term is sometimes used for larger particles, that is, up to 500 nm (Adams and Barbante 2013). Based upon their origin, the nanostructures can be divided into three major groups: natural, incidental, and engineered. Natural nanomaterials are the results of various (bio)geochemical or mechanical processes occurring in the nature (Baker et al. 2017b). The incidental nanomaterials are nanomaterials produced unintentionally under direct or indirect influence of anthropogenic processes (Sharma et al. 2020). The engineered nanomaterials are nanostructures designed and produced by humans for their benefits in different areas. However, on the basis of their sale as regulated commercial products, the nanoscale materials have been classified into four main groups (Chaudhary et al. 2015). The first and biggest group of nanoscale material contains the application of metal oxide (oxides of cerium, iron, zirconium, and zinc), semiconducting nanomaterials as a chemical polishing agent, sunscreens, and cosmetics agents (Zhang et al. 2018). Another group consists of nanoclays, which are naturally occurring particles in the environment. Nanoclays are naturally occurring claylike particles, which have the tendency to augment the harness, strength, heat resistance, and flame-resistant properties to material (Esmaeili et al. 2020). The different types of nanomaterials have further been used for the manufacturing of paper cartons, plastic bottles, and tennis balls (Sharma et al. 2018). The third category of particles consists of nanotubes that are fundamentally used to reduce the static electricity level in hard disk handling trays and fuel lines (Jagusiak et al. 2021). They are mainly utilized in industrial sector especially in field emitter sources, flame reduction fillers, and car electrostatically paintable machineries. The last and fourth group has comprised quantum dots and has widely been applied in the medical therapeutic and diagnostic sector (Solaimani 2020).

Nano"risk"ology: Issues in Nanotechnology Including Environmental Health Safety

The potential usages and benefits offered by nanomaterials to humans in the area of medical diagnostics, pollution monitoring (Chaudhary et al. 2016), remediation (Chaudhary et al. 2017), and energy solution cannot be replaced by any other technology. However, the toxicity imparted by the nanoparticles to the environment cannot be ignored by the researchers. Table 11.1 explains influential impact of different types of nanomaterials on environmental species.

Being the most integral part of our survival, the safety of environment is the top most priorities of researchers (Kumar et al. 2020a, b).

The above brief table is enough to sketch the risks related to nanotechnology, that is, nano-riskology. The need of risk governance is the only way of dealing with the toxic impact of nanoparticles. The main aim of the chapter is to understand the associated risks and to find the answer to "What we want from nanotechnology and what are we getting?"

Table 11.1 Nanomaterials with their intended use and expected exposure to the environment

S. no.	Nanomaterial	Intended application	Chances of exposure to the environment	Toxicity profile (toxic to)	References
1.	TiO_2	Supercapacitor	Through leaching from dumping yards after disposal	Algae, aquatic animals, plants	Wu et al. (2014)
2.	ZnO	Industrial level	Through industrial waste effluent and water contamination	Environmental species	Buerki-Thurnherr et al. (2012)
3.	CdSe	Semiconductor	Through water streams	Cells, tissue culture stem cells	Wang et al. (2008)
4.	CuO	Semiconductor	Through metallurgy industries	Aquatic organisms	Rohilla et al. (2020)
5.	CdS	Mineral	Through incineration of municipal waste	Bacterial species	Hussain et al. (2013)

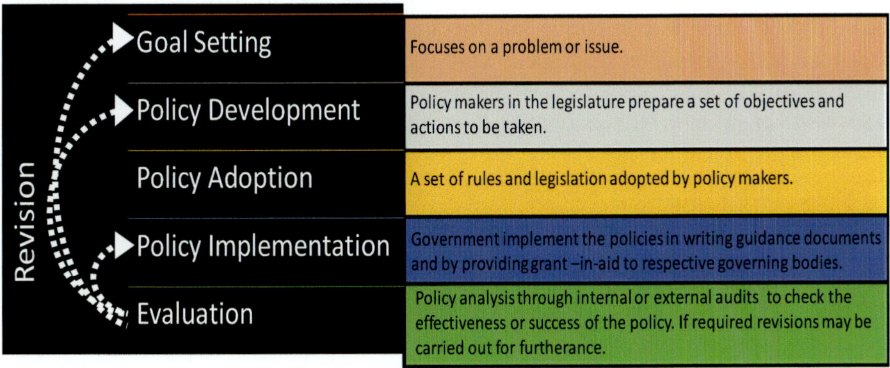

Fig. 11.1 An overview of governance working on policy-making

Risk Governance

Risk governance here refers to the application of the governing approach to tackle the issues of risks associated with the excessive release of nanomaterials in the environment. In particular, the governance works in portions, that is, firstly, the risk governance recognizes that decisions about issues of risks are not perspectives of the group of people rather those are scientifically evident facts (Stone et al. 2018). After the validation and assessment of the risks, the preventive or safety frameworks are designed (Fig. 11.1). The designing of the framework mainly involved the scientific, aesthetic, and administrative factors. In addition, all the financial feasibility aspects need to be taken care for the proper management of environmental impact of hazardous materials. Despite of it, developing a framework for nanotechnology has many implications; for instance, the dealing governance must be anticipatory and flexible as risks are inconclusive, whereas the threats to the environment and the human health have also not been taken jokily.

11.2 Key Requirements for Sustainable Use of Nanomaterials

"Sustainable development" refers to the production and consumption of goods and resources in a manner that the needs of the current generations can be satisfied without compromising, limiting, or threatening the needs and environmental conditions of future generations (Bardos et al. 2016). For nanotechnology, a wealth of applications has been proposed. For instance, nanotechnology enables the manufacturing process with lesser energy consumption and waste generations, thus saving the expenditure on carbon trading (Chauhan et al. 2020). Nanotechnology is not restricted to the production of nanoparticles but can also be a decisive step of complex productions (Bhatio 2016). Such ways lead to the macro productions of products. For instance, the potential application of nano-based catalysts in the sector of energy production is well documented in literature. On a bigger approach, certain converging technologies involve the macro-production and result in "meta-technologies" such as computing and biochemical analytics (Fig. 11.2). The interdependence of these technologies with each other has shown that attributing sustainability to any associated technology is very tricky due to the entangled network of production protocols (Chakraborty et al. 2019).

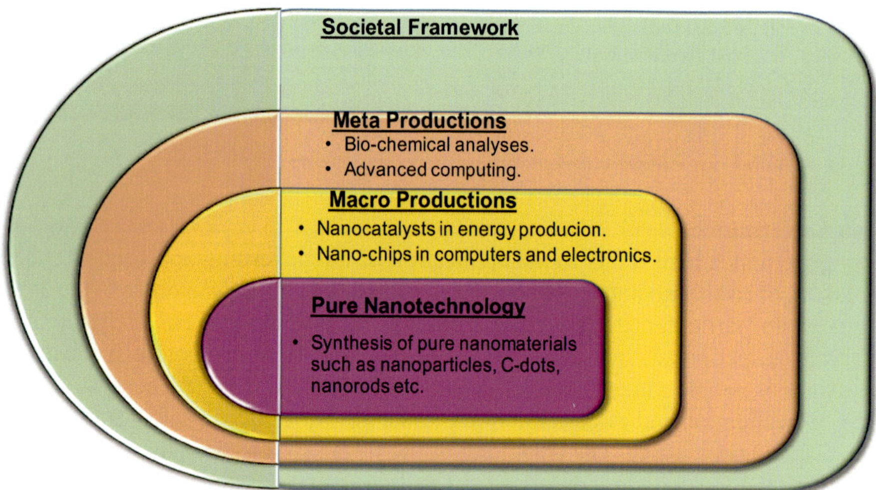

Fig. 11.2 Illustration of interdependence among production processes and societal network

11.3 Nanotechnology: Policy Implications

David Rejeski (Woodrow Wilson International Center for Scholars) said that nanotechnology is considered as the safer approach for consumer products (Vance et al. 2015). Before the development of nanotechnology in earlier era, various governmental regulatory frameworks were present to analyze the specific and unique properties of material. Rejeski mentions (Rejeski and Lekas 2008) that the population around the world does not always convict the laws and regulations implemented by government for the fortification of human health and environmental surrounding (Baun et al. 2010). Impacts and advantages of nanotechnology are not restricted to one area; it is related to various areas for risk prevention in genetically modified sectors. From the last few years, various nano-products have been entering into the market for commercialization (Lo et al. 2012). He also proposed that development of nanotechnology depends upon four plausible scenarios, which would provide meaning to the nanotechnology in terms of liability and asset (Kuang et al. 2020). The possible scenarios include: "tipping scales," "nano Bhopal," "Hollywood wins," and "old Europe."

According to "tipping scales" and "nano Bhopal," the field of nanotechnology is considered as the search for novelty. If new properties and performances can be obtained from the already existing chemical substances, then we should be careful while using old knowledge for future emergence (Michelson 2008). While making policies, it is mandatory to understand that nanotechnology is not confined to the industries, and it is applicable at global level for conventional methodology (Soltani and Pouypouy 2019). In 1984, Bhopal chemical plant disaster happens in India to provide global press coverage and public backlash against nanotechnology (Peterson 2009). Small companies had received large orders to fulfill the availability of different products. During the relative processes, different mistakes and negative impacts would arise in terms of public perception. One way to protect respective methodology is to encourage the utilization of nanotechnology at translational corporation level (Baker et al. 2017a). Rejeski said that public reviews of nanotechnology could be variable because they have not fully esteemed their coverage (Rejeski and Lekas 2008).

According to "Hollywood wins," the policy implantation teams should understand the fact that society receives different messages and results from most common ways such as books, games, and films. Mad scientists have been considered as the major object of movie fiction, and nanotechnology helps in understanding science fictions in a better way. In Hollywood, many films (Agent Cody Banks and Spiderman II) and games (PlayStation 2) have been produced on the basis of nanotechnology. The produced things have suggested that the nanotechnology can bite back while integrated into our lives.

According to "old Europe," Rejeski said European environmental NGOs and different activists have tried to utilize the new technological assessment better than American counterparts (Rejeski and Lekas 2008). He also stated that European Union has manufactured and tuned many preventive principles over the years. The respective models have provided a message to the society to "learn and act." Meanwhile, the United States follows the rule of "act and then learn" in terms of the usage of new technologies for the betterment of the nation. In European countries, Union and Commission have already established various kinds of technology assessment analytical system at different places. In 1995, US congress had eradicated Office Technology Assessment (OTA) for checking the perceptive view of nanotechnology. According to Rejeski, OTA had played a significant characteristic while evaluating the risk factors associated with nanotechnology. Therefore, it should not be removed out from the system (Davies 2008). In the end, the developed policies should provide voluntary agreements for the development and usage of nanotechnology in a better way. It is very important to start this process today not two years from now, said Rejeski (Rejeski and Lekas 2008).

11.4 Risk Assessment

The "sustainable potential" of nanotechnology needs to be explored thoroughly for its future applications and implications (Falkner et al. 2012). In general, the applications and implications of safer use of nanomaterials are made on the basis of three principles of sustainability: that is, economy, society, and the environment or informally called profit, people, and planet (Hristozov et al. 2016). But the concept of "sustainability potential" about nanoparticles has found to be inadequate in current state through information about the effect of external conditions over the safer use of nanoparticles, which is not well document in literature. Therefore, such projections are usually beneficial in various ideological purposes and safer use of nanomaterials (Gajewick et al. 2012; Trybula and Newberry 2014).

Many scientific methods have been developed to estimate the sustainability of nanomaterials (Dixit et al. 2018). These methods mainly focus on understanding of the environmental and the societal and economic dimensions of chosen particles. The common risk assessment methods for safer use of nanomaterials have been discussed below:

11.4.1 Life Cycle Analysis (LCA)

Life cycle assessment or life cycle analysis is one of the most common methods to assess the overall environmental impacts associated with nanomaterials, from its formation to use and its effect after disposal (Senan-Salinas et al. 2020). This assessment is also known as "cradle-to-grave" approach in risk management. LCA has mainly comprised of a set of ISO standards parameters. LCA, unlikely other methods, has offered the longer-term assessment and can be used for assessing existing and expected developments in nanotechnology (Visentin et al. 2021). According to literature, Lloyd et al. examined the features of nanoscale particles of platinum-group metal in the field of automotive catalysts (Fleischer and Grunwald 2008).

Their results displayed the reduction in the metal loading capacity in automotive catalysts. The outcomes have supported the environmental eminence with the disposal of harmful toxins. However, the energy consumptions and excessive usage of sustainable resources have been enhanced with the application of nanoparticles. Further, Osterwalder et al. utilized improved LCA approach for the comparison of energy production with the nanoparticles (Fleischer and Grunwald 2008).

The major limitation of LCA is the applicability of chosen methods to the selected technologies. It basically deals with the transformation of a prevailing technique into the advanced and innovative method through equivalent substitutions (Rashid and Liu 2020), which has further restricted its wide spread usage to all the methods. In addition, this method of analysis has only been applied to existing commercial products (Atia et al. 2020). Up to now, only scientifically recognized and adapted products have been taken up for the environmental assessment under LCA analysis. It also permits conversion of sustainable requirements into technical routine feature for the "relative sustainable evaluation." The respective substitution approach exists in the tradition of efficiency strategies as an analytical classic (Fleischer and Grunwald 2008). It is majorly employed for the research and development field where existing techniques can be modified in terms of application framework to support the decrement of ecological technologies (Xiong et al. 2020). The LCA recognize the ecological blockage of prevailing items and provide evidences with the objectives of technological innovation. The comprehensive LCAs for nanomaterials would be useless until the complete information about the environmental impact has not been provided for all the existing materials in nanometric range (Hosseinzadeh-Bandbafah et al. 2020).

In addition, life cycle impact assessment (LCIA) is the most crucial step of LCA because it handles large database extracted from inventory analysis. Thus, the choice of an appropriate inventory is very important for getting through information about the LCA (Yao et al. 2020). LCIA's main role is to estimate and understand the magnitude and significance of the environmental impact of the understudy product or services during the entire course of life cycle of advanced nanomaterials (Salehpour et al. 2020). The results are obtained into understandable impact indicators such as climate change, ozone depletion, eutrophication, acidification, human toxicity (cancer and noncancer related), respiratory inorganics, ionizing radiation, ecotoxicity, photochemical ozone formation, land use, and resource depletion (Hossain et al. 2020). The emissions and resources are assigned to each of these impact categories, which are then converted into indicators using impact assessment models (Recipe, USEtox, Traci, Caltox, etc.).

11.4.2 Future Technology Analyses (FTA)

For the sustainable development of nanotechnology, it is not appropriate to collect information only about the capabilities, threats, and promises. In literature, eco-assessment method especially LCA has been reported in comprehensive way for the certainty and accessibility of data for developing nanotechnology (Fleischer and Grunwald 2008). Various ideas have been recommended for safer utilization of nanotechnology. In order to simplify the respective entitlements about the safe use of nanomaterials, it becomes a rational way to calculate the intellectual and administrative process for permitting the different ranges of procedures and strategies for the betterment of nanotechnologies through sustainable valuation (Krajnik et al. 2016). The major task associated with these objectives is to provide much clean and transparent scheme to design biocompatible range of nanomaterials with minimum impact over environment (Kamali et al. 2019). In order to discriminate among assumptions for sustainable effect of nanoworld, the possible solutions with beneficial probability and emergence of possible future situations through organized foresight methodologies with better control over the growth and expansion of advanced materials in real world have been discussed in detail (Lu and Ozcan 2015). From the future point of view, it is essential to understand the philosophy of knowledge for nanomaterials. According to this, properties and assumptions of the different potential and menace of nanoworld should be analyzed to validate the sustainable development. The sustainable development of nanotechnology for the future projection necessitates two major facts such as evaluate the valid future estimations for sustainable development and accomplish the aptitude and pressures of nanotechnology applications in terms of sustainability (Nazarko 2017). The philosophy of knowledge for the sustainable development of nanotechnology depends upon three basic criteria action-guiding knowledge, explanatory knowledge, and orientation knowledge. Therefore, formatting the patterns of nanotechnology in terms of sustainable development has further led the enhancement in the basic knowledge through decision-making in synthetic methodologies. Further, to provision the process of decision-making sustainable influential behavior for the usage, construction and removal effects of nanotechnology should be provided for safe use (Eyo et al. 2020).

11.4.3 Nanotechnology Assessment

11.4.3.1 Technology Assessment

The technology assessment (TA) has basically provided the information and alignment for the policy-making about the chosen technology and its employment for developing better civilization. It improves the society, political, and diffusion possibility for future technology (Buckley et al. 2017). It can be carried out via the combination of research with supplying evidences for decision-making about the safe use of technology and promoting social communication. It majorly involves shaping and finishing of technology in terms of social context such as funding of public methodologies, their diffusion, and regulatory matters (Millard et al. 2019). It has also been classified in the field of nanotechnology as it provides the information about the associated risk and their side effects on living beings. Secondly, it has provided the strategized procedure for the evaluation of benefits of nano-technological outcomes. From the last few decades, TA has provided information about the safe usage of different technologies (Oomen et al. 2018). Different trials have been performed for the fabrication of sustainable and innovative methods of nanotechnology under TA methods. The most recent and developed examples are constructive technology assessment (CTA) and real-time technology assessment (RTTA), which have been implemented throughout the European countries for sustainable development of nanotechnology (Fleischer and Grunwald 2008; Sia 2017).

11.4.3.2 Technology Assessment for Sustainable Use of Nanotechnology: Task and Challenges

Different challenges are associated with nanotechnology during sustainable development such as application or progress of nanotechnology for the sustainable product, which can be organized or not has been assessed in a comparative manner under TA. The formatting of procedural methods for the safer use of nano-technological application for the development of sustainable nation has to be developed under TA. Why does the potential of formatting nanoworld in R & D according to technology evaluation has to be checked in proper manner? Before finding out the answers of the respective questions for future benefits, it is mandatory to understand the task and challenges associated with these goals of sustainable development (Ridsdale and Noble 2016; Sanchez et al. 2011).

11.4.3.3 Co-Evaluation of Technology, Society, and Their Inseparability Issue

It is mandatory to understand the sustainability of particles for assessing the production and usage of nanotechnology in effective manner. The development in nanofield is decided on the basis of societal benefits and their impact on environment (Helland and Kastenholz 2008). It has been observed that the nanotechnology will not be helpful in the social development. Their intention for the coevolution of technology for societal benefits is not thoroughly investigated till date. Further, there should be specific methodologies for better understanding of the mutual prompting factors for society in reference to nanomaterials. During sustainable development for nanotechnology, prolonged societal aspects need to be established for the safe usage (Purohit et al. 2017). The effect of sustainable assessment for the technological purpose can be investigated through structures, technique, resources, and emissive performance of chosen particles during their safe utilization in technological applications. The contribution of social development and nanoworld depends upon the technologies and societal mechanism (Corsi et al. 2018). The possible ability of nanotechnology with systematic approach leads to sustainable development along with modification in values, custom, and structures of chosen materials. In addition, the sustainable development of nanotechnology has mainly based upon two factors, that is, improving technology in terms of sustainability standards and social personification of technology for maintaining relationship between environments (Ittipanuvat et al. 2014). The distribution and economics of nanotechnology depends upon the political background of state and social surroundings. Therefore, the major task is to recognize the contribution credited by nanotechnology in terms of new energies and facts and provided their comprehensive scenario for road mapping (Kamarulzaman et al. 2020).

11.4.3.4 Life Cycle Issue and Prediction Changes

The sustainable development in terms of exaggerating the nanotechnology should require a clear discrepancy among more or less sustainable technology. The preventive methodology for nanoworld has needed a complete and transparent life cycle of technological system (Durante et al. 2015). The life cycle has included the detailed information about the progressive integration and balancing of all the sustainable possessions of chosen nanomaterials. During the contribution of LCA toward shaping of nanotechnology, it should consider various parameters such as environmental consequences, new technologies, political and economic framework, development lifestyles, health benefits, market analysis, and sustainable assessment of particles for performing reliable life cycle (Bauer et al. 2008). The use of nano titanium dioxide in SPF sunscreens is considered as the primary example of nanotechnology around the world market to replace the toxic organic constituents. The accumulation of nanoparticles on food chain has to be investigated out for preserving human health and ecosystem. The measurement of the development probability parameters for the predictive knowledge of nanoparticles in atmosphere has been discussed in detail. The quantitative usage of sunscreen for human benefits is considered as the foremost part for discussing the safe use of nanomaterials (Ma et al. 2020). The augmented utilization of LCA is considered as the most important implemented factor for long usage of strategically employed processes (Oliveira et al. 2017). The prepared hybrid methodology has been further utilized for exploring various technologies under greener approach program for eco-potential and developed products in terms of nanotechnology (Stucki and Woerter 2019).

11.4.3.5 Integration Issue and Incommensurability Problem

The different aspects of sustainable usage of advanced nanomaterials should be measured and modified as per the reference of their environmental effect. The minor issue has arisen due to the heterogenous and scattered knowledge about the methodology involving the formation of single and unique sustainable program for developing the "master sustainability indicator" for nanotechnology. Till date, the insufficient and invariant knowledge has restricted their wide spread exploration in health, water sanitation, and environmental remediation applications (Younis et al. 2018). The different programs and policies of sustainable development for nanotechnology have not yet organized into single analyses, which involve the sustainability index. The formed program has caused many barriers in the quantitative utilization of sustainable valuation (Subramanian et al. 2014). In that case, all the experimentation and results should be carried out in qualitative manner. One other issue arises due to heterogenous sustainability program and causes conflicting and inconsistent outcomes in the area of nanotechnology. The sustainable indicators have both positive and negative contribution in different fields on the feature of their application in terms of technology (Rossi et al. 2014). For example, sustainable nanotechnology in the field of science especially in chemistry results into positive influence due to their ecological and green innovation and found very competitive as compared to other methodology (Nabipour and Hu 2020). During the green procedure, the complete replacement of tedious technologies involving the usage of hazardous chemical and solvents has been done for saving ecological factors. The less usage of hazardous and polluted solvents all over the country has further affect the market price of old products and protect the environment (Part et al. 2015). The developed means have suggested that the sustainable technology has provided a comprehensive way to understand their explicit and undesired results.

11.4.3.6 Reflexive Sustainability Assessment for Nanotechnology

The incomplete knowledge about the developed nanomaterials has possessed several challenges for providing sustainable valuation about nanotechnological providences. Out of different challenges, many methods are completely known for the future technical assessment. The complexity of developed schemes gets enhanced in the presence of TA. The uncertain nature, insufficient evaluation, undiscovered challenges, and noncomprehensive ideas about nanoparticles have further possessed hindrance to understand their behavior to large extent. TA is considered as most basic assessment, and its different perspective should be augmented to meet the raised challenges (Bystrzejewska-Piotrowska et al. 2009). To maintain the future sustainable goals and discipline programs for assessing various points should be keep in mind for developing technologies. The manner of anticipation program in terms of nanotechnology for improving sustainable development according R & D performa has been thoroughly investigated for each type of nanomaterials. The roadmaps and line scheme for application purpose should be analyzed properly and optimized different indicator system for predictive sustainable assessment development of nano-programs. The life cycle assessment (LCA) program should be enhanced for increasing the prospective knowledge and substitute the past traditional methodologies in reference to the developed nano schemes along with exploring their side effects (Salieri et al. 2018). The development of different national and international programs for the safe usage of nanotechnology in industrial sector by incorporating greener approach at different sources resolves the associated conflicts (Khataee et al. 2020). Further, it intricates the problem arising during sustainable assessment in terms of limits of availability and shaping their requirement under prescribed conditions. The nanotechnology development in terms of sustainable development would be purposeful only when all the ambiguities, rawness, and knowledge hitched would be tackled carefully. Therefore, careful measurement and analysis should be performed to measure the implemented program for normative, empirical, theoretical, and political monitoring of nanotechnology (Oke et al. 2017).

11.4.4 International and National Scenario

The nanotechnology safety programs have emerged out as a special assessment program during the last many years for the sustainable usage of particles. The amendments in these programs in reference to systematic and appropriate analysis of contributing factors need to be explored more for the betterment of society (Hansen and Baun 2012). Various drawbacks such as different qualms, reservations, and knowledge gap results from nanotechnology for environmental surroundings and human health need to be explored in an excessive manner. In the United States, multiagency governance has been appointed at different levels such as nanoscale science engineering, technology subcommittee, and practices for overcoming the knowledge gaps. The Nanotechnology Research and Development Act (2003) has been introduced for the safe usage of nanoparticles. It works at different level such as food, drugs, health, and R & D sectors. In the United Kingdom, Nanotechnology Issue Dialogue Group works for coordinating society, environment, health hazards, rural and urban affairs, food safety, and minerals (Karjalainen et al. 2017). It deals with the safe handling of nanomaterials. In Japan, no particular committees and technical policies are present for chemical screening and law regulations. In China, National Steering Committee for Nanoscience and Nanotechnology works for education and science programs.

Indian Scenario

Nanotechnology is considered as the most rapidly growing technology in India. In 2007, government of India had established a project on Nano Science and Technology. It is considered as the most promising and enhanced project of Nano Science and Technology Initiative (NSTI). These sectors have identified the nano success mission in different sectors. The Union Cabinet has also encouraged the role of nano mission in Phase II (12th plan period) with allotment of Rs. 650 crores. The Department of Science and Technology, India, is considered as the most appropriate platform for applying projects under nano mission scheme for getting highest rate of public funding (Kumar et al. 2020a, b). While keeping the data for the entire risk assessment governance program, India has come a developing nation in the field of nanotechnology. From the 1980s, the policies of science and technology have started their main focus on the application part of research work in higher fields. Various programs such as five-year plans (1980–1985) and intensification of Research in High Priority Areas (IRHPA) program were established to qualify and demonstrate the scientific research and their ability for technological expansion. The success in respective program was obtained through the development of scientific proficiency, technical programs, and interdisciplinary programs (Nawale et al. 2021).

11.5 Risk Treatment

In general, nanoparticles should be prepared with high biocompatibility and are nontoxic for the environmental surrounding. Simultaneously, particles can be recovered easily and could be utilized again at a greater extent. Further, advanced technological systems should be used, which can deduct the usage of nanoparticles, increase productivity of materials, and release less waste by-products. In industrial sectors, various effluents have been released in the atmosphere, which is highly toxic to the aquatic flora and fauna species (Iavicoli et al. 2017). Those effluents should be treated before releasing in the system by including environmentally viable manner without affecting the operating behavior of atmosphere. Additionally, during photocatalysis activity, the catalyst and nanomaterials are released at a higher risk after the completion of reaction. The contaminated sites should be tackled carefully to avoid the toxic impact of advanced nanomaterials (Jayanthi et al. 2012).

11.5.1 Involvement of Social, Economic, and Biophysical Factors

It is mandatory to evaluate the social, economic, and biophysical impact of developed nanotechnology for the sustainable development. At social level, it is essential to investigate whether the imposing nanotechnology would not go to harm any environmental species (human, plants, and aquatic animal). Secondly, in the developed nation, it is urgent to calculate the economic cost of the nanomaterial (Hegde et al. 2016). Finally, biophysical factor should be inspected out whether the developed system is biologically or physically affecting the ecosystem or not. The amalgamation of development and environmental goals must be followed by "a triple bottom-line" approach (TBL). It includes splitting of sustainable remediation (SR) into components in accordance to the government and industrial strategies and their different practice tools.

11.5.2 Guidance on How to Use and Deal

The amendment of different trademark is very important in the primary phase of remediation process. In terms of future use and remedial solution to endorse sustainability, it is primary important to develop the methodologies for producing safer particles (Chaudhary et al. 2018). A clear approach for addressing the trademarks and tradeoffs along with their explicit examination would be required. It will provide the outline and steps involve during the procedure of nanomaterial development and their utilization for sustainable nature. The term "risk treatment" can be cured majorly during the respective step. If the message of less utilization of nanomaterials, their side effects, and effluent property gets introduced during the usage, then the society will use the material in safer way (Musee 2011).

11.5.3 Consider Future Cycle for the Remediation

The remediation events basically reduced the impact of different kinds of toxins and re-established the ecological functions for meeting community needs (Nowack 2017). All the resources, experimentation, and activities should be performed via keeping in mind that they should preserve the quality and quantity of future cycles or products for the future generation in terms of sustainable remediation. The remedial design should be known to answer the questions such as how the lands will be utilized after meeting continuous industrial effluents and after post remediation planning (Chaudhary et al. 2019). It is our foremost duty to preserve all the resources for future cycle so that they can also enjoy the natural environmental processes in similar way as we are enjoying.

11.5.4 Participation and Contribution to the Assurance

For preserving the natural resources and cycles for future, it is highly mandatory to participate and contribute toward the different safety data sheets issued by the government. It would provide meaningful information about the risk treatment, resource conservation, and side effects of nanotechnology. The active involvement and collaboration of society in designing and decision-making for different activities takes place for sustainable remediation (Kanagaraj et al. 2015). It is considered as the meaningful engagement to evaluate and substantiate the potential benefits for better land management decisions. The good amount of participation would provide new effects to the remediation projects. Additionally, it gives a new and innovation ideas for the remediation options, new sense of ownership, and commitment along with new end strategies.

11.6 Preparation of Nanomaterial Safety Data Sheets

Safety Data Sheets (SDS) have been considered as the important platform for analyzing the safer use of various types of chemical components. It provides the knowledge of hazardous substances and different guidelines for their safe use in supply chain in trained manner. It basically deals with the ecoliability (both plants and animals) and toxicity evaluation of material. It is highly beneficial for the workers, employers, professionals, vendors, companies, and producers (Karim 2020). So, it is very much necessary to develop high standards. SDS analog for the safer, ecolabeling, and responsible usage of nanomaterials all around the world is one of the prime measures to deal the associated risk of new technologies. In nano world, SDS will provide accurate, reliable, current, and toxicity profiling information about nanomaterials to fulfil the gaps arise during the development. It should follow "TOP Procedure Principle," where T, O, and P stands for technical, organizational, and personal protective measures.

11.6.1 Perilous Evaluation of SDS for Nanomaterials

In Australia, a national policy body known as Safe Work Australia (SWA) was established in 2010 for investigating out the major points of SDS for nano range materials such as silicates, carbon dots, paints, pigments, and metal oxides (Vishwakarma and Samal 2010). SWA has evaluated that the information provided by SDS was not sufficient to explore the different risk assessment approaches such as information about bulk material, toxicity, and exposure condition along with their safety measures. Further, in 2013, Lee et al. reported different SDS for variety of nanoparticles (Schulte et al. 2014), which belongs to the Organization for Economic Co-operation and Development (OECD), which includes the following:

- Deficiency of evidences for the toxicological profiling, exposure condition, and physicochemical properties (size distribution, solubility, stability, redox capacity, and potential ability)
- Utilization of ambiguous information such as use of same CAS number for different forms of carbon
- Removal of various information regarding the risk factor of material

Further, National Institute for Occupational Safety and Health (NIOSH) has analyzed that SDS has formatted during the year of 2007 to 2011 and explored their disadvantages (Lu et al. 2016). In similar manner, Lee et al. also studied the same and found insufficient information for various nanomaterials (Song et al. 2010).

11.6.2 Guidelines for the Preparation of SDS

In 2010, Swiss State Secretariat for Economic affairs (SECO) has established various guidelines for formatting the available SDS (Germann et al. 2016). In 2012, they have been updated for the better utilization of nanotechnology by including the following:

- Important and obligatory information about the documentation, threat, and composition of substance
- Essential precautions for treatment, storage, exposure, protection, risk, safety, and disposal of substance
- Major awareness against the accidental and first aid measures, stability, ecological evidences, transportation, and regulatory information

Table 11.2 Representation of different priorities for the maintenance of SDS for nanomaterial

S. no.	SDS information	Significances for the development of nanomaterials
1.	Identification of substance	Necessary
2.	Hazardous information	Necessary
3.	Physical and chemical properties	Necessary and precautionary
4.	Report of firefighting and first aid information	Important
5.	Toxicological and ecological information	Preferable
6.	Transport and regulatory data	Preferable

Further, various guidelines were introduced to amend the SDS of nanomaterials (data and source). In 2010, SECO has provided the application of SDS guidelines for two nanomaterials such as NANO-BLOGGO (surface finisher) and SECOKAT (photocatalyst). The international organization for standardization includes safety data sheets for the synthesis of nanoparticles, utilization, and their practical analysis (Hossain and Mukherjee 2013; Lee et al. 2011). Following are the key points that are included for the formatting of SDS:

- The marking of SDS data should be very fast, due to enlargement in data availability.
- Transparent and strong information should be provided for the nanoparticles when CAS number of bulk materials is used.
- Statement and assertion should be provided when toxicological evidence is not accessible.
- Declaration of statement provided for the exposure limits and application of bulk and nanomaterials (Table 11.2).

11.7 Enhancement of Reusability and Recovery of Nanomaterials After Application

Due to high potential value of nanotechnology, they have been receiving large amount of interest around the manufacturers and users due to their recovery and reusage during the nanoproduct cycle. The recycling of nanomaterials and nanotechnology has been divided into three categories such as health benefits at occupational level in terms of recycling, environmental influences of various deposits produced during the synthesis and introduction, and utilization of recovered nanomaterials during product manufacturing (Boldrin and Hansen 2014; Rahman et al. 2020). The advantage and efficiency of recovered products are mainly dependent upon the technology of nano-products, precursor of nanomaterial, type of suspension, and matrices used during fabrication of advanced materials. In past era, separation, centrifugation, and solvent evaporation were used as a conventional method for the recycling of nano waste materials. But in recent years, the recovery of nanomaterials has mainly included pH, magnetic field, biological analysis, nanostructured colloidal solvent, and molecular antisolvent methods. The respective methodologies are cost-effective in nature and require very less time and energy (Ramchandran and Gernand 2020). Further, chemical, mechanical, physical, and thermal properties of nanomaterials have required the efficient separation and recycled processing of material. In literature, cloud point extraction (CPE) technique has been reported for the separation of various aqueous nanomaterials (Ag, Cu, Zn, Au, and CuO) and utilized at a greater extent. The separation phenomena are found very difficult in case of emulsion where two layers are present together. For this reason, nanomaterials have been heated at higher temperature and allowed to centrifuge for complete separation. The reusability and recyclability of nanomaterials have further overcome the release of the hazardous nanomaterials in the environment. The process of recycling nanomaterials should pay major attention during the manufacturing step such as simple chemical substances, which can individually separate, and no aggregation takes place in recycling analysis. The toxic nature of material can be controlled from the physicochemical properties of parent material. Only those nanomaterials should be used, which can be recycled again and again and have the ability for the environmental remediation application. The nature of nanomaterials should be biocompatible and nontoxic for the sustainable development around the society (Li et al. 2008; Dutta et al. 2018).

11.8 Future Prospects

Nanotechnology is considered as the most influential and applied methodology all around the world. The development of urgent risk governance system is required for monitoring the advantages and disadvantages of nanomaterials in technological world. The governance of nanotechnology has required special company and agency to overlook the research network, infrastructure, human resources, hazards nature, and imparted toxicity. The facets of nanomaterials would directly influence the sustainable development and various species present around atmospheric environment. It is the high time to applicate standard operating parameters for the development and utilization of nanomaterials. Nanotechnology could replace various materials significantly without utilizing much energy resources. By creating a governance scheme, the potential utilization of nanomaterials can be made a success step. The polices should be based upon the scientific understanding of nanomaterials. With the ongoing globalization of nanomaterials, international coordination, and harmonization, there is an urgent need for making the sound regulatory approaches for the safer use of nanomaterials. This chapter reviews and outlines the risk associated with the nanomaterials, current regulatory approaches, and the efforts in developing the sustainable nanotechnology.

Multiple Choice Questions

Question 1. The issues of risks associated with the excessive release of nanomaterials in the environment are recognized through

(a) Risk governance
(b) Standard operating parameters
(c) Sustainable development
(d) Risk treatment

Question 2. Sustainable development of nanomaterials related to the

(a) Risk treatment
(b) Computing and biochemical analytics
(c) Consumption of the goods and resources
(d) Risk governance

Question 3. According to "tipping scales" and "nano Bhopal," the field of nanotechnology is considered as the search for

(a) Stagnation
(b) Novelty
(c) Utilization of resources
(d) Risk governance

Question 4. The common risk assessment methods for safer use of nanomaterials depend upon

(a) LCA
(b) FTA
(c) Nanotechnology assessment
(d) All of above

Question 5. Safety Data Sheets (SDS) have been considered as the important platform for analyzing the

(a) Safer use of chemical components
(b) Utilization of nanotoxicity
(c) Knowledge of hazardous substances
(d) All of above

Question 6. In 2010, SECO has provided the application of SDS guidelines for nanomaterials such as

(a) SECOKAT (photocatalyst)
(b) NANO-BLOGGO (surface finisher)
(c) a & b (both)
(d) None of above

Question 7. In Australia, a national policy body known as Safe Work Australia (SWA) was established for investigating out the major points of SDS for nano range materials in the year

(a) 2012
(b) 2018
(c) 2010
(d) None of above

Question 8. To qualify and demonstrate the scientific research and their ability for technological expansion, which proposal was established

(a) Five-year plans (1980–1985)
(b) IRHPA
(c) All of above
(d) None of above

Question 9. National Institute for Occupational Safety and Health (NIOSH) has analyzed that SDS has formatted during the year of

(a) 2008–2018
(b) 2007–2011
(c) 2015–2017
(d) None of above

Question 10. The toxic nature of material can be controlled from the

(a) Sustainability use
(b) Precursor
(c) Physicochemical properties of parent material
(d) All of above

Short Questions

Question 1. Explain the concept of risk governance in terms of sustainability.
Question 2. Explain sustainable development in the field of nanotechnology and upon what factor it depends upon.
Question 3. Explain different factors upon which risk assessment of nanomaterial sustainability depends.
Question 4. Explain different parameters that are related to risk treatment in nanotechnology.
Question 5. What are the key points for the development of nanomaterials according to safety data sheets?

Multiple Choice Question Answers

Answer 1. (a)
Answer 2. (c)
Answer 3. (b)
Answer 4. (d)
Answer 5. (d)
Answer 6. (c)
Answer 7. (c)
Answer 8. (c)
Answer 9. (b)
Answer 10. (d)

Short Question Answers

Answer 1. Risk governance here refers to the application of the governing approach to tackle the issues of risks associated with the excessive release of nanomaterials in the environment. In particular, the governance works in portions, that is, firstly, the risk governance recognizes that decisions about issues of risks are not perspectives of the group of people rather those are scientifically evident facts. After the validation and assessment of risks, the preventive or safety frameworks are designed. The designing of the framework mainly involved the scientific, aesthetic, and administrative factors. In addition, all the financial feasibility aspects need to be taken care for the proper management of environmental impact of hazardous materials.

Answer 2. "Sustainable development" refers to the production and consumption of goods and resources in a manner that the needs of the current generations can be satisfied without compromising, limiting, or threatening the needs and environmental conditions of future generations. For nanotechnology, a wealth of applications has been proposed. For instance, nanotechnology enables the manufacturing process with lesser energy consumption and waste generations, thus saving the expenditure on carbon trading. Nanotechnology is not restricted to the production of nanoparticles but can also be a decisive step of complex productions. Such ways lead to the macro productions of products. For instance, the potential application of nano-based catalysts in the sector of energy production. On a bigger approach, certain converging technologies involve various macro-production and result in "meta-technologies" such as computing and biochemical analytics. The interdependence of these technologies with each other has shown that attributing sustainability to any technology is very tricky due to the intertwined network of production protocols.

Answer 3. Many scientific methods have been developed to estimate the sustainability of the nanomaterials. These methods mainly focus on the understanding of environmental and the societal and economic dimensions of chosen particles. The common risk assessment methods for safer use of nanomaterials have been discussed below:

(a) Life cycle assessment
(b) Future technology analyses
(c) Nanotechnology assessment

Answer 4. The concept of risk treatment for the sustainable development of nanotechnology majorly depends upon four factors such as:

(a) Involvement of social, economic, and biophysical factors
(b) Guidance on how to use and deal
(c) Consider future cycle for the remediation
(d) Participation and contribution to the assurance

Answer 5. The international organization for standardization includes safety data sheets (SDS) for the synthesis of nanoparticles, utilization, and their practical analysis. Following are the key points that are included for the formatting of SDS:

- The marking of SDS data should be very fast, due to enlargement in data availability.
- Transparent and strong information should be provided for the nanoparticles when CAS number of bulk materials is used.
- Statement and assertion should be provided when toxicological evidence is not accessible.
- Declaration of statement provided for the exposure limits and application of bulk and nanomaterials.

References

Adams FC, Barbante C (2013) Nanoscience, nanotechnology and spectrometry. Spectro Chim Acta B At Spectrosc 86:3–13. https://doi.org/10.1016/j.sab.2013.04.008

Atia NG, Bassily MA, Elamer AA (2020) Do life-cycle costing and assessment integration support decision-making towards sustainable development? J Clean Prod 267:122056. https://doi.org/10.1016/j.jclepro.2020.122056

Baker JR, Ward BB, Thomas TP (2017a) Chapter 11: Nanotechnology in clinical and translational research. In: Clinical and translational science (2nd ed), pp 195–205. https://doi.org/10.1016/B978-0-12-802101-9.00011-9

Baker S, Volova T, Prudnikova SV, Satish S, Prasad MN (2017b) Nanoagroparticles emerging trends and future prospect in modern agriculture system. Environ Toxicol Pharmacol 53:10–17. https://doi.org/10.1016/j.etap.2017.04.012

Bardos RP, Bone BD, Boyle R, Evans F, Harries ND, Howard T, Smith JW (2016) The rationale for simple approaches for sustainability assessment and management in contaminated land practice. Sci Total Environ 563:755–768

Bauer C, Buchgeister J, Hischeir R, Poganietz WR, Schebek L, Warsen J (2008) Towards a framework for life cycle thinking in the assessment of nanotechnology. J Clean Prod 16:910–926. https://doi.org/10.1016/j.jclepro.2007.04.022

Baun A, Sayre P, Steinhauser KG, Rose J (2010) Regulatory relevant and reliable methods and data for determining the environmental fate of manufactured nanomaterials. NanoImpact 8:1–10. https://doi.org/10.1016/j.impact.2017.06.004

Bhatio S (2016) Chapter 2: Nanoparticles types, classification, characterization, fabrication methods and drug delivery applications. Springer, Cham, pp 33–93. https://doi.org/10.1007/978-3-319-41129-3_2

Boldrin A, Hansen SF (2014) Environmental exposure assessment framework for nanoparticles in solid waste. J Nanopart Res 16:2394

Buckley JA, Thompson PB, Whyte KP (2017) Collingridge's dilemma and the early ethical assessment of emerging technology: the case of nanotechnology enabled biosensors. Technol Soc 48:54–63. https://doi.org/10.1016/j.techsoc.2016.12.003

Buerki-Thurnherr T, Xiao L, Diener L, Arslan O, Hirsch C, Maeder-Althaus X, Grieder K, Wampfler B, Mathur S, Wick P, Krug H (2012) *In vitro* mechanistic study towards a better understanding of ZnO nanoparticle toxicity. Nanotoxicology 7:402–416

Bystrzejewska-Piotrowska G, Gollimowski J, Urban PL (2009) Nanoparticles: their potential toxicity, waste and environmental management. Waste Manage 29:2587–2595. https://doi.org/10.1016/j.wasman.2009.04.001

Chakraborty D, Chauhan P, Kumar S, Chaudhary S, Chandrasekaran N, Mukherjee A, Ethiraj KR (2019) Utilizing corona on functionalized selenium nanoparticles to design doxorubicin delivery system. J Mol Liq 296:11864

Chaudhary S, Sharma P, Kumar R, Mehta SK (2015) Nanoscale surface designing of Cerium oxide nanoparticles for controlling growth, stability, optical and thermal properties. Ceram Int 41:10995–11003. https://doi.org/10.1016/j.ceramint.2015.05.044

Chaudhary S, Sharma P, Renu, Kumar R (2016) Hydroxyapatite doped CeO_2 nanoparticles: impact on biocompatibility and dye adsorption properties. RSC Adv 6:62797–62809

Chaudhary S, Sharma P, Singh D, Umar A, Kumar R (2017) Chemical and pathogenic cleanup of wastewater using surface-functionalized CeO_2 nanoparticles. ACS Sustain Chem Eng 5:6803–6816

Chaudhary S, Sharma P, Kumar S, Alex SA, Kumar R, Mehta SK, Mukherjee A, Umar A (2018) A comparative multi-assay approach to study the toxicity behaviour of Eu_2O_3 nanoparticles. J Mol Liq 269:783–795

Chaudhary S, Rohilla D, Umar A, Kaur N, Shanavas A (2019) Synthesis and characterizations of luminescent copper oxide nanoparticles: toxicological profiling and sensing applications. Ceram Int 45:15025–15035

Chauhan P, Dogra S, Chaudhary S, Kumar R (2020) Usage of coconut coir for sustainable production of high-valued carbon dots with discriminatory sensing aptitude toward metal ions. Mater Today Chem 16:100247

Corsi I, Winther-Nielsen M, Sethi R, Punta C, Torre Della C, Libralato G et al (2018) *Ecofriendly* nanotechnologies and nanomaterials for environmental applications: key issue and consensus recommendations for sustainable and ecosafe nanoremediation. Ecotoxicol Environ Saf 154:237–244. https://doi.org/10.1016/j.ecoenv.2018.02.037

Davies JC (2008) Nanotechnology oversight: an agenda for the new administration. Project on Emerging Nanotechnologies is supported by THE PEW CHARITABLE TRUSTS. http://emerginglitigation.shb.com/Portals/f81bfc4f-cc59-46fe-9ed5-7795e6eea5b5/pen13.pdf

Dixit AM, Shrestha SN, Guragain R, Pandey BH, Oli KS, Adhkari SR et al (2018) Chapter 5: Risk management, response, relief, recovery, reconstruction, and future disaster risk reduction. Impacts and insights of Gorkha earthquake, pp 95–134

Durante S, Comoglio M, Ridgway N (2015) Chapter 34: Life cycle assessment in nanotechnology, materials and manufacturing. In: Micromanufacturing engineering and technology (2nd ed), pp 775–804. https://doi.org/10.1016/B978-0-323-31149-6.00034-7

Dutta T, Kim KH, Deep A, Szulejko JE, Vellingiri K, Kumar S (2018) Recovery of nanomaterials from battery and electronic wastes: a new paradigm of environmental waste management. Sust Energ Rev 82:3694–3704. https://doi.org/10.1016/j.rser.2017.10.094

Esmaeili A, Sbarufatti C, Casati R, Jimenez-Suarez A, Urena A, Hamouda AMS (2020) Effective addition of nanoclay in enhancement of mechanical and electromechanical properties of SWCNT reinforced epoxy: strain sensing and crack-induced piezoresistivity. Theor Appl Fract Mech 110:102831. https://doi.org/10.1016/j.tafmec.2020.102831

Eyo EU, Abbey SJ, Ngambi S, Ganjian E, Coakley E (2020) Incorporation of a nanotechnology-based product in cementitious binders for sustainable mitigation of sulphate-induced heaving of stabilised soils. Eng Sci Tech Int J. https://doi.org/10.1016/j.jestch.2020.09.002

Falkner R, London School of Economics and Jasperse N, Free University Berlin (2012) Regulating nanotechnologies: risk, uncertainty and the global governance gap. Glob Environ Polit 12: 30–35

Fleischer T, Grunwald A (2008) Making nanotechnology developments sustainable. A role for technology assessment? J Clean Prod 16:889–898

Gajewick A, Rasuley B, Dinadayalane TC, Urbaszek P, Puzyn T, Leszczynska D, Leszczynski J (2012) Advancing risk assessment of engineered nanomaterials: application of computational approaches. Adv Drug Deliv Rev 64:1663–1693. https://doi.org/10.1016/j.addr.2012.05.014

Germann, Mora IG, Alfonso (2016) Sovereign disaster risk finance in middle income countries: a partnership with the Swiss state secretariat for economic affairs (SECO). 1:107922. World Bank Group, Washington, DC. http://documents.worldbank.org/curated/en/978941471933369121/Sovereign-disaster-risk-finance-in-middle-income-countries-a-partnership-with-the-Swiss-state-secretariat-for-economic-affairs-SECO

Hansen SF, Baun A (2012) European regulation affecting nanomaterials. Rev Limit Future Recomm 10:364–368. https://doi.org/10.2203/dose-response.10-029.Hansen

Hegde K, Brar SK, Verma M, Surampalli RY (2016) Current understandings of toxicity, risks and regulations of engineered nanoparticles with respect to environmental microorganisms. Nanotechnol Environ Eng 1:5. https://doi.org/10.1007/s41204-016-0005-4

Helland A, Kastenholz H (2008) Development of nanotechnology in light of sustainability. J Clean Prod 16:885–888. https://doi.org/10.1016/j.jclepro.2007.04.006

Hossain SK, Mukherjee SK (2013) Toxicity of cadmium sulfide (CdS) nanoparticles against *Escherichia coli* and HeLa cells. J Hazard Mater 260:1073–1082

Hossain MU, Wang L, Chen L, Tsang DW, Thomas Ng S, Poon CS et al (2020) Evaluating the environmental impacts of stabilization and solidification technologies for managing hazardous wastes through life cycle assessment: a case study of Hong Kong. Environ Int 145:106139. https://doi.org/10.1016/j.envint.2020.106139

Hosseinzadeh-Bandbafah H, Tabatabei M, Aghbashlo M, Khanali M, Khalife E, Shojaei TR et al (2020) Consolidating emission indices of a diesel engine powered by carbon nanoparticle-doped diesel/biodiesel emulsion fuels using life cycle assessment framework. Fuel 267:117296. https://doi.org/10.1016/j.fuel.2020.117296

Hristozov D, Gottardo S, Semenzin E, Oomen A, Bos P, Peijnenburg W (2016) Frameworks and tools for risk assessment of manufactured nanomaterials. Environ Int 95:36–53. https://doi.org/10.1016/j.envint.2016.07.016

Hussain S, khaliq A, Matllob A, Wahid A M, Afzal I (2013) Germination and growth response of three wheat cultivars to NaCl salinity. Soil Environ 32:36–43

Iavicoli I, Leso V, Beezhold DH, Shvedova AA (2017) Nanotechnology in agriculture: opportunities, toxicological implications, and occupational risks. Toxicol Appl Pharmacol 329:96–111. https://doi.org/10.1016/j.taap.2017.05.025

Ittipanuvat V, Fujita K, Sakata I, Kajikawa Y (2014) Finding linkage between technology and social issue: a literature based discovery approach. J Eng Tech Mang 32:160–184. https://doi.org/10.1016/j.jengtecman.2013.05.006

Jagusiak A, Goclon J, Panczyk T (2021) Adsorption of Evans blue and Congo red on carbon nanotubes and its influence on the fracture parameters of defective and functionalized carbon nanotubes studied using computational methods. Appl Surf Sci 539:148236. https://doi.org/10.1016/j.apsusc.2020.148236

Jayanthi AP, Beumer K, Bhattacharya S (2012) Nanotechnology: risk governance in India. Econ Polit Week 4:34–40

Kamali M, Persson KM, Costa ME, Capela I (2019) Sustainability criteria for assessing nanotechnology applicability in industrial wastewater treatment: current status and future outlook. Environ Int 125:261–276. https://doi.org/10.1016/j.envint.2019.01.055

Kamarulzaman NA, Lee KE, Sio KS, Mokhtar M (2020) Public benefit and risk perceptions of nanotechnology development: psychological and sociological aspects. Technol Soc 62:101329. https://doi.org/10.1016/j.techsoc.2020.101329

Kanagaraj J, Senthivelan T, Panda RC, Kavitha S (2015) Eco-friendly waste management strategies for greener environment towards sustainable development in leather industry: a comprehensive review. J Clean Prod 89:1–17

Karim ME (2020) Chapter 31: Functional nanomaterials: selected occupational health and safety concerns. In: Handbook of functionalized nanomaterials for industrial applications, pp 995–1006. https://doi.org/10.1016/B978-0-12-816787-8.00031-4

Karjalainen T, Hoeveler A, Draghia-Akli R (2017) European Union research in support of environment and health: building scientific evidence base for policy. Environ Int 103:51–60. https://doi.org/10.1016/j.envint.2017.03.014

Khataee A, Pouran SR, Hassani A (2020) Editorial note-special issue on "Ultrasonic nanotechnology: new insights into industrial and environmental applications". Ultrason Sonochem 65:104878. https://doi.org/10.1016/j.ultsonch.2019.104878

Krajnik P, Rashid A, Pusavec F, Remskar M, Yui A, Nikkam M et al (2016) Transitioning to sustainable production – Part III: Developments and possibilities for integration of nanotechnology into material processing technologies. J Clean Prod 112:1156–1164. https://doi.org/10.1016/j.jclepro.2015.08.064

Kuang L, Burgess B, Cuite CL, Tepper BJ, Hallman WK (2020) Sensory acceptability and willingness to buy foods presented as having benefits achieved through the use of nanotechnology. Food Qual Pref 83:103922. https://doi.org/10.1016/j.foodqual.2020.103922

Kumar H, Kumari N, Sharma R (2020a) Nanocomposites (conducting polymer and nanoparticles) based electrochemical biosensor for the detection of environment pollutant: its issues and challenges. Environ Impact Assess Rev 85:106438. https://doi.org/10.1016/j.eiar.2020.106438

Kumar SB, Padhi RK, Mohanty AK, Satpathy KK (2020b) Distribution and ecological- and health-risk assessment of heavy metals in the seawater of the southeast coast of India. Marine Poll Bull 161:111712. https://doi.org/10.1016/j.marpolbul.2020.111712

Lee JH, Kuk WK, Kwon M, Lee JH, Lee KS, Yu J (2011) Evaluation of information in nanomaterial safety data sheets and development of international standard for guidance on preparation of nanomaterial safety data sheets. Nanotoxicology 7:338–345. https://doi.org/10.3109/17435390.2012.658095

Li SQ, Zhu RR, Zhu H, Xue M, Sun XY, Yao SD, Wang SL (2008) Nanotoxicity of TiO_2 nanoparticles to erythrocyte in vitro. Food Chem Toxicol 46:3626–3631. https://doi.org/10.1016/j.fct.2008.09.012

Lo CC, Wang C, Chien PY, Hung CW (2012) An empirical study of commercialization performance on nanoproducts. Technovation 32:168–178. https://doi.org/10.1016/j.technovation.2011.08.005

Lu Y, Ozcan S (2015) Green nanomaterials: on track for a sustainable future. Nanotoday 10:417–420

Lu ML, Putz-Anderson V, Garg A, Davis KG (2016) Evaluation of the impact of the revised national institute for occupational safety and health lifting equation. Human Factors 58-667-682. https://doi.org/10.1177/0018720815623894

Ma DM, Xian Y, Ding A, Guo H, Qian DJ (2020) Interfacial self-assembled thioxathone monolayers on the surfaces of silica nanoparticles as efficient heterogeneous photocatalysts for the selective oxidation of aromatic thioethers under air atmosphere. Colloid Surf A: Physicochem 125856. https://doi.org/10.1016/j.colsurfa.2020.125856
Michelson ES (2008) Globalization at the nano frontier: the future of nanotechnology policy in the United States, China, and India. Technol Soc 30:405–410. https://doi.org/10.1016/j.techsoc.2008.04.018
Millard DC, Nicolinj MA, Arrowood CA, Hayes HB, Bradley JA et al (2019) Evaluating the use of microelectrode array technology and cell-based neuronal culture models for proconvulsant risk assessment: progress from the HESI NeuTox Consortium. J Pharmacol Toxicol Methods 99:106595
Musee N (2011) Nanotechnology risk assessment from a waste management perspective: are the current tools. Hum Exp Toxicol:1–16
Nabipour H, Hu Y (2020) 4: Sustainable drug delivery systems through green nanotechnology. Nanoeng Biomater Adv Drug Deliv:61–89. https://doi.org/10.1016/B978-0-08-102985-5.00004-8
Nawale VP, Malpe DB, Marghade D, Yenkie R (2021) Non-carcinogenic health risk assessment with source identification of nitrate and fluoride polluted groundwater of Wardha sub-basin, central India. Ecotoxicol Environ Safe 208:111548. https://doi.org/10.1016/j.ecoenv.2020.111548
Nazarko T (2017) Future-oriented technology assessment. Procedia Eng 182:505–509. https://doi.org/10.1016/j.proeng.2017.03.144
Nowack B (2017) Evaluation of environmental exposure models for engineered nanomaterials in a regulatory context. NanoImpact 8:38–47. https://doi.org/10.1016/j.impact.2017.06.005
Oke AE, Aigbavboa AV, Semenya K (2017) Energy savings and sustainable construction: examining the advantages of nanotechnology. Energy Procedia 142:3839–3843. https://doi.org/10.1016/j.egypro.2017.12.285
Oliveira L, Messagie M, Rangaraju S, Hernandez M, Sanfelix J, Mierlio JV (2017) Chapter 7: Life cycle assessment of nanotechnology in batteries for electric vehicles. In: Emerging nanotechnologies in rechargeable energy storage systems, pp 231–251. https://doi.org/10.1016/B978-0-323-42977-1.00007-8
Oomen AG, Steinhauser KG, Bleeker EAJ, Broekhouzen F, Slip A, Dekkers S et al (2018) Risk assessment frameworks for nanomaterials: scope, link to regulations, applicability, and outline for future directions in view of needed increase in efficiency. NanoImpact 9:1–13. https://doi.org/10.1016/j.impact.2017.09.001
Part F, Zecha G, Causon T, Sinner EK, Humer EH (2015) Current limitations and challenges in nanowaste detection, characterisation and monitoring. Waste Manag 43:407–420. https://doi.org/10.1016/j.wasman.2015.05.035
Peterson MJ (2009) Bhopal plant disaster – situation summary. International dimensions of ethics education in science and engineering case study series. https://scholarworks.umass.edu/cgi/viewcontent.cgi?article=1004&context=edethicsinscience
Purohit R, Mittal A, Dalela S, Warudkar V, Purohit K, Purohit S (2017) Social, environmental and ethical impacts of nanotechnology. Mater Today Proc 4:5461–5467. https://doi.org/10.1016/j.matpr.2017.05.058
Rahman MM, Awual MR, Asiri AM (2020) Preparation and evaluation of composite hybrid nanomaterials for rare-earth elements separation and recovery. Sep Purif Technol 253:117515. https://doi.org/10.1016/j.seppur.2020.117515
Ramchandran V, Gernand JM (2020) Examining the *in vivo* pulmonary toxicity of engineered metal oxide nanomaterials using a genetic algorithm-based dose-response-recovery clustering model. Comput Toxicol 13:100113. https://doi.org/10.1016/j.comtox.2019.100113
Rashid SS, Liu YQ (2020) Comparison of life cycle toxicity assessment methods for municipal wastewater treatment with the inclusion of direct emissions of metals, PPCPs and EDCs. Sci Total Environ 143849. https://doi.org/10.1016/j.scitotenv.2020.143849

Rejeski D, Lekas D (2008) Nanotechnology field observations: scouting the new industrial west. J Clean Prod 16:1014–1017. https://doi.org/10.1016/j.jclepro.2007.04.014
Ridsdale DR, Noble BF (2016) Assessing sustainable remediation frameworks using sustainability principles. J Environ Manag 184:36–44
Rohilla D, Chaudhary S, Singh N, Batish DR, Singh HP (2020) Agronomic providences of surface functionalized CuO nanoparticles on *Vigna radiata*. Environ Nanotechnol Monit Manag 14:100338
Rossi M, Cubadda F, Dini L, Terranova ML, Aureli F, Sorbo A et al (2014) Scientific basis of nanotechnology, implications for the food sector and future trends. Trends Food Sci Technol 40:127–148. https://doi.org/10.1016/j.tifs.2014.09.004
Salehpour T, Khanali M, Rajabipour A (2020) Environmental impact assessment for ornamental plant greenhouse: life cycle assessment approach for primrose production. Environ Pollut 266:115258. https://doi.org/10.1016/j.envpol.2020.115258
Salieri B, Turner DA, Nowack B, Hischier R (2018) Life cycle assessment of manufactured nanomaterials: where are we? NanoImpact 10:108–120. https://doi.org/10.1016/j.impact.2017.12.003
Sanchez A, Recillas S, Font X, Casals E, Gonzalez E, Puntes V (2011) Ecotoxicity of, and remediation with, engineered inorganic nanoparticles in the environment. Trends Anal Chem 30:507–516. https://doi.org/10.1016/j.trac.2010.11.011
Schulte PA, Geraci CL, Murashov V, Kuempel ED, Zumwalde RD, Castranova V et al (2014) Occupational safety and health criteria for responsible development of nanotechnology. J Nanopart Res 16:2153. https://doi.org/10.1007/s11051-013-2153-9
Senan-Salinas J, Landaburu-Aguirre J, Blanco A, Garcia-Pacheco R, Garacia-Calvo E (2020) Data of the life cycle impact assessment and cost analysis of prospective direct recycling of end-of-life reverse osmosis membrane at full scale. Data Brief 33:106487. https://doi.org/10.1016/j.dib.2020.106487
Sharma P, Kaun S, Chaudhary S, Umar A, Kumar R (2018) Bare and nonionic surfactant-functionalized praseodymium oxide nanoparticles: toxicological studies. Chemosphere 209:1007–1020. https://doi.org/10.1016/j.chemosphere.2018.06.041
Sharma P, Chaudhary S, Kumar R (2020) Assessment of biotic and abiotic behaviour of engineered SiO_2 nanoparticles for predicting its environmental providence. NanoImpact 17:100200. https://doi.org/10.1016/j.impact.2019.100200
Sia PD (2017) Nanotechnology among innovation, health and risks. Procedia Soc Behav Sci 237:1076–1080. https://doi.org/10.1016/j.sbspro.2017.02.158
Solaimani M (2020) Binding energy and diamagnetic susceptibility of donor impurities in quantum dots with different geometries and potentials. Mater Sci Eng B262:114694. https://doi.org/10.1016/j.mseb.2020.114694
Soltani AM, Pouypouy H (2019) Chapter 19: Standardization and regulations of nanotechnology and recent government policies across the world on nanomaterials. In: Advances in phytonanotechnology, pp 419–446. https://doi.org/10.1016/B978-0-12-815322-2.00020-1
Song W, Zhang J, Guo J, Zhang J, Ding F, Li L et al (2010) Role of the dissolved zinc ion and reactive oxygen species in cytotoxicity of ZnO nanoparticles. Toxicol Lett 199:389–397. https://doi.org/10.1016/j.toxlet.2010.10.003
Stone V, Fuhr M, Feindt PH, Bouwmeester H, Linkov I, Sabella S et al (2018) The essential elements of a risk governance framework for current and future nanotechnologies: perspective. Risk Anal 38:1321–1331. https://doi.org/10.1111/risa.12954
Stucki T, Woerter M (2019) The private returns to knowledge: a comparison of ICT, biotechnologies, nanotechnologies, and green technologies. Technol Forecast Soc Change 145:62–81. https://doi.org/10.1016/j.techfore.2019.05.011
Subramanian V, Semenzin E, Hristozov D (2014) Sustainable nanotechnology: defining, measuring and teaching. NanoImpact 9:6–9. https://doi.org/10.1016/j.nantod.2014.01.001
Trybula W, Newberry D (2014) Nanotechnology risk assessment. Nanotechnol Saf:195–206
Vance ME, Kuiken T, Vejerano EP, McGinnis SP, Hochella MF Jr, Rejeski D et al (2015) Nanotechnology in the real world: redeveloping the nanomaterial consumer products inventory Marina. Beilstein J Nanotechnol 6:1769–1780. https://doi.org/10.3762/bjnano.6.181

Visentin C, Silva Trentin AWD, Braun AB, Thome A (2021) Life cycle sustainability assessment of the nanoscale zero-valent iron synthesis process for application in contaminated site remediation. Environ Pollut 268:115915. https://doi.org/10.1016/j.envpol.2020.115915

Vishwakarma V, Samal SS (2010) Safety and risk associated with nanoparticles – a review. J Miner Mater Chara Eng 9:455–459

Wang L, Nagesha K, Selvarasah DS, Dokmeci RM, Carrier M (2008) Toxicity of CdSe nanoparticles in Caco-2 Cell cultures. J Nanobiotech 6:1–15

Wu H, Li D, Zhu X, Yang C, Liu D, Chen X, Song Y, Lu L (2014) High-performance and renewable supercapacitors based on TiO_2 nanotube array electrodes treated by an electrochemical doping approach. Electrochim Acta 116:129–136. https://doi.org/10.1016/j.electacta.2013.10.092

Xiong J, Zhu J, He Y, Ren S, Huang W, Lu F (2020) The application of life cycle assessment for the optimization of pipe materials of building water supply and drainage system. Sustain Cities Soc 60:102267. https://doi.org/10.1016/j.scs.2020.102267

Yao X, Cao Y, Zheng G, Devlin TA, Yu B, Hou X et al (2020) Use of life cycle assessment and water quality analysis to evaluate the environmental impacts of the bioremediation of polluted water. Sci Total Environ 143260. https://doi.org/10.1016/j.scitotenv.2020.143260

Younis SA, El-Fawal MA, Serp M (2018) Nano-wastes and the environment: potential challenges and opportunities of nano-waste management paradigm for greener nanotechnologies. In: Handbook of environmental materials management, pp 1–65. https://doi.org/10.1007/978-3-319-58538-3_53-1

Zhang J, Chen J, Ren J, Guo W, Li X, Chen R et al (2018) Biocompatible semiconducting polymer nanoparticles as robust photoacoustic and photothermal agents revealing the effects of chemical structure on high photothermal conversion efficiency. Biomaterials 181:92–102. https://doi.org/10.1016/j.biomaterials.2018.07.042

Chapter 12
Misconceptions in Nanotoxicity Measurements: Exploring Facts to Strengthen Eco-Safe Environmental Remediation

Chitven Sharma, Deepika Bansal, and Sanjeev Gautam

Abstract Nanotechnology has quickly increased its employment in various fields, be it sensors, paints, cosmetics, food packaging, drug delivery, or wastewater treatment. The nanotechlogical advances are due to its nanosize and surface area properties, but at the same time, another problem named "nanotoxicity" comes into the discussions. Properties that prove to help treat certain diseases may possess toxicity simultaneously. Although some nanomaterials are toxic, some misunderstandings are related to toxicity in general. These misconceptions mainly arise due to a lack of accurate data about the toxicities. This chapter deals with the properties of nanomaterials that make them toxic and how these properties can be altered to produce a less negative impact. Also, the commonly used nanomaterials and their myths have been discussed. Some nanoparticles may prove toxic due to their synthesis techniques; hence, green synthesis can be used as an alternative method. Also, the toxic NPs should be disposed of carefully using various methods, as explained. The future of nanotoxicity involves the better learning of nano-bio interactions.

Keywords Nanotoxicity · Nano-waste · Nano-bio interface · Misconceptions · Nanotoxicology

C. Sharma · D. Bansal · S. Gautam (✉)
Advanced Functional Materials Lab., Dr. S. S. Bhatnagar University Institute of Chemical Engineering & Technology, Panjab University, Chandigarh, India
e-mail: sgautam@pu.ac.in

R. Kumar et al. (eds.), *Advanced Functional Nanoparticles "Boon or Bane" for Environment Remediation Applications*, Environmental Contamination Remediation and Management, https://doi.org/10.1007/978-3-031-24416-2_12

12.1 Introduction

Nanotechnology has emerged as a revolutionary field for the modern era research areas of science and technology, leading to the fastest growing sector worldwide. Nanomaterials (NMs) have gained popularity in diverse applications ranging from biochemistry to electrochemistry. Their unique properties, such as the nanometer size of particles and greater surface area, are some of those aspects that have potentially resolved various present-day concerns. These concerns include drug delivery, multidrug-resistant bacteria, wastewater treatment, energy storage, industrial catalysts, biodegradable materials, UV protection, etc. However, at times, the same attributes that prove beneficial might give appalling results of nanomaterials, which is termed nanotoxicity.

12.1.1 What Is Nanotoxicity?

Ever since NMs have come into existence, the researchers have always been astonished by the unexpected results they exhibit. With the rising application of nanotechnology in various fields, the concerns related to the toxic effects of these materials are also rising as it is directly linked with the safety of the consumers. Toxic here generally means the materials recommended as poisonous or hazardous for any living cell, and nanotoxicology is the study that deals with the toxicity of the NMs. The study provides information about the interactions of nanomaterials with living systems, where main stress is laid upon the interactivity of physical (e.g., size, shape, etc.) and chemical properties (e.g., surface chemistry, composition, etc.) of nanostructures with toxic responses of the biological system (Fischer and Chan 2007). Factors such as the chemistry of nanoparticles, morphological traits, particle aggregation, and biological aspects such as genetics and existing diseases also greatly determine the harmful effects NMs might pose on human health. These materials can also cause contamination of water, air, and soil resources after their disposal. Since these things surround us, hence it is likely that we might get exposed to them either through inhaling them, ingesting via the food we consume (Armstead and Li 2016) or through the dermal route (skin absorption) (Gaiser et al. 2009). Studies have revealed that the removal of NMs if inhaled is much more complicated than particles of greater size, leading to chronic health problems (Buzea et al. 2007).

Nevertheless, in some instances, these toxic properties may prove valuable to fight against diseases at the cellular level. The most common includes the medical treatment of cancer wherein NMs such as magnetic, polymeric, and lipids are used for the targeted drug delivery against cancer-causing cells. They are also used as unique cancer-detecting sensors and tumor-imaging agents (Amiri et al. 2019; Fernandez-Fernandez et al. 2011; Minelli et al. 2010). Apart from this, they have shown promising effects in food technology as antibacterial agents (Kumar et al. 2020) (Fig. 12.1).

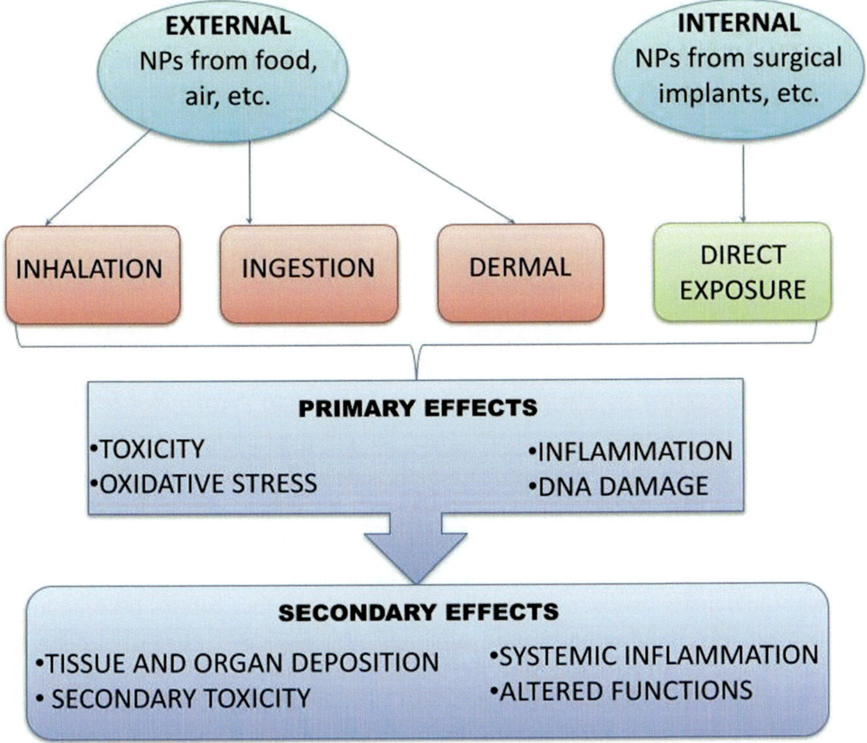

Fig. 12.1 Possible external and internal routes for the entry of NPs and the risks associated with them

The surface area plays a vital role in determining the toxic effects of nanoparticles, as it is the first line of interaction with the biological system (Fard et al. 2015). Since NMs are known for their greater surface-to-volume ratio, this property is one of the primary reasons for increased chemical reactivity at the nanoscale, which easily triggers the production of reactive oxygen species (ROS). These species may cause oxidative stress, inflammation in cell tissues, and damage the cell components (Fischer and Chan 2007). Also, NMs, due to their physical attributes, possess remarkable magnetic, optical, and electrical properties. So, there is a possibility that such nanostructures' breakdown might result in toxic effects. Apart from this, heavy metals, carbon-based materials, and Al-based nanostructures are already known for their toxicity. The breakdown of composites containing these materials might risk living organisms. The reason nanoparticles are thought to be more prone to toxicity is their size, which is much smaller than the size of cell and cell organelles. As a result, these particles easily pave their way inside these biological structures, causing disruption in the regular activity, thereby leading to serious health issues (Buzea et al. 2007). However, many researchers have regarded this as a common misconception, discussed in Sect. 12.2 of the chapter.

Although many theories have been put forward related to the toxicity of nanoparticles, the ground reality lies in the fact that the understanding of risks and measures for preventing them usually stands at a lower priority in the world of technology development (Maynard et al. 2006). Due to this, misconceptions related to the NMs have risen recently. People might lose confidence in this technology due to its natural and perceived fears, which can cause a major setback to this industry. Applying these particles with incomplete knowledge can prove even more hazardous for people who are producing and using them. Therefore, a proper insight into the risk factors, the strategies, and the targeted research to combat the present-day problems due to the NMs is the need of the hour.

12.1.2 Alteration in Properties from Macro to Nano Scale

The transition from bulk to nanoscale range causes many modifications from morphological characteristics to properties of particles. The leading factors that provoke NMs to act differently from macro-size materials include the surface and quantum effects (Buzea et al. 2007). As the particle size is reduced to 100 nm, the surface area vastly increases, as discussed earlier. Hence, the surface effects become more prominent at this stage, significantly affecting chemical and physical properties (Julien et al. 2011). In addition to it, the quantum effects also influence the behaviour, actions, reactivity, etc. of nanoparticles (Takei et al. 2011). The free electrons in the crystal lattice get confined as the size is reduced to the nanometer range, leading to remarkable changes in the properties of particles. For example, opaque substances become transparent (copper), insulators become conductors (silicon), solids turn to liquids at room temperature (gold), etc. Also, for instance, the NM, which seems to be nontoxic, might turn to give harmful results or vice versa owing to its characteristic reactivity (Naseer et al. 2018). Therefore, such vast types of alterations are sometimes unpredictable, and this is probably why no proper theory related to the toxicity of NMs has been put forth.

12.2 Misconceptions About Nanotoxicity

Several misconceptions are prevailing in society regarding the toxicity of nanoparticles that need to be addressed at the earliest. The main reason is the lack of proper knowledge of handling these particles, which eventually gives people common assumptions. Some of them have been discussed below:

(a) The first and most frequently discussed belief is that their small size permits them to penetrate the tissues, cell organelles, and other biological structures due to their similarity in physical dimensions with biological molecules such as proteins, antibodies, etc. This leads to the damage of vital biological systems

(Fischer and Chan 2007). However, some studies have reported that they cannot just enter freely inside the biological systems despite their usual small size. Instead, it is the functional group attached to the surface of these particles that determines their activity. For example, a study was carried out using citrate-based gold nanoparticles where it was observed that NMs could make their way inside the mammalian cell but could not get into the cytoplasm or nucleus (Chithrani et al. 2006). Thus, by modifying their surface chemistry, the reactivity of NMs and their access to specific cells or cell organelles can be controlled (Chen and Gerion 2004).

(b) NMs are always more toxic than their bulk counterparts with similar compositions. However, this is a common myth that does not have proper evidence. Since properties change drastically from bulk to nanoscale, there's no linearity in their behavior, and this has probably forced us to consider nanoparticles to be toxic. For example, CNTs are consistently recognized as harmful; however, this is not the case in all conditions. Their morphological characteristics and synthesis techniques play a vital role (Allegri et al. 2016). Apart from this, the amount of dose used also dramatically determines the consequences in general (Naseer et al. 2018).

(c) Surface area and size are the critical factors influencing their toxicity. Nevertheless, these two factors alone cannot be held responsible; chemical properties also play their role. As one can always engineer NMs according to the requirement, their physical aspects can still be controlled. To reduce the toxic effects of these particles, they can be further doped with biocompatible materials such as chitosan (Naskar et al. 2019), starch (Famá et al. 2012), etc. to make nanocomposite (Fig. 12.2).

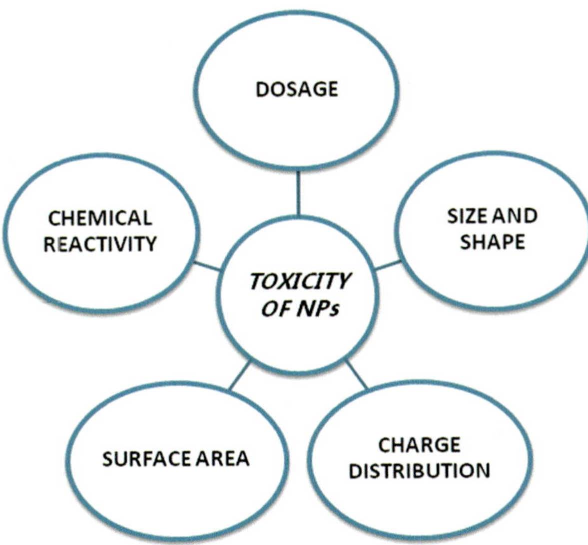

Fig. 12.2 The typical factors affecting toxicity of nanoparticles in general

In general, the toxic effects of nanoparticles are unpredictable. They are complex to understand as they involve various factors, including particle-cellular interactions, physicochemical attributes, routes and degrees of exposure, and several others. Hence, it is difficult to analyze the aftermaths of nanoparticle application in various fields. However, many researchers are working toward an effective solution to overcome this problem. Many methods have been proposed to check the toxic effects of these particles, including in vivo, in vitro, in silico, etc., which is elaborated in the next section. Studies have shown that the nanocomposite of these particles with biodegradable and biocompatible materials is an effective solution. The implementation of green synthesis over the conventionally used chemical synthesis route has also demonstrated remarkable effects to lessen the toxic effects. Thus, a proper study and integrated approach together can surely solve the existing problems, which eventually help in eradicating the common myths prevailing in society.

12.3 Nanotoxicity Measurements

12.3.1 In Vitro Studies

Just as in cellular biology, experiments are conducted outside the living organism; likewise, in vitro is carried out. The investigation includes cell cultures performed in a test tube and on petri dishes. These cultures serve as models for various animal and human tissues exposed to the NMs whose cytotoxicity must be checked. The cells selected for the tests are determined by the potential route through which NPs may enter the body. For example, tumor cell lines were used to check the toxic effects in cancer chemotherapy, whereas intestinal epithelium cells were used for estimating the toxicity of ingested NPs (Sukhanova et al. 2018). The nanoparticles must be supplied as aerosols instead of suspensions. In the latter case, the culture cells denied the nutrients from the apical pole because it faces the air, whereas the basal side gets exposure to nutrients. Thus, before the exposure to NPs, the cells must be cultured similarly from each side.

The advantage of this method is that it requires no concerns related to cross-species correlation, economic restrictions, and ethical considerations (Fröhlich and Salar-Behzadi 2014). They reduce the use of living organisms such as mice to a large extent for conducting the study, and due to this, it can be done several times at a large scale as cultures can be prepared easily. In addition, one does not have to master laboratory experience in animal handling. However, the major drawback of this technique is its failure to replace the actual cellular conditions existing in the living organisms. It may show results that do not relate to the circumstances occurring in the human body. Moreover, this cellular model cannot incorporate transport mechanisms such as blood, lymph, and bile present in the living system. Due to these concerns, the toxicity analysis was earlier carried out in vivo.

12.3.2 In Vivo Studies

Unlike in vitro studies, in vivo studies are performed directly on the whole living body or cells, especially on laboratory animals such as mice or humans and plants (Deng et al. 2007). These studies mainly include animal testing and clinical trials for carrying out the experiments. The two major routes through which nanoparticles are exposed to the organism are dermal or oral (direct respiration) routes. Since the study is carried out for NPs to be employed in biomedical applications, the main focus is on the toxic effects of any of these materials on the living body. The major limitation of this study is that it includes rodent species for conducting toxicity tests known to be obligatory nose breathers. This does not represent the model for human inhalation exposure as they breathe via both nose and oral (mouth) routes. Moreover, it is a comparatively expensive and time-consuming approach linked with ethical and regulatory issues. Nevertheless, above all these disadvantages, the plus point of this study is that it provides an actual living body condition, is clinically relevant, and can be tested for checking the chronic effects of NPs, which in vitro study otherwise fails to predict.

There are mainly three common pathways of inhalation exposure for the testing species: whole-body exposure, nose/head-only exposure, or lung-only exposure.

12.3.2.1 Whole-Body Exposure

This type of exposure aerosol is generated by compressing the filtered air, then heated, and added to dry filtered room air. A charge neutralizer is placed inside the exposure chamber to limit the electrostatic interaction with the chamber, and the aerosol is then directed inside it. The critical thing to be kept in mind while carrying out this procedure is that the aerosol concentration should be steady for the entire exposure time and free from contaminants and particle aggregation. However, this can be a significant challenge for NM-containing aerosols as they tend to agglomerate and make it difficult to break down easily. Animals receive a highly variable amount of doses in this method, mainly attributed to exposure via other routes such as mouth, eyes, etc. But animals might avoid getting exposed by gathering together or covering their nose in others fur or the corners of cages.

12.3.2.2 Nose/Head-Only Exposure

The chamber that holds the animal species is very small in this exposure method. A tube is generally attached to the section where aerosols are generated. The tube is directed toward the animal's nose, extending the chamber. A restraint is provided at the back end of the tube to prevent the animal from running away and seal the ending completely, which helps prevent the leakage of air around the testing system. However, due to the chamber's size constraints, the animal movement gets restricted and might result in causing discomfort. Some may even try to escape from testing conditions due to the suffocation, which can cause hindrance in carrying out further study. So to avoid ventilation issues, the minimum flow through the chamber should be 2.5 times that of the animal's minute volume.

12.3.2.3 Lung-Only Exposure

The method is also known as intratracheal instillation. It is carried out by injecting a delivery device into the animal's trachea and projecting its tip close to the bifurcation of the trachea. As soon as the animal takes in the material, it gets distributed to the lungs. For example, the method was used to distribute polystyrene NPs and beryllium oxide particles throughout the lungs (Rao et al. 2003). This approach might be risky as it may cause organ damage by dehydrating the trachea.

12.3.3 *In Silico Studies*

In silico study is a relatively new approach to toxicity testing (Shityakov et al. 2017). Its implementation can probably resolve the present-day limitations of in vitro and in vivo techniques. The experiment is entirely carried out on a computer or via computer simulation, particularly in biological experiments. Two models come under the deposition model, namely, empirical and mechanistic models. The practical model deals with mathematical equations that fit with experimental data, taking the pathways in the respiratory tract as identical, that is, having linear dimensions.

On the other hand, mechanical models calculate the deposition rate in the respiratory tract based on a realistic description of lung structure and physiology while considering the different breathing scenarios and parameters. The empirical model is generally used to calculate the deposition in the whole lung. In contrast, the mechanical models can deposit either across the entire lung or in a specific organ region. Both models are based on idealized descriptions of lung morphology and physiology and computational fluid dynamics (CFD) or estimation of inhaled aerosol flow by taking into account the fate of a large number of particles or an individual particle as per the Eulerian-Lagrangian models, respectively (Table 12.1).

According to the Marshall Protocol, there are varieties of in silico techniques

(a) Bacterial Sequencing Techniques: It has been developed as an alternative to in vitro methods for identifying bacteria through DNA and RNA sequence analysis. The technique uses polymerase chain reaction (PCR), which takes a single or few copies of a piece of DNA and replicates it to several orders of magnitude, leading to the generation of millions or more copies of a particular DNA sequence. This technique has allowed for the detection of bacteria that are responsible for a variety of conditions.

Table 12.1 Comparison between in vitro and in vivo methods based on general criteria

Measurement techniques	In vitro	In vivo
Testing species	Dead organisms or incubated cell cultures are used	A whole living organism is used, such as mice, rabbits, apes, etc.
Experiment conditions	Conducted under controlled laboratory conditions	Conducted under physiological conditions
Cost efficiency	It is a comparatively less expensive method	It is a comparatively more costly method
Time consumption	It is a fast process	It is comparatively time-consuming
Experimental results	Less precise results	More precise results
Limitations	In vitro methods have fewer restrictions	In vivo method have more restrictions

(b) Molecular Modeling: This design helps demonstrate how drugs and other substances interact with the nuclear receptors of cells. For example, computer-based results were reported for 25-D, a vitamin D metabolite, and capnine, a substance produced by bacteria that turns the vitamin D receptor off. The conclusions drawn are clinically approved. Another study was conducted to understand the interactions between single-walled CNT and the ATP entry point of the human mitochondrial voltage-dependent anion-selective channel where molecular docking and molecular dynamics simulations were performed (González-Durruthy et al. 2020).

(c) Whole-Cell Simulation: Researchers have devised a computer model of the crowded interior of a bacterial cell that accurately simulates the behavior of living cells in response to sugar in its environment.

These studies can even be carried out simultaneously to understand better the biological responses to various nanomaterials (Bai et al. 2018), as shown in Fig. 12.3. The sequence of performing these techniques can always be customized according to specific conditions.

12.4 Commonly Used Nanomaterials

12.4.1 Carbon Nanotubes

Carbon nanotubes, both single-walled and multiwalled, are cylindrical structures made up of graphene sheets. CNTs are highly tensile and about a hundred times stronger than steel. Also, CNTs can bend easily due to their elastic nature. However, several methods can be used to prepare CNTs; these are generally produced using chemical vapor deposition (CVD) method. CNTs show their use in several areas like energy storage, air and water purification, fabrics and fibers, etc. (Wilson et al.

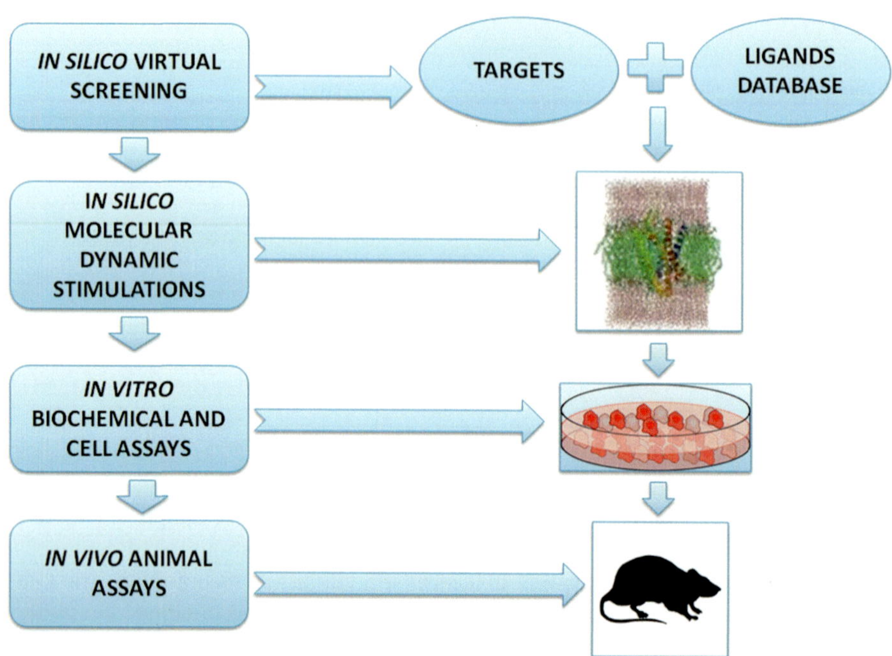

Fig. 12.3 Simultaneous in silico, in vivo, and in vitro experimental procedures for better results

2002). Also, CNTs are widely used in the field of medicine and pharmacy. These NPs can be used in cancer therapy, infection therapy, tissue generation, neurodegenerative diseases, and gene therapy in drug delivery. The usage of CNTs has been experiencing an ever-increasing rise, but its toxicity has been identified as a major barrier in its use. This is generally due to their tiny nanoparticle size. CNTs can be cancer-causing or carcinogenic due to their size or fiber length. These NPs may cause health issues similar to those caused by asbestos (Francis and Devasena 2018). The most crucial toxic issue caused by CNTs is pulmonary toxicity. It targets the lungs and the respiratory tract. The poisonous effects of CNTs can be classified into pulmonary toxicity, cardiovascular outcomes, and reproductive and developmental toxicity. Talking in detail about pulmonary toxicity, the respiratory tract is the entry point for NPs. These nanoparticles displace the air particles in the tract. Although the track possesses various reaction mechanisms that keep such foreign particles away, however, due to their nanosize, they easily travel to other organs. Once these NPs get deposited in the tract, they can travel to other parts like the nervous or cardiac system. The deposition of CNTs on the inner walls of the body may cause inflammation (Kayat et al. 2011). As per research, SWCNTs proved to be more toxic than MWCNTs. This is because SWCNTs have a smaller surface area, and due to this, they cause more apoptosis, which is a form of programmed cell death that occurs in multicellular organisms. Apart from this, carbon nanotubes can also cause damage to the central nervous system. When CNTs interact with the

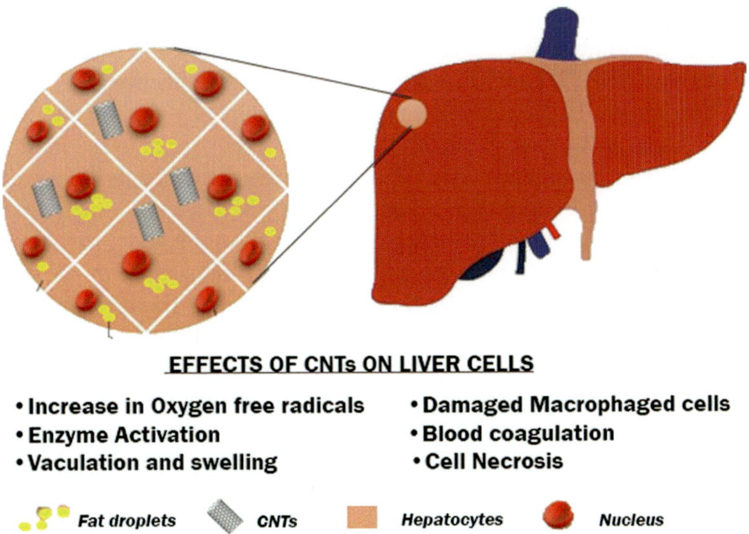

Fig. 12.4 Effects of carbon nanotubes on liver cells resulting in various damages (Mohanta et al. 2019)

brain cells, certain chemicals are released to cause inflammation or apoptosis. When CNTs contact cardiac cells, there may be some damage to the heart, which causes cell proliferation, muscle damage, hindrance in the blood flow, and vascular atherosclerosis (Mohanta et al. 2019). Only MWCNTs cause some toxicity that may result in mild irritation for the eyes.

On the other hand, SWCNTs do not prove toxic to the eyes. In addition to these body parts, CNTs may cause adverse effects on the spleen, kidneys, and liver. Due to all these reasons, CNTs are not being used for medical and other purposes on a large scale, even when they are excellent candidates in various fields. Particular methods like changing their size or modifying them into nanocomposites may help reduce their toxicity (Fig. 12.4).

12.4.2 Zinc Oxide

Zinc oxide (ZnO) nanoparticles serve as excellent materials for biomedical use. They show antibacterial and antimicrobial properties. These NPs are highly used in sunscreens because of their superb ultraviolet (UV)-absorbing properties and transparency for visible light. Most of the uses of ZnO come into play due to its ability to generate reactive oxygen species (ROS). For example, the mechanism of ZnO that helps it show excellent antibacterial, antimicrobial, and antifungal properties is the production of ROS, which results in cell death or apoptosis (Mishra et al. 2017). However, during in vivo and in vitro studies, it was found out that zinc oxide NPs can cause obstructions in development and egg hatching in zebrafish. When ZnO particles enter the lungs, they can cause severe inflammation and related responses. However, as most NPs contact the skin, the related skin reactions have not been reported. Apart from this, ZnO NPs can reach various body organs like the heart, spleen, liver, lung, and kidney. Zinc oxide NPs can prove toxic to the neurological system as they can cause apoptosis. When experimented on rats, these particles showed interference in spatial learning and memory by altering synaptic plasticity. They may generally hinder many biological processes in the body, including enzymatic actions. Usually, when proteins are deposited at the surface of the ZnO NPs, a layer called corona is formed, which may then seem like foreign particles to the immune system, thereby causing reactions (Sruthi et al. 2018).

There are specific methods in which their toxicity can be of great use too. One of such methods is to destroy the cancer cells. But this can destroy both good and bad cells. However, ZnO can be doped with certain materials or associated with another material to create a nanocomposite. As ZnO causes high oxidative stress in general, it can be coated with SiO_2 to decrease oxidative stress and cell toxicity. When ZnO NPs were doped with magnetic materials, their toxicity gets reduced. However, much research has not been performed on it as of now. Also, its stability increases on doping it with manganese. Hence, such methods can reduce toxicity and prove beneficial in different areas (Xue et al. 2011).

12.4.3 *Titanium Dioxide*

Titanium dioxide NPs show excellent optical properties, along with electrical properties (Nyamukamba et al. 2018). In areas of wastewater treatment, TiO_2 can be used to degrade organic and inorganic waste from wastewater. TiO_2 nanotubes help in the field of photocatalysis too. It finds uses in cosmetics, toothpaste, paints, enamels, paper industries, etc. Various chemical methods like sol-gel, chemical vapor deposition, physical vapor deposition, or hydrothermal methods are used to synthesize these NPs. Alternatively, green synthesis can also be used to prepare TiO_2 in an environment-friendly way. The entire process for green synthesis using *Echinacea purpurea* herba extract has been explained in Fig. 12.5 (Dobrucka 2017). Using this method, the average size of the obtained nanoparticles was observed to be around 120 nm, measured using SEM-EDS.

Considering the toxic effects of titanium dioxide, these NPs usually prove to be carcinogenic and cause genotoxicity and oxidative stress. These NPs generally enter through the respiratory tract. They then travel into the blood, but according to a few studies, they could also enter through skin tissues and get carried away to the other parts of the body (Shakeel et al. 2016). TiO_2 NPs can cause damage to microorganisms, plants, and humans. Their most common mechanism of damage includes rupturing the cell membrane. However, they prove to be highly useful, but research about their toxic effects is still being done on a priority basis (Grande and Tucci 2016).

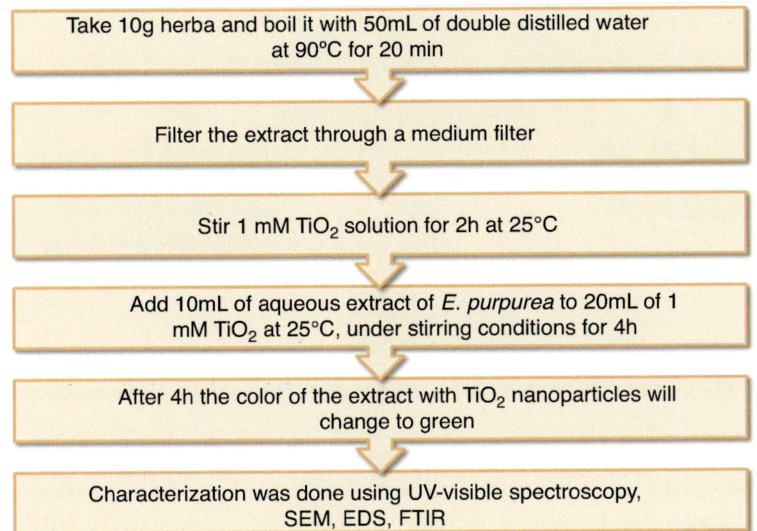

Fig. 12.5 Green synthesis of titanium dioxide NPs using *Echinacea purpurea* herba extract

12.4.4 Copper Oxide

CuO nanoparticles exhibit their use in various materials, including storage media, sensors, catalysts in rocket propellants, etc. (Ren et al. 2009). CuO NPs have excellent optical properties as well as antimicrobial properties. Usually, the sol-gel method is used to synthesize CuO as it is relatively faster and easier than the other methods available. NPs with sizes around 10–40 nm are obtained using this method. These NPs are highly toxic; the reason for this might be their extremely small size. Smaller CuO NPs are more harmful than larger ones. A positive charge indicates enhanced toxicity and dependence on pH and temperature. The toxicity of CuO may also include the mechanism of ROS generation, and these NPs may decrease levels of reduced glutathione (GSH). In various culture tests, high toxic behavior was observed for human lung cultured cells and human skin organ cultures. In an experiment, Chibber et al. treated neurons with CuO NPs that resulted in the disintegration of the membrane of the neurons. Once NPs overcome the resistance of the cell membrane, they enter the cell and generate ROS, which eventually causes cell death, as shown in Fig. 12.6 (Grigore et al. 2016).

12.4.5 Other Commonly Used Nanomaterials

The nanomaterials above are the most used ones in various industrial spheres. Apart from these, NPs like gold, silver, etc. also prove toxic at times.

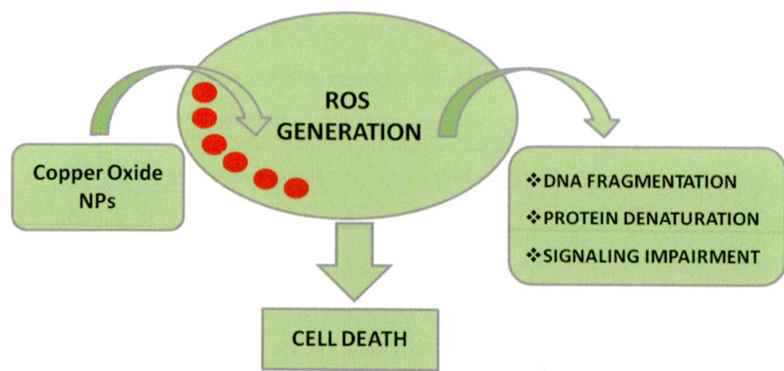

Fig. 12.6 Mechanism of ROS generation by copper oxide NPs resulting in damage to cells

Silver NPs possess antimicrobial properties that help produce bandages and medical devices to control bacterial growth. Due to their optical properties, Ag NPs can also harvest light. Various chemical methods can synthesize Ag NPs, but the green synthesis method is an eco-friendly option. Talking about its toxicity, Ag NPs may somehow get accumulated in the food chain, thus causing toxic effects to many organisms. When Ag NPs enter the cell by any means, reactive oxygen species are usually released, and dysfunctioning of mitochondria occurs. This can further damage the cell. The toxicity of Ag NPs sized less than 10 nm is the highest. Out of many tests performed to determine the toxicity of silver NPs, in vitro methods are more rapid, whereas in vivo methods take a longer time; however, in vivo display better results. As an alternative to both, in silico methods are rapid and produce accurate results. Also, they help achieve another motive of toxicity studies: prevention of animal sacrificing (Vazquez-Muñoz et al. 2017; Budama-Kilinc et al. 2018).

Gold NPs find their significant use in sensors, catalysis, electronics, diagnosis, etc. In food technology, Au NPS can be used in sensors that display whether some food is suitable for consumption or not. These are also used to diagnose heart disease infectious agents and are commonly used in pregnancy testing kits. Au NPs are being considered for use in fuel-cell application areas (Aldrich 2015). However, Au particles possess some toxic properties as well. In vivo and in vitro tests were performed on gold NPs to test their toxic behavior. As per the results of the in vitro tests, it was concluded that gold NPs induce ROS production, which may increase oxidative stress on living cells and cause cell damage due to apoptosis and necrosis. In in vivo methods, the intravenous method was the most commonly used. Hence, some research pertaining to inhalation, absorption through the skin, and oral absorption is still required to understand the toxicity of gold NPs better. According to a specific study, it was concluded that cationic gold nanoparticles could cause damage to the cell membrane when they come into contact with the negatively charged cellular membrane. Au NPs with a size around 1.4 nm can cause high oxidative stress on cells and damage them as a result. In contrast, those NPs with a length of about 15 nm did not damage the cell wall. In conclusion, many suggest that gold nanoparticles' toxicity may depend on their size (Alkilany and Murphy 2010; Jia et al. 2017).

Another common nanomaterial is silicon dioxide. Silicon dioxide NPs are highly stable and used in drug delivery systems. These NPs find their use in drug delivery systems, DNA detection, protein adsorption, humidity sensors, etc. (Rao et al. 2005). As the use of silica/silicon dioxide has been increasing rapidly, more and more humans are being exposed to this nanomaterial. SiO_2 possesses some toxic properties that may be fatal to live organisms (Murugadoss et al. 2017). Toxicity in SiO_2 is both size- and dose-dependent. These NPs generally attack the spleen and liver area. They may cause damage to the cell membrane due to ROS generation. Hence, SiO_2 may be useful generally but may prove fatal in high doses (Table 12.2).

Table 12.2 Typically determining toxic effects of commonly used nanomaterials on living systems

Commonly used nanomaterials	Toxic effects on living systems	References
Carbon nanotubes	Damages DNA	Simate et al. (2012) and Tejral et al. (2009)
	Carcinogenic	
	Skin irritation	
	Lung toxicity	
	Diminishing cellular activities	
Zinc oxide	Cytotoxicity	Vandebriel and De Jong (2012) and Sahu et al. (2013)
	Oxidative stress	
	Mitochondrial dysfunction	
	Toxicity in human lung cells	
Titanium dioxide	Disturb glucose and lipid homeostasis in mice and rats	Baranowska-Wójcik et al. (2020)
	Oxidative stress	
	Apoptosis or chromosomal instability	
	Oxidative damage	
	Impairment of antioxidative capacity	
Silica	Lung cancer	Merget et al. (2002)
	Silicosis	
	Chronic obstructive pulmonary disease (COPD)	
	Kidney disease	
Iron oxide	Acute cytotoxicity	Yarjanli et al. (2017)
	Oxidative stress	
	Protein aggregation in the neural cells	
	Iron accumulation	
Gold nanoparticles	Hepatotoxicity (liver)	Alkilany and Murphy (2010)
	Nephrotoxicity (kidney)	
	Immunogenicity	
	Hematological toxicity (blood)	
	Inflammatory and oxidative responses	
Silver nanoparticles	Histopathologic abnormalities	Korani et al. (2015)
	Toxicity in the muscle	
	Genotoxicity and carcinogenicity	
	Immune system toxicity	
	Reproductive and developmental toxicity	

(continued)

Table 12.2 (continued)

Commonly used nanomaterials	Toxic effects on living systems	References
Quantum dots	Bioaccumulation in organs and tissues	Hardman (2006) and Wu and Tang (2014)
	Cytotoxicity	
	Genetic material damage	
	Disturbed cell viability	
	Disordered immune cell	
Aluminum-based nanoparticles	Disturb the cell viability	Bahadar et al. (2016)
	Alter mitochondrial function	
	Increase oxidative stress	
	Alter tight junction protein expression of the blood-brain barrier (BBB)	
Copper oxide	Liver damage	De Jong et al. (2019)
	Histopathological alterations	
	Ulceration and degeneration	
	Morphological alterations in various organs	

12.5 Nano-Bio Interface

This is an era of engineered nanomaterials. There has been an increase in the liking of NPs in almost all fields, including medicines, biosensors, drug delivery, wastewater treatment, energy storage, etc. But as the use of engineered nanomaterials increases, their toxicity becomes more noticeable. As studied, most nanomaterials react with living cells and produce toxic results. The most common ways of entering NPs into living organisms are skin or inhalation. By both ways, it may get carried to different body organs and thus damage them. To understand the effects of nanomaterial toxicity on living organisms, it is crucial to understand a nano-bio interface. According to its definition, a nano-bio interface is between nanomaterials and biological systems that includes a dynamic series of interactions between nanomaterial surfaces and biological nanoscale or nanostructured surfaces (Foroozandeh et al. 2019).

When the NPs contact the organism's surface, they interact with proteins and a layer called protein-corona is formed (Lujan and Sayes 2017). Nano-bio interface usually involves the redox reaction mechanism leading to the formation and rummaging of reactive oxygen species (ROS). However, more research is required on understanding the mechanism of redox reactions. Several knowledge gaps lead to misconceptions in understanding the nano-bio interface. Critical information is needed as the study of nano-bio interface serves as an essential factor in drug delivery systems and biocompatibility (Tian et al. 2020).

12.6 Loopholes

Despite several studies being conducted on nanotechnology and its possible health hazards, there is still a wide knowledge gap between the survey of nanotoxicology and nanomaterial safety (Hu et al. 2016). It is imperative that the studies performed to analyze nanotoxicity of several NMs and record the biological responses for the same should be relevant to humans and the natural environment. However, in reality, the amount of dose of NMs used for carrying out the procedure in vivo and in vitro studies are very high, which makes it difficult to estimate the effects of the same materials on living organisms and the environment (Donaldson et al. 2013; Singh et al. 2019). In addition to this, there is an urgent need to address all those data gaps related to nanomaterials ranging from their synthesis techniques to characterization methods (Szakal et al. 2014), which will eventually eradicate all those myths prevailing in society related to their toxic effects. Also, a better insight into the biological responses to any foreign particles is a must to eliminate any sought of risk factors attached with the biomedical applications of nanotechnology.

12.7 Methods to Control Toxicity

12.7.1 Green Synthesis

Most of the nanoparticles used at a larger scale show some toxic properties due to the by-products formed or their properties. There is a need to find alternatives to the conventional methods that produce toxic products. One of the emerging synthesis techniques is the green synthesis method, which involves using plant/microorganism extract. Green synthesis helps save time and energy and is also cost-effective (Devatha and Thalla 2018). The materials prepared using this method will possess an environment-friendly nature and be less toxic to nature. Through this technique, toxic by-products can be reduced, morphology may be varied, and overall safety can be enhanced (Singh et al. 2018).

12.7.1.1 Green Synthesis Using Plants

Silver nanoparticles can be prepared using various chemical methods. However, as we talk about alternatives to such chemical processes that may produce toxic by-products, green synthesis becomes a viable choice (Verma and Mehata 2016). Silver nanoparticles can be successfully synthesized using Neem (*Azadirachta indica*) plant leaves. A few neem leaves were taken and washed thoroughly under tap water, followed by double-distilled water in a study. It was then dried at room temperature.

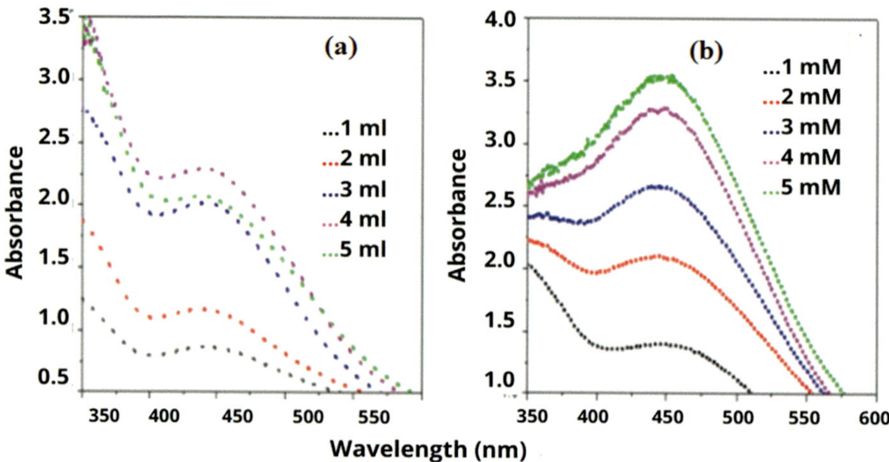

Fig. 12.7 UV-vis spectroscopy data of silver NPs synthesized using Neem leaves (Ahmed et al. 2016)

Approximately 20 g of chopped leaves were boiled for half an hour and 200 mL double distilled water. Afterward, the product was cooled and filtered using Whatman filter paper no.1. It was then stored at 4 °C for its use in the future. Silver nitrate ($AgNO_3$) was taken, and in 100 mL, 1 mM solution was prepared. Then 1, 2, 3, 4, and 5 mL of plant extract was poured separately to 10 mL of $AgNO_3$ solution, maintaining its concentration at 1 mM. Ag nanoparticles were also prepared by changing the concentration of $AgNO_3$ (1–5 mM), keeping extract concentration constant (1 mL). To reduce the photo-activation of AgNo3, the setup was incubated in a dark chamber. The color change of the solution from colorless to brown confirmed the reduction of Ag NPs. Further, its characterization was done using UV-vis spectroscopy (Ahmed et al. 2016). The results are shown in Fig. 12.7.

12.7.1.2 Green Synthesis Using Microorganisms

Apart from plants like neem, hibiscus, and tulsi, some microorganisms have also been used for the synthesis of silver nanoparticles. A few microorganisms, including algae, bacteria, and fungi, are being used. For this particular experiment involving the synthesis of Ag nanoparticles, *Escherichia coli* has been used. Mainly, *E. coli* does not cause any harm and keeps our digestive system healthy. These are generally found in the intestinal tract and also guts of some animals. *E. coli* supernatant was taken as per the experiment, and 1 mM $AgNO_3$ of varying concentrations was added. It was incubated for about 24 h. The varying concentrations S1, S2, S3, and S4 are defined as below, where C means culture supernatant and Solution A means 1 nM $AgNO_3^-$

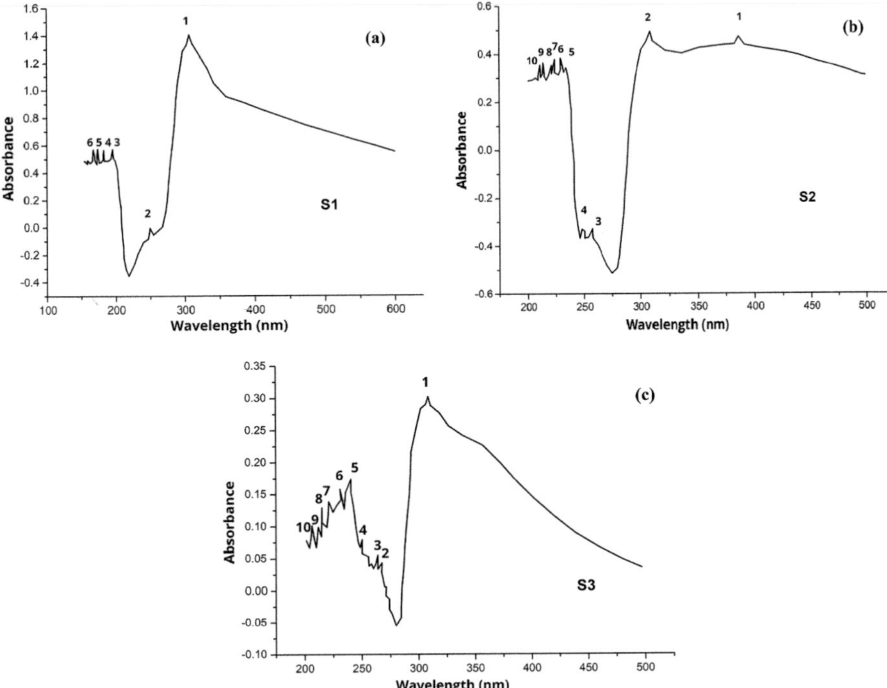

Fig. 12.8 UV-vis spectroscopy data of Ag NPs at different concentrations S1, S2, and S3, synthesized using microorganisms. (Divya et al. 2016)

- S_1- 10 ml C + 40 ml A
- S_2- 20 ml C + 30 ml A
- S_3- 30 ml C + 20 ml A
- S_4- 40 ml C + 10 ml A

After this, characterization was performed, and UV readings were obtained. S4, one of the resultant solutions, didn't show any color change, which indicated the nonavailability of Ag NPs. On the other hand, S3 showed an absorbance of 0.29 at 307 nm. S2 had an absorbance of 0.46 at 309 nm, whereas S1 had maximum absorbance of 1.35 at 302 nm. Hence, S1 was considered to be the standard ratio. The UV-visible spectrum of Ag NPs is shown in Fig. 12.8. The increase in culture filtrate causes a decrease in absorbance. The decline suggests a reduction in nanoparticles size. It seems that by increasing filtrate, the amount and size of nanoparticles become smaller (Divya et al. 2016).

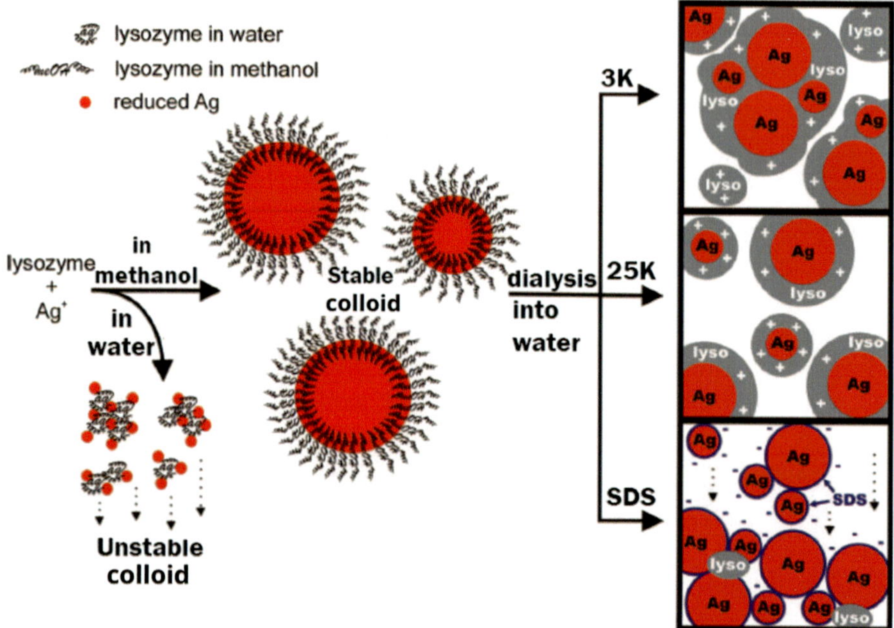

Fig. 12.9 Ag-lysozyme nanoparticles formation mechanism under different reaction conditions. (Reprinted with permission from Eby et al. 2009)

12.7.1.3 Green Synthesis Using Enzymes

Enzymes, in general, are compounds that help catalyze an inevitable biochemical reaction. Enzymes are amino acids that are connected through peptide bonds. Enzymes cause reactions to occur rapidly at ambient temperatures with a high degree of specificity. Some of these enzymes can be even used for conducting the green synthesis of nanoparticles (Talens-Perales et al. 2016). Enzymes like immobilized NADH-dependent nitrate reductase obtained from *Fusarium oxysporum*, fibrinolytic enzyme obtained from *Bacillus cereus* NK1, and keratinase from *B. safensis* LAU 13 have been previously used for synthesizing Ag NPs in an eco-friendly manner (Adelere and Lateef 2016). Using the keratinase enzyme at 30 °C ± 2 °C, silver NPs of about 8 nm were obtained (Lateef et al. 2015). Another enzyme used to prepare Ag NPs is the hen egg-derived lysozyme. The formation mechanism of Ag-lyso NPs is depicted in Fig. 12.9 (Eby et al. 2009). It clearly shows that lysozyme and Ag do not show colloidal stability when dispersed in water. On the other hand, when dispersed in methanol, stable colloid formation occurs.

12.7.2 Magnetic Nanocomposites

Materials that possess magnetic properties are relatively easier to separate from their respective solutions. For example, as carbon nanotubes have toxicity, we can prepare a magnetic nanocomposite of CNTs and spinel ferrites. This makes it easy to regenerate the nanoparticles after using them in wastewater treatment and dye removal processes. Bare spinel ferrite nanoparticles can be used as adsorbents, but modified spinel ferrites, along with multiwalled carbon nanotubes, work as better adsorbents, both in terms of efficiency and cost-effectiveness (Briceno et al. 2013).

One of the methods that can be used to synthesize the above nanocomposite is by adding 100 mg of 95% pure MWCNTs in 30 mL of TREG (triethylene glycol, 99%). The MWCNTs used were without any surface modifications, having a thickness of about 20–30 mm and a length of 5–15 micrometres. This mixture was then sonicated for 5 min. Fe(acac)3 (Iron(III) Acetylacetonate, 99%) was used as a magnetic precursor. 200 mg of this precursor was added under magnetic mixing. The resultant mixture was then provided with a 278 °C temperature at a rate of 3 °C per minute with continuous mixing and argon protection. It was kept at reflux for 30 min. After letting it cool down to room temperature, some water was added to the solution. Also, 30 mL ethanol (C_2H_5OH) was added to it. Then the resultant nanocomposites were segregated using a commercial magnet and rinsed with C_2H_5OH many times and dried in the absence of air (Wan et al. 2007).

Separation of spinel ferrite nanoparticles is chosen because it is relatively simpler and faster than the other conventional methods. The advantage of using ferrites for wastewater treatment is that these nanoparticles can be separated through magnetic separation (Guillot 2006). After the entire process is completed, spinel ferrites can be easily recovered using an external magnet. The recovery of SFNPs through the magnetic separation method saves a considerable amount of time and energy. Moreover, the same material can be used over and over again. It should be noted that the main motive of the recovery of SFNPs is to restore their adsorption capacity and make it available for further use.

12.8 Disposal of Nanowaste

As discussed in the chapter earlier, nanomaterials are being used in different sectors. These areas include sensors, dye removal, wastewater treatment, and food packaging. It is widely known that nanomaterials have brought a wave of change in the technology areas. But these areas do generate nano-waste at some point in time. Nanowaste is the kind of waste generated through nanomaterials after their use in a particular reaction. The twist is that the nano waste is still reactive after being termed "waste." The response within it has not finished yet. These materials are more reactive than their bulk counterparts. They need to be deactivated or made inert before dumping them off. Here are some techniques that are being used to get rid of nano waste. However, these techniques sometimes produce results that can worsen the situation.

12.8.1 Incineration

In general terms, incineration is a method that involves burning the waste to get rid of it. Incineration is also known as "thermal treatment" as a high temperature is provided inside the incinerator. There are various types of incinerators, including water wall, modular, multiple hearths, catalytic combustion, waste-gas, direct-flame, liquid injection, fluidized bed, rotary kiln, and grate incinerators (moving and fixed). The choice of incinerator type depends on the type, volume, and hazard of the waste to be destroyed (Holder et al. 2013). Incineration destroys most of the waste through burning. Also, at some stage, energy can be gathered from the waste. However, the main disadvantage of incineration is that it releases extremely small particles at the end after burning that may cause serious human health issues. These particles, when inhaled, can cause pulmonary diseases and heart problems. Although not much information is present about the incineration of nano waste, it is known that there should be specific guidelines for the disposal of nano waste. At the end step, flue gas filters must be there to deactivate the nanoparticles produced at the end (Walser et al. 2012). Incineration may burn most of the waste and prove to be useful, but now the task lies with flue gas and slag handling. The particles are not released in the atmosphere but get stuck onto the final residues using flue gas filters. An important point to be noticed is that they do not lose their chemical and physical properties and may prove toxic while disposing of the final products in landfills. Hence, necessary precautions must be taken at this stage, and also there is a need to create nanoparticles that are less toxic and degradable (Walser et al. 2012). Also, in some cases, the heat released from burning can be used as an energy source, mainly electric energy. The total greenhouse gas emissions are lesser from incinerators as compared to landfills. According to certain research, metal oxide NPs increased combustion efficiency in an incinerator and reduced the emission of toxic polycyclic aromatic hydrocarbons (PAHs). On the other hand, Ag nanoparticles reduced combustion efficiency and increased PAH emissions (Vejerano et al. 2014).

According to recent research, pulse jet bag filters were prepared to collect NPs even from the flue gas. This could help in the proper disposal of the nano waste, thus causing no harm to the environment and the people who come in contact with this waste while transporting. The efficiency of these bag filters ranged from 98% to 99.98% (Boudhan et al. 2018). These bags can be used more than once, increasing efficiency as per results. The experimental setup for this has been shown in Fig. 12.10.

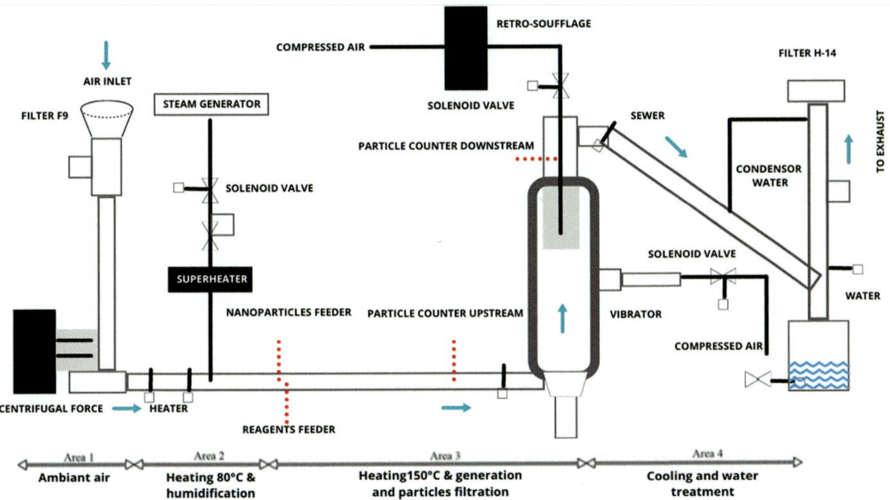

Fig. 12.10 Incineration setup for nanowaste disposal using pulse jet filters for better results

12.8.2 Landfill

As discussed above, the fate of incineration's end products, namely, flue gas and slag, lies in the landfill. In simple terms, a landfill is an area created to dispose of waste. There have been instances when these landfills went unnoticed and were not taken care of, which led to contamination and insect breeding. However, now the municipalities of that very place take care of the landfills. The incinerators can burn a significant portion of the nano waste provided if it contains the mechanism for flue gas filtration. This will prevent the nanoparticles from escaping into the air. However, the remaining part of the nano waste gets released into the landfills (Kim 2014). Many studies have been done on the issue of landfill leaching. Leaching is a common phenomenon in landfills, resulting from rainfall penetrating the landfill and reaching the degrading waste. This is an issue of great concern because it causes contamination, which may take no time to get into the neighboring soil and water streams.

To find a solution to solve the issue of leachates, research was performed using tungsten-doped titanium dioxide (Azadi et al. 2017). This material could treat the landfill leachate using the principle of photocatalytic degradation. The doped compound was prepared through the sol-gel technique. These particles were characterized through XRD, PSA, FESEM, and EDS. The results of the XRD data suggest that the average crystallite size was changed with the calcination temperature. When the photocatalytic activity of the W-doped TiO_2 was studied, it was seen that doped nanoparticles show better performance than undoped nanoparticles under visible light irradiation. That is also a method to check whether the doping was correctly done or not. The best contact time of the nanoparticles and leachate was found out

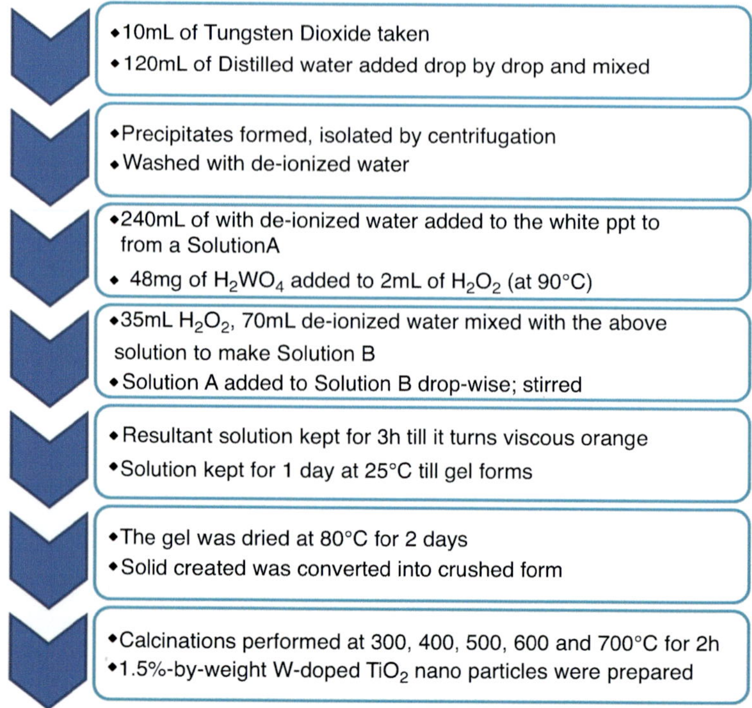

Fig. 12.11 Preparation process of tungsten-doped titanium dioxide NPs for treatment of landfill leachate

to be 34 h. After this time, the removal rate becomes very low, and after this, the process becomes costly.

It was concluded that augmenting the tungsten value to an optimal value increases the process efficiency. However, the further increase may block some reactive surface sites for photocatalytic activity and negatively affect the process efficiency. Moreover, if the engineered nanoparticles (ENPs) are thermally treated before their disposal, they may lose some part of their endurance in the environment (Part et al. 2020). As per another research, the landfill should be monitored carefully, and methods of leachate treatment such as nanofiltration or precipitation technologies should be adopted (Fig. 12.11).

12.9 Recycle and Reuse

As mentioned earlier in the chapter, nano-wastes make up a substantial portion of the total collected annually. With the increase in engineered nanomaterial (ENM) production, a lot of nano-waste is generated every year. This may include waste from dye and textile industries, sensor manufacturing units, and food technology sectors. The endpoint of this nano-waste is landfills most often. In these times, the obvious question that arises in mind is, "Are nanoparticles recyclable?" or "Can nanoparticles be re-used?" Some studies have been performed to find a solution to these questions.

Although nanomaterials are a boon to society in many different ways, these materials also compromise the environment. The landfill leachate causes significant pollution and health hazards. Not many recycling or reuse techniques are known because they are difficult to recover due to the small size of the nanoparticles. The ideal methods for this purpose need to be cost-effective and simple. The most used way to recover magnetic nanoparticles is by using an ordinary magnet. In the case of spinel ferrites like manganese ferrites, zinc ferrites, cobalt ferrites, nickel ferrites, etc., the magnetic recovery is effortless, economical, and time-saving (Gupta et al. 2020). The SFNPs can work in both acidic and basic conditions. Hence, they can be desorbed easily to serve for reuse over many times. In the process of dye removal using methyl orange (MO) as a dye, the adsorbent can be removed with the help of an external magnet, followed by rinsing with ethanol and drying (Guillot 2006). As per experimental data, the reusability did not decrease the successive adsorption efficiency much. A simple magnetic separation process of gold nanoparticles has been shown in Fig. 12.12 (Chen and Gerion 2004; Oliveira et al. 2010).

Researchers from Bristol University suggested a method to recover NPs using the microemulsion technique (Myakonkaya et al. 2010). They took the example of cadmium and zinc NPs. Most importantly, there was no change in their chemical properties and shape. This made them suitable for reuse. According to the experiment performed, using cadmium and zinc nanoparticles, it was seen how the oil and water in the microemulsion got separated into two layers when provided with temperature. One layer contained the nanoparticles that could be recovered, and the other had no nanoparticles.

Fig. 12.12 A simple magnetic separation process of Au nanoparticles using commercially available external magnet. (Reprinted with permission from Oliveira et al. 2010)

During the COVID-19 pandemic, an ever-increasing demand for face masks was seen. But these masks possessed some limitations too. Hence, the Korea Advanced Institute of Science and Technology (KAIST) recently created reusable nano filter masks (Liu et al. 2020). As per data, these show 95% efficiency and can be used up to 20 times of washing. For these masks, the researchers have used orthogonal nanofibers. The structure of nanofibers minimizes air filter pressure and enhances the filtration process. A team at the Queensland University of Technology (QUT) has created nanotechnology-based biodegradable face masks made up of degradable plant waste products (Talebian et al. 2020). This mask can prevent virus-sized particles from entering and does not cause fatigue in those wearing the mask so that they can wear it for a longer time. Also, people with breathing issues can wear it easily. There has been another kind of reusable nanoparticle-based mask in the market. These are infused with metal oxide nanoparticles that kill germs. These can be washed about 100 times and still not lose their properties (Bose 2020).

However, more guidelines are required for better recycling and reuse of nanoparticles so that no harm is caused to nature. Also, there is a need to develop materials that are less toxic and have recyclable properties.

12.10 Future Perspectives

Nanotoxicity has been studied for over a decade now. As the chapter discusses various limitations in the nanotechnology field, there is an immediate need to find ways to reduce toxicity and save the environment and its components, including human beings. Ignored aspects of nanomaterials like bandgap and leaching problems should be considered (Singh et al. 2019). There is also a need to study in detail the properties of nanomaterials due to their extremely small size. The misconceptions associated with it should also be studied thoroughly. As methods of recycling nanowaste have been found, further research can be done to modify them to increase their efficiency and make them suitable to recycle every kind of nano-waste. The lack of proper data accounts for some of the research problems about nanotoxicity. Apart from using nanocomposites with magnetic properties, other nanocomposite alternatives may be found so that they can be recovered easily.

Nanotoxicity is not only limited to physics and chemistry but may also apply to pharmacology and biological studies at the molecular level and also computer studies. The combination of all these may help create an extensive database that can help trace the origin of nano-waste and assist in deciding the future actions to be taken for it (Fadeel 2019). Moreover, Health Safety Standards and other regulations should be updated as per today's requirements. The disposal of nano-waste is an important area of concern as it may cause harm to nature as well as the people involved in the disposal process. Not only this but also students working at small levels should be provided with proper knowledge about the toxic nanomaterials and the precautions to be taken while working with such materials. Nanotechnology is an emerging field and can't be ignored at any cost. The toxicity-related data should be executed in various areas before using (Saini and Srivastava 2018). Hence, better synthesis, characterization, and data should be extracted to help reduce toxicity.

12.11 Conclusions

This chapter deals with the toxic behavior of commonly used nanomaterials in various fields. The importance of nanomaterials in different areas has been mentioned as an introductory portion, along with the changes in properties that occur in material from macro to nanoscale. As the topic suggests, the misconceptions associated with nanotechnology and nanotoxicity are also emphasized. It is essential to understand the methods used to define the toxic properties of widely employed nanomaterials. Various methods in context to this, including characterization and synthesis, have been discussed in detail. There may be specific knowledge gaps that may cause such misconceptions to develop. Hence, such gaps have been emphasized. These gaps include knowledge gaps and gaps that may instigate the invention of better synthesis and characterization techniques. The in vivo and in vitro methods have been mentioned, and alongside, in silico method has also been discussed for certain compounds. An important term used to understand nanotoxicity is the nano-bio interface.

The main focus is laid on the toxic properties of certain compounds and the misconceptions associated with such materials. The properties and uses of various materials are thoroughly discussed. The barriers in the uninterrupted use of these materials due to their toxicity are explained in detail. The diseases or issues caused due to toxicity have been compared in tabular form. Much emphasis is laid on the generation of nano-waste and its disposal. The pros and cons of disposal processes like incineration and landfills have been mentioned. As a solution to the harm caused by nano-waste, the option of recycling and reuse has been identified as an important area of research. Particular examples of the recycling and reuse of some nanomaterials have been discussed. Moreover, methods that can be used to produce less toxic nanomaterials have been stressed upon. Also, the future perspectives related to the topic have been explained.

References

Adelere IA, Lateef A (2016) A novel approach to the green synthesis of metallic nanoparticles: the use of agro-wastes, enzymes, and pigments. Nanotechnol Rev 5(6):567–587

Ahmed S, Saifullah, Ahmad M, Swami BL, Ikram S (2016) Green synthesis of silver nanoparticles using Azadirachta indica aqueous leaf extract. J Radiat Res Appl 9(1):1–7

Aldrich S (2015) Gold nanoparticles: properties and applications. Sigma-Aldrich, St. Louis

Alkilany AM, Murphy CJ (2010) Toxicity and cellular uptake of gold nanoparticles: what we have learned so far? J Nanopart Res 12(7):2313–2333

Allegri M, Perivoliotis DK, Bianchi MG, Chiu M, Pagliaro A, Koklioti MA, Charitidis CA (2016) Toxicity determinants of multiwalled carbon nanotubes: the relationship between functionalization and agglomeration. Toxicol Rep 3:230–243

Amiri M, Salavati-Niasari M, Akbari A (2019) Magnetic nanocarriers: evolution of spinel ferrites for medical applications. Adv Colloid Interf Sci 265:29–44

Armstead AL, Li B (2016) Nanotoxicity: emerging concerns regarding nanomaterial safety and occupational hard metal (WC-Co) nanoparticle exposure. Int J Nanomedicine 11:6421

Azadi S, Karimi-Jashni A, Javadpour S (2017) Photocatalytic treatment of landfill leachate using W-doped TiO 2 nanoparticles. Int J Environ Eng 143(9):04017049

Bahadar H, Maqbool F, Niaz K, Abdollahi M (2016) Toxicity of nanoparticles and an overview of current experimental models. Iran Biomed J 20(1):1

Bai Q, Li L, Liu S, Xiao S, Guo Y (2018) Drug design progress of *in silico*, *in vitro* and *in vivo* researches. In-vitro In-vivo In-silico J 1(1):16

Baranowska-Wójcik E, Szwajgier D, Oleszczuk P, Winiarska-Mieczan A (2020) Effects of titanium dioxide nanoparticles exposure on human health – a review. Biol Trace Elem Res 193(1):118–129

Bose P (2020) Nanotechnology in washable and reusable face masks, AzoNano. Available online via: https://www.azonano.com/article.aspx?ArticleID=5529. Accessed 28 Nov 2020

Boudhan R, Joubert A, Gueraoui K, Durécu S, Venditti D, Tran DT, Le Coq L (2018) Pulse-jet bag filter performances for treatment of submicronic and nanosized particles from waste incineration. Waste Biomass Valorization 9(5):731–737

Briceno S, Sanchez Y, Braemer-Escamilla W, Silva P, Rodriguez JP, Ramos MA, Plaza E (2013) Comparative study of preparation methods of ferrites nanoparticles cofe2o4. Acta Microsc 22(1):62–68

Budama-Kilinc Y, Cakir-Koc R, Zorlu T, Ozdemir B, Karavelioglu Z, Egil AC, Kecel-Gunduz S (2018) Assessment of nano-toxicity and safety profiles of silver nanoparticles. In: Khan M (ed) Silver nanoparticles-fabrication, characterization and applications. Intech, p 185

Buzea C, Pacheco II, Robbie K (2007) Nanomaterials and nanoparticles: sources and toxicity. Biointerphases 2(4):MR17–MR71

Chen F, Gerion D (2004) Fluorescent CdSe/ZnS nanocrystal– peptide conjugates for long-term, nontoxic imaging and nuclear targeting in living cells. Nano Lett 4(10):1827–1832

Chithrani BD, Ghazani AA, Chan WC (2006) Determining the size and shape dependence of gold nanoparticle uptake into mammalian cells. Nano Lett 6(4):662–668

De Jong WH, De Rijk E, Bonetto A, Wohlleben W, Stone V, Brunelli A, Cassee FR (2019) Toxicity of copper oxide and basic copper carbonate nanoparticles after short-term oral exposure in rats. Nanotoxicology 13(1):50–72

Deng X, Jia G, Wang H, Sun H, Wang X, Yang S, Liu Y (2007) Translocation and fate of multi-walled carbon nanotubes *in vivo*. Carbon 45(7):1419–1424

Devatha CP, Thalla AK (2018) Green synthesis of nanomaterials. In: Synthesis of inorganic nanomaterials. Woodhead Publishing, pp 169–184

Divya K, Kurian LC, Vijayan S, Manakulam Shaikmoideen J (2016) Green synthesis of silver nanoparticles by *Escherichia coli*: analysis of antibacterial activity. J Water Environ Nanotechnol 1(1):63–74

Dobrucka R (2017) Synthesis of titanium dioxide nanoparticles using Echinacea purpurea herba. IJPR 16(2):756

Donaldson K, Schinwald A, Murphy F, Cho WS, Duffin R, Tran L, Poland C (2013) The biologically effective dose in inhalation nanotoxicology. Acc Chem Res 46(3):723–732

Eby DM, Schaeublin NM, Farrington KE, Hussain SM, Johnson GR (2009) Lysozyme catalyzes the formation of antimicrobial silver nanoparticles. ACS Nano 3(4):984–994

Fadeel B (2019) The right stuff: on the future of nanotoxicology. Front Toxicol 1:1

Famá L, Rojo PG, Bernal C, Goyanes S (2012) Biodegradable starch based nanocomposites with low water vapor permeability and high storage modulus. Carbohydr Polym 87(3):1989–1993

Fard JK, Jafari S, Eghbal MA (2015) A review of molecular mechanisms involved in toxicity of nanoparticles. Adv Pharm Bull 5(4):447

Fernandez-Fernandez A, Manchanda R, McGoron AJ (2011) Theranostic applications of nanomaterials in cancer: drug delivery, image-guided therapy, and multifunctional platforms. Biotechnol Appl Biochem 165(7-8):1628–1651
Fischer HC, Chan WC (2007) Nanotoxicity: the growing need for *in vivo* study. Curr Opin Biotechnol 18(6):565–571
Foroozandeh P, Aziz AA, Mahmoudi M (2019) Effect of cell age on uptake and toxicity of nanoparticles: the overlooked factor at the nanobio interface. ACS Appl Mater Interfaces 11(43):39672–39687
Francis AP, Devasena T (2018) Toxicity of carbon nanotubes: a review. Toxicol Ind Health 34(3):200–210
Fröhlich E, Salar-Behzadi S (2014) Toxicological assessment of inhaled nanoparticles: role of *in vivo*, ex vivo, *in vitro*, and *in silico* studies. Int J Mol Sci 15(3):4795–4822
Gaiser BK, Fernandes TF, Jepson M, Lead JR, Tyler CR, Stone V (2009) Assessing exposure, uptake and toxicity of silver and cerium dioxide nanoparticles from contaminated environments. Environ Health 8(1):S2
González-Durruthy M, Giri AK, Moreira I, Concu R, Melo A, Ruso JM, Cordeiro MND (2020) Computational modeling on mitochondrial channel nanotoxicity. Nano Today 34:100913
Grande F, Tucci P (2016) Titanium dioxide nanoparticles: a risk for human health? Mini-Rev Med Chem 16(9):762–769
Grigore ME, Biscu ER, Holban AM, Gestal MC, Grumezescu AM (2016) Methods of synthesis, properties and biomedical applications of CuO nanoparticles. Pharmaceuticals 9(4):75
Guillot M (2006) Magnetic properties of Ferrites. Mater Sci Technol
Gupta NK, Ghaffari Y, Kim S, Bae J, Kim KS, Saifuddin M (2020) Photocatalytic degradation of organic pollutants over MFe_2O_4 (M= Co, Ni, Cu, Zn) nanoparticles at neutral pH. Sci Rep 10(1):1–11
Hardman R (2006) A toxicologic review of quantum dots: toxicity depends on physicochemical and environmental factors. Environ Health Perspect 114(2):165–172
Holder AL, Vejerano EP, Zhou X, Marr LC (2013) Nanomaterial disposal by incineration. Environ Sci Process Impacts 15(9):1652–1664
Hu X, Li D, Gao Y, Mu L, Zhou Q (2016) Knowledge gaps between nanotoxicological research and nanomaterial safety. Environ Int 94:8–23
Jia YP, Ma BY, Wei XW, Qian ZY (2017) The *in vitro* and *in vivo* toxicity of gold nanoparticles. Chin Chem Lett 28(4):691–702
Julien CM, Mauger A, Zaghib K (2011) Surface effects on electrochemical properties of nano-sized $LiFePO_4$. J Mater Chem 21(27):9955–9968
Kayat J, Gajbhiye V, Tekade RK, Jain NK (2011) Pulmonary toxicity of carbon nanotubes: a systematic report. Nanomedicine 7(1):40–49
Kim Y (2014) Nanowastes treatment in environmental media. Environ Health Toxicol 29
Korani M, Ghazizadeh E, Korani S, Hami Z, Mohammadi-Bardbori A (2015) Effects of silver nanoparticles on human health. Eur J Nanomed 7(1):51–62
Kumar P, Mahajan P, Kaur R, Gautam S (2020) Nanotechnology and its challenges in the food sector: a review. Mater Today Chem 17:100332
Lateef A, Adelere IA, Gueguim-Kana EB, Asafa TB, Beukes LS (2015) Green synthesis of silver nanoparticles using keratinase obtained from a strain of *Bacillus safensis LAU* 13. Int Nano Lett 5(1):29–35
Liu Z, Ramakrishna S, Liu X (2020) Electrospinning and emerging healthcare and medicine possibilities. APL Bioeng 4(3):030901
Lujan H, Sayes CM (2017) Cytotoxicological pathways induced after nanoparticle exposure: studies of oxidative stress at the 'nano–bio' interface. Toxicol Res 6(5):580–594
Maynard AD, Aitken RJ, Butz T, Colvin V, Donaldson K, Oberdörster G et al (2006) Safe handling of nanotechnology. Nature 444(7117):267–269
Merget R, Bauer T, Küpper H, Philippou S, Bauer H, Breitstadt R, Bruening T (2002) Health hazards due to the inhalation of amorphous silica. Arch Toxicol 75(11-12):625–634

Minelli C, Lowe SB, Stevens MM (2010) Engineering nanocomposite materials for cancer therapy. Small 6(21):2336–2357

Mishra PK, Mishra H, Ekielski A, Talegaonkar S, Vaidya B (2017) Zinc oxide nanoparticles: a promising nanomaterial for biomedical applications. Drug Discov Today 22(12):1825–1834

Mohanta D, Patnaik S, Sood S, Das N (2019) Carbon nanotubes: Evaluation of toxicity at biointerfaces. J Pharm Anal 9(5):293–300

Murugadoss S, Lison D, Godderis L, Van Den Brule S, Mast J, Brassinne F et al (2017) Toxicology of silica nanoparticles: an update. Arch Toxicol 91(9):2967–3010

Myakonkaya O, Guibert C, Eastoe J, Grillo I (2010) Recovery of nanoparticles made easy. Langmuir 26(6):3794–3797

Naseer B, Srivastava G, Qadri OS, Faridi SA, Islam RU, Younis K (2018) Importance and health hazards of nanoparticles used in the food industry. Nanotechnol Rev 7(6):623–641

Naskar S, Kuotsu K, Sharma S (2019) Chitosan-based nanoparticles as drug delivery systems: a review on two decades of research. J Drug Target 27(4):379–393

Nyamukamba P, Okoh O, Mungondori H, Taziwa R, Zinya S (2018) Synthetic methods for titanium dioxide nanoparticles: a review. In: Yang D (ed) Titanium dioxide – material for a sustainable environment, pp 151–175

Oliveira RL, Kiyohara PK, Rossi LM (2010) High performance magnetic separation of gold nanoparticles for catalytic oxidation of alcohols. Green Chem 12(1):144–149

Part F, Zaba C, Bixner O, Zafiu C, Lenz S, Martetschläger L et al (2020) Mobility and fate of ligand stabilized semiconductor nanoparticles in landfill leachates. J Hazard Mater:122477

Rao GVS, Tinkle S, Weissman D, Antonini J, Kashon M, Salmen R et al (2003) Efficacy of a technique for exposing the mouse lung to particles aspirated from the pharynx. J Toxicol Environ Health Part A 66(15–16):1441–1452

Rao KS, El-Hami K, Kodaki T, Matsushige K, Makino K (2005) A novel method for synthesis of silica nanoparticles. J Colloid Interface Sci 289(1):125–131

Ren G, Hu D, Cheng EW, Vargas-Reus MA, Reip P, Allaker RP (2009) Characterization of copper oxide nanoparticles for antimicrobial applications. Int J Antimicrob Agents 33(6):587–590

Sahu D, Kannan GM, Vijayaraghavan R, Anand T, Khanum F (2013) Nanosized zinc oxide induces toxicity in human lung cells. Int Sch Res Notices 2013

Saini B, Srivastava S (2018) Nanotoxicity prediction using computational modelling-review and future directions. IOP Conf Ser Mater Sci Eng 348:012005

Shakeel M, Jabeen F, Shabbir S, Asghar MS, Khan MS, Chaudhry AS (2016) Toxicity of nano-titanium dioxide (TiO 2-NP) through various routes of exposure: a review. Biol Trace Elem Res 172(1):1–36

Shityakov S, Roewer N, Broscheit JA, Förster C (2017) *In silico* models for nanotoxicity evaluation and prediction at the blood-brain barrier level: a mini-review. Comput Toxicol 2:20–27

Simate GS, Iyuke SE, Ndlovu S, Heydenrych M, Walubita LF (2012) Human health effects of residual carbon nanotubes and traditional water treatment chemicals in drinking water. Environ Int 39(1):38–49

Singh J, Dutta T, Kim KH, Rawat M, Samddar P, Kumar P (2018) 'Green' synthesis of metals and their oxide nanoparticles: applications for environmental remediation. J Nanobiotechnol 16(1):84

Singh AV, Laux P, Luch A, Sudrik C, Wiehr S, Wild AM et al (2019) Review of emerging concepts in nanotoxicology: opportunities and challenges for safer nanomaterial design. Toxicol Mech Methods 29(5):378–387

Sruthi S, Ashtami J, Mohanan PV (2018) Biomedical application and hidden toxicity of Zinc oxide nanoparticles. Mater Today Chem 10:175–186

Sukhanova A, Bozrova S, Sokolov P, Berestovoy M, Karaulov A, Nabiev I (2018) Dependence of nanoparticle toxicity on their physical and chemical properties. Nanoscale Res Lett 13(1):44

Szakal C, Roberts SM, Westerhoff P, Bartholomaeus A, Buck N, Illuminato I, Rogers M (2014) Measurement of nanomaterials in foods: integrative consideration of challenges and future prospects. ACS Nano 8(4):3128–3135

Takei K, Fang H, Kumar SB, Kapadia R, Gao Q, Madsen M et al (2011) Quantum confinement effects in nanoscale-thickness InAs membranes. Nano Lett 11(11):5008–5012

Talebian S, Wallace GG, Schroeder A, Stellacci F, Conde J (2020) Nanotechnology-based disinfectants and sensors for SARS-CoV2. Nat Nanotechnol 15(8):618–621

Talens-Perales D, Marín-Navarro J, Polaina J (2016) Enzymes: functions and characteristics

Tejral G, Panyala NR, Havel J (2009) Carbon nanotubes: toxicological impact on human health and environment. J Appl Biomed (De Gruyter Open) 7(1)

Tian X, Chong Y, Ge C (2020) Understanding the nano–bio interactions and the corresponding biological responses. Front Chem 8:446

Vandebriel RJ, De Jong WH (2012) A review of mammalian toxicity of ZnO nanoparticles. Nanotechnol Sci Appl 5:61

Vazquez-Muñoz R, Borrego B, Juárez-Moreno K, García-García M, Morales JDM, Bogdanchikova N, Huerta-Saquero A (2017) Toxicity of silver nanoparticles in biological systems: does the complexity of biological systems matter? Toxicol Lett 276:11–20

Vejerano EP, Leon EC, Holder AL, Marr LC (2014) Characterization of particle emissions and fate of nanomaterials during incineration. Environ Sci Nano 1(2):133–143

Verma A, Mehata MS (2016) Controllable synthesis of silver nanoparticles using Neem leaves and their antimicrobial activity. J Radiat Res Appl 9(1):109–115

Walser T, Limbach LK, Brogioli R, Erismann E, Flamigni L, Hattendorf B, Rossier M (2012) Persistence of engineered nanoparticles in a municipal solid-waste incineration plant. Nat Nanotechnol 7(8):520–524

Wan J, Cai W, Feng J, Meng X, Liu E (2007) In situ decoration of carbon nanotubes with nearly monodisperse magnetite nanoparticles in liquid polyols. J Mater Chem 17(12):1188–1192

Wilson M, Kannangara K, Smith G, Simmons M, Raguse B (2002) Nanotechnology: basic science and emerging technologies. CRC Press

Wu T, Tang M (2014) Toxicity of quantum dots on respiratory system. Inhal Toxicol 26(2):128–139

Xue F, Liang J, Han H (2011) Synthesis and spectroscopic characterization of water-soluble Mn-doped ZnOxS1− x quantum dots. Spectrochim Acta A 83(1):348–352

Yarjanli Z, Ghaedi K, Esmaeili A, Rahgozar S, Zarrabi A (2017) Iron oxide nanoparticles may damage to the neural tissue through iron accumulation, oxidative stress, and protein aggregation. BMC Neurosci 18(1):51

GPSR Compliance

The European Union's (EU) General Product Safety Regulation (GPSR) is a set of rules that requires consumer products to be safe and our obligations to ensure this.

If you have any concerns about our products, you can contact us on ProductSafety@springernature.com

In case Publisher is established outside the EU, the EU authorized representative is:

Springer Nature Customer Service Center GmbH
Europaplatz 3
69115 Heidelberg, Germany

Batch number: 10372213

Printed by Printforce, the Netherlands